Oats Nutrition and Technology

Oats Nutrition and Technology

Edited by

YiFang Chu
Quaker Oats Center of Excellence,
PepsiCo R&D Nutrition, Barrington,
Illinois, USA

WILEY Blackwell

Library of Congress Cataloging-in-Publication Data

Oats nutrition and technology / edited by YiFang Chu.
        pages cm
    Includes index.
    ISBN 978-1-118-35411-7 (cloth)
    1. Oats.   2. Oats as food.   3. Oats–Analysis.   4. Oats–Processing.   I. Chu, YiFang, editor of compilation.
    TX558.O3.O28 2013
    641.3′313–dc23

                                                                                            2013018937

A catalogue record for this book is available from the British Library.

Wiley also publishes its books in a variety of electronic formats. Some content that appears in print may not be available in electronic books.

Cover image ©iStockphoto/OliverChilds
Cover design by Meaden Creative

Set in 10/12pt Times Ten by Aptara Inc., New Delhi, India
Printed and bound in Singapore by Markono Print Media Pte Ltd

2   2014

Contents

List of Contributors   xi

Preface   xv

Acknowledgements   xvii

PART I: INTRODUCTION

**1 Introduction: Oat Nutrition, Health, and the Potential Threat of a
Declining Production on Consumption**   **3**
Penny Kris-Etherton, Chor San Khoo, and YiFang Chu

1.1   A landmark health claim   3
1.2   The growing interest in oats and health   4
1.3   Declining production poses threats to the growth of oat intake   5
      References   6

PART II: OAT BREEDING, PROCESSING, AND PRODUCT
PRODUCTION

2 Breeding for Ideal Milling Oat: Challenges and Strategies   **9**
Weikai Yan, Judith Frégeau-Reid, and Jennifer Mitchell Fetch

2.1   Introduction   9
2.2   Breeding for single traits: Genotype-by-environment interactions   11
2.3   Breeding for multiple traits: Undesirable trait associations   19
2.4   Strategies of breeding for an ideal milling oat   25
2.5   Discussion   28
      Acknowledgements   32
      References   32

3 Food Oat Quality Throughout the Value Chain   **33**
Nancy Ames, Camille Rhymer, and Joanne Storsley

3.1   Introduction: Oat quality in the context of the value chain   33
3.2   Physical oat quality   36

3.3 Nutritional oat quality   41
3.4 Agronomic factors affecting physical and nutritional quality   46
3.5 Oat end-product quality   47
3.6 Mycotoxins   58
3.7 Summary   59
     Acknowledgements   60
     References   60

PART III: OAT NUTRITION AND CHEMISTRY

4 Nutritional Comparison of Oats and Other Commonly Consumed Whole Grains   **73**
Apeksha A. Gulvady, Robert C. Brown, and Jenna A. Bell

4.1 Introduction to oats as a cereal grain   73
4.2 Overview of the nutritional composition of oats   75
4.3 Conclusion   91
     References   91

5 Oat Starch   **95**
Prabhakar Kasturi and Nicolas Bordenave

5.1 Introduction   95
5.2 Native oat starch organization: From the molecular to the granular level 96
5.3 Starch minor components, isolation, and extraction   104
5.4 Beyond native starch granule: Gelatinization, pasting, retrogradation, and
     interactions with other polysaccharides   107
5.5 Industrial uses   115
5.6 Conclusion and perspectives   116
     References   116

6 Oat β-Glucans: Physicochemistry and Nutritional Properties   **123**
Madhuvanti Kale, Bruce Hamaker, and Nicolas Bordenave

6.1 Introduction   123
6.2 Molecular structures and characteristics   124
6.3 Extraction   131
6.4 Solution properties   135
6.5 Oat β-glucan nutritional properties   144
6.6 Conclusion and perspectives   158
     References   159

7 Health Benefits of Oat Phytochemicals   **171**
Shaowei Cui and Rui Hai Liu

7.1 Introduction   171
7.2 Oat phytochemicals   172
7.3 Health benefits of oat phytochemicals: Epidemiological evidence   185
7.4 Summary   189
     References   189

8   Avenanthramides: Chemistry and Biosynthesis   195
Mitchell L. Wise

8.1   Introduction   195
8.2   Nomenclature   196
8.3   Synthesis   197
8.4   Chemical stability   197
8.5   Antioxidant properties   199
8.6   Solubility of avenanthramides   200
8.7   Analysis of avenanthramides   201
8.8   Biosynthesis of avenanthramides   201
8.9   Victorin sensitivity   206
8.10 Environment effects on avenanthramide production   207
8.11 Hydroxycinnamoyl-CoA: Hydroxyanthranilate N-hydroxycinnamoyl
      transferase (HHT)   209
8.12 Cloning HHT   211
8.13 Metabolic flux of avenanthramides   214
8.14 Localization of avenanthramide biosynthesis   216
8.15 Plant defense activators   218
8.16 False malting   219
8.17 Conclusion   221
      References   222

PART IV: EMERGING NUTRITION AND HEALTH RESEARCH

**9   The Effects of Oats and Oat-β-Glucan on Blood Lipoproteins and
     Risk for Cardiovascular Disease   229**
Tia M. Rains and Kevin C. Maki

9.1   Introduction   229
9.2   Hypocholesterolemic effects of fiber   230
9.3   Hypocholesterolemic effects of oats and oat β-glucan   231
9.4   Summary/Conclusions   233
      References   233

**10   The Effects of Oats and β-Glucan on Blood Pressure and
      Hypertension   239**
Tia M. Rains and Kevin C. Maki

10.1 Introduction   239
10.2 Dietary patterns and blood pressure   240
10.3 Oats and oat β-glucan: Effect on blood pressure and hypertension   246
10.4 Conclusion   251
      References   251

**11   Avenanthramides, Unique Polyphenols of Oats with Potential Health
      Effects   255**
Mohsen Meydani

11.1 Introduction   255
11.2 Avenanthramides, the bioactive phenolics in oats   256

11.3 Anti-inflammatory and antiproliferative activity of avenanthramides   258
11.4 Summary and conclusion   261
      Acknowledgements   261
      References   261

12   Effects of Oats on Obesity, Weight Management, and Satiety   265
      Chad M. Cook, Tia M. Rains, and Kevin C. Maki

12.1 Introduction   265
12.2 Effects of oats and oat β-glucan on body weight   266
12.3 Effects of oats on appetite   271
12.4 Possible mechanisms of action   274
12.5 Summary   276
      References   276

13   Effects of Oats on Carbohydrate Metabolism   281
      Susan M. Tosh

13.1 Introduction   281
13.2 Epidemiology   281
13.3 Mechanisms of postprandial blood glucose reduction   282
13.4 Clinical studies using whole oat products   284
13.5 Clinical studies using oat bran products   286
13.6 Clinical studies using oat-derived β-glucan preparations   289
13.7 Dose response   289
13.8 Longer-term glucose control   291
13.9 Summary   292
      References   293

14   Effects of Oats and β-Glucan on Gut Health   299
      Renee Korczak and Joanne Slavin

14.1 Oats and β-glucan   299
14.2 Digestive health   299
14.3 Short chain fatty acids and fiber fermentability   301
14.4 Large bowel effects of whole grains   302
14.5 Fermentation of individual dietary fibers   303
14.6 Prebiotics   303
14.7 Other mechanisms underlying the effect of oats on gut function   306
14.8 Conclusion   306
      References   307

15   Oats and Skin Health   311
      Joy Makdisi, Allison Kutner, and Adam Friedman

15.1 History of colloidal oatmeal use   311
15.2 Oat structure and composition   312
15.3 Clinical properties   313
15.4 Clinical applications of oats   318
15.5 Side effects of oats   323
15.6 Conclusions   326
      References   326

PART V: PUBLIC HEALTH POLICIES AND CONSUMER RESPONSE

16 Health Claims for Oat Products: A Global Perspective **335**
Joanne Storsley, Stephanie Jew, and Nancy Ames

16.1 Introduction 335
16.2 Definition of health claims 336
16.3 Substantiation of health claims 338
16.4 Health claims and dietary recommendations for oat products 339
16.5 Benefits of health claims 346
16.6 Nutritional information and health claims: How can health claims ensure clarity
versus confusion? 348
16.7 Considerations in conducting research for health claim substantiation 349
References 351

**17 Oh, What Those Oats Can Do: Quaker Oats, the US Food and Drug
Administration, and the Market Value of Scientific Evidence
1984–2010** **357**
Robert Fitzsimmons

17.1 Introduction 357
17.2 Wild oats: The oat bran craze 1988–1990 363
17.3 Brantastic voyage: Oats through dietetic history 364
17.4 Gruel intentions: The NLEA and Quaker's health claim 1990–1997 382
17.5 Cash crop: Leveraging scientific evidence 1997–2010 395
17.6 Conclusions 413
References 420

PART VI: FUTURE RECOMMENDATIONS

18 Overview: Current and Future Perspectives on Oats and Health **429**
Penny Kris-Etherton

18.1 Chapter summaries 429
18.2 Relevance to the nutrition and dietetic communities and the medical profession 433
18.3 Future needs and recommendations 434
References 436

Index **439**

List of Contributors

Nancy Ames, PhD Research Scientist, Agriculture and Agri-Food Canada, Winnipeg, MB, Canada

Jenna A. Bell, PhD, RD Chair-Elect (2012–2013), The Sports, Cardiovascular and Wellness Nutrition Dietetic Practice Group, Academy for Nutrition and Dietetics, Chicago, IL, USA

Nicolas Bordenave, PhD Associate Principal Scientist, Global R&D Technical Insights – Analytical Department, PepsiCo Inc., Barrington, IL, USA

Robert C. Brown, R&D Nutrition Senior Director, Global R&D Nutrition, PepsiCo Inc., Barrington, IL, USA

YiFang Chu, PhD Senior Manager, Quaker Oats Center of Excellence, PepsiCo R&D Nutrition, Barrington, IL, USA

Chad M. Cook, PhD Senior Scientist/Medical Writer, Biofortis Clinical Research, Addison, IL, USA

Shaowei Cui, MPS Technician, Department of Food Science, Cornell University, Ithaca, NY, USA

Jennifer Mitchell Fetch, Research Scientist (oat breeding), Cereal Research Centre, Agriculture and Agri-Food Canada, Winnipeg, MB, Canada

Robert Fitzsimmons, Harvard College, Cambridge, MA, USA

Judith Frégeau-Reid, PhD Research Scientist (grain quality), Eastern Cereal and Oilseed Research Center, Agriculture and Agri-Food Canada, Ottawa, ON, Canada

Adam Friedman, MD, FAAD Assistant Professor of Medicine (Dermatology)/Physiology and Biophysics, Director of Dermatologic Research, Associate Residency Program Director, Division of Dermatology, Department of Medicine, Montefiore Medical Center, Bronx, New York, USA, Department of Physiology and Biophysics, Albert Einstein College of Medicine, Bronx, New York, USA

Apeksha A. Gulvady, R&D Nutrition Senior Scientist, Global R&D Nutrition, PepsiCo Inc., Barrington, IL, USA

Bruce Hamaker, Whistler Center for Carbohydrate Research, Purdue University, West Lafayette, IN, USA

Stephanie Jew, RD Sector Specialist – Regulation, Agriculture and Agri-Food Canada, Ottawa, ON, Canada

Madhuvanti Kale, Whistler Center for Carbohydrate Research, Purdue University, West Lafayette, IN, USA

Prabhakar Kasturi, Global R&D Technical Insights – Analytical Department, PepsiCo Inc., Barrington, IL, USA

Chor San Khoo, PhD Nutritionist, Mt. Laurel, NJ, USA

Renee Korczak, MS Department of Food Science and Nutrition, University of Minnesota, St. Paul, MN, USA

Penny Kris-Etherton, PhD RD Distinguished Professor, Department of Nutritional Sciences, The Pennsylvania State University, University Park, PA, USA

Allison Kutner, MS IV Research Fellow, Division of Dermatology, Department of Medicine, Montefiore Medical Center, Bronx, New York, USA

Rui Hai Liu, MD, PhD Professor, Department of Food Science, Cornell University, Ithaca, NY, USA

Joy Makdisi, Research Fellow, Division of Dermatology, Department of Medicine, Montefiore Medical Center, Bronx, New York, USA

Kevin C. Maki, PhD Chief Science Officer, Biofortis Clinical Research, Addison, IL, USA

Mohsen Meydani, DVM, PhD, FAAA, FASN Professor of Nutrition, Friedman School of Nutrition Science and Policy, Tufts University Senior Scientist and Director of Vascular Biology Laboratory, Jean Mayer USDA Human Nutrition Research Center on Aging at Tufts University, Boston, MA, USA

Tia M. Rains, PhD Principal Scientist, Biofortis Clinical Research, Addison, IL, USA

Camille Rhymer, MSc Research Assistant, Agriculture and Agri-Food Canada, Winnipeg, MB, Canada

Joanne Slavin, PhD, RD Department of Food Science and Nutrition, University of Minnesota, St. Paul, MN, USA

Joanne Storsley, MSc Cereal Research Biologist, Agriculture and Agri-Food Canada, Winnipeg, MB, Canada

Susan M. Tosh, PhD Research Scientist, Guelph Food Research Centre, Agriculture and Agri-Food Canada, Guelph, ON, Canada

Mitchell L. Wise, PhD Research Chemist, United States Department of Agriculture, Agricultural Research Service, Cereal Crops Research, Madison, WI, USA

Weikai Yan, PhD Research Scientist (oat breeding), Eastern Cereal and Oilseed Research Center, Agriculture and Agri-Food Canada, Ottawa, ON, Canada

Preface

Why a book on the life cycle of oats?

To our knowledge, a book that discusses the life cycle of oats from on-farm production to finished product to health and policy has not previously been presented. As a result, we felt that such a compendium of articles from multidisciplinary fields would be interesting and educational.

Oats Nutrition and Technology presents a comprehensive and integrated overview of the coordinated activities of plant scientists, food scientists, nutritionists, policy makers, and the private sector in developing oat products for optimal health. Many areas of expertise are integrated, necessarily so, to create the continuum that we know as the contemporary food system (i.e., from "farm to fork"). Readers will gain a good understanding of the value of best agricultural production and processing practices that are important in the oats food system, as well as of all other aspects of today's food system. The book reviews plant agricultural practices for the production of oat products, the food science involved in the processing of oats, and nutrition science aimed at understanding the health effects of oats and how they can affect nutrition policies. There are individual chapters that summarize oat breeding and processing, the many bioactive compounds that oats contain, and their health benefits. With respect to the latter, the health benefits of oats and oat constituents on chronic diseases, gut health, and skin health are reviewed. The book concludes with a global summary of food labeling practices that are particularly relevant to oats.

The book is framed from the perspective of multiple disciplines: plant breeding and processing, the nutritional value of oats (i.e., nutrients and bioactive components) and related health effects, and nutrition policies related to food labeling and health claims. There is much we have learned about the oat food system, but the reality is that much remains to be learned about all of these areas and the advances that are needed to develop the best and most cost-effective oat products for farmers and processing companies in a way that benefits consumers' health as much as possible. In addition, oats and oat products must meet consumer quality expectations that relate to both their sensory preferences and nutritional expectations. Oats must be acceptable to consumers with respect to appearance, texture, flavor, and aroma.

The contents of this book are deliberately organized to familiarize the readers with the various stages of the oat product life cycle. This approach underscores an appreciation for building on the scientific discoveries and knowledge contributed by each discipline, and how important this process is to the development and validation of future oat products for human health.

The eighteen chapters in this book are divided into six sections, with an introductory section (Chapter 1) on oat nutrition research and production. The five remaining sections include Part II: Oat Breeding, Processing, and Product Production; Part III: Oat Nutrition and Chemistry; Part IV: Emerging Nutrition and Health Research; Part V: Public Health Policies and Consumer Response; and Part VI: Future Recommendations. Each section provides readers with an overview on current insights into research, issues, and opportunities.

Part II: Oat Breeding, Processing, and Product Production: This section consists of two chapters (Chapters 2 and 3) that focus on the importance of oat breeding and current challenges in farming and agriculture. Readers will gain a good understanding of the value of best agricultural production and processing practices that are important in the oat food system, and also an appreciation of all other aspects of today's complexity of food production, farming challenges, and product developments.

Part III: Oat Nutrition and Chemistry: This section comprises five chapters (Chapters 4–8) covering chemical and nutritional compositions of whole oats. Discussions also include recently discovered bioactive compounds/phytochemicals in oats, such as avenanthramides, which have strong antioxidative properties and potential health effects. Biosynthesis of bioactive compounds is also discussed.

Part IV: Emerging Nutrition and Health Research: This section consists of seven chapters (Chapters 9–15) that cover emerging research on lipid and lipoprotein metabolism, blood pressure, weight and satiety, diabetes and carbohydrate metabolism, gut health, and skin health. Current insights on studies related to the effects of oats and whole grains on disease and health are presented.

Part V: Public Health Policies and Consumer Response: This section comprises two chapters (16 and 17) that provide global insights into regulatory claims, substantiation requirements, and health policies in the USA, Canada, and the European Union. These chapters also discuss the impact of health claims on government public educational programs (food labeling and advertising), food industry innovation in oat products and sales, and consumer and professional responses to oat products.

Part VI: Future Recommendations: In this section, summaries of the previous 17 chapters are discussed in a single chapter. Future research needs and recommendations are discussed as well. There are many opportunities to expand our knowledge of oats and their development to optimize nutrition, as well production and sustainability.

This book is intended to offer scientists and health practitioners interested in this field in-depth information about the life cycle of oats. It is intended to be thought provoking and stimulate readers to address the many research challenges associated with the oat life cycle and food system.

YiFang Chu

Acknowledgements

In the field of nutrition, having the opportunity to work with oats and to edit a book about them is a true blessing. I am deeply grateful to Marianne O'Shea and Richard Black for this privilege. I am also thankful for colleagues and mentors who made the process fun and rewarding: Yuhui Shi, Alan Koechner, Yongsoo Chung, Sarah Murphy, Debbie Garcia, and Maria Velissariou. Special thanks go to Chor San Khoo, who challenged and drove us to bring this book to a much better place.

My sincere gratitude goes to the contributors who took the time to provide excellent reviews of the current science and to help bring awareness to the challenges facing oats. Your diligent efforts come through in each chapter brilliantly.

I would also like to acknowledge colleagues, friends, and collaborators who helped with various aspects of this book: Andrea Bruce, John St. Peter, Prabhakar Kasturi, Jan-Willem van Klinken, Debra Kent, Gary Carder, Laura Harkness, Mike Morello, John Yen, John Schuette, Nancy Moriarity, Jeanette Ramos, Ellen Moreland, Michelle Slimko, Bonnie Johnson, Steve Bridges, Tia Bradley, Tiffany Richardson, Chris Visconti, Lori Romano, and Renuka Menon from PepsiCo; Boxin Ou from International Chemistry Testing Inc.; and David McDade from Wiley-Blackwell.

I am profoundly indebted to my parents, Pi-Chi and Li-Chiu, whose unwavering love has been the constant anchor in my life. Finally, thank you April, Winston, and Isis – you have filled my life with simple joy and made me about the happiest person on earth every single day.

Part I
Introduction

1

Introduction: Oat Nutrition, Health, and the Potential Threat of a Declining Production on Consumption

Penny Kris-Etherton[1], Chor San Khoo[2], and YiFang Chu[3]

[1]Department of Nutritional Sciences, The Pennsylvania State University, University Park, PA, USA
[2]Nutritionist, Mt. Laurel, NJ, USA
[3]Quaker Oats Center of Excellence, PepsiCo R&D Nutrition, Barrington, IL, USA

1.1   A landmark health claim

The landmark approval of a health claim for oats in 1997 by the United States Food and Drug Administration (FDA) marked the first food specific health claim. The FDA had concluded that an intake of at least 3 g β-glucan from oats as part of a diet low in saturated fats could help reduce the risk of heart disease (Chapter 17). Of importance is that the oat health claim signifies for the first time recognition by a public health agency that dietary intervention could be beneficial in disease prevention, and that certain foods or food components, when consumed as part of a healthy diet, may reduce the risk of certain diseases. It is, therefore, not surprising that the first food-related health claim was approved for reducing the risk of cardiovascular disease (CVD), the leading cause of death in the United States and many western countries, including Canada (Health Canada, 2010). Often under communicated is that CVD is the leading cause of death among women in the United States (Roger *et al.*, 2012). The FDA approval of a health claim elevated the role of diet in overall health, adding emphasis to disease prevention in addition to treatment. For example, many of the risk factors

Oats Nutrition and Technology, First Edition. Edited by YiFang Chu.
© 2014 John Wiley & Sons, Ltd. Published 2014 by John Wiley & Sons, Ltd.

associated with CVD are preventable by dietary interventions, including high blood pressure, high total serum cholesterol, low-density lipoprotein-cholesterol (LDL-C) and very low density lipoprotein-cholesterol, and high blood glucose associated with type 2 diabetes, and obesity.

1.2   The growing interest in oats and health

The oat health claim that underwent extensive scientific review for approval by the FDA sparked great interest in the scientific community. For the first time, health practitioners (dietitians, nutritionists, and physicians) had the option to recommend that a specific food be incorporated into a diet for an adjunct intervention in the management and prevention of disease.

The unique chemistry and nutritional composition of oats suggest that the benefits of oats may not be confined to just a cholesterol-lowering effect but, as demonstrated by further research, that they may also have other favorable health benefits. As of 2010, ischemic heart disease (number 1 ranking) and stroke (number 3 ranking) were two of the top 12 world health problems that could be favorably affected by oat consumption (Cohen, 2012; Lim *et al.*, 2012). Important risk factors recently highlighted by the Global Burden of Disease Study that could be affected by oats include high blood pressure, high body mass index, and high fasting blood glucose levels (Cohen, 2012; Lim *et al.*, 2012), as well as an elevated LDL-C level as noted by the American Heart Association (Roger *et al.*, 2012).

The oat health claim has sparked interest in developing a better understanding of oats, from breeding for the best oat cultivar, processing, nutrition research on oats and health, as well as public health education and policy. It has become clear that the challenges to improving the quality of oats are not just yield but rather a combination of three possible dependent traits—yield, groat percentage, and β-glucan level (Chapter 2).

Recent advances in research have focused on oat chemistry and nutrition with the goal of demonstrating the mode of action of oats on lipid and glucose metabolism. Of interest is the form of β-glucan in oats, which differs from other whole grain soluble fibers. In oats, the majority of the soluble fibers are β-glucan, accounting for 3–6% of whole groat weight. Although β-glucan also exists in barley and wheat, the β-glucan in oats differ in many physicochemical properties, such as solubility, gelation, and molecular weight, all of which affect physiological functions in the gastrointestinal tract, for example, bile acid binding, colonic viscosity accumulation, and fermentation. These differences in β-glucan structure may explain the reduction in cholesterol and postprandial blood glucose levels with oat consumption (Chapter 5)

The health benefits of oats can be attributed largely to their unique chemistry and nutrient profile. Recent efforts have focused on isolating, identifying, and characterizing the bioactive constituents unique to oats. Compared to other whole grains such as corn, wheat, and rice, oat nutrition profiles are uniquely "complete" across many constituents, ranging from nutrients to phytochemicals and bioactive compounds. Nutritionally, oats provide many essential nutrients.

On a 100 g basis, oats are a significant source of dietary fiber, soluble fiber mostly as β-glucan, thiamin, folate, iron, magnesium, copper, and zinc. Additionally, oats are an excellent source of potassium and are low in sodium, with a Na:K ratio less than one (Chapter 4).

Avenanthramides are phytonutrients in oats known to have anti-inflammatory and antioxidative activity, and may be involved in some of the health effects unique to oats. Avenanthramides are emerging as an interesting class of chemicals that may be beneficial for skin health, including treatment for atopic dermatitis, contact dermatitis, pruritic dermatoses, sunburn, drug eruptions, and other conditions. Colloidal oatmeal has also been used to relieve skin irritation and itching, and for cleansing and moisturizing. The flavonoids in oats may also protect against ultraviolet A radiation.

More recently, research has focused on the impact of oat intake on other health outcomes beyond the lipid lowering effect, such as blood pressure, body mass index and weight, glucose metabolism and type 2 diabetes, as well as caloric regulation and satiety. These studies are ongoing and the data are still preliminary. A consistent finding is that oat β-glucan lowers serum cholesterol, and although the magnitude of cholesterol lowering varies, it correlates to the amount of β-glucan consumed.

1.3   Declining production poses threats to the growth of oat intake

Although oat and health research have advanced significantly, a very different picture is emerging on the global scene with respect to oat production and consumption. Since the approval of the health claim for oats in 1997, there has been a steep growth in the demand for hot breakfast cereals and oats sales have soared. This positive trend developed in North America was also observed in eastern and western Europe over the same period. On the other hand, world production of oats has declined and is at a record low rate. In 2011, world oat production lagged behind wheat, corn, and barley, dropping to its lowest level since 1960, from 6.8 to 0.8% of the world's crop production. In the United States, oats are fading from a commodity to a specialty crop. The worldwide drop in production may be attributed to several factors, including more land devoted to growing more profitable crops for foods, feeds, biofuels, and vegetable oils; low amounts of funding for research, little innovation in production techniques; and a weak demand for oats as a feed source (Strychar, 2011). Today, oats are considered an orphan crop, receiving little research investment from either government or industry.

If the trend of decreased oat production continues, oats will become so expensive that affordable and widely accessible oat products for the public may be limited. Reversing this trend will require programs that involve both public and private collaborations to assure an adequate level of research investment for advancing the understanding and securing the accessibility of this important crop.

References

Cohen, J. (2012) A controversial close-up of humanity's health. *Science* **338**, 1414–1416.

Health Canada (2010) Cardiovascular Disease Morbidity, Mortality and Risk Factors Surveillance Information. Public Health Agency of Canada (www.publichealth.gc.ca; last accessed 14 May 2013).

Lim, S., *et al.* (2012) A comparative risk assessment of burden of disease and injury attributable to 67 risk factors and risk factor clusters in 21 regions, 1990–2010: A systematic analysis for the Global Burden of Disease Study 2010. *Lancet* **330**, 2224–2260.

Roger, V.L., *et al.* (2012) Executive summary: Heart disease and stroke statistics – 2012 update: A report from the American Heart Association. *Circulation* **125**, 188–197.

Strychar, R. (2011) The Future of Oats. Presentation at the Nordic Oat Days conference, 10 October 2011, Helsinki, Finland.

Part II
Oat Breeding, Processing, and Product Production

2

Breeding for Ideal Milling Oat: Challenges and Strategies

Weikai Yan[1], Judith Frégeau-Reid[1], and Jennifer Mitchell Fetch[2]

[1]*Eastern Cereal and Oilseed Research Center, Agriculture and Agri-Food Canada, Ottawa, ON, Canada*
[2]*Cereal Research Centre, Agriculture and Agri-Food Canada, Winnipeg, MB, Canada*

2.1   Introduction

Both acreage devoted to oats and oat production have dramatically decreased worldwide since the 1960s, as working horses have been replaced by modern farm machinery. The introduction of short-seasoned and more profitable corn and soybean cultivars to the northern regions of the United States and southern areas of Canada in the recent decade is another major reason for reduced oat production. However, the oat acreage in Canada has more or less stabilized at around 1.5 million hectares in recent years (Agriculture and Agri-Food Canada, 2010). This is partially due to the need for growing oats as a rotation crop and the use of oats as a forage crop, oat grains as feed, and oat straw for animal bedding. However, more important are the increased purchase and processing of oat grains by the milling industry and increased awareness and human consumption of oat products as healthy food.

Oats are a minor crop compared with other cereal crops and oilseeds. In addition, it is a self-pollinated crop, obviating the need for purchasing hybrid seed every year. Because of its lower profitability, relatively little breeding and research on oats are carried out. The limited breeding and research effort has been supported primarily through government funding with support from the oat milling industry and growers of oat seed and grain. As a result, breeding for superior milling oats has become a main driving force for oat breeding and related research. Although there are some differences in the specifications for oats used as feed or fodder, oats that are excellent for milling are also suitable for forage and feed. In this chapter, an attempt is made to define the ideal milling oat cultivar and the challenges and strategies in breeding such an oat cultivar to discuss.

2.1.1   What is an ideal milling oat?

An ideal milling oat cultivar must be defined from the perspective of the oat value chain, which starts with the oat growers and ends with consumers of the oat product, with the oat processors serving as the key link between the two. An ideal oat cultivar must benefit each of these stakeholders. A reliably high yield, along with supporting agronomic traits (good resistance to important diseases and pests, lodging resistance, and proper maturity), is the number one consideration of oat growers when choosing a crop cultivar. The second factor they consider is whether the quality of their oat grains meets the requirements of potential buyers (i.e., millers), because selling to millers is often more profitable than using or selling the oats as feed. The requirements of the millers include higher groat percentage, so that more oat product can be produced per unit weight of purchased oat grains, uniform grains and easy dehulling to reduce the energy cost for processing, and better compositional quality so their oat products meet consumers' expectations. Consumers consider oat products to be nutritious and especially healthy because of the dietary fiber contained in the oat groat (β-glucan in particular). Oat products must contain a minimum level of β-glucan and total dietary fiber to be labeled as healthy food (Chapter 6). The traits of an ideal milling oat cultivar are listed in Table 2.1.

Despite the tremendous effort of oat breeders and great progress made in improving oat cultivars throughout the world, a cultivar with all the desired traits has not yet been developed. Why is this so? Is it even possible to achieve such a goal? What are the challenges for developing the ideal cultivar? What strategies should be used in breeding towards such a cultivar? These are the questions this chapter attempts to answer.

Table 2.1   Trait compositions for an ideal milling oat cultivar

For growers	For millers	For consumers
High and stable grain yield for the target environment	High groat percentage (milling yield)	High β-glucan and dietary fiber content
Good lodging resistance	Easy dehulling	High protein content
Proper maturity	Uniform kernels	High levels of essential amino acids
Good resistance to relevant diseases and pests	Low groat breakage during dehulling	Low oil content
Good tolerance to relevant abiotic stresses	White groat color	High antioxidant content
High test weight		Other desirable ingredients
Large kernels		
High straw yield		

To facilitate the discussion, a real data set from the 2011 "Nationwide Oat Test" is examined in this chapter. The breeding lines tested at seven locations across Canada included 45 new covered oat breeding lines developed from the oat breeding program at the Eastern Cereal and Oilseed Research Center (ECORC) of Agriculture and Agri-Food Canada (AAFC or AAC) located in Ottawa, Ontario, and 45 lines from the Cereal Research Center (CRC) of AAFC located in Winnipeg, Manitoba, plus six official check cultivars for the Prairies, Ontario, and Quebec. These locations were: Lacombe (AB), Saskatoon (SK), Portage (MB), Ottawa (ON), New Liskeard (ON), Normandin (QC), and Harrington (PE). The experimental design was randomized incomplete blocks with three replications at each location. Grain yield and important quality characteristics (e.g., test weight, kernel weight, groat percentage, and concentrations of β-glucan, oil, and protein) were determined for each location.

The data analysis method used in this chapter is GGE biplot analysis (Yan *et al.*, 2000; Yan and Kang, 2003). A GGE biplot summarizes the information of genotype main effect (G) and genotype-by-environment (location in this case) interaction effects (GE) in a genotype-by-environment two-way data set. G and GE are the two pieces of information pertinent to cultivar and test environment evaluations. The biplot was first developed by Gabriel (1971) to graphically display the principal component analysis results of a two-way data set, such as the yield data of a set of genotypes in a set of environments. It is so named because it displays both genotype names and location names in the same plot. The unique features of the GGE biplot allow visual examination of the data to answer the important questions a plant breeder needs to ask.

2.2   Breeding for single traits: Genotype-by-environment interactions

Breeding for a single trait is limited by two factors: the availability of genetic variation, (i.e., availability of germplasm with desired levels of that trait) and its heritability. Germplasm collection, preservation, evaluation, and utilization have always been the key components of plant breeding that set the ultimate limit of crop improvement. However, for the current discussion, it is assumed that sufficient genetic variation exists for each trait within the breeding lines tested and discussion focuses on the second factor, trait heritability. Ignoring experimental errors at individual test locations, the heritability of a trait in the multilocation scenario is a matter of the relative magnitude of genetic variance versus genotype-by-location interaction variance (i.e., the G/[G+GE] ratio), which can also be expressed as genetic correlations among test locations. A high heritability across environments (high G/[G+GE] ratio) or a close genetic correlation among test environments means that the test environments (or locations) are relatively homogeneous; therefore, selection for general adaptation for the whole region based on mean yield across all environments is feasible and effective. Otherwise, the target environments must be divided into subregions or mega-environments, and specific adaptation to each subregion must be sought (Yan *et al.*, 2007a).

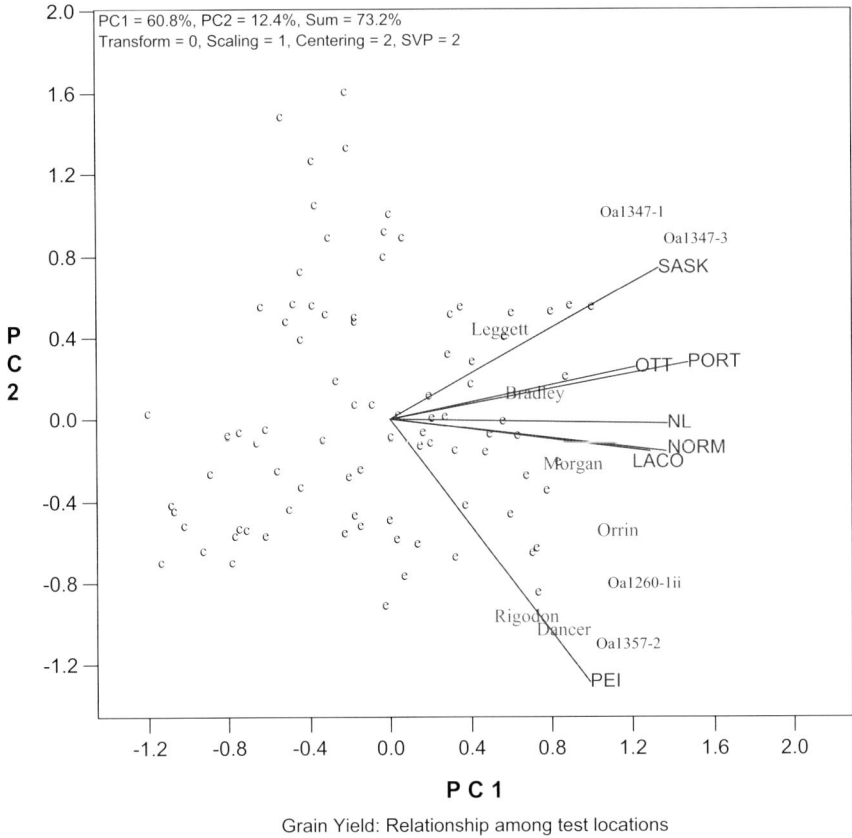

Grain Yield: Relationship among test locations

Figure 2.1   The "environment association" view of the GGE biplot for grain yield. (Biplot based on location-standardized data and location-focused singular value partition, SVP.) (For color version, see color plate section.)

2.2.1   Grain yield

The yield data for each of the 96 genotypes (90 breeding lines plus six check cultivars) at each of the seven locations in the 2011 Nationwide Oat Test are summarized in the form of a GGE biplot (Figure 2.1).[1]

A GGE biplot can be viewed in many different ways by adding supplementary lines to the biplot to explore specific aspects of the two-way data. The biplot shown in Figure 2.1 is the "environmental relationship" view, which is useful for visualizing genetic correlations among the test locations. The biplot explains

[1] In all figures, locations are indicated in upper case: LACO: Lacombe AB; NL: New Liskeard ON; NORM: Normandin QC; OTT: Ottawa ON; PEI: Harrington PE; PORT: Portage MB; and SASK: Saskatoon SK. Breeding lines from the CRC oat breeding program are labeled as "c" and those from the ECORC program as "e." The names of the six check cultivars and a few breeding lines are spelled out.

Table 2.2 Genetic correlations among test locations for grain yield

Locations	LACO	NL	NORM	OTT	PEI	PORT	SASK	Mean
LACO (Lacombe)		0.56	0.62	0.39	0.44	0.62	0.57	0.60
NL (New Liskeard)	0.56		0.58	0.62	0.47	0.69	0.56	0.64
NORM (Normandin)	0.62	0.58		0.45	0.52	0.61	0.68	0.64
OTT (Ottawa)	0.39	0.62	0.45		0.36	0.66	0.51	0.57
PEI (Harrington)	0.44	0.47	0.52	0.36		0.43	0.17	0.48
PORT (Portage)	0.62	0.69	0.61	0.66	0.43		0.74	0.68
SASK	0.57	0.56	0.68	0.51	0.17	0.74		0.61

The threshold correlation value is 0.206 for $P < 0.05$ and 0.265 for $P < 0.01$.

73% of the G+GE of the yield data and is adequate for displaying the main patterns of the data. The cosine of the angle between any two locations approximates the genetic correlation between them. The locations appear to be positively correlated to each other, because the angles between them are all smaller than 90°, except the angle between locations PEI (Harrington, PE) and SASK (Saskatoon, SK), which is close to 90°. This biplot presentation of genetic correlations among test locations can be verified by using the numerical correlation matrix of test locations (Table 2.2). The correlation matrix shows that all locations are positively correlated with each other, except for PEI, which is uncorrelated with SASK and is also less correlated with the other locations. The biplot presentation is much easier to comprehend.

Lack of positive genetic correlation between any two test locations is due to the presence of a large GE; a large GE relative to G can cause significant crossover GE (i.e., obvious rank change of genotypes at different locations), which in turn can lead to differentiation of subregions or mega-environments. Indeed, the "which-won-where" view of the same biplot (Figure 2.2) reveals that although breeding line OA1347-3 appeared to be the highest yielding line at most locations, the highest yielding line at PEI was OA1357-2. The "which-won-where" view of the GGE biplot contains an irregular polygon, which is formed by connecting the genotypes farthest from the biplot origin at various directions, such that all genotypes are either on the sides of the polygon or enclosed within the polygon. This biplot view also contains a set of straight lines that originate from the biplot origin and are perpendicular to each side of the polygon, dividing the biplot area into sectors. Each of the environments inevitably falls into one of the sectors. For example, the location PEI falls into one sector, and all other locations into another. An interesting property of the "which-won-where" view is that the genotype placed at the vertex of the polygon in a sector is nominally the one with the highest values for all environments falling into that respective sector. Thus, the highest yielding genotype for PEI was OA1357-2, whereas the highest yielding genotype for the other six locations was OA1347-3.

Figure 2.2 suggests that the seven test locations may be divided into two subregions or mega-environments. However, this cannot be considered conclusive

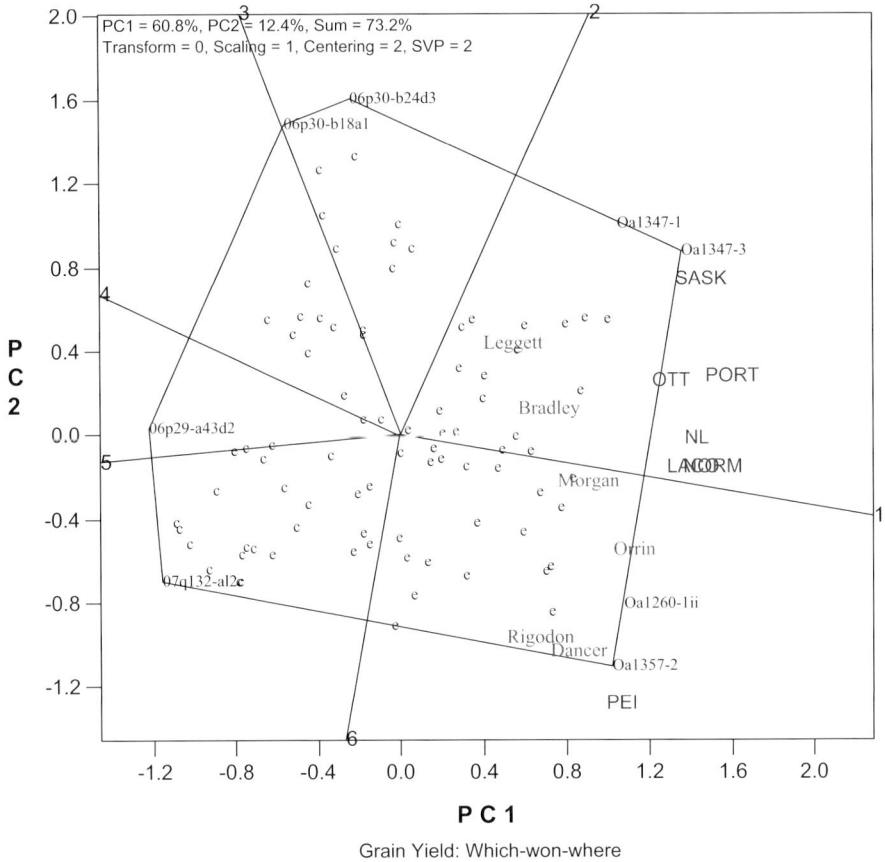

Figure 2.2 The "which-won-where" view of the GGE biplot for grain yield. (Biplot based on location-standardized data and location-focused singular value partition, SVP.) (For color version, see color plate section.)

because the biplot was based on data from a single year, and the suggestion contradicts previous reports that PEI belonged to the same mega-environment with New Liskeard and Normandin (Yan *et al.*, 2010). Given that most locations were positively correlated, the test locations may be relatively homogenous in terms of yield response. Indeed, the G/(G+GE) ratio for this data set was 57.5%, and the heritability across test location was 0.887, supporting this idea. Accepting that all test locations belong to the same mega-environment simplifies cultivar evaluation. It means that genotypes can be evaluated based on their mean yields across test locations. The "mean-versus-stability" view (Figure 2.3) was designed for this purpose. The red line with a single arrow points to a higher mean yield across all environments and is called the average environment axis. It is drawn to pass through the biplot origin and the small circle that represents the average environment. Thus, genotypes are ordered in terms of their mean yields across

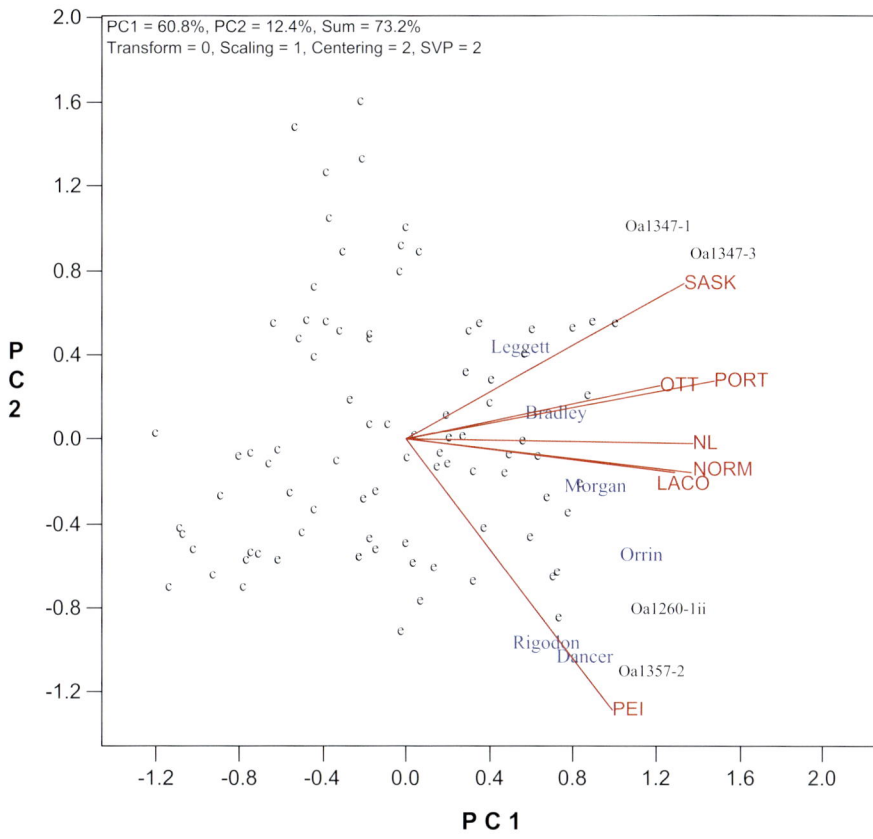

PC1 = 60.8%, PC2 = 12.4%, Sum = 73.2%
Transform = 0, Scaling = 1, Centering = 2, SVP = 2

Oa1347-1
Oa1347-3
SASK
Leggett
OTT PORT
Bradley
NL
NORM
Morgan LACO
Orrin
Oa1260-1ii
Rigodon
Dancer
Oa1357-2
PEI

P C 2 — P C 1

Grain Yield: Relationship among test locations

Plate 2.1 The "environment association" view of the GGE biplot for grain yield. (Biplot based on location-standardized data and location-focused singular value partition, SVP.)

Oats Nutrition and Technology, First Edition. Edited by YiFang Chu.
© 2014 John Wiley & Sons, Ltd. Published 2014 by John Wiley & Sons, Ltd.

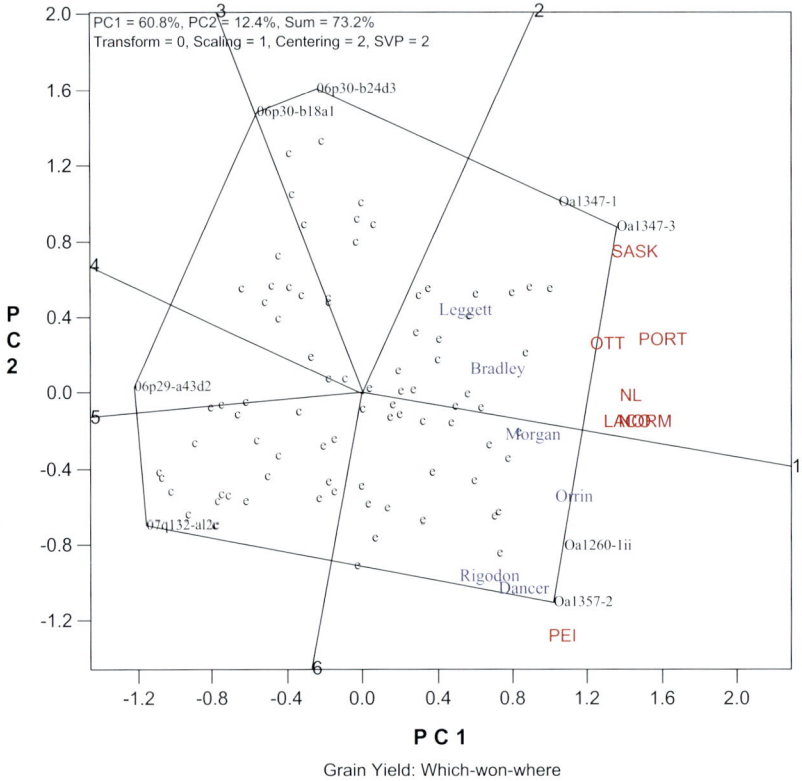

Grain Yield: Which-won-where

Plate 2.2 The "which-won-where" view of the GGE biplot for grain yield. (Biplot based on location-standardized data and location-focused singular value partition, SVP.)

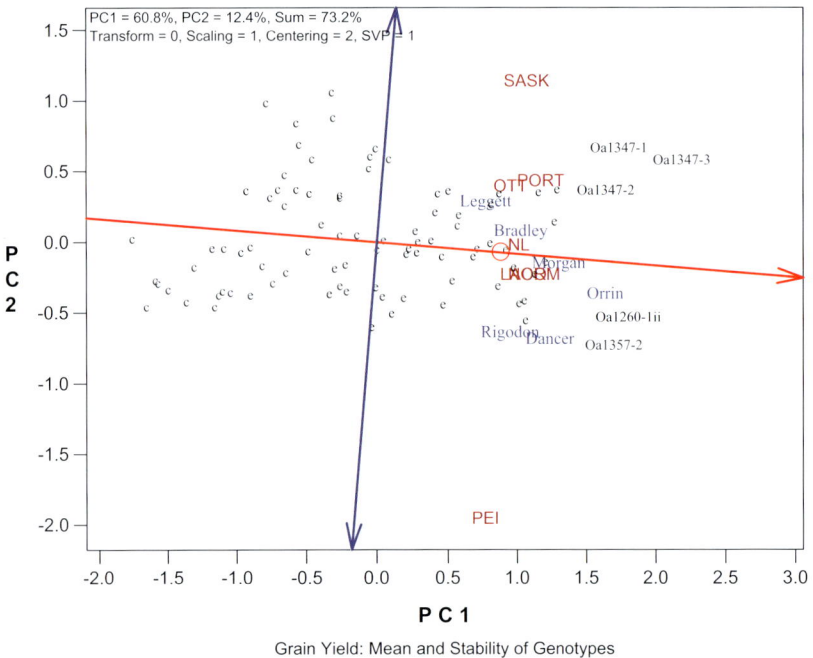

Grain Yield: Mean and Stability of Genotypes

Plate 2.3 The "mean-versus-stability" view of the GGE biplot for grain yield. (Biplot based on location-standardized data and genotype-focused singular value partition, SVP.)

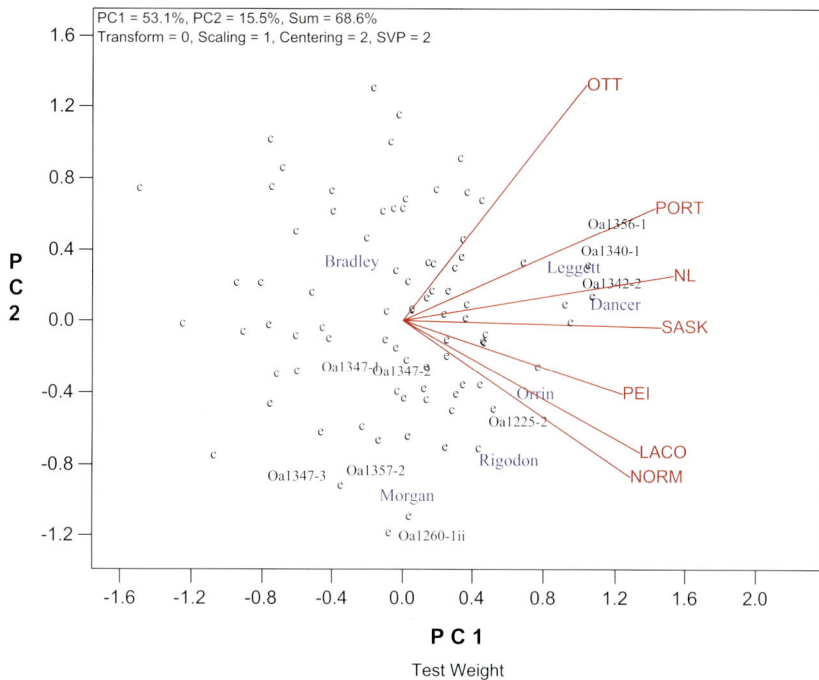

Plate 2.4 The "environment association" view of the GGE biplot for test weight. (Biplot was based on location-standardized data and location-focused singular value partition, SVP.)

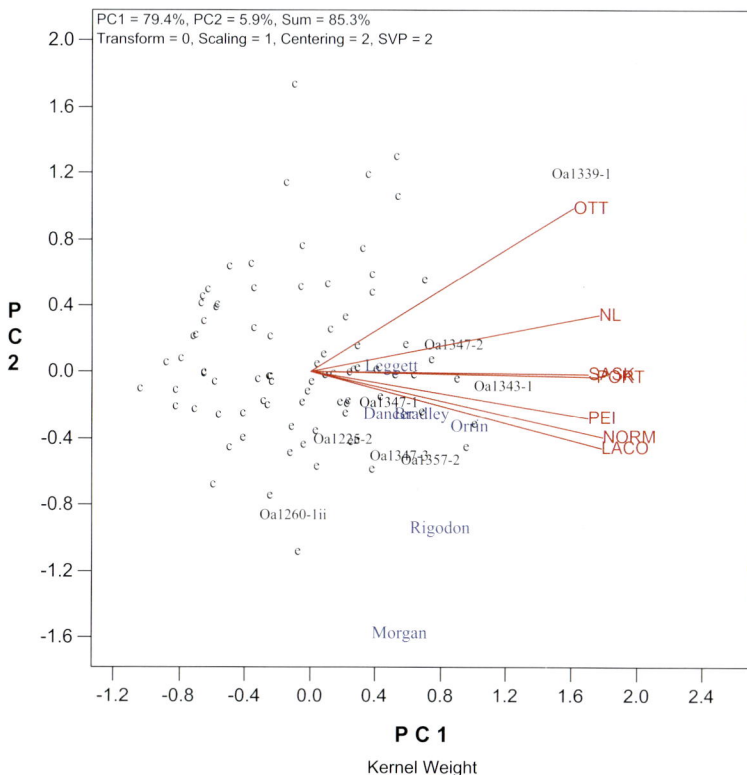

Plate 2.5 The "environment association" view of the GGE biplot for thousand-kernel weight. (Biplot based on location-standardized data and location-focused singular value partition, SVP.)

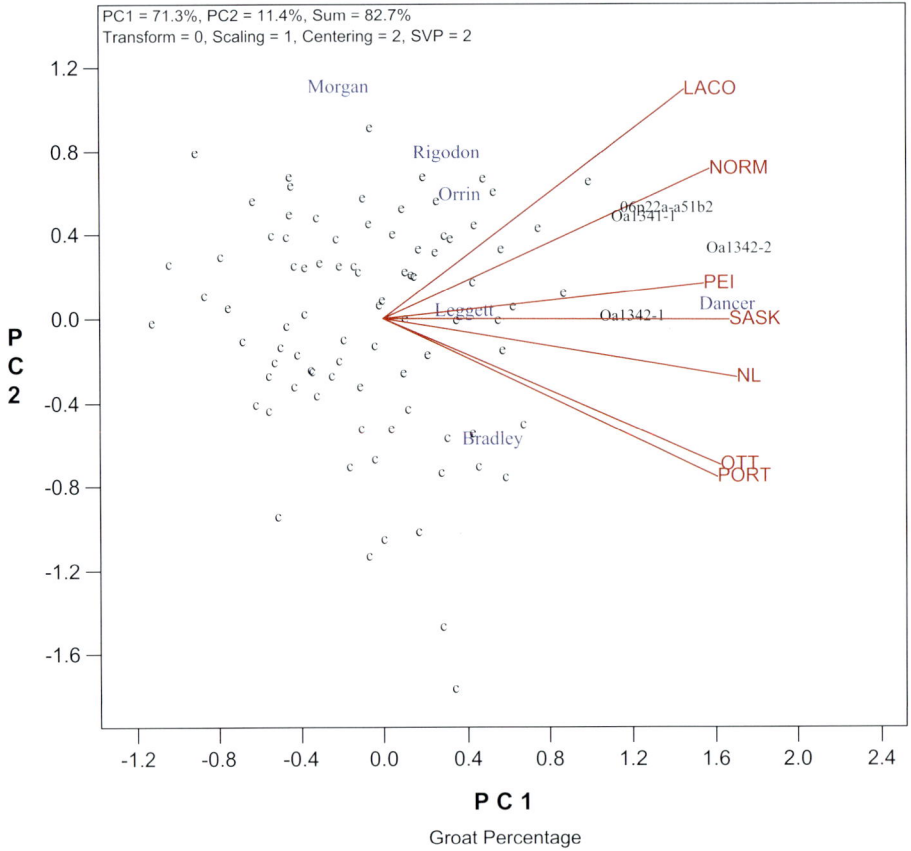

Plate 2.6 The "environment association" view of the GGE biplot for groat percentage. (Biplot based on location-standardized data and location-focused singular value partition, SVP.)

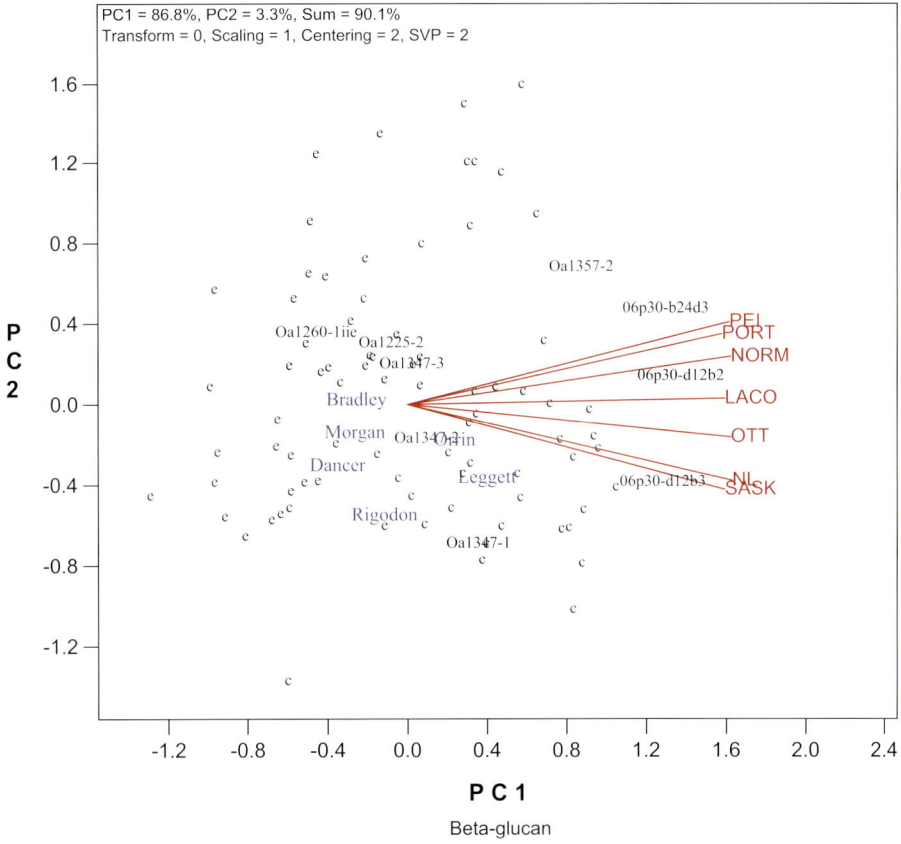

Plate 2.7 The "environment association" view of the GGE biplot for β-glucan concentration in the groat. (Biplot was based on location-standardized data and location-focused singular value partition, SVP.)

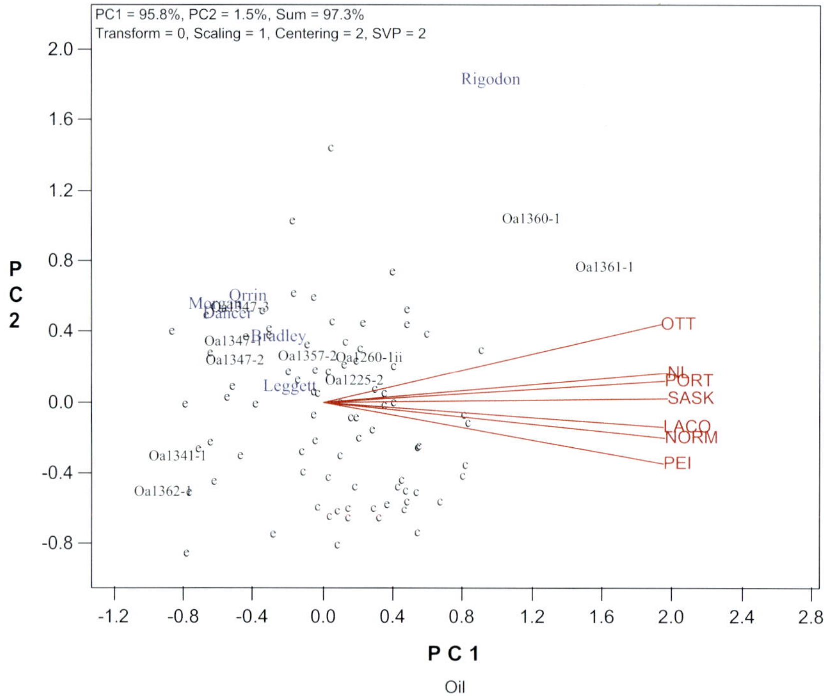

Plate 2.8 The "environment association" view of the GGE biplot for oil concentration in the groat. (Biplot based on location-standardized data and location-focused singular value partition, SVP.)

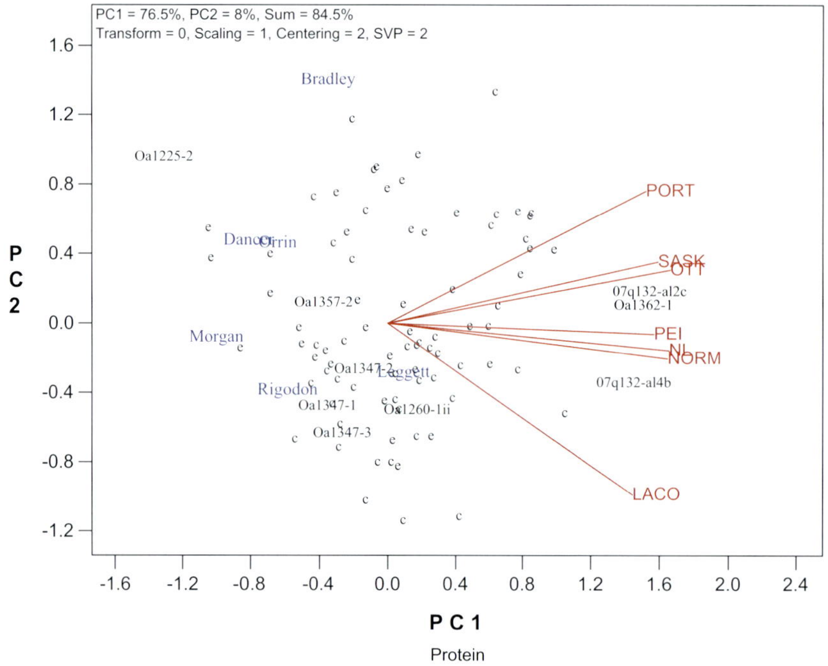

Plate 2.9 The "environment association" view of the GGE biplot for protein concentration in the groat. (Biplot based on location-standardized data and location-focused singular value partition, SVP.)

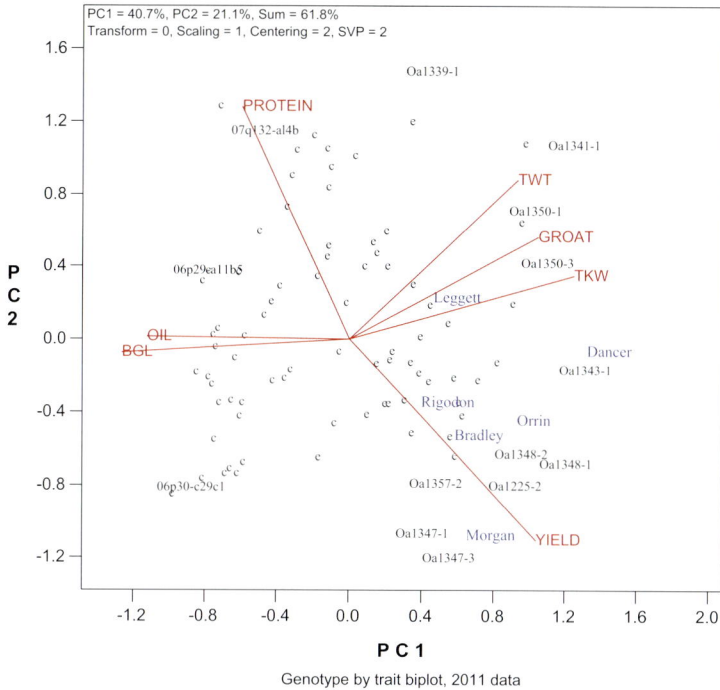

Genotype by trait biplot, 2011 data

Plate 2.10 The genotype-by-trait biplot involving seven traits: grain yield (YIELD), groat percentage (GROAT), β-glucan concentration (BGL), oil concentration (OIL), protein concentration (PROTEIN), test weight (TWT), and thousand-kernel weight (TKW). (Biplot based on trait-standardized data and trait-focused singular value partition, SVP.)

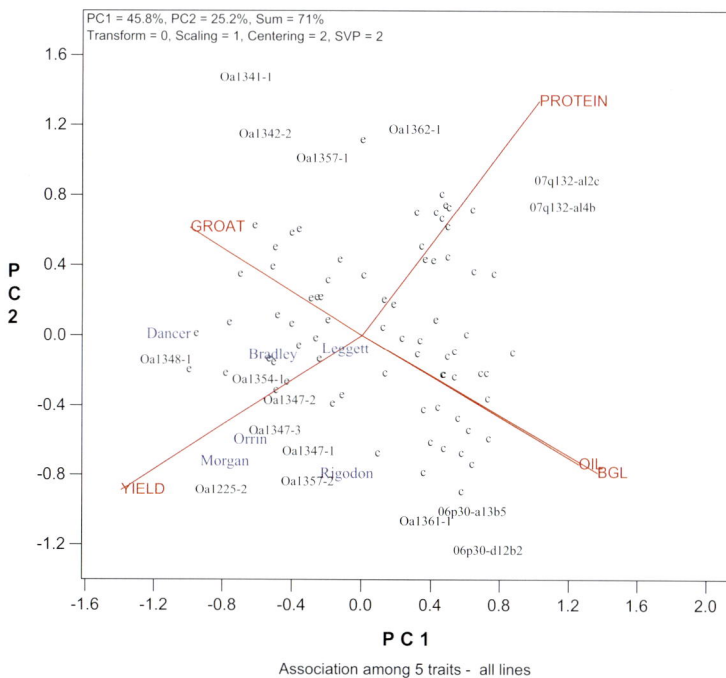

Association among 5 traits - all lines

Plate 2.11 The genotype-by-trait biplot involving five traits: grain yield (YIELD), groat percentage (GROAT), β-glucan concentration (BGL), oil concentration (OIL), and protein concentration (PROTEIN). (Biplot based on trait-standardized data and trait-focused singular value partition, SVP.)

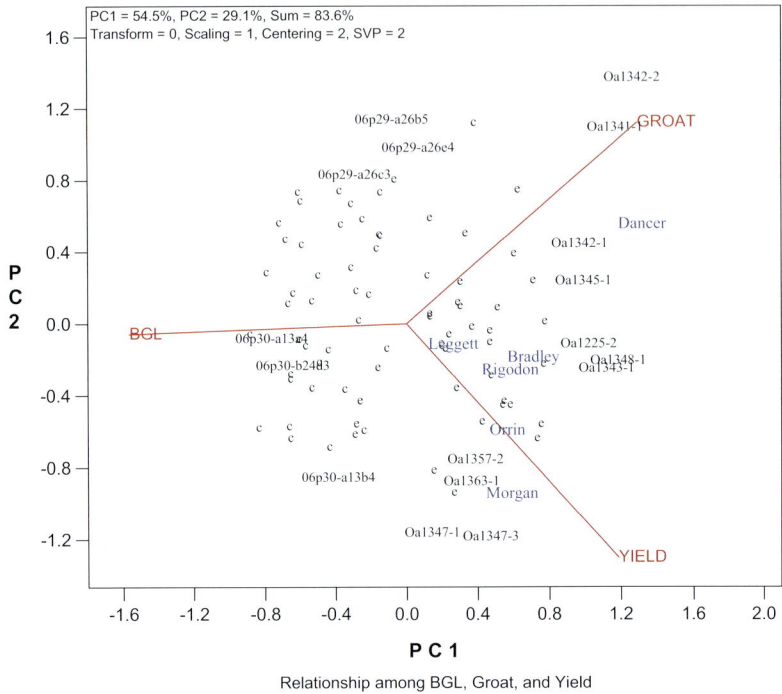

Relationship among BGL, Groat, and Yield

Plate 2.12 The genotype-by-trait biplot involving three traits: grain yield (YIELD), groat percentage (GROAT), and β-glucan concentration (BGL). (Biplot based on trait-standardized data and trait-focused singular value partition, SVP.)

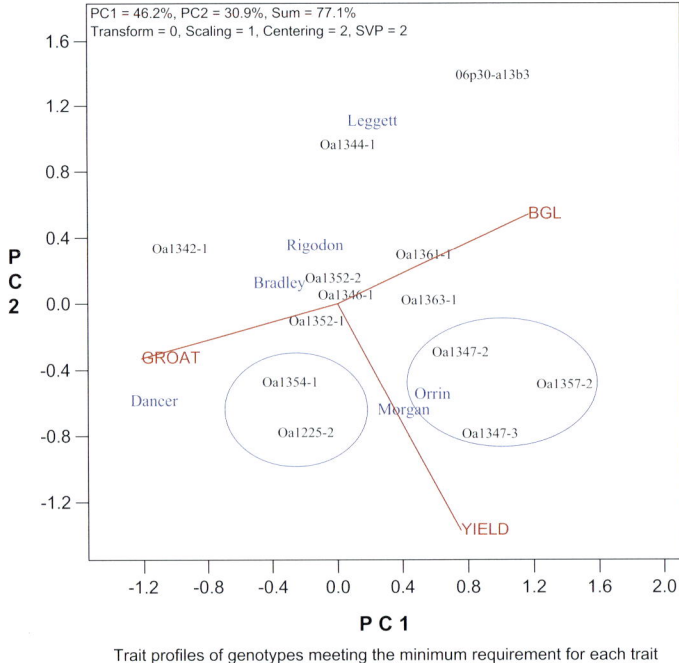

Trait profiles of genotypes meeting the minimum requirement for each trait

Plate 2.15 The genotype-by-trait biplot involving three traits and 19 selected genotypes, which approximates the mean trait levels of the six check cultivars and 13 promising genotypes for grain yield (YIELD), groat percentage (GROAT), and β-glucan concentration (BGL). (Biplot based on trait-standardized data and trait-focused singular value partition, SVP.)

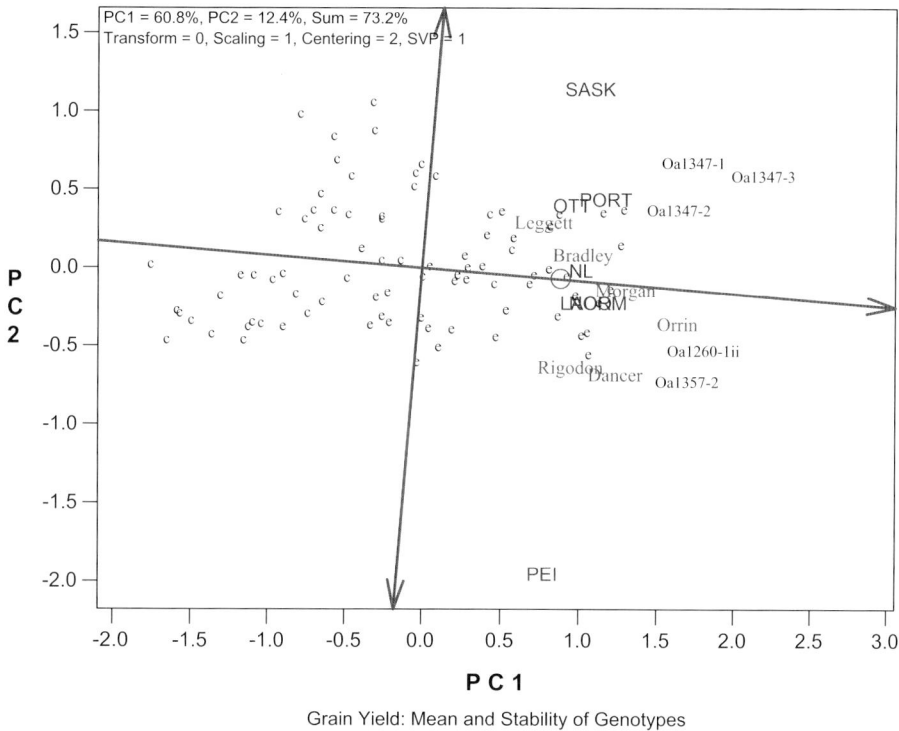

Grain Yield: Mean and Stability of Genotypes

Figure 2.3   The "mean-versus-stability" view of the GGE biplot for grain yield. (Biplot based on location-standardized data and genotype-focused singular value partition, SVP.) (For color version, see color plate section.)

the seven locations on the biplot: OA1347-3 > OA1260-1II > OA1357-2 ≈ Orrin > OA1347-1 >... The line with two arrows pointing outwards represents genotype instability. The closer the placement of a genotype to the red line, the more stable it is in yield performance. The biplot shows that the check cultivar Morgan is highly stable and that the check cultivar Orrin is more stable than the new breeding lines with higher mean yields.

In conclusion, high-yielding genotypes can be easily selected based on their mean yields across environments. The genotype-by-location interaction did not seem to constitute a main challenge in selecting high-yielding genotypes.

2.2.2   Test weight

Like the yield data, the GGE biplot for test weight data (Figure 2.4) shows significant positive genetic correlations among the test locations, although the Ottawa site (OTT) was less correlated with the locations Lacombe and Normandin. Ottawa is the southernmost location of the seven test sites, and resistance to crown rust is usually an important genetic factor for traits such as yield and test weight. The heritability among locations was 0.846 and the G/(G+GE)

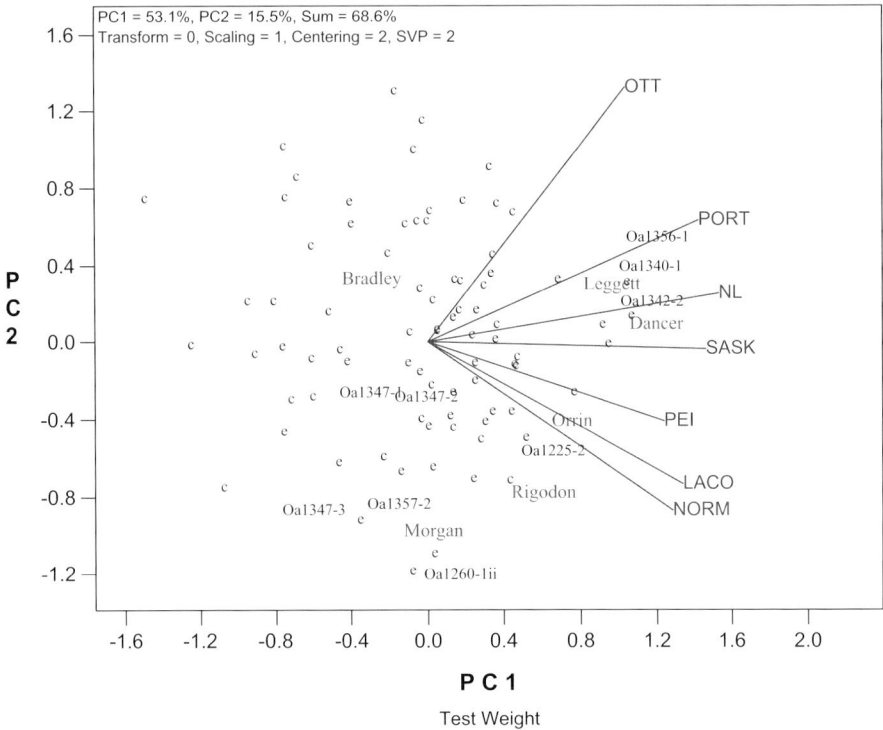

Figure 2.4 The "environment association" view of the GGE biplot for test weight. (Biplot was based on location-standardized data and location-focused singular value partition, SVP.) (For color version, see color plate section.)

was 52%, which are considered relatively high. As a result, genotypes with high test weight (e.g., Dancer, OA1356-1, and OA1342-2) can be easily selected.

2.2.3 Kernel weight

Kernel weight has even higher heritability (0.956) and G/(G+GE) ratio (79%) than grain yield and test weight. This is reflected in the close genetic correlations among the test locations (Figure 2.5). Consequently, genotypes with high kernel weight (e.g., OA1339-1 and OA1343-1) can be easily selected from any single test location.

2.2.4 Groat percentage

The genetic correlations among test locations for groat percentage (Figure 2.6) were not as high as those for kernel weight but were higher than those for yield or test weight. All test locations were positively correlated, although to varying degrees. Its heritability across locations was 0.926 and its G/(G+GE) ratio was 69%. As a result, genotypes with high groat percentage (e.g., Dancer and OA1342-2) can be easily identified at any single location.

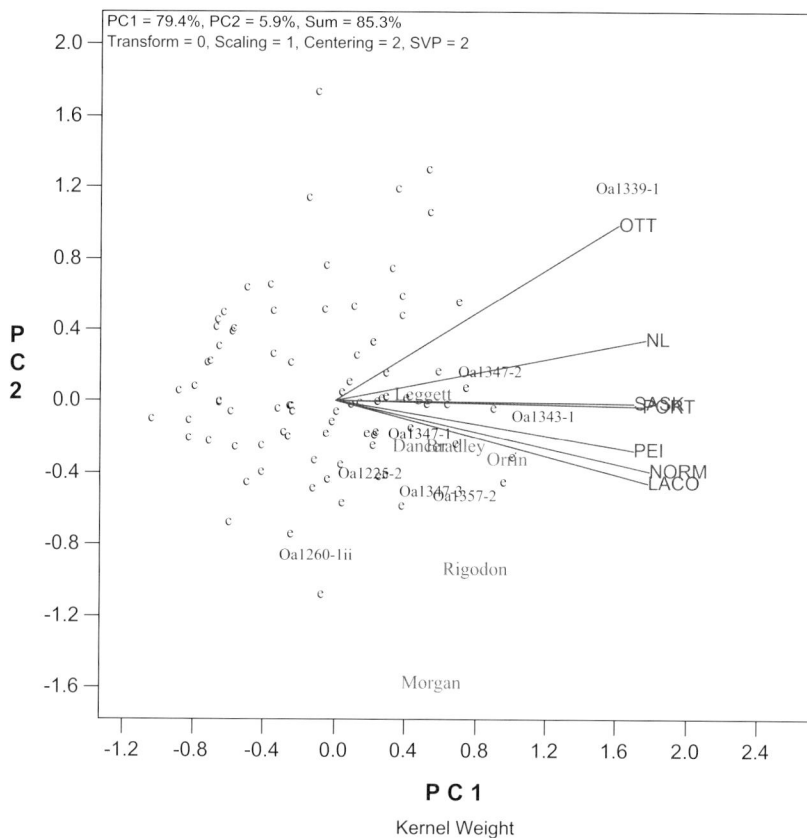

Figure 2.5 The "environment association" view of the GGE biplot for thousand-kernel weight. (Biplot based on location-standardized data and location-focused singular value partition, SVP.) (For color version, see color plate section.)

2.2.5 β-glucan concentration

β-glucan had very high heritability (0.957) and was only slightly affected by genotype-by-location interaction, with a G/(G+GE) ratio of 87%, as reflected by the narrow angles among locations (Figure 2.7). The genetic correlation between any two locations was higher than 0.88. As a result, genotypes with high β-glucan levels (e.g., several CRC lines from the 06p30 family) can be easily selected at any location. High heritability for β-glucan has also been reported in other studies (Holthaus *et al.*, 1996; Cervantes-Martinez *et al.*, 2001; Yan *et al.*, 2011).

2.2.6 Oil concentration

Oil concentration had the highest heritability among most quantitative traits in oat. Across-location heritability was 0.989 and G/(G+GE) ratio was 93%, as reflected by the very acute angles between locations (Figure 2.8). As a result, genotypes with high oil concentrations (e.g., OA1361-1) or low oil concentrations

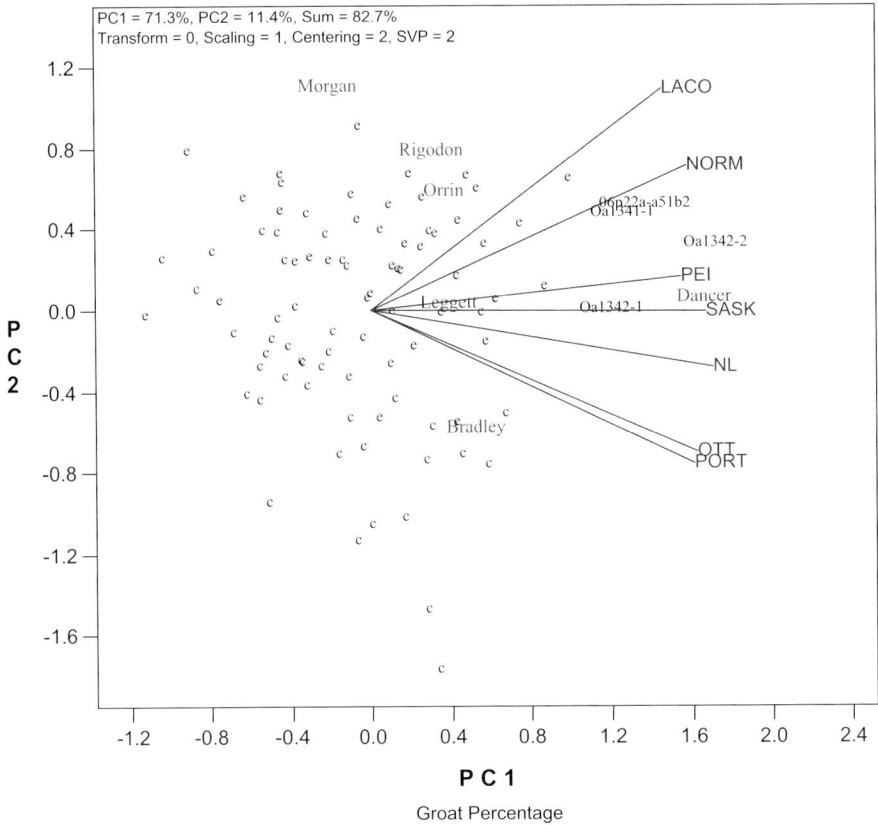

Figure 2.6   The "environment association" view of the GGE biplot for groat percentage. (Biplot based on location-standardized data and location-focused singular value partition, SVP.) (For color version, see color plate section.)

(e.g., OA1362-1) can be easily identified at any location. High heritability for oat oil concentration was reported as early as the 1970s (Baker and McKenzie, 1972; Frey and Hammond, 1975).

2.2.7   Protein concentration

The magnitude of heritability (0.943) and G/(G+GE) ratio (75%) for protein concentration were similar to those of groat percentage and test weight. All locations were positively correlated (Figure 2.9) such that high-protein genotypes (e.g., OA1362-1 and 07q132-al2c) can be easily identified.

To summarize this section, β-glucan and oil concentrations were highly heritable and the rank of genotypes for these traits was similar across locations. As a result, selection for these traits can be conducted at a few locations. Grain yield, test weight, groat percentage, and protein concentration were somewhat affected by genotype-by-location interactions. Nevertheless, most test locations were positively correlated and no negative correlations among locations were found. This

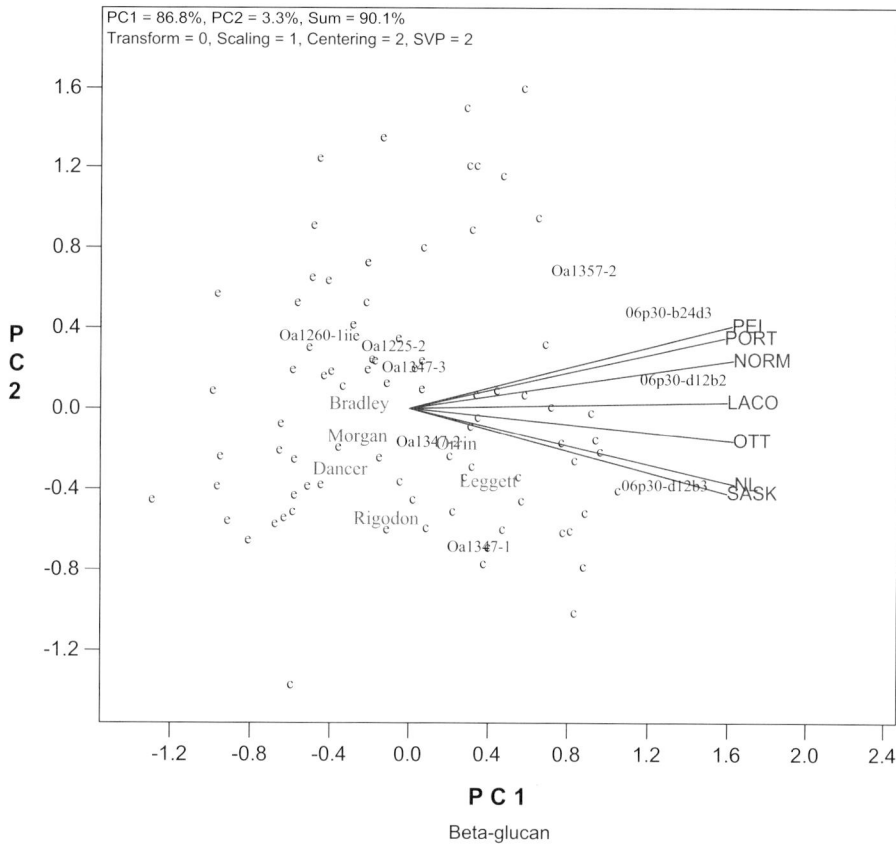

Figure 2.7   The "environment association" view of the GGE biplot for β-glucan concentration in the groat. (Biplot was based on location-standardized data and location-focused singular value partition, SVP.) (For color version, see color plate section.)

means that each of these traits can be improved relatively easily based on data from multiple representative locations. Improvement for any single trait does not generally constitute a major challenge in breeding for an ideal milling oat cultivar.

2.3   Breeding for multiple traits: Undesirable trait associations

2.3.1   Pairwise associations

Successfully combining the desired levels of two traits in a single genotype depends on the nature of the genetic association between them. A positive correlation or lack of correlation means that they can be combined easily, whereas a negative correlation means they cannot be combined easily. Interrelationships among the measured traits in the 2011 Nationwide Oat Tests are summarized

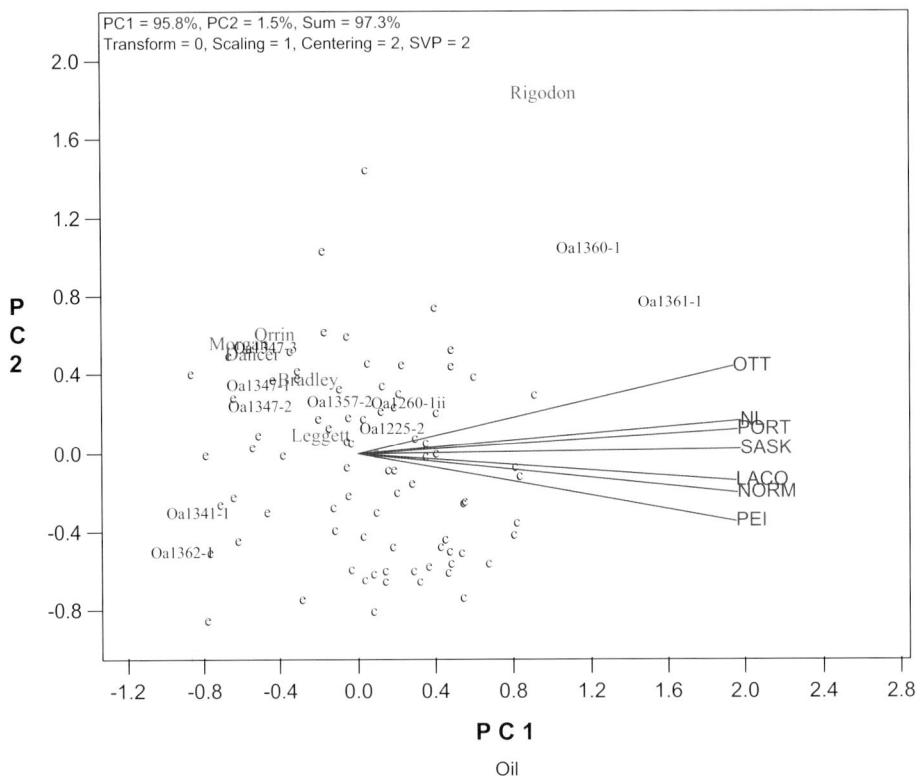

Figure 2.8   The "environment association" view of the GGE biplot for oil concentration in the groat. (Biplot based on location-standardized data and location-focused singular value partition, SVP.) (For color version, see color plate section.)

graphically in Figure 2.10 and numerically in Table 2.3. The biplot revealed positive correlations (acute angles) among groat percentage, thousand-kernel weight, and test weight and a positive correlation between oil and β-glucan concentrations. However, these two groups of traits were negatively correlated (obtuse angles). The biplot also revealed a negative correlation between grain yield and protein concentration.

In Figure 2.10 and Table 2.3, β-glucan concentration behaved like a "troublemaker" among the traits and in the breeding for an ideal milling oat. It was negatively correlated with groat concentration, grain yield, test weight, and thousand-kernel-weight but positively correlated with oil concentration. All these associations are undesirable. Deleting test weight and kernel weight from Figure 2.10 led to the biplot in Figure 2.11. This biplot best summarizes the key undesirable associations in milling oat breeding as follows: (i) negative association between β-glucan concentration and groat percentage (obtuse angle); (ii) negative association between β-glucan concentration and grain yield (obtuse angle); (iii) positive correlation between β-glucan and oil concentration (acute angle); and (iv) negative correlation between protein concentration and grain yield (obtuse angle).

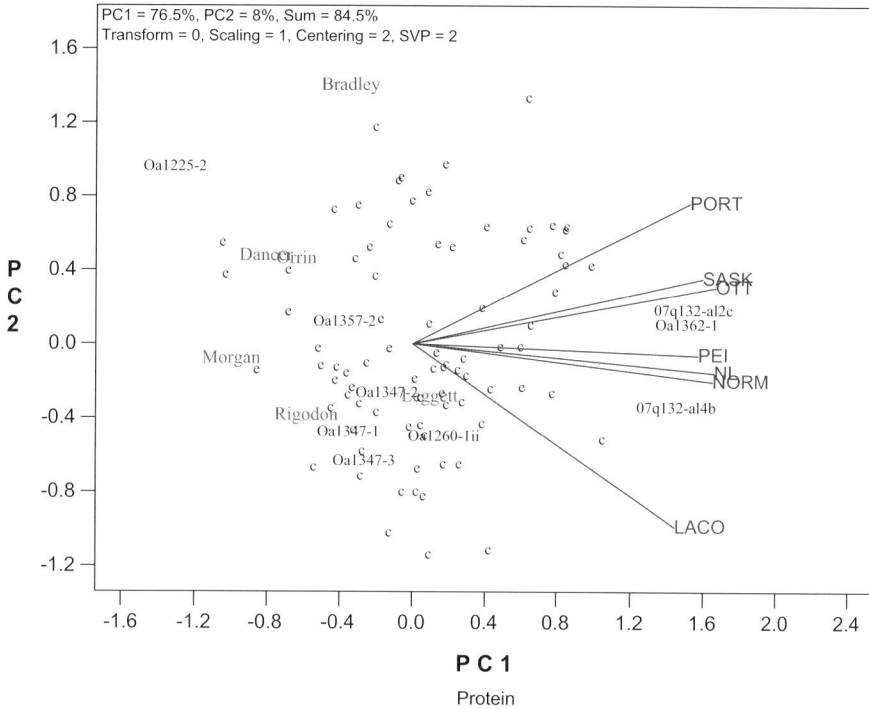

PC1 = 76.5%, PC2 = 8%, Sum = 84.5%
Transform = 0, Scaling = 1, Centering = 2, SVP = 2

Figure 2.9   The "environment association" view of the GGE biplot for protein concentration in the groat. (Biplot based on location-standardized data and location-focused singular value partition, SVP.) (For color version, see color plate section.)

These associations are consistent with previous observations. For example, Yan and colleagues reported a relatively consistent negative association between protein concentration and grain yield and a positive association between β-glucan and oil concentrations from the Quaker Uniform Oat Nursery data obtained at seven to nine locations across Canada and the United States during 1996 to 2003

Table 2.3   Genetic correlations among oat grain traits

Traits	BGL	GROAT	PROTEIN	TWT	TKW	YIELD	OIL
BGL		−0.408	0.112	−0.268	−0.454	−0.387	0.561
GROAT	−0.408		−0.185	0.546	0.352	0.086	−0.241
PROTEIN	0.112	−0.185		−0.001	-0.026	−0.651	0.058
TWT	−0.268	0.546	−0.001		0.461	0.100	−0.123
TKW	−0.454	0.352	−0.026	0.461		0.357	−0.483
YIELD	−0.387	0.086	−0.651	0.100	0.357		−0.365
OIL	**0.561**	−0.241	0.058	−0.123	−0.483	−0.365	

The threshold correlation value is 0.206 for $P < 0.05$ and 0.265 for $P < 0.01$.
BGL: groat β-glucan concentration; GROAT: groat percentage; PROTEIN: groat protein concentration; TWT: test weight; TKW: thousand-kernel weight; YIELD: grain yield; OIL: groat oil concentration.

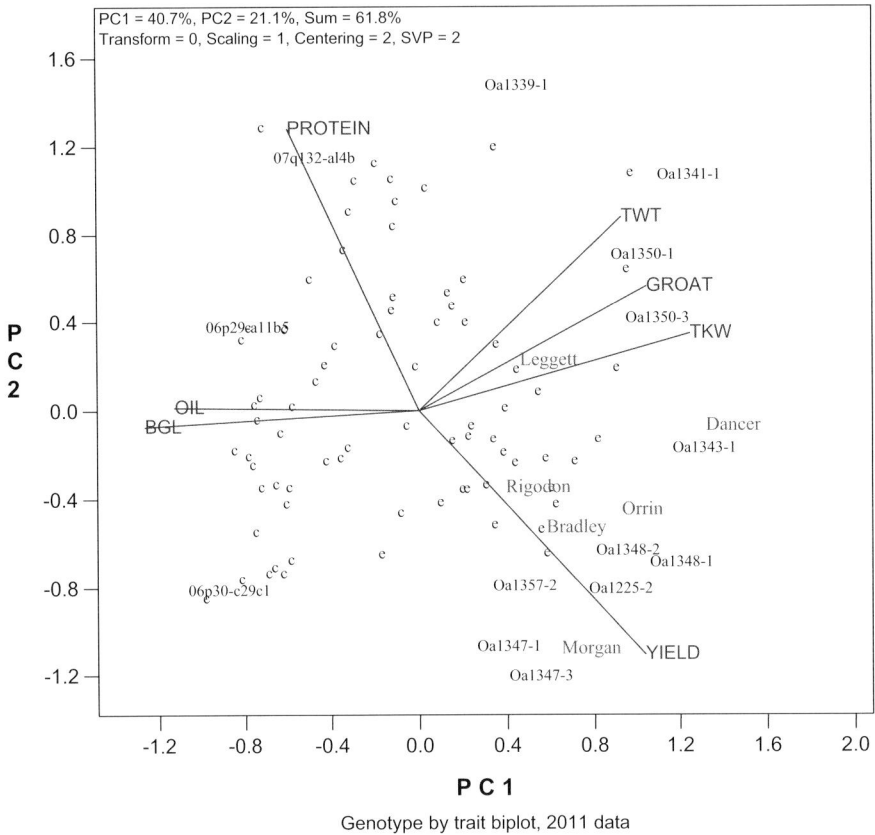

Figure 2.10   The genotype-by-trait biplot involving seven traits: grain yield (YIELD), groat percentage (GROAT), β-glucan concentration (BGL), oil concentration (OIL), protein concentration (PROTEIN), test weight (TWT), and thousand-kernel weight (TKW). (Biplot based on trait-standardized data and trait-focused singular value partition, SVP.) (For color version, see color plate section.)

(Yan *et al.*, 2007b). Yan and Frégeau-Reid (2008) reported a negative correlation between β-glucan and groat percentage and a positive correlation between oil and β-glucan concentrations for an oat breeding population. However, other studies reported the opposite findings. Kibite and Edney (1998) reported a negative correlation between oil and β-glucan concentrations, and Peterson and colleagues reported a positive correlation between β-glucan concentration and groat percentage in the American oat nurseries (Peterson *et al.*, 1995).

Among the undesirable trait associations observed in this example data set, the first two are most challenging, because they involve the three most important traits for an ideal milling oat. Grain yield is the trait oat growers care most about; groat percentage is the trait millers care most about; and β-glucan concentration is the trait consumers care most about. Therefore, discussion here focuses on associations among these three traits.

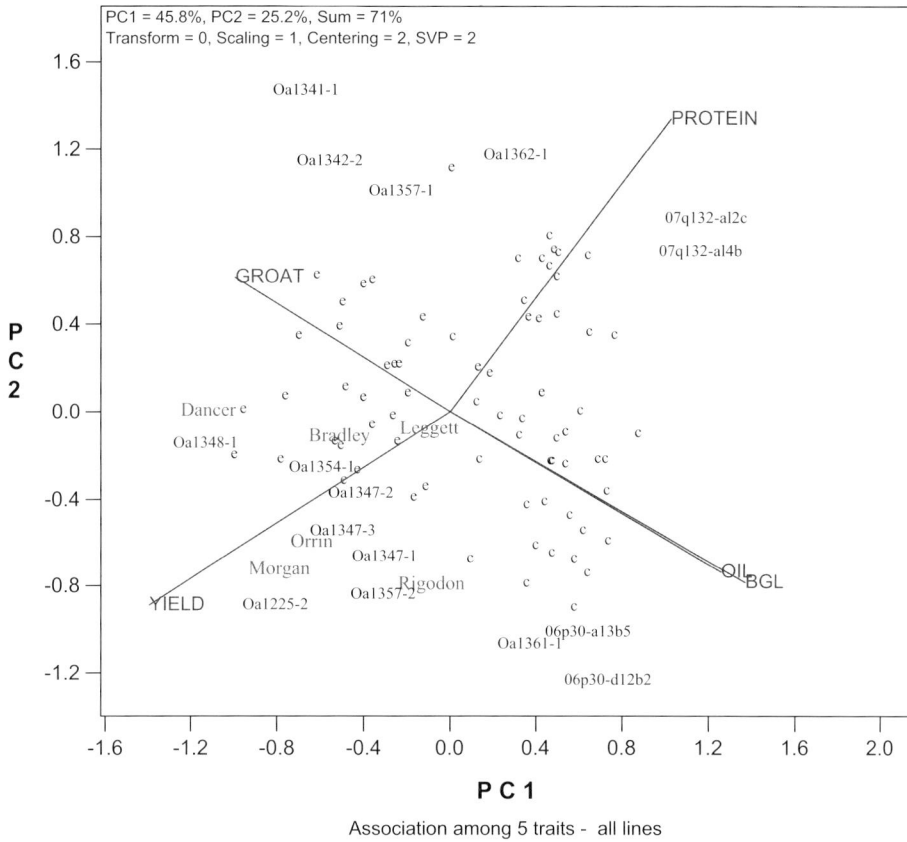

Figure 2.11   The genotype-by-trait biplot involving five traits: grain yield (YIELD), groat percentage (GROAT), β-glucan concentration (BGL), oil concentration (OIL), and protein concentration (PROTEIN). (Biplot based on trait-standardized data and trait-focused singular value partition, SVP.) (For color version, see color plate section.)

2.3.2   The three-way association

Figure 2.12 is the biplot containing only grain yield, groat percentage, and β-glucan concentration as traits. As in Figures 2.10 and 2.11, this biplot shows a modest negative association between β-glucan concentration and grain yield, a modest negative association between β-glucan and groat percentage, but a near-zero association between grain yield and groat percentage. The r-squared values between any two traits did not exceed 16%, suggesting that a reasonable combination of any two of the three traits is not an impossible task. The real challenge, however, is to combine high levels of all three traits.

Among the check cultivars, Dancer had an excellent groat percentage but a low β-glucan level. Morgan had an excellent yield potential but below average groat and β-glucan levels. Leggett was a well-rounded cultivar; it is positioned near the biplot origin, meaning that it had an average level for each of the three

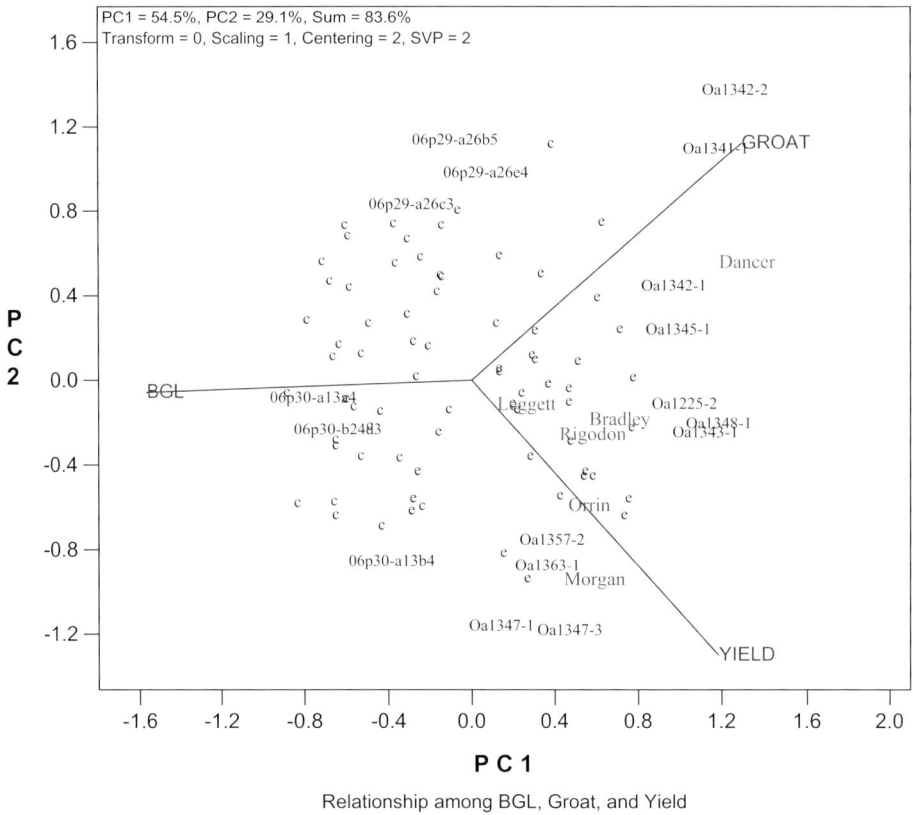

Figure 2.12 The genotype-by-trait biplot involving three traits: grain yield (YIELD), groat percentage (GROAT), and β-glucan concentration (BGL). (Biplot based on trait-standardized data and trait-focused singular value partition, SVP.) (For color version, see color plate section.)

traits. Among the breeding lines, OA1225-2, OA1343-1, and OA1348-1 exhibited a combination of high grain yield and high groat percentage. Unfortunately, and as expected, they also had low β-glucan levels. In contrast, genotypes with high levels of β-glucan (e.g., 06p30-a13a4 and many of its "sisters") resulted in low grain yield, low groat percentage, or both. Some genotypes produced modest groat percentage and β-glucan concentration (e.g., 06p29-a26b5 and 06p29-a26e4) but also produced the lowest grain yields in the test. Some genotypes produced a combination of modest β-glucan and grain yield (e.g., 06p30-a13b4) but nearly the lowest groat percentages. Some genotypes produced very high groat percentages (OA1341-1 and OA1342-2) but only average grain yields and nearly the lowest levels of β-glucan. No single genotype produced a good combination of all three traits.

Therefore, the real challenge in breeding for an ideal milling oat is not lack of genetic variation for any single trait nor undesirable associations between any two traits but the three-way association between grain yield, groat percentage,

and β-glucan concentration, as depicted in Figure 2.12. This three-way association was repeatedly observed in the Canadian Prairies trials (Yan *et al.*, 2011).
These three traits are interconnected in such a way that improving the level of
any one leads to the decrease of one or both of the other two. Thus, combining
any two traits at a higher level would almost certainly lead to the lowering of the
level of the third. Admittedly, groat percentage and grain yield were not negatively correlated; however, their simultaneous improvement was accompanied by
lower β-glucan levels. Adding other traits to this picture, such as oil and protein
concentrations and other traits listed in Table 2.1, would add further complexity
to the breeding task.

2.4 Strategies of breeding for an ideal milling oat

Given the three-way association among grain yield, groat percentage, and β-
glucan concentration (Figure 2.12), a two-step selection strategy is proposed here
for breeding ideal milling oats. This strategy consists of independent culling followed by comprehensive selection based on an integrated index.

2.4.1 Step 1: Independent culling to select for promising genotypes

Independent culling is conducted using check cultivars as a reference to set a
bar (minimum required level) for each key trait. Although the check cultivars
differ from each other in various ways, they are all considered milling oats and
meet the minimum requirements for each of the key traits. For each trait, the
check cultivar that shows the lowest level of that trait was used as a bar to reject
breeding lines. All breeding lines performing below this bar for any single trait
were discarded, no matter how well they performed for other traits. Thus, only
those that exceeded the bar for all three traits were retained for the second step
of selection.

In the data set discussed here, there were six check cultivars (Bradley, Dancer,
Leggett, Morgan, Orrin, and Rigodon). The poorest check cultivars for β-glucan,
groat percentage, and grain yield were Dancer, Morgan, and Leggett, respectively, so they were used to set the bar for each respective trait (Figure 2.13).
Using these criteria, only 13 of the 90 new breeding lines were tentatively selected
(Table 2.4). Figure 2.13 is a snapshot of the "multitrait selection against checks"
tool in the GGE biplot software package (www.ggebiplot.com) used in this work,
which allows easy selection of the appropriate check cultivar for each trait and
setting of the bar relative to the check for each trait. However, independent
culling can be conducted with other software packages with the same functionality; even a spreadsheet will do the job.

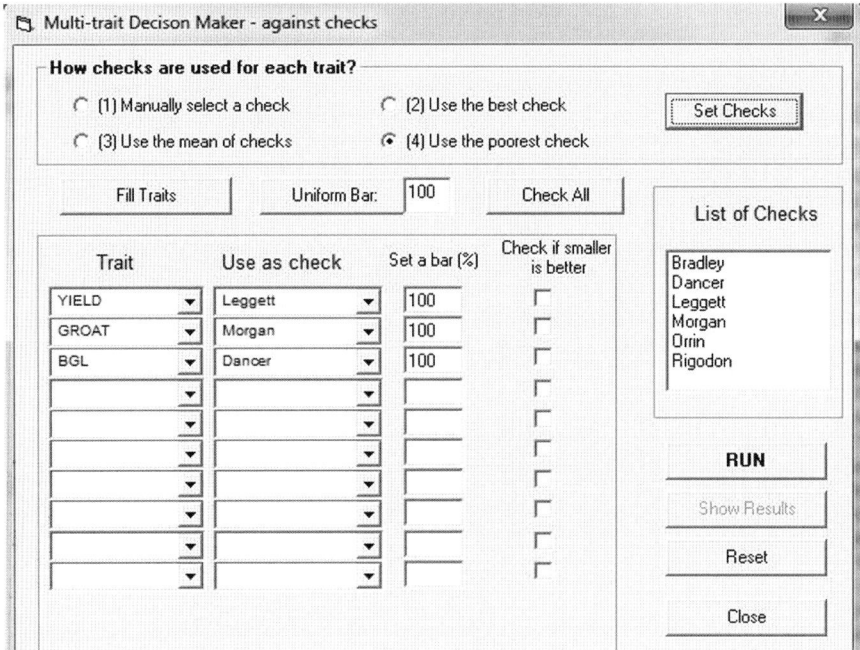

Figure 2.13 A snapshot of the "multitrait selection against checks" tool in the GGE biplot software. This tool offers flexibility for choosing traits to be used in selection, check cultivars to be used as references for each trait, and the cut-off value (bar) to be used to reject genotypes.

Table 2.4 Trait values of the genotypes retained after independent culling.

Genotypes	β-glucan concentration (%)	Groat percentage	Grain yield (kg/ha)
06p30-a13b3	5.0	70.6	4720
Bradley (check)	4.3	73.2	4870
Dancer (check)	4.2	77.4	4991
Leggett (check)	4.8	72.8	4700
Morgan (check)	4.3	71.0	5203
OA1225-2	4.4	75.0	5191
OA1342-1	4.3	75.6	4730
OA1344-1	4.7	72.7	4708
OA1346-1	4.3	71.8	4952
OA1347-2	4.6	70.9	5174
OA1347-3	4.5	70.5	5345
OA1352-1	4.2	72.3	4968
OA1352-2	4.4	72.9	4926
OA1354-1	4.3	73.9	5066
OA1357-2	5.2	72.9	5387
OA1361-1	4.5	70.5	4956
OA1363-1	4.5	70.6	5042
Orrin (check)	4.7	73.0	5256
Rigodon (check)	4.4	72.7	4838

2.4.2 Step 2: Index selection to identify promising genotypes

Breeding lines that survive independent culling may not be better than the current check cultivars. Given the negative correlations among traits, it is essential to develop an integrated index, so that genotypes can be compared for overall superiority. This involves standardizing the data by trait, assigning a weight to each trait, and then applying weights to calculate a superiority index for each genotype. The weights are subjective and reflect the researcher's understanding of the relative importance of each trait. For example, on the basis of independent culling, weights may be given to the three traits as grain yield (1.0), groat percentage (0.8), and β-glucan concentration (0.6) (Figure 2.14). A superiority index can then be calculated and the genotypes ranked accordingly (Table 2.5).

Although 13 breeding lines were accepted as potential cultivars, only one line (OA1357-2) was ranked better than the best ranked check cultivar (Orrin) and only four lines (OA1357-2, OA1225-2, OA1347-2, and OA1347-3) were ranked better than the second best ranked check cultivar (Dancer). These lines deserve more attention in future tests. Table 2.5 was generated by the "multitrait decision maker" of the GGE biplot software package (Figure 2.14) but any other software with similar functionality can accomplish this task.

Another way to compare the accepted lines with the check cultivars is to display the data from Table 2.4 in a biplot (Figure 2.15). Similar to the biplots shown in Figures 2.10, 2.11, and 2.12, this biplot shows a negative correlation between groat percentage and β-glucan concentration across the 19 genotypes (6 check cultivars and 13 breeding lines). Grain yield did not correlate with either of these two traits. These relationships are reflected in the trait

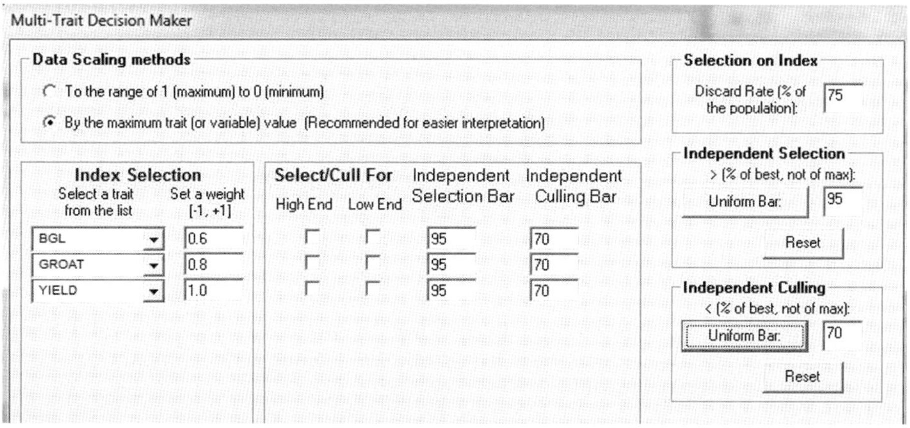

Figure 2.14 A snapshot of the "multitrait decision maker" in the GGE biplot software. This tool combines three selection strategies: independent selection based on any trait to select useful parents, independent culling based on key traits to reject inferior genotypes, and index selection to rank genotypes based on an integrated index. The index selection component was used to generate Table 2.5.

Table 2.5   Ranking of the genotypes after independent culling.

Trait	BGL	GROAT	YIELD	Superiority Index
Weight	0.6	0.8	1.0	
OA1357-2	1.000	0.942	1.000	0.981
Orrin	0.911	0.944	0.976	0.949
OA1225-2	0.851	0.969	0.964	0.937
OA1347-3	0.868	0.911	0.992	0.934
OA1347-2	0.880	0.917	0.960	0.926
Dancer	0.812	1.000	0.926	0.922
OA1354-1	0.824	0.955	0.940	0.916
Morgan	0.826	0.918	0.966	0.915
Leggett	0.930	0.942	0.872	0.910
OA1352-2	0.859	0.943	0.914	0.910
06p30-a13b3	0.963	0.912	0.876	0.910
OA1363-1	0.863	0.912	0.936	0.910
OA1361-1	0.868	0.911	0.920	0.904
OA1344-1	0.899	0.940	0.874	0.902
OA1346-1	0.834	0.928	0.919	0.901
OA1352-1	0.818	0.934	0.922	0.900
Rigodon	0.847	0.939	0.898	0.899
Bradley	0.826	0.946	0.904	0.898
OA1342-1	0.822	0.977	0.878	0.897

Note: Values are relative to the highest value for each trait.
BGL: groat β-glucan concentration; GROAT: groat percentage; YIELD, grain yield.

profiles of the genotypes. OA1357-2, which ranked first in Table 2.5, had a good combination of grain yield and β-glucan level. Unfortunately, but expectedly, it had a below-average groat level and is, therefore, not an ideal milling oat cultivar as defined earlier in this chapter. OA1225-2, ranked third in Table 2.5, had a combination of above-average groat percentage and grain yield but below-average β-glucan level. Therefore, it is not an ideal milling oat, either. A truly ideal milling oat cultivar would combine the characteristics of OA1357-2 and Dancer. Are such cultivars obtainable?

2.5   Discussion

2.5.1   Identification of the main challenges

There are only three types of challenges for any plant breeding program: (i) insufficient genetic variability for each of the key traits that make up the ideotype; (ii) any genotype-by-environment interaction for each of the key traits; and (iii) undesirable associations among the key traits. Although germplasm availability sets the ultimate limitation to the improvement of any crop type, the breeder must assume that the current germplasm pool has sufficient genetic variability to make further progress, even though the second and third types of challenges may be rooted in the first challenge and can be solved only by the introduction of new germplasm.

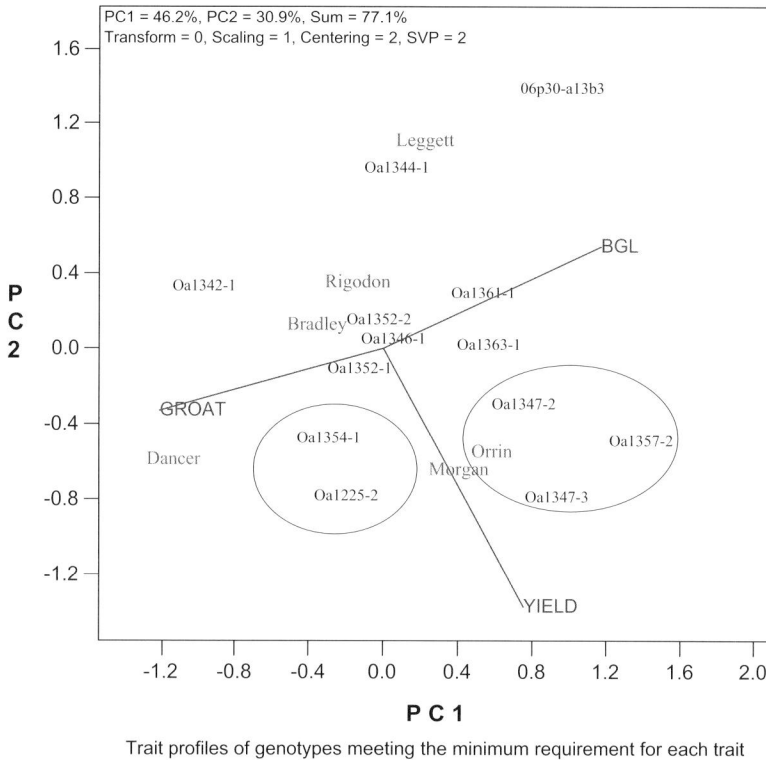

PC1 = 46.2%, PC2 = 30.9%, Sum = 77.1%
Transform = 0, Scaling = 1, Centering = 2, SVP = 2

Trait profiles of genotypes meeting the minimum requirement for each trait

Figure 2.15 The genotype-by-trait biplot involving three traits and 19 selected genotypes, which approximates the mean trait levels of the six check cultivars and 13 promising genotypes for grain yield (YIELD), groat percentage (GROAT), and β-glucan concentration (BGL). (Biplot based on trait-standardized data and trait-focused singular value partition, SVP.) (For color version, see color plate section.)

The genotype-by-location interactions observed for each of the key traits in the 2011 Nationwide Oat Test were small to moderate and, therefore, did not constitute a great challenge. Of course, this conclusion is based on data from only a single year and the genotype-by-year interaction may constitute a greater challenge. Another important reason for this result may be that the test locations belong to a relatively homogeneous mega-environment. By definition, a mega-environment is a subregion for the production of a given crop within which the same genotype(s) perform best across years at all representative locations. Very large genotype-by-location interactions can occur by including a wider range of test locations. However, once mega-environments are well defined, and selections are confined to a single mega-environment, genotype-by-location interactions become a minor challenge to breeding progress. It is necessary to base decisions on data from multiple representative test locations. Very large genotype-by-year interactions can occur within a mega-environment. There is not much the researcher can do other than to base selections on data from multiple years, which slows breeding progress. The challenge imposed by strong genotype-by-year interactions can be relieved only by developing higher yielding and more stable genotypes, which may depend on introducing new germplasm.

Undesirable associations among key traits are the greatest challenge identified in this chapter. Four undesirable pairwise associations were identified, but the three-way association among grain yield, groat percentage, and β-glucan concentration was the most important. Among the 90 new breeding lines, some were identified to have a good combination of grain yield and β-glucan (e.g., OA1357-2) or a good combination of grain yield and groat percentage (e.g., OA1225-2). However, no lines exhibited a good combination of all three traits. Given the persistent negative association between groat percentage and β-glucan, it is fair to ask whether it is possible to develop such a cultivar.

2.5.2   The possibility of developing a truly ideal milling oat cultivar

To tackle this question, it is necessary to examine how groat percentage and β-glucan concentration are defined. Groat percentage is calculated as:

$$Groat\% = \frac{Groat\ yield}{Grain\ yield}$$

which predetermines a negative correlation between groat percentage and grain yield. In fact, it is necessary to ask why the correlation was positive although nonsignificant (0.086) rather than the expected -1 (Table 2.3). The explanation is that the large genetic variability in groat yield among the genotypes overcame the negative relationship between its two components, grain yield and groat percentage.

Similarly, β-glucan concentration is calculated as:

$$BGL\% = \frac{BGL\ yield}{Groat\ yield}$$

which can be expressed as:

$$BGL\% = \frac{BGL\ yield}{Grain\ Yield \times Groat\%}$$

From this formula, it is not surprising that β-glucan was negatively correlated with both grain yield and groat percentage. On the contrary, it is surprising that the correlations were not stronger (Table 2.3). The reason is that large genetic variability exists among the genotypes in terms of β-glucan yield per unit area of land. Therefore, the only way to combine these three traits in a single genotype is to increase the genetic potential of oats in terms of β-glucan yield per unit area of land. This is similar to the idea of Cervantes-Martinez and colleagues, who proposed that improving β-glucan yield could simultaneously improve both β-glucan content and grain yield (Cervantes-Martinez *et al.*, 2002). Although achieving this goal is a challenging task, there is no concrete evidence that the genetic potential of β-glucan yield in oats has been reached and cannot be further improved.

In other words, it may be possible to develop an ideal milling oat that combines all three traits at a high level. However, this possibility would again lie in the introduction of new germplasm.

2.5.3 Long-term goals and current strategies

The reality is that it is difficult to combine all three traits at a high level, but it is relatively easy to combine two of the three traits at relatively high levels. Therefore, it may be meaningful to define subideal breeding goals that are more achievable. There may be three types of subideal oat cultivars:

Type I: High grain yield + high groat percentage

Type II: High grain yield + high β-glucan level

Type III: High groat percentage + high β-glucan level

Currently, there are examples of Type I (e.g., OA1225-2) and Type II (e.g., OA1357-2), but an example for Type III is still lacking. Currently all known high β-glucan cultivars or breeding lines exhibit only intermediate groat percentage at best. Type III does not yet exist, but it is an essential step toward developing a truly ideal milling oat.

Are the subideal Type I and Type II genotypes acceptable to the producer–miller–consumer oat value chain? They probably are, with a condition. Type I genotypes must have an acceptable β-glucan level, so that the oat products derived from their grains can be labeled as healthy food. Type II genotypes must have an acceptable groat level, so that oat millers can draw a profit. This can be achieved through independent culling using acceptable cultivars as references, as described earlier. Breeding lines that survive independent culling will surpass existing cultivars. Breeding lines with merits relative to check cultivars can be selected based on a single superiority index, as described earlier. These lines will inevitably fall into the Type I or Type II subideal group or somewhere in between.

Classifying promising genotypes into proper subideal groups can be beneficial to breeders, producers, and oat processors. It can help the breeders choose parent cultivars, formulate new crosses, and select among the progenies. It may help the producer choose cultivars according to their intended end use. Finally, it may help processors to select cultivars to purchase oat grains from. All stakeholders in the oat value chain have important roles in shaping the breeding programs. They can help the breeder choose the cultivars to be used as checks in independent culling and decide the weights for each of the key traits to determine the superiority index.

As a final note: the data set discussed in this chapter was used only to demonstrate concepts and methods. Although the comments on the check cultivars are consistent with long-term observations, the comments on the new breeding lines should be considered tentative, because they are based on data from only a single year.

Acknowledgements

We would like to thank the following colleagues who contributed to obtaining the 2011 Nationwide Oat Test data: Richard Martin, Allan Cummiskey, Denis Pageau, Isabelle Morasse, John Roswell, John Kobler, Dorothy Sibbitt, Brad DeHaan, Steve Thomas, Aaron Beattie, Tom Zatorski, Kim Stadnyk, and Wes Dyck. The Nationwide Oat Test project was funded by Agriculture and Agri-Food Canada (AAFC) and the Prairie Oat Growers' Association (POGA).

References

Agriculture and Agri-Food Canada. (2010) *Oats: Situation and Outlook. Market Outlook Report* [Online]. Available: http://www.agr.gc.ca/pol/mad-dam/pubs/rmar/pdf/rmar_02_03_2010-08-03_eng.pdf (last accessed 18 April 2013).

Baker, R.J. and McKenzie, R.I.H. (1972) Heritability of oil content in oats *Avena sativa* L. *Crop Science* **2**, 201–202.

Cervantes-Martinez, C.T., *et al.* (2001) Selection for greater β-glucan content in oat grain. *Crop Science* **41**, 1085–1091.

Cervantes-Martinez, C.T., *et al.* (2002) Correlated responses to selection for greater β-glucan content in two oat populations. *Crop Science* **42**, 730–738.

Gabriel, K.R. (1971) The biplot graphic display of matrices with application to principal component analysis. *Biometrika* **58**, 453–467.

Holthaus, J.F., *et al.* (1996) Inheritance of β-glucan content of oat grain. *Crop Science* **36**, 567–572.

Frey, K.J. and Hammond, E.G. (1975) Genetics, characteristics, and utilization of oil in caryopses of oat species. *Journal of the American Oil Chemists' Society* **52**, 358–362.

Kibite, S. and Edney, M.J. (1998) The inheritance of β -glucan concentration in three oat (*Avena sativa* L.) crosses. *Canadian Journal of Plant Science* **78**, 245–250.

Peterson, D.M., *et al.* (1995) β-Glucan content and its relationship to agronomic characteristics in elite oat germplasm. *Crop Science* **35**, 965–970.

Yan, W. and Frégeau-Reid, J.A. (2008) Breeding line selection based on multiple traits. *Crop Science* **48**, 417–423.

Yan, W. and Kang, M.S. (2003) *GGE Biplot Analysis: A Graphical Tool for Breeders, Geneticists, and Agronomists*. CRC Press, Boca Raton, FL.

Yan, W., *et al.* (2000) Cultivar evaluation and mega-environment investigation based on the GGE biplot. *Crop Science* **40**, 597–605.

Yan, W., *et al.* (2007a) GGE Biplot vs. AMMI analysis of genotype-by-environment data. *Crop Science* **47**, 641–653.

Yan, W., *et al.* (2007b) Associations among oat traits and their responses to the environment in North America. *Journal of Crop Improvement* **20**, 1–30.

Yan, W., *et al.* (2010) Identifying essential test locations for oat breeding in eastern Canada. *Crop Science* **50**, 504–551.

Yan, W., *et al.* (2011) Genotype × location interaction patterns and testing strategies for oat in the Canadian prairies. *Crop Science* **51**, 1903–1914.

3
Food Oat Quality Throughout the Value Chain

Nancy Ames, Camille Rhymer, and Joanne Storsley
Agriculture and Agri-Food Canada, Winnipeg, MB, Canada

3.1   Introduction: Oat quality in the context of the value chain

Oat (*Avena sativa* L.) quality is key to a successful food oat value chain, culminating in an oat product that meets the needs of the end user. Understanding the consumer trends and issues that shape the food oat market is an important aspect of the processing and marketing of good quality oat products. For example, approval of health claims for oats in the United States (US FDA, 1997) and, more recently, in Canada (Health Canada, 2010), together with demonstrated health benefits of daily consumption of whole grains, has led to increased consumer demand for oats and oat products in North America. Along with this strengthened demand comes the expectation that both the nutritional and sensory qualities of the oat products will be high. Quality factors of the end product at the consumer level will ultimately determine the value and marketability of oats, but the entire value chain must be considered to secure a reliable and consistent source of oats that fulfills these specifications. Each member of the value chain, which includes plant breeders, growers, grain companies, exporters, processors, and food companies, has unique capabilities and challenges that warrant attention to different aspects of oat quality (Figure 3.1). Although quality may mean different things to different sectors of the value chain, all members must be motivated towards the development of a high-quality oat product that meets consumer quality expectations.

Oats Nutrition and Technology, First Edition. Edited by YiFang Chu.
© 2014 John Wiley & Sons, Ltd. Published 2014 by John Wiley & Sons, Ltd.

Breeders	Producers	Grain Handlers	Millers	Food Processors	Consumers
Consider all quality factors	Market opportunities	Preferred varieties for processing	Nutrient content (high fibre; low fat)	Nutrition labels and claims	Health benefits
	Grain color; no sprouting; no frost damage	Kernel color & soundness	No groat discoloration; granulation; functional properties; low enzyme activity	Functional specifications; sensory attributes; shelf stability	End-product color, flavor/aroma, texture & appearance
	Grain yield & grade	Grain supply; grain physical properties	Milling yield; kernel uniformity	Ingredient supply; end-product pricing	Cost; convenience
	Free from disease; low dockage	Grain appearance; free from contaminants	Clean grain; free from microorganisms & mycotoxins	Free from microorganisms & mycotoxins	Food safety

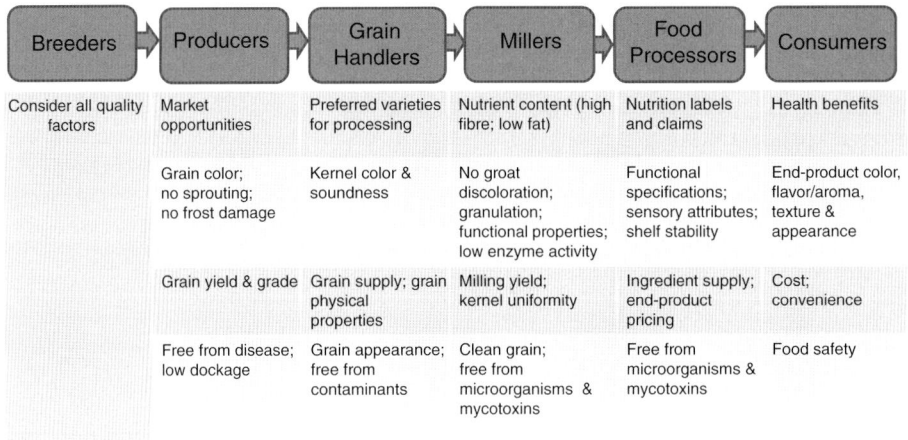

Figure 3.1   Some quality factors important to key participants in the food oat value chain.

To the producer (or oat grower), grain quality is anything that affects yield or the value of the grain at the elevator or point of sale. Physical quality characteristics of the grain, particularly those related to grain weight and size of kernels, are of primary concern, as they affect yield and profitability. Factors such as kernel weathering, disease, or frost damage are also paramount because they degrade overall quality and limit market opportunities. For the producer growing and selling oats into various market situations, quality will be determined by the specifications of the buyer. Markets for oats can be separated into the food oat milling market or the feed market; the feed market in turn is separated into livestock feed or quality performance feed (equine) markets (Agriculture and Agri-Food Canada, 2010). Although the same quality characteristics may be measured for all three markets, each has its own acceptable limits for these characteristics, which will determine the grade standard and the potential market opportunities. For example, the Canadian Grain Commission Inspection Division Grade Standards separate oat grades into Canada Western (CW) No. 1, CW No. 2, CW No. 3, and CW No. 4, depending on test weight (kg/hL), degree of soundness, percentage hulled or hull-less, and percentage damage due to frost (Canadian Grain Commission, 2012a). Based on grade standards alone, a good quality food oat destined for the milling industry would have a high test weight, high degree of purity (free from damage and foreign material, including dirt and stones), plump and uniform kernel size, good color, moisture and soundness, and meet the remaining standards for a CW No. 1 grade. Additional criteria for specialty food oat markets may be stipulated by processors, prompting grain handlers to source oats with specific characteristics on a contract basis (e.g., hull-less or high β-glucan content). Processors often adopt "preferred variety" lists to facilitate the sourcing of oats with desired end-use quality (Manitoba Co-operator, 2013).

Food oats are typically processed into several milled products that fall within the categories of flakes, flour, steel cut groats, and bran. The processing of oats requires specific quality criteria to obtain the best quality end product at a reasonable cost. Oat properties important to the millers include physical characteristics that impact the yield of high-value oat fractions, such as percentage groat and percentage plumps and thins. The oats must also perform well during industrial processing steps and exhibit ease of hulling as well as low susceptibility for groat breakage. Millers sell their oat products to food manufacturers or directly into retail markets; therefore, they also must satisfy the quality needs of the end user. This includes meeting specifications for achieving food safety and product labeling regulations (e.g., low microbial content, fiber levels for nutrient and health claims). Specialized quality requirements may be defined by the millers and food manufacturers according to in-house specifications developed for optimizing end-product processing. For example, specifications may stipulate target β-glucan content, degree of toasted flavor, granulation, or pasting behavior that impact the desired quality and consistency of the specific food application.

Ultimately, end products made from food oats are consumed by humans, therefore quality of the end product depends on both consumer sensory preferences and nutritional demands or expectations. Some quality characteristics may be unanimously important for all end products and form the basis for acceptance or rejection of oats from entering food processing markets. Consumers readily associate oat products with the wholesomeness and healthfulness of the oat grain, and thus the perception of a natural, safe, and nutritional whole grain oat must be maintained. Characteristics of the raw oat that would influence these perceptions include: clean grain, free from disease, insects, and dirt; oat grain and groats (grain minus hull) with a light uniform color that is unstained or unmarked; free from pesticide residues, natural toxins, or fungal/ bacterial contamination; compositional quality to provide the nutritional balance associated with whole grain recommendations in national food guides, such as Canada's Food Guide and the US MyPlate; and correct quantity of specific functional nutritional components, such as soluble fiber or fat, which meet the local health claims provided on the product label. Often, defining "good end-product quality" is further challenged by considering personal preferences. For example, preference mapping of muesli oat flakes has shown that consumer age and dental condition influenced which muesli flavor and texture attributes were considered most important in evaluating sample pleasantness, suggesting that product quality could be tailored to different consumer groups (Kälviäinen *et al.*, 2002).

Quality assessment of oats begins during the cultivar development stage. Plant breeders must be concerned with screening germplasm for a number of breeding criteria important for economic and sustainable production. Through communication with all participants in the food oat value chain, plant breeders are well informed of consumer quality demands and consider processing and nutritional quality traits among the breeding criteria when developing new cultivars/varieties. The screening criteria applied for improved food oat quality depend on the heritability of the trait as well as on the ease of measurement. Several test procedures are routinely used to screen oat germplasm throughout the cultivar development process. Early in the breeding process, the small sample sizes of experimental breeding lines limit the number of quality tests that can be performed. The food quality criteria of significant interest in early generation breeding material include levels of nutritional components, such as protein, oil, and β-glucan; these can be measured on whole seeds using near infrared technology to facilitate rapid analysis without compromising the seed itself. Early generation quality screening also typically includes a number of physical or anatomical quality traits that can affect the milling yield, such as test weight, seed size and shape (percentage plumps and thins), and percentage hull. In the later stages of cultivar development, breeding lines are planted in large plots at multiple locations so that both genotypic and environmental influences on quality can be assessed before a variety is released for production. At this point, seed samples are large enough to be mechanically dehulled for assessment of milling performance (e.g., groat breakage) and ground for further screening via chemical or functional analysis (e.g., dietary fiber and flour properties). Cultivar improvement of quality attributes involves assessment throughout the breeding process. In Canada, any new potential cultivars must undergo 2–3 years of comparative testing with commonly grown check cultivars at several locations to ensure that their quality characteristics meet or exceed the current acceptable level of quality. The quality characteristics assessed are determined by representatives from the oat value chain, including producers, breeders, researchers, as well as the grain distribution, milling, and processing industries.

This chapter addresses oat quality attributes of importance at each step in the value chain from pre-production quality through to specific end-product characteristics.

3.2   Physical oat quality

Physical characteristics of the kernel are important factors affecting the value and use potential of oat cultivars. The main physical traits of concern include grain appearance, kernel shape and size, test weight, kernel weight, percentage groat/percentage hull, and breakage. The oat kernel is surrounded by a hull. When the hull is removed, the part remaining is called the groat. The groat can be divided into three main structural components: the bran, the germ or embryo, and the starchy endosperm. The embryo accounts for 3%, the bran 40%, and the starchy endosperm 57% of the groat by weight (Lásztity, 1998). The structure and chemistry of the oat kernel impact oat quality and have been thoroughly reviewed elsewhere (Miller and Fulcher, 2011).

Although many of these traits are of particular interest to millers and oat breeders, several traits are highly correlated with characteristics of interest to other members of the value chain. For example, in a study of 120 oat genotypes, Buerstmayr *et al.* (2007) reported a significant correlation between grain yield, of great importance to the producer, and groat yield, of great importance to the miller.

3.2.1   Oat and groat color

Oat color can influence quality from a number of perspectives. Varieties are often classified according to their characteristic hull color (e.g., white, tan, or light grey) (Manitoba Co-operator, 2013), which can be used as selection criteria for end use.

For example, performance feed markets for race horses demand very high quality white oats (Agriculture and Agri-Food Canada, 2010). There are no reports in the literature indicating a relationship between oat hull color classification and the color of milled products, and significant variation in groat color has been observed among genotypes within the same hull color type (Ames, unpublished data). Good groat color free from staining is a requirement for oats entering processing markets (Agriculture and Agri-Food Canada, 2010) and problems with groat discoloration have been cause for millers to reject oats. Environmental factors such as excessive moisture and presence of fungal pathogens are linked to groat discoloration; some genotypes were found to be more susceptible (Newton et al., 2003; Tekauz et al., 2004). There are also reports of oat cultivars being considered unsuitable for food processing on the basis of rolled oat properties including inferior color, as was described for the Australian cultivar Yarran (Zhou et al., 1999a). Color changes are expected to occur during heat processing, as an increase in yellow and red colors are concurrent with the development of toasting (Cenkowski et al., 2006). Significant genotypic variation in processed oatmeal color suggests that investigation into the propensity for some cultivars to form greater toasted color upon processing is warranted (Lapveteläinen and Rannikko, 2000). Incorporation of oat flour into a traditionally wheat-based end product, such as noodles, generally results in a darker color (Aydin and Gocmen, 2011; Zhou et al., 2011; Majzoobi et al., 2012; Mitra et al., 2012). Darker end-product color can reduce a product's appeal for some consumers, whereas others find it acceptable and may associate it with increased fiber and healthfulness (Aydin and Gocmen, 2011; Mitra et al., 2012). In any case, color is an important contributor to the quality of food oats and should be taken into consideration when breeding varieties, especially for the milling and specialty markets.

3.2.2 Milling yield

The objective of the milling process is to obtain a maximum yield of sound oat kernels, free from husks and other extraneous matter. Milling yield refers to the units of undehulled grain required to produce 100 units of finished product or the grain weight from which 100 kg of rolled oats is obtained upon milling (Groh et al., 2001). A higher milling yield would suggest that more grain would be required to produce 100 units; therefore, the higher the number, the lower the quality. The physical characteristics of the grain are the major quality factors affecting milling yield. Doehlert et al. (1999) calculated milling yield by subtracting the mass of hulls remaining and broken groats from the mass of the crude groat, dividing this by the mass of the oat starting material, and multiplying the inverse of this value by 100.

3.2.3 Hull and groat content

The oat hull or husk refers to the outer portion of a whole oat kernel that serves as a protective covering for the seed during growth, harvest, and storage, but is considered inedible and, therefore, is removed prior to milling. Although both hulled and hull-less oat types exist, the more common milling oat type consists of a kernel enclosed within a hull. Hulls are a low-value product in milling oats; therefore, high levels will reduce economic return. Since the hull does not contribute to the edible portion of the grain, using hull-less oats may appear to be a practical alternative. However, problems with hull retention, disease, preharvest sprouting, threshability, handling, and storage have limited their acceptance (Forsberg and Reeves 1992; Ronald et al., 1999; Kirkkari et al., 2004). When the whole oat has the hull removed, it is referred to as a groat (oat minus the hull or husk). The hull or husk typically comprises 25–30% of the grain weight, depending on cultivar, location, and year of production (Rhymer, 2002). Hull content is measured as the ratio of the weight of hull compared to the weight of the whole grain and is expressed as a percentage. The groat percentage (reverse of hull percentage) and groat:hull ratio are other measures used to express the amount of hull relative to groat and are used throughout the value chain as indicators of grain quality.

The whole oat grain typically undergoes the dehulling operation prior to heat treatment (or conditioning) and is carried out using mechanical dehullers. After preliminary cleaning, whole oats are fed into an impact dehuller, where the grains are thrown outwards by centrifugal force. The combination of high velocity and impact detach the hull from the kernel. The speed of the rotor is adjustable, allowing optimization of the process according to the physical quality characteristics of the oat lot (Kent and Evers, 1994).

Both the quantity of hull relative to grain and ease of hull removal are important quality characteristics for the miller. Reducing hull percentage in oats is an important goal in breeding programs, since lower hull percentage or higher groat content improves milling yield and increases the value to the miller. The ease of hull removal also impacts milling efficiency and grain quality. Hulls that are difficult to remove are not desirable; oats that are not dehulled or separated out will contaminate the purified groat product. However, hulls that come off too easily often do not provide the kernel protection and may result in more damaged or rancid kernels. The term "hulling degree" is defined as the percentage of the grain that is successfully dehulled during milling and is used to describe the efficiency of the dehulling process (Browne et al., 2002).

Hullability is the term used to describe the ease of hull removal from its enclosed kernel; it impacts milling efficiency (Browne et al., 2002). Some studies have reported a correlation between hullability and kernel content (Ganßmann and Vorwerck, 1995; Doehlert et al., 1999). However, according to Browne et al. (2002), physical and morphological characteristics of the husk and kernel, other than simply the proportion by weight of husk and kernel, will determine hullability. There is currently no standard milling test to assess the hullability of oat lots and varieties, which limits the ability to develop and select varieties with a high value for milling. Higher impact speeds are necessary to achieve a satisfactory hulling degree in oat lots with poor hullability, but these may result in greater kernel breakage and the production of fines (small fragments of kernel), leading to a loss in kernel yield in the subsequent processing steps.

In breeding programs, where reducing hull percentage and increasing the groat:hull ratio are important objectives, the most accurate way to measure hull and groat weight is through hand separation, with careful removal of the lemma and palea (comprising the hull). This is a tedious process, so mechanical dehulling methods involving compressed air are typically used to evaluate the hull and groat content of oat lines. A number of laboratory and industrial-scale dehulling processes are reported but generally, the objective involves removal of the hull with the least amount of groat loss or damage. Examples of laboratory dehulling equipment used include the Codema LLC laboratory dehuller LH5095 (Rhymer *et al.*, 2005) and the Streckel & Schrader KG Laboratory hulling machine BT459 (Browne *et al.*, 2002). Although several laboratory-scale dehullers and proto-cols are available, they do not directly duplicate the results of industrial impact dehullers. In addition, variation among laboratory-scale equipment suggests that although results may be correlated, absolute values will differ (Doehlert and McMullen, 2001).

3.2.4 Groat breakage

The impact that occurs during dehulling causes groat breakage, an undesir-able result of the milling process that increases with higher rotational speeds (Peltonen-Sainio *et al.*, 2004), a higher number of passes through an impact dehuller, or increased air pressure in a compressed air dehuller (Doehlert and McMullen, 2001). The degree of breakage also varies with cultivar and growing location (Doehlert *et al.*, 1999; Rhymer 2002). Doehlert and McMullen (2000) reported higher breakage in locations with heavy infections of rust and lower breakage in oats with a high hull content. The trend between low breakage and high hull content was also observed by Rhymer (2002). Although this is a con-tradiction in terms of quality, it suggests that thicker hulls may protect groats from damage during dehulling. Reducing groat breakage is one of the cultivar improvement objectives in many breeding programs. Breakage is most accu-rately measured by separating out the broken kernels by hand after the mechan-ical dehulling operation, weighing, and expressing as a percentage of the total dehulled groats. However, hand sorting is very time consuming and impractical for breeding programs. Ames evaluated rapid methods for screening germplasm for susceptibility to groat breakage and found that a visual assessment and scor-ing method correlated moderately to hand sorting, and that a mechanical sieving method was highly correlated (unpublished data). Engleson and Fulcher (2002) suggested differences in some oat components can confer "toughness" charac-teristics to kernels and, therefore, decrease dehulling impact damage. Compo-nents such as moisture, protein, and β-glucan were found to increase tough-ness, decreasing impact damage, whereas higher starch levels increased damage (Engleson and Fulcher, 2002).

3.2.5 Kernel size and shape

The size and shape of oat kernels, commonly expressed as "plumps and thins," is a quality factor important to breeders and the milling industry. In general, large and uniform seed size is preferred, particularly for millers producing large oat flakes. Groat size affects the maximum size of the flake that can be made (Doehlert *et al.*, 2006). Kernel size and uniformity also affect the efficiency of the dehulling operation, since large and small kernels require different rotor speeds and energy input (Doehlert and Wiessenborn, 2007). Larger, plumper seeds are associated with better milling yields. Plumps and thins are measured in oat breeding programs with the goal of increasing the plumps and reducing thins. Plump seeds comprise the portion of a 50 g sample of oats that remain on top of a 5.5/64″ × 0.75″ slotted screen, whereas thin seeds are those that pass through a 5/64″ × 0.75″ screen (Prairie Recommending Committee for Oat and Barley, 2010). Seed size and shape vary significantly within and among oat cultivars (Doehlert *et al.*, 2006, 2008; Hu *et al.*, 2009) and are related to other quality factors as well. For example, the relationship between kernel size/shape and lipase activity was studied by Hu *et al.* (2009), who showed large cultivar differences for kernel lipase activity. Within each cultivar, large and small kernels (over and through a 5/64″ × 0.75″ sieve, respectively) exhibited variable lipase activity, which was negatively correlated with increased kernel area, width, length, and thousand kernel weight (Hu *et al.*, 2009). This observation is likely explained, at least in part, by the fact that the majority of lipase activity exists on the surface of the groat and, therefore, within cultivars—the larger the surface area (as is the case with the smaller kernels), the greater the lipase activity. This is another reason that larger kernels are preferred by industry.

3.2.6   Test weight

Test weight refers to the weight of a specific volume of grain, expressed as kg/hL. Test weight is determined using a 0.5- or 1.0-L measuring cylinder filled with grain in a consistent and uniform manner prior to weighing in grams and converting to kg/hL (Canadian Grain Commission, 2012b). A higher test weight is desirable because it is associated with higher quality. In Canada, the Canadian Grain Commission specifications for a No. 1 CW Grade include a minimum test weight of 56 kg/hL and for a No. 4 CW Grade include a minimum test weight of 48 kg/hL (Canadian Grain Commission, 2012a). In general, the specifications of the Canadian milling industry require a minimum test weight of 52 kg/hL (Saskatchewan Ministry of Agriculture, 2011). Similar requirements have been identified in Australia, where a premium oat grade (Oat 1) must have a minimum hectolitre weight of 51 kg/hL (Winfield *et al.*, 2007).

Test weight reflects packing properties of individual grains in a container and is influenced by physical factors such as grain size and shape, hull properties, and moisture content, as well as environmental factors like preharvest weathering (Bayles, 1977). This trait is of specific interest to growers and grain buyers because it is a determining factor for grain grading and plays a role in efficient storage and transportation. Test weight is often correlated with other quality characteristics, such as kernel size and shape, and is, therefore, impacted by factors during plant development. Thin kernels generally have lower test weights, which may suggest disease or moisture stress during grain filling. Kernel length is determined at the time of anthesis but kernel width is an indicator of degree of grain filling. Since test weight is affected by dry matter accumulation during grain filling, factors such as rust, drought, lodging, late planting, or high seeding rates can reduce test weight. Test weight is also affected by groat density.

Factors positively affecting density, such as protein, are positively correlated with test weight. Starch content reduces density, and therefore is negatively correlated with test weight (Doehlert and McMullen, 2000, 2008).

3.2.7   Thousand kernel weight

Kernel weight is an indicator of kernel size and density. It is determined by counting and weighing 100 or 1000 kernels using automated equipment such as an Agriculex (Guelph, Canada) seed counter or Numigral grain counter (Zhou *et al.*, 1999b). The kernel weight is expressed as grams per thousand kernels, or thousand kernel weight (TKW). Alternatively, it can be expressed on a single-kernel basis in milligrams (Mitchell Fetch *et al.*, 2003a). A TKW greater than 30 g is considered the standard in many Canadian breeding programs and is used as a minimum for screening potential new cultivars (Prairie Recommending Committee for Oat and Barley, 2010).

3.3   Nutritional oat quality

The superior nutritional value of oats compared to other cereals has long been recognized, since they contain naturally high amounts of valuable nutrients, such as soluble fibers (β-glucans), proteins, and unsaturated fatty acids; they also contain vitamins, minerals, and antioxidants (Lásztity, 1998). β-glucan soluble fiber in oats has been associated with cholesterol lowering backed by numerous studies, and has resulted in health claims for oat β-glucan/soluble fiber and cholesterol lowering found on many oat product labels. Additionally, β-glucan has been associated with reductions in glycemic response, which have led to health claim approval for some countries (European Food Safety Authority, 2011), and has the potential to result in FDA- and Health Canada-approved health claims in the future, provided there is sufficient clinical evidence to substantiate the claim. Details on health claims for oat products are found in Chapter 16.

Because of health claims and advertising, most consumers are aware of antidotal information about the "goodness" of oatmeal or other oat-based breakfast products. Food manufacturers have certainly used this to their advantage, since even an oatmeal cookie may be associated with healthy snacking due to the inclusion of rolled oats. In some countries, oat food processors are able to advertise the "goodness" of oats directly on the package label as long as the product contains whole oat or oat bran with minimum of 0.75 g oat soluble fiber as β-glucan.

Consumer demand for oat products is largely dependent on the nutritional properties and associated health benefits. β-glucan, the bioactive component for lowering cholesterol and reducing glycemic response in humans, is considered one of the most important quality characteristics when developing new oat cultivars for the milling and food industry, and for encouraging consumer health. Plant breeders have responded to the needs of these markets by selecting for desired nutritional composition, especially high β-glucan, high total dietary fiber, and low oil content. The major components of the oat groats include carbohydrates (starch and fiber), protein, and fat. The content and characteristics of these components represent quality attributes for millers, food processors, and consumers. In addition, several minor components, such as antioxidants, are a source of bioactive compounds and represent added nutritional benefits or quality factors for oats used in nutraceuticals or functional foods. However, the major oat components are the focus of this chapter.

3.3.1   β-glucan

The importance of β-glucan content in the milling of oats is well understood by all participants in the value chain, since it is the basis for many of the well-documented health benefits (Beer *et al.*, 1995; Andersson *et al.*, 2002; Beck *et al.*, 2009; Andersson and Hellstrand, 2012). Consumers recognize the value of β-glucan through label claims on products and marketing/advertising carried out by the milling and food processing industries. These industries in turn work closely with oat producers and plant breeders to communicate the levels of β-glucan required to meet the health claims specified in their products. The health claims allowed internationally are covered in Chapter 16 but, in general, health claims are associated with minimum amounts of β-glucan per serving and per day. In most cases, the level of β-glucan is 3 g/day with a minimum of 0.75–1.0 g/serving (US FDA, 1997; Health Canada, 2010). To meet this level in food products, millers prefer to process oat cultivars containing higher levels of β-glucan, with the level of 4% as a minimum.

Breeding and growing cultivars with higher levels of β-glucan is one of the selection criteria of most food oat breeding programs. The groat β-glucan content of domestic oats from international sources has been reported to range from as low as 3.7% to as high as 7.5% (Peterson, 1991; Miller *et al.*, 1993). Registered oat cultivars grown in Canada typically contain 4.3–5.5% β-glucan in the groat (Mitchell Fetch *et al.*, 2006, 2007, 2009, 2011a, 2011b). Five years of genotype-by-location data from the Western Canadian Cooperative Oat Registration trials showed that groat β-glucan content ranged from 3.6 to 7.2%, depending on the breeding line and growing location (Prairie Recommending Committee for Oat and Barley, 2012). Analysis of data from these trials also showed that both β-glucan and total dietary fiber were more heritable than grain yield (Yan *et al.*, 2011). Variations in β-glucan content and/or molecular characteristics can change due to genotype (Doehlert *et al.*, 2001; Andersson and Börjesdotter, 2011), environment (Doehlert *et al.*, 2001; Yan *et al.*, 2007; Andersson and Börjesdotter, 2011), agronomics (Tiwari and Cummins, 2009), and processing (Beer *et al.*, 1997; Gutkoski and El-Dash, 1999; Kerckhoffs *et al.*, 2003; Regand *et al.*, 2009; Hu *et al.*, 2010; Immerstrand, 2010; Tosh *et al.*, 2010; Gujral *et al.*, 2011; Yao *et al.*, 2011a; Brummer *et al.*, 2012).

Although efforts to improve the nutritional quality of oats have led to the development of cultivars with higher β-glucan levels (McMullen *et al.*, 2005), some recent studies suggest that health benefits of β-glucan also depend on the quality of β-glucan, specifically: solubility, viscosity, and/or molecular weight (Wood *et al.*, 2000; Wood, 2002; Lan-Pidhainy *et al.*, 2007; Wolever *et al.*, 2010). A method for the continuous measurement of β-glucan viscosity with a Rapid Visco Analyzer (RVA) was recently developed to overcome the complexity of the common protocols based on *in vitro* digestion methods (Gamel *et al.*, 2012).

Enzymes similar to the *in vitro* method developed by Beer and colleagues (1997) are added to the sample (to digest starch, protein, and fat) along with a buffer, placed in the RVA, and a final viscosity is achieved over a defined period of time, with the main contributor to this viscosity being β-glucan. The method gives an indication of how the food may behave during digestion in the gut, and since high viscosity of β-glucan (a product of solubility and molecular weight) is believed to be associated with lowering cholesterol and blood glucose levels, this method can predict the relative healthfulness of the food being tested or whether the characteristics of the β-glucan are changing as a result of food formulation or processing.

Cultivar and environmental effects on β-glucan quality with respect to molecular characteristics were studied by Doehlert and Simsek (2012), and although environment influenced flour slurry viscosity and extractability, no effect of genotype was observed. Genotype was reported to have significant effects on degree of polymerization (DP) of the cellulosic regions (β-1-4 linked glucose) in the β-glucan chains, where higher β-glucan cultivars were identified as having lower frequency of DP3 fragments and higher frequencies of DP4 and DP6 compared to other cultivars tested (Doehlert and Simsek, 2012).

3.3.2 Total dietary fiber

Total dietary fiber (TDF) is a major constituent of oat groats, making up approximately 10–15% of the proximate analysis (Manthey *et al.*, 1999; Welch, 2006). Total dietary fiber comprises the soluble fiber and insoluble fiber portions, together with the low-molecular weight oligosaccharides (AACC International, 2009, 2011).

The soluble fiber portion in oats, which makes up approximately 40–50% of the TDF (Manthey *et al.*, 1999) is comprised mainly of β-glucan (well known for its viscous properties and health benefits) but also contains smaller quantities of other soluble fibers, such as arabinoxylan and arabinogalactan (Manthey *et al.*, 1999, Doehlert *et al.*, 2012) that may contribute to the viscous properties and/or health benefits of oat. Research specifically on oat soluble arabinoxylan and its contribution to viscosity and health is limited. However, Doehlert *et al.* (2012) examined the monosaccharide composition of soluble fiber extracts from the oat genotype HiFi, and confirmed the presence of both arabinose and xylose; assuming all xylose and arabinose originated from arabinoxylan, the quantity of extractable arabinoxylan was much less than that of extractable β-glucan (approximately one tenth). Mannose and galactose were present (approximately 2 and 14% abundance compared to β-glucan, respectively), as was glucose (derived mostly from β-glucan, but some from starch contamination of the extract).

The insoluble fiber portion in oat is made up of cellulose and other noncellulosic polysaccharides (Englyst *et al.*, 1989) such as arabinoxylan. The majority of arabinoxylan in oats is in the insoluble portion of dietary fiber (Manthey *et al.*, 1999; Shewry *et al.*, 2008). The CODEX Alimentarius definition of dietary fiber includes all nondigestible carbohydrate polymers with a degree of polymerization of three or more if they show health benefits (Jones, 2013). Several analytical methods are currently available for the measurement of dietary fiber in plant and food products, depending on the portion of fiber of interest and definition adopted (Howlett *et al.*, 2010). The American Association of Cereal Chemists International (AACCI) recommends method AACC 32-21.01 for measuring insoluble and soluble fiber in oat products (AACC International, 1989). Recently, new methods have been suggested that include measurement of low-molecular weight dietary fiber (soluble in water and ethanol): AOAC Methods 2009.01/AACC Method 32-45.01 and AOAC Method 2011.25/AACC Method 32.50.01 (AACC International, 2009, 2011; AOAC International, 2009, 2011). Method 2009.01 is the recommended reference method in Canada (Health Canada, 2012b) and is under consideration in other countries (McCleary *et al.*, 2013).

Dietary fiber is a component of primary interest to consumers, particularly since promoting its adequate consumption is a recommendation of national dietary guidelines due to various health benefits (US Department of Agriculture & US Department of Health and Human Services, 2010; Health Canada, 2012c). Oat is recognized as a source of dietary fiber. For example, one serving of oatmeal will provide approximately 10–15% of the daily fiber requirement, a fact that consumers are familiar with, based on nutrition facts tables present on product packages. The importance of fiber to the consumer flows down through the value chain and is considered a main quality trait to oat processors and millers, aiming to meet the requirements for nutrient content claims for dietary fiber (Health Canada, 2012b) and whole grain-related health claims approved in the United States (US FDA, 2009). TDF has become an important breeding criterion in Canadian breeding programs, since both genotype and environment vary significantly for this quality trait. TDF has ranged from 6.7 to 13.8% in oat groats from Canadian breeding lines grown at multiple locations over four years (Prairie Recommending Committee for Oat and Barley, 2012); reports of registered cultivars range from 10.0 to 14.3% (Mitchell Fetch *et al.*, 2006, 2007, 2009, 2011a, 2011b). Correlations observed between TDF and β-glucan content suggest selection for improvements in β-glucan, an easier trait to measure, will also result in higher TDF (Yan *et al.*, 2011). However, TDF comprises several distinct fiber components that should be taken into consideration to achieve higher levels of TDF. Effects of genotype, environment, and their interaction can also play roles in variation in TDF levels. The content and composition of soluble and insoluble dietary fiber were shown to vary with oat genotype but not year, suggesting oats could be bred for specific fiber composition (Manthey *et al.*, 1999). Comparisons of water extractable and unextractable (i.e., soluble and insoluble) fiber components (e.g., arabinoxylan) among five cultivars of oats suggest that it is possible to breed for lines with high dietary fiber and other specific phytochemicals (Shewry *et al.*, 2008).

3.3.3   Starch

Starch comprises a large percentage of the oat and varies with cultivar and growing conditions. Starch content ranges from 39 to 55% in the whole oat and from 39 to 65% in the groat (Paton, 1977; Lásztity, 1998; Zhou *et al.*, 1998a; Rhymer *et al.*, 2005). Aspects of oat starch that set it apart from other grains such as wheat are that the starch granules are smaller (Hoover and Vasathan, 1992) and have a high lipid content (Hartunian-Sowa and White, 1992). Physiochemical properties of oat starch have been recently reviewed (Sayar and White, 2011), and examples of its importance to end-product quality are discussed later in this chapter.

3.3.4 Protein

Oats are distinct among cereals due to their considerably higher protein concentration and superior protein composition (Peterson and Brinegar, 1986; Klose and Arendt, 2012). The protein content of oats is considered a quality factor, although specific advantages of higher or lower content in traditional oat products like flakes have not been well documented. Use of oat flour in bread products suggests that lower rather than higher protein levels result in a superior bread formulation (Hüttner et al., 2010a). Protein content in oats varies, depending on the oat product, cultivar, and growing environment. Registered cultivars in Canada typically contain 11.1–13% protein in whole oats (Mitchell Fetch et al., 2003a, 2003b) and 12.3 to 16.3% in groats (Mitchell Fetch et al., 2006, 2007, 2009, 2011a, 2011b). A wider range in groat protein content was observed (10.6–22.6%) among Canadian oat breeding lines grown at several locations between 2008 and 2012 (Prairie Recommending Committee for Oat and Barley, 2012). There has been disagreement on the proportion of globulins (salt-water-soluble proteins) in oats. Quantitative data published on the proportion of globulins vary widely, from 40–50 to 70–80% (Lásztity, 1996). However, most studies agree that the globulin fraction of oats accounts for the majority of oat storage protein. Because the hull contains a high amount of cellulose and a very low amount of protein, there is a significant difference in composition between oat kernels and oat groats. The protein content of oat groats may even range from 12.4 to 24.5%, which is the highest among cereals (Lásztity, 1996). The protein concentration of grain is typically obtained by multiplying its total nitrogen content (measured by combustion) by a nitrogen-to-protein conversion factor calculated from the amino acid composition of the grain. There is some controversy regarding the conversion factor for oat protein; studies report the use of conversion factors of 6.25, 5.83, and 5.4, with the latter considered most accurate (Mariotti et al., 2008). Most cereals (wheat, barley, and rye) have a high percentage of prolamins, the alcohol-soluble fraction, which usually contains most of the storage proteins, but oats are an exception. Their major storage proteins belong to the salt-water-soluble globulin fraction, whereas in oats prolamins are a minor component. The consequence of the high-globulin, low-prolamin composition of oat seed is that oat protein compared to other cereal protein provides a better balance of the amino acids essential for humans and other monogastric animals (Shotwell et al., 1990).

Protein composition is not commonly measured in typical oat quality analysis, yet the amino acid composition of oats has set it apart from other cereal grains in terms of nutritional value. The higher lysine content and excellent functional properties suggest oat protein concentrates and isolates are ideal ingredients for protein enrichment (Lásztity, 1998). Tryptophan, another essential amino acid, is found in high amounts in oats compared to other cereals (Wieser et al., 1983).

Protein concentrations differ considerably among cultivars and species, and even among identical cultivars exposed to different environments (Lásztity, 1998). The effect of the level of nitrogen fertilizers on protein concentration is well known. However, the amino acid composition of oats is not altered as extensively by the increase in nitrogen as that of other cereals (Lásztity, 1996).

Recent studies have shown that oats can be tolerated by most people suffering from celiac disease (Hoffenberg *et al.*, 2000; Janatuinen *et al.*, 2002; Peräaho *et al.*, 2004). Consequently, oats could improve the nutritional quality of the gluten-free diet and reduce the risk of nutrient deficiencies. Oats have long been known for their benefits and new oat products are being developed and are emergent in the functional food market (Goulet *et al.*, 1986; Angelov *et al.*, 2006; Guan *et al.*, 2007). However, according to Codex Alimentarius, oats are still considered a gluten-containing cereal and government regulations regarding the use of oats for celiacs varies among countries (Klose and Arendt, 2012).

3.3.5   Fat

Oats have a relatively high fat or oil content compared to other cereals. High oil content can be a desirable quality attribute for animal feed or for industrial oil extraction. Selection for high oil oat genotypes with up to 15% oil has been successful (Branson and Frey, 1989). However, oats intended for human food should be low in oil, for two reasons. Firstly, in food applications, higher oil concentrations are deleterious because of their potential for rancidity and production of off-flavors (Doehlert, 2002). Secondly, although several countries have a health claim for cholesterol lowering with oat-soluble fiber, the claim also stipulates maximum oil levels. Oat oil content varies greatly among genotypes, ranging from 3% to over 10%. Oil composition consists of triglycerides, free fatty acids, plant sterols, glycolipids, and phospholipids. Although low oat oil content is preferred by processors, the nutritional quality of the oil varies, depending on the fatty acid composition. Oat oil contains both linoleic and linolenic acids, which are essential fatty acids for human nutrition, and palmitic acid, which plays a role in oil stability (Lásztity, 1998).

3.4   Agronomic factors affecting physical and nutritional quality

There are reports that modification of agronomic management techniques, such as seeding date, nitrogen fertilizer, and cultivar selection, could result in improved physical and compositional seed quality (May *et al.*, 2004). Delayed seeding date has been shown to affect kernel size (May *et al.*, 2004) and influence protein content (Humphreys *et al.*, 1994). Oat quality is somewhat resilient compared to other grains, with respect to maintaining quality under variable agronomic conditions. For example, weed competition can significantly reduce oat grain yield, but studies have shown the effects of high weed populations on oat quality are relatively negligible. Physical oat kernel quality, measured as kernel weight and percentage plump kernels, was only mildly reduced with high wild oat weed densities present during cultivation with no resulting grade reductions (Willenborg *et al.*, 2005). Rivera-Reyes *et al.* (2008) studied agronomic factors affecting oat seed quality and showed that to increase yield and seed quality in oats, a plant density of 40 kg/ha and nitrogen fertilization of 60 kg/ha are necessary. Phosphate fertilization of 80–120 kg/ha increased both yield and seed quality (Rivera-Reyes *et al.*, 2008). An analysis of 15 years of data from Finnish trials suggests reduced grain quality was partially due to reduced inputs of nitrogen and phosphorus fertilizer (Salo *et al.*, 2007). In general, reduced inputs in oat production for a variety of reasons—from low grain prices to environmental concerns—could negatively impact oat quality. Increased levels of nitrogen resulted in lower test weights but did not affect seed weight and positively affected protein content (Ohm, 1976). Higher levels of nitrogen fertilizer were shown to result in increased protein accompanied by a reduction in plump kernels and lower oil content (May *et al.*, 2004). Güler (2011) reported that higher levels of nitrogen fertilizer increased β-glucan content along with thousand kernel weight and test weight. Analysis of oat yield and quality data from nine locations over 7 years in Canada revealed a negative association between protein content and yield, and a positive association between protein and β-glucan content (Yan *et al.*, 2007). This suggests that selection for high levels of nutritional characteristics like protein and β-glucan may be difficult when higher yielding cultivars are desired.

3.5 Oat end-product quality

In general, food oats should impart qualities to their end products that are acceptable to consumers with respect to appearance, texture, flavor, and aroma. Milled oat products are traditionally used in numerous food applications, including cooked cereals (e.g., oatmeal), muesli, ready-to-eat breakfast cereals, bakery items (e.g., cookies and muffins), and snack products (e.g., granola bars), each with its own quality requirements. Oat attributes that influence end-use quality are discussed further by milled product type with examples of several major food applications.

3.5.1 Oat flakes

Flakes represent an important oat milling product that can be marketed as an ingredient for bakery products and granola bars, or as an end product that is consumed cold (muesli) or hot (oatmeal). Several flake products are commonly available, including those made from whole groats (large or "old fashioned" flakes) and steel cut groats (quick and instant flakes). Flake quality is typically described by parameters related to size (thickness, granulation, and specific gravity), texture (strength/durability), and water absorption. Flake size specifications are an important means for controlling end-product quality, since thickness can play a role in product functionality and the consumer's sensory perception.

Thickness is one of the most important factors affecting sensory properties of oat flakes. A study of muesli quality showed that decreased flake thickness resulted in weaker taste intensity, increased adhesion to teeth, and decreased requirement for mastication (Kälviäinen *et al.*, 2002). Thinner flakes were also perceived by panelists as more fragile (Kälviäinen *et al.*, 2002), a finding that is reiterated by an instrumental assessment, showing greater force was required to rupture thick flakes compared to thin flakes (Gates *et al.*, 2004). Flake strength not only has implications for sensory properties but also relates to durability of the flakes during packing and handling. Flake size affects absorption, as demonstrated in a study that tested the water absorption capacity of commercial flakes representing a range of granulation types (baby, instant, quick, and large) (Ames and Rhymer 2003). Samples with significantly smaller flake thickness absorbed significantly more water and vice versa for the significantly thicker flakes. The large flake sample carried this trend through to the cooked oatmeal, where instrumental analysis of the texture revealed significantly less work of penetration, adhesiveness, and stringiness, all characteristics of a more "runny" oatmeal (Ames and Rhymer, 2003). A sensory panel also perceived that thicker muesli flakes absorbed less milk (Kälviäinen *et al.*, 2002), and associations between oat flake properties (including thickness and water-binding capacity) and oatmeal sensory properties were found using principal component analysis (Lapveteläinen *et al.*, 2001).

Flake quality parameters are largely a function of processing, and therefore can be directly controlled by the processor. Variations in heat moisture treatments that groats undergo prior to flaking have been shown to alter flake and end-product quality. For example, in a study by Gates and colleagues (2008), kilned groats compared to raw groats were found to produce flakes with a significantly greater specific gravity; varying the tempering conditions just prior to flaking effected the most differences. Specific weight, thickness, and water absorption were all influenced by significant interactions between tempering time and temperature, but clear trends have suggested that increased heat treatment at this stage of processing would result in flakes with higher absorption capacity (Gates *et al.*, 2008). Similarly, in another study, subjecting heat-treated oat groats to an additional high-temperature kiln drying process prior to flaking resulted in a perceptible increase in milk absorption by sensory panelists, provided the flakes were of a thin- or medium-size classification (Kälviäinen *et al.*, 2002). Studies also suggest that flake durability can be increased by using appropriate tempering times prior to flaking (Gates *et al.*, 2008) and storing flakes at low water activity (Gates *et al.*, 2004). Cooking procedures used for preparing oatmeal, which are typically controlled by the consumer, can also significantly impact oatmeal texture, color, and flavor (Lapveteläinen and Rannikko, 2000).

Production of high-quality flakes could be aided by selecting oats with superior traits for end-product processing. However, there are relatively few published reports showing genetic variation in oat flake and cooked oatmeal quality. Rhymer *et al.* (2005) studied five Canadian oat genotypes grown in six environments. Genotype was the main source of variation influencing flake granulation but some interactions with growing environment were observed. Flake water absorption was largely affected by genotype-by-environment interactions, indicating the need for multiple growing sites when breeding for these flake characteristics. Cooked oatmeal texture, as measured by an instrumental method, was significantly affected by oat genotype. Similarly, a comparison of Finnish and Swedish genotypes revealed genotypic and environmental effects on numerous flake measurements and cooked oatmeal sensory properties (Lapveteläinen and Rannikko, 2000; Lapveteläinen et al., 2001). For example, flake thickness was significantly affected by a main effect of genotype, and damaged flake particles were significantly affected by growing year (Lapveteläinen et al., 2001). These researchers also found water-binding capacity to be affected by significant genotype-by-environment interactions, as well as several cooked oatmeal characteristics evaluated by a sensory panel (Lapveteläinen et al., 2001). In addition to texture, genotypic variations were observed for odor, flavor, and color properties of cooked oatmeal (Lapveteläinen and Rannikko, 2000). Evidence of genetic control over flake and end-product traits indicate potential for breeding oats with improved quality. Furthermore, correlations between groat and flake properties have been reported (Lapveteläinen et al., 2001; Gates et al., 2008), warranting further investigation of the use of groat measurements to predict certain qualities after processing. As more research progresses, clear quality definitions for specific end-use markets may be developed for breeding and cultivar registration purposes.

3.5.2   Steel cut groats

Steel cut groats are a unique milling product made by cutting processed groats into small pieces, which can be cooked into a hot cereal. Due to the large particle size, longer cooking times are required but the resulting porridge texture is desirable (Caldwell et al., 2000). Little information is available regarding steel cut product quality, other than that related to its lower glycemic response (Gonzalez and Stevenson, 2011) and formation of an undesirable green color when cooked in the presence of iron, which could rarely occur in freshly pumped well water (Doehlert et al., 2009).

3.5.3   Oat flour

Traditionally, oat flour is a whole grain product made from heat-treated groats or flakes, but it can also encompass milling fractions remaining after bran removal (Caldwell et al., 2000). Oat flour is commonly used in ready-to-eat breakfast cereals (Tahvonen et al., 1998; Fast and Caldwell, 2000; Holguin-Acuña et al., 2008; Núñez et al., 2009; Sandoval et al., 2009; Yao et al., 2011a) and infant cereals (Fernández-Artigas et al., 1999a, 1999b, 2001) and has increasingly been investigated as a means of adding fiber to products that are traditionally not made with oats, including pasta and noodles (Sgrulletta et al., 2005; Aydin and Gocmen, 2011; Wang et al., 2011; Zhou et al., 2011; Majzoobi et al., 2012; Mitra et al., 2012), bread (Flander et al., 2007, 2011; Tiwari et al., 2012), beverages (Angelov et al., 2006), and extruded snacks (Liu et al., 2011). Recent interest in developing gluten-free products for the celiac niche market has initiated more research into using oats as a main ingredient in staple products such as bread, pasta, and noodles (Chillo et al., 2009; Hüttner et al., 2010a, 2010b, 2011; Renzetti et al., 2010; Mastromatteo et al., 2012; Hager et al., 2012a,2012b). Given the wide range of food products and the relatively early stage of research, it is challenging to define what inherent oat flour characteristics are required to optimize end-product quality. When added to a food application, oat flour takes the role of an ingredient, and its functionality in the food system is influenced by its composition (starch, protein, lipid, and fiber), water absorption capacity, and pasting properties. Oat flour is unique in that it contains high amounts of β-glucan and fat compared to many other cereal grains, generally has high water absorption, and lacks gluten, which can present a technological challenge for some product applications.

Measuring the pasting properties of oat flour is one of the main tools used to assess quality. Pasting refers to the changes in viscosity that a flour slurry undergoes when subjected to a specified heating profile with continuous stirring. Viscosity can be measured by a number of instruments and methodologies. For example, a typical standard procedure involves heating a flour slurry to a temperature greater than that at which starch gelatinizes and holding at the elevated temperature for a short period of time before subsequent cooling (Zhou *et al.*, 1998a). A lot of information can be gained from plotting the resulting viscosity changes pertaining to the peak viscosity attained during heating, the subsequent drop in viscosity due to stirring during the holding period, and the final viscosity reached upon cooling (Zhou *et al.*, 1998a). There is also an AACC International Approved Method (76-22.01) specific to measuring the pasting properties of oat flour for use by industry as a predictor of processing quality. This method differs in that it heats the slurry to 64°C and holds this temperature constant for the remainder of the test (AACC International, 2007), which typically results in a constant increase in viscosity that is more rapid at the onset of the test and slows or plateaus towards the end-point viscosity, which is measured at 20 minutes. This method was developed to maximize discrimination between oat flour samples with different pasting properties due to variations in processing and composition (Ames, unpublished data). A similar type of heating profile was used to accentuate differences in viscosity resulting from novel hydrothermal processing of oat groats compared to commercial kilning (Cenkowski *et al.*, 2006). Spindle-type viscometers have been used to measure the viscosity of oat flour slurries incubated at constant temperatures (20, 30, or 40°C) over several hours and have resulted in a hyperbolic increase in viscosity over time (Zhang *et al.*, 1997).

Several researchers have studied the role of different components in oat flour pasting by observing changes in viscosity resulting from selective enzymatic breakdown. Starch is an important contributor to pasting due to its capacity to swell, gelatinize, and form gels, as seen by the pasting properties of purified starch slurries. Pasting properties of oat starch can be influenced by starch composition and also vary with genotype (Zhou *et al.*, 1998a; Rhymer *et al.*, 2005; Šubarić *et al.*, 2011). Although starch is the most abundant component in oat flour, it is not the necessarily the main contributor to pasting. β-glucan, which exhibits a high viscosity at relatively low concentrations, also plays a significant role (Doehlert *et al.*, 1997; Yao *et al.*, 2007; Liu *et al.*, 2010; Kim and White, 2012). Furthermore, Zhou *et al.* (2000b) found that viscograms of oat flour slurries that were allowed to pre-soak were greatly affected by the presence of endogenous β-glucanase enzymes, which also explained some genotypic variations observed. β-glucan is recognized as such a major contributor to pasting properties that oat flour viscosity measurements have been used to successfully estimate β-glucan content, provided endo-β-glucanase enzymes are inhibited (Doehlert *et al.*, 1997; Colleoni-Sirghie *et al.*, 2004). In addition to content, β-glucan molecular weight and structure can impact pasting (Yao *et al.*, 2007; Liu and White, 2011). In contrast, protein plays a minor role in pasting compared to β-glucan and starch; however, its interaction with these components does have some impact on viscosity (Zhou *et al.*, 2000b; Liu *et al.*, 2010; Kim and White, 2012). Similarly, both the quantity of lipid and its fatty acid composition have been shown to influence oat flour pasting properties (Zhou *et al.*, 1999a).

Since the underlying basis for viscosity comes from a number of components and their interactions, pasting properties of oat flour are easily influenced. For example, effects of genotype, growing environment, and crop management practices on oat flour viscous properties have been documented (Zhou *et al.*, 1998b, 1999b; Rhymer, 2002; Yao *et al.*, 2007; Doehlert and Simsek, 2012). Heat treating the grain, such as during industrial conditioning processes, also changes flour pasting properties and the degree of change can be cultivar-dependent (Zhou *et al.*, 1999c).

There are multiple mechanisms proposed by which heat treatment alters the viscous properties of oat flour. The first is related to the presence of endogenous enzymes, mainly β-glucanases, which break down β-glucan. Thus, untreated or insufficiently heated oat flour can exhibit low viscosity and/or rapid decline in viscosity. The extent of this effect is related to the amount of intrinsic enzyme activity and the nature of the heat treatment (Doehlert *et al.*, 1997; Zhang *et al.*, 1997; Zhou *et al.*, 2000b). For example, moist steam was found to be more effective in maintaining oat flour viscosity than dry roasting (Doehlert *et al.*, 1997). In addition, there is evidence that heat treatments impact β-glucan polymer properties, independent of enzyme activity, that are reflected in changes in flour viscosity. This was suggested by Zhang and colleagues (1997), who observed higher flour slurry viscosities for autoclaved and steamed oats compared to those that were sterilized to inactivate enzymes. They also found that dry roasting at a very high temperature was successful at inactivating enzymes but still resulted in reduced viscosity. Furthermore, subsequent roasting of steamed oats decreased the flour slurry viscosity compared to steaming alone, suggesting that the hydration capacity of the β-glucan was impeded by roasting (Dohelert *et al.*, 1997). Further insight can be gained from studies on the effects of heat treatments on structural properties of extracted β-glucan polymers, but these are not discussed here.

Processing also has the potential to alter starch properties, and thus impact flour pasting. However, changes in thermal properties upon commercial processing practices as measured by differential scanning calorimetry, indicated that disorganization of the starch crystalline structure occurred but without complete gelatinization (Zhou *et al.*, 2000b). Changes in flour slurry viscosity due to roasting and steaming were also attributed mainly to changes in β-glucan rather than starch (Doehlert *et al.*, 1997; Zhang *et al.*, 1997).

Other factors that have been shown to influence the pasting properties of oat flour include storage (untreated samples in the ground state can change within days) (Zhou *et al.*, 1999c) and physical factors such as the flour particle size (finer particle size results in higher viscosity) (Zhang *et al.*, 1997). To apply practical significance to these variations, relationships between pasting properties of oat flour and end-product quality have been investigated for a number of food applications, including oat bread (Hüttner *et al.*, 2010a, 2011) and noodles (Zhou *et al.*, 2011; Mitra *et al.*, 2012).

3.5.4   Oat pasta and noodles

Oat flour has been studied for use in pasta and noodle applications. In general, adding oat flour to wheat formulations imparts a softer, stickier, and less elastic texture to noodles and pasta (Chillo *et al.*, 2009; Majzoobi *et al.*, 2012; Mitra *et al.*, 2012) due to the lack of gluten-forming proteins that provide structure and viscoelasticity. The cooking quality of oat noodles is compromised by susceptibility to breakage and high cooking losses (Aydin and Gocmen, 2011; Zhou *et al.*, 2011). The high cooking loss observed in oat noodles is attributed in part to the presence of water-soluble β-glucan, as well as to the lack of gluten network that acts to trap starchy material, and reduce leaching and surface adhesiveness (Zhou *et al.*, 2011; Majzoobi *et al.*, 2012). The unique structure of oat noodles is evident from electron scanning micrographs that show reduced surface uniformity, increased cracks and holes, and differences in protein matrix and starch granules compared to wheat (Zhou *et al.*, 2011; Majzoobi *et al.*, 2012). In addition to unique textural properties, oat noodles and pasta are characterized by a darker, more reddish color (likely due to the inclusion of bran layers) and a greater tendency for the color to change over time (Zhou *et al.*, 2011; Majzoobi *et al.*, 2012; Mitra *et al.*, 2012). Sensory panelists have also detected differences in noodle and pasta quality due to oat flour addition. In some cases, oat pasta was less acceptable overall (Aydin and Gocmen, 2011), specifically with respect to low firmness and high adhesiveness (Chillo *et al.*, 2009). In one study, panelists found no difference in noodle firmness or flavor with increasing oat addition up to 30% but the degree of liking decreased for color and appearance (Mitra *et al.*, 2012). Although the darker color and presence of specks were the most common reasons for low sensory scores, a subset of the panelists valued these characteristics and associated them with increased fiber and healthfulness (Mitra *et al.*, 2012). Panelists in another study evaluated oat noodles added to a soup and found that color and flavor only had a slight negative impact on sensory ratings when noodles contained more than 30% oat flour (Majzoobi *et al.*, 2012).

Oat flour is often only used as a minor ingredient in wheat-based formulations, due to its significant impact on product quality, but there are strategies for improving its functionality in noodle and pasta applications. Both Hager *et al.* (2012a) and Chillo *et al.* (2009) reported that extruded pasta could not be successfully produced from oat flour alone, but extrusion of 100% oat pasta using unique processing parameters has been accomplished (Ames, unpublished data; Sgrulletta *et al.*, 2005). Where oat is used as a major ingredient, efforts have focused on compensating for the lack of gluten by adding other high-protein ingredients, such as egg and wheat gluten, which have successfully improved texture, cooking quality, and sensory properties (Wang *et al.*, 2011; Zhou *et al.*, 2011; Hager *et al.*, 2012a). By adding carboxymethylcellulose or a portion of pre-gelatinized oat flour as structural agents, Chillo and colleagues (2009) enabled the extrusion of oat spaghetti and improved the dried product's resistance to break-ing. Similarly, pre-gelatinization of starch with steam during the dough mixing stage was used by Sgrulletta and colleagues (2005) to produce 100% oat pasta. The positive impact of incorporating these structural aids and applying processes that cause gelatinization reiterate the important role that protein and starch play in noodle and pasta quality.

There may also be opportunities to improve oat noodle and pasta quality through cultivar selection. Pasta made from five different naked oats showed variation in cooking time, water uptake, and flavor (Sgrulletta *et al.*, 2005). In addition, the effect of adding different Australian oat genotypes to white salted noodles was investigated by Mitra and colleagues (Mitra *et al.*, 2012); although wheat was replaced with oat only up to 30% in the formulation, differences in noodle color and texture were observed based on the source of oat flour. Cooked noodle texture was softer with increasing oat addition but selected genotypes were added up to 30% with no significant difference in firmness compared to the wheat control (Mitra *et al.*, 2012). These authors also identified superior geno-types with regard to ease of processing and maintenance of β-glucan levels after processing and cooking. Genotypic differences may be related to pasting proper-ties as relationships between oat flour RVA viscosity measurements and noodle qualities, such as processability and firmness, were observed (Mitra *et al.*, 2012). These findings support the need for further work to identify oat flour properties that lead to improved noodle texture, color, and β-glucan content, which are all important for future product development.

3.5.5 Oat bread

Similar to oat noodle and pasta applications, oat flour use in breads is challenged by the lack of gluten network that provides superior leavening in wheat breads. Despite this, oat flour has shown the most potential for bread baking among six other gluten-free grains based on analysis of dough development properties, loaf volume, crumb softness and springiness, and aroma (Hager *et al.*, 2012b). It is apparent that in the absence of a gluten network, the formation of a starch gel during oat bread processing is essential to provide the structural base for the gas retention required to achieve good bread quality. In general, oat bread formula-tions require greater amounts of water than wheat, resulting in batter rather than dough (Hager *et al.*, 2012b; Renzetti *et al.*, 2010). Measurements of batter rheol-ogy have been useful in predicting bread quality, with low batter viscosity/high deformability being preferred, as this allows for maximum gas cell expansion (Hüttner *et al.*, 2010a, 2011).

Overall, factors that support the starch gel structure and reduce batter vis-cosity are favorable for the production of good quality oat bread; however, the interaction between various influences is complex. For example, an optimum protein content of approximately 12% is needed for good quality oat bread, as suggested by Hüttner and colleagues (2010a). When they used oat flour with higher levels of protein (17%), the resulting bread had significantly lower specific loaf volume and increased crumb density, likely due to disruption of the starch gel and increased water hydration (Hüttner *et al.*, 2010a). Good quality bread has been obtained from oat flour with high protein content when high amounts of fat are also present (Hüttner *et al.*, 2011). It is hypothesized that the fat compensates by acting to retard starch water absorption, consequently reducing hydration and increasing batter deformation (Hüttner *et al.*, 2011). Furthermore, recent research has identified that oats contain specific foam-promoting proteins (tryptophanins) as well as foam-inhibiting nonpolar lipids, which warrant further investigation as to their role in oat bread structure (Kaukonen *et al.*, 2011). Dietary fiber, including β-glucan, is another component of oat flour that increases water hydration and is associated with poor baking quality (Hüttner *et al.*, 2010a, 2011). An enzyme that promotes protein polymerization was found to increase the hardness of oat bread (Renzetti *et al.*, 2010), whereas batter viscosity was reduced and bread texture was greatly improved by the addition of enzymes that hydrolyze protein and β-glucan, thus confirming their role in the baking potential of oat flour (Hüttner *et al.*, 2010a; Renzetti *et al.*, 2010).

Other factors besides composition should be considered when optimizing oat-based breads. For example, oat milling conditions can significantly impact bread quality. Oat flour with too fine a particle size and high amounts of starch damage exhibits poor baking performance (Zhang *et al.*, 1998; Hüttner *et al.*, 2010a), due to excessive water hydration and reduced batter deformation (Hüttner *et al.*, 2010a). The larger number of smaller bran pieces is also thought to create more sites for gas to escape from cells within the batter, thus resulting in the lower specific loaf volumes observed for fine oat flours (Zhang *et al.*, 1998). Hydrothermal grain treatments, such as those used to inactivate endogenous enzymes in oats prior to milling, can also influence bread baking quality. Zhang and colleagues (1998) found that grain treatments involving steam or a combination of steam and roasting resulted in good mixing and bread characteristics, whereas oat flour made from dry roasted grain (where enzymes were not fully inactivated) produced bread with low specific volume and poor quality scores. Other reports, where oats were not heat-treated prior to milling, found that higher α-amylase activity in oat flour was detrimental to bread quality, likely due to weakening of the starch gel structure (Hüttner *et al.*, 2011).

Research suggests that oat cultivar selection and improvement for superior bread baking performance is possible when variations due to milling were minimized, compositional differences between six oat genotypes were sufficient to impact bread crumb structure and hardness (Hüttner *et al.*, 2011). Furthermore, modifications to the baking process itself (e.g., longer proofing time and higher temperature) and ingredients (water and gluten) can be optimized to improve the specific volume and reduce the hardness of oat-based breads (Salmenkallio-Marttila *et al.*, 2004; Flander *et al.*, 2007). In oat breads containing wheat, β-glucan molecular weight is reduced during the bread baking process, particularly during the fermentation step, due to endogenous enzymes present in the wheat ingredient (Flander *et al.*, 2007; Tiwari *et al.*, 2012). Similar molecular weight reductions have been observed in straight dough and sourdough processes (Flander *et al.*, 2011). Novel bread processes, such as subjecting a portion of the oat batter to low levels of hydrostatic pressure, can be used to weaken proteins that interfere with starch gel structure and improve oat bread quality (Hüttner *et al.*, 2010b).

3.5.6   Extruded oat products

Whole oat flour properties have been related to the quality of ready-to-eat extruded cereal products. Two separate studies by Yao and colleagues demonstrated that the oat genotype used made a significant difference to the physical and sensory properties of extrudates (Yao *et al.*, 2006, 2011a). For example, in the first study, two genotypes were compared; the one containing higher starch with lower protein and β-glucan resulted in extrudates with a superior expansion ratio, which subsequently impacted water hydration and hardness properties (Yao *et al.*, 2006). However, in the second study, extruded products made from this same cultivar showed the least expansion ratio compared to three other genotypes with lower starch and similar or higher protein and β-glucan contents, indicating that other components, possibly high insoluble dietary fiber, or other factors are involved (Yao *et al.*, 2011a). It was also shown that oat genotypes can respond differently to changes in extrusion processing parameters (Yao *et al.*, 2006). These studies show the importance of oat flour composition and suggest that genotypic (and possibly environmental) variation has the potential to be used to select sources of oats for improved extruded cereal quality. Other studies have acknowledged the key role of starch in extrusion and how other components such as lipids and proteins can impact functionality during processing (Núñez *et al.*, 2009).

3.5.7   Oat bran

Oat bran became available as a commercial product in the 1980s; before that time, traditional oat milling did not include any separation except that of hulls (Fast and Caldwell, 2000). Unlike the wheat kernel, the morphology of the groat does not allow for separation of a distinct bran layer. However, a fraction of ground oats rich in oat bran can be attained by a combination of grinding, screening and aspiration (Wood *et al.*,1989). The following definition of oat bran was recommended by the American Association of Cereal Chemists (AACC, 1989):

Oat bran is the food that is produced by grinding clean oat groats or rolled oats and separating the resulting oat flour by sieving, bolting, and/or other suitable means into fractions such that the oat bran fraction is not more than 50% of the starting material, and has a total β-glucan content of at least 5.5% (dry weight basis) and a total dietary fiber content of at least 16.0% (dry weight basis), and such that at least one third of the total dietary fiber is soluble fiber.

Oat bran yield can vary significantly due to a variety of factors. For example, tempering oats to 12% moisture for 20 minutes improved (nearly doubled) bran yield from roller milling compared to no tempering (Doehlert and Moore, 1997). Genotype and environmental differences affecting oat β-glucan and oil levels have been found to influence bran yield, with higher β-glucan and oil levels being associated with higher yield (Doehlert and McMullen, 2000).

Health benefits associated with β-glucan from oat bran, such as lowering of cholesterol and subsequent reduction of coronary heart disease risk (Berg *et al.*, 2003), have attracted research into the development of oat fractionation processing strategies to produce various value added products with increased functionality. For example, oat bran has successfully been incorporated into porridge (Yao *et al.*, 2011b), noodles (Reungmaneepaitoon *et al.*, 2006), pasta (Åman *et al.*, 2004; Bustos *et al.*, 2011), and baked products (Åman *et al.*, 2004), illustrating the potential of oat bran to enrich various cereal products for the production of foods with high dietary fiber (β-glucan) and protein content, resulting in nutritional benefits. Porridge can be made with 100% oat bran, thereby delivering a high concentration of β-glucan (Yao *et al.*, 2011b). Reungmaneepaitoon *et al.* (2006) found that instant fried noodles could be made with 10–15% oat bran concentrate (OBC), which contained enough β-glucan (0.80–1.27 g/serving) to meet the FDA-approved health claim that requires 0.75 g per serving. Kaur and colleagues (2012) found that addition of oat bran at up to a 15% level did not reduce overall acceptability scores of pasta, whereas incorporation of other brans, such as barley, had significantly lower acceptability scores beyond 10% (although all brans were successfully added to pasta at lower levels without adversely affecting the physicochemical, cooking and sensory quality). However, according to Bustos and colleagues (2011), oat bran could only be incorporated into pasta at 5% without adversely affecting cooking properties. Åman *et al.* (2004) incorporated OBC into fresh pasta (7% OBC), macaroni (10.2% OBC), muffins (4.2 and 9% OBC), and yeast-leavened soft bread (4.5% OBC). They measured the molecular weight distribution of β-glucan in these foods and found that oat raw materials like groats and bran (produced by dry processing) contained intact β-glucan with high average molecular weight, whereas baking including a fermentation step, fresh pasta preparation, and production of fermented soup and pancake batter all resulted in extensive degradation of the oat β-glucan. Large oat bran particles and short fermentation time helped to reduce β-glucan degradation during bread making (Åman *et al.*, 2004).

3.5.8   Oat product aroma and flavor

Aroma and flavor characteristics are important contributors to the overall quality of processed oats and their end products. Processed oats are noted for their characteristic sweet, toasted cereal aromas and flavors that are generally mild and highly desirable. For example, high intensities of "creamy" and "oat" flavors were identified as important to the consumer acceptability of cooked oatmeal (Zhou *et al.*, 2000a). Sensory panels have evaluated processed oat products on the basis of overall aroma and flavor intensity, as well as by rating individual attributes. Common terms used to describe the dominant aroma and flavor properties include toasted, roasted, sweet, cereal, and oat. Less desirable odor and flavor attributes, including aftertaste, that have been measured in processed oats include metallic, bitter, musty, and yeasty (Lapveteläinen and Rannikko, 2000; Zhou *et al.*, 2000a; Heiniö *et al.*, 2001; Sides *et al.*, 2001).

The aroma and flavor of oat products are influenced by oat source, industrial processing, storage, and further end-product processing and preparation techniques. Zhou and colleagues (2000a) reported significant genotypic variations in the sensory properties of cooked oatmeal for several aroma and flavor attributes that could impact consumer acceptability. They found that fewer oatmeal attributes were also significantly impacted by growing site, particularly the less desirable properties metallic, bitter, and starch. Genotypic variations in aroma and flavor properties of processed oat products were also reported by other researchers (Molteberg *et al.*, 1996; Lapveteläinen & Rannikko, 2000) as well as genotype-by-environment interactions (Lapveteläinen *et al.*, 2001).

Processing contributes significantly to flavor and aroma development in oat products; differences can be distinguished via sensory evaluation and electronic nose technology based on stage of processing and type of hydrothermal condition employed to stabilize enzymes (Sides *et al.*, 2001; Klensporf and Jeleń, 2008; Head *et al.*, 2011; Ruge *et al.*, 2012). Subjecting raw groats to traditional kilning processes (steaming and drying) imparts toasted and cereal aromas and flavors; however further steaming to facilitate the flaking process has resulted in the detection of an additional yeasty attribute (Sides *et al.*, 2001). Other unique aroma properties of flakes have been identified as nutty, bread, and floury (Klensporf and Jeleń, 2008). Some processing conditions can lead to negative sensory properties. For example, performing heat treatments prior to removal of hulls was associated with rancid and bitter properties (Molteberg *et al.*, 1996). Furthermore, germination of oats, which is used in some processing applications, has resulted in musty and earthy properties, but these can be eliminated upon drying (Heiniö *et al.*, 2001). The drying conditions also have an impact; higher temperatures and quick drying has led to positive results, including increased intensities of roasted, nutty, and sweet attributes (Heiniö *et al.*, 2001). Storage of oats can result in a loss of sweetness (Molteberg *et al.*, 1996) as well as increased musty, earthy, bitter, and rancid sensory attributes, due to an increase in free fatty acids and volatile compounds associated with lipid oxidation; however, processing generally increases shelf life (Heiniö *et al.*, 2002). Further processing used by food manufacturers, such as extrusion cooking (Parker *et al.*, 2000), and even end-product preparation steps usually performed by the consumer (e.g., cooking oatmeal) can impact flavor properties (Lapveteläinen and Rannikko, 2000).

Specific compounds responsible for aroma and flavor in oat products have been investigated. Many flavor and aroma-active volatiles have been extracted from raw oats, and the number of compounds is substantially greater for heat processed oats (Heydanek and McGorrin, 1981; Zhou *et al.*, 2000a; Ren and Tian, 2012), although in some cases the amount or concentration of volatiles was reduced with processing (Heiniö *et al.*, 2001; Sides *et al.*, 2001; Klensporf and Jeleń, 2008). It is important to note that volatile analysis is greatly influenced by extraction and analytical methodology, as well as the oat sample preparation technique (Zhou *et al.*, 1999d, 2000a; Klensporf and Jeleń, 2008; Cognat *et al.*, 2012). Despite experimental differences, it is clear that a combination of volatile compounds is responsible for the complex aroma and flavor profiles exhibited by processed oat products, although key compounds have been identified (Parker et al., 2000; Zhou et al., 2000a; Heiniö et al., 2001; Sides et al., 2001; Klensporf and Jeleń, 2008; Ren and Tian, 2012). For example, (E,E,Z)-2,4,6-nonatrienal has been identified as a key component responsible for the characteristic "oatmeal-like, sweet" aroma noted in flakes (Schuh and Schieberle, 2005). Other non-volatile components, particularly phenolic acids, may also play a role (Molteberg et al., 1996; Heiniö et al., 2001).

3.5.9 Shelf stability of oat products

The shelf stability of the oat groat, flour, or flake product is an important quality attribute for the processor and consumer; it depends to a large extent on the high lipid content in oat and the potential for lipid derived rancidity. Oxidation of lipids is a chemical reaction that results in undesirable flavors and aromas. Traditional conditioning or kilning processes, which result in a toasty oat aroma and taste associated with oatmeal, were initially implemented to reduce oxidation of lipids and reduce rancidity. Various types of heat moisture treatments continue to be an essential part of oat processing resulting in inactivation of lipolytic enzymes such as lipase, while minimizing lipid breakdown and oxidation (Head et al., 2011). Lipase hydrolyzes oat lipids to release free fatty acids, which are further degraded to hydroperoxides by other lipolytic enzymes in oats (lipoxygenase and lipoperoxidase). An increase of free fatty acids during storage can be an indicator of hydrolytic rancidity. The hydroperoxides are precursors for secondary lipid oxidation that produces volatile aldehydes. Aldehydes such as hexanal are well known to be associated with oxidation of polyunsaturated fatty acids and are often used as an indicator of oat rancidity (Heiniö et al., 2002; Lehto et al., 2003). The aldehydes and alcohols are more abundant in rancid samples, whereas alkanes and furans are more prevalent in fresh samples (Cognat et al., 2012). The levels of these compounds can help assess the flavor attributes of oats and provide a means of potentially reducing rancidity through processing studies or genotypes with higher antioxidant activity.

3.6 Mycotoxins

Some fungal diseases have an indirect effect on the health and safety of food oats through the secondary toxins they produce, so a brief discussion of some of these toxins is warranted in the context of oat quality. During the pre- and/or postharvest (storage) stages of oat production, infection by fungal diseases such as *Fusarium* and *Penicillium*, respectively, can result in contamination of the oat grain. These contaminants can produce potentially toxic metabolites, or mycotoxins, making the oats unsuitable for human consumption. These toxins are now being tested at the end-product level and are, therefore, a concern to millers and processors (Roscoe et al., 2008).

The European Union (EU) has developed formalized acceptable maximum limits for mycotoxins in food, especially the toxins deoxynivalenol (DON) and zearalenone (ZON) produced by *Fusarium* (European Mycotoxin Awareness, 2012; Scudamore et al., 2007). Although DON may not constitute a significant threat to public health, it is known to be immunotoxic, and the need to set standards with regards to the exposure to DON from the food supply has been recognized internationally (Sobrova *et al.*, 2010). In general, lower tolerance levels are set for processed cereal products, depending on the product and what is known about the reduction of DON levels through processing. The EU standard maximum limit range is from 1750 ppb for raw oats to 200 ppb for infant cereals (European Mycotoxin Awareness, 2012). In a recent survey of 18 infant oat cereals, levels of DON from 0 to 19 ppb were reported (Dombrink-Kurtzman *et al.*, 2010). DON concentrations in harvested grain vary, depending on the cultivar, location of growth, and if grain is dehulled (Tekauz *et al.*, 2008; Slikova *et al.*, 2010). Industrial processing of raw oats into flakes was effective in reducing the levels of *Fusarium* mycotoxins by 90–95% in a UK study (Scudamore *et al.*, 2007). The majority of the mycotoxin loss observed in oat processing occurs during the dehulling stage, since DON accumulation is highest in hulls (Adler *et al.*, 2003; Slikova *et al.*, 2010). DON accumulation was 34% lower in the groats of hulled cultivars compared to hull-less cultivars (Slikova *et al.*, 2010). Additional heat treatments applied throughout the various processing steps have also been shown to reduce mycotoxin levels. Reductions in DON concentrations of up to 52% were achieved with superheated steam at 185°C (Cenkowski *et al.*, 2007).

Ochratoxin A, a potent renal carcinogen and nephrotoxic agent, is a naturally occurring fungal metabolite produced by *Penicillium verrucosum* (Canadian Grain Commission, 2011; Health Canada, 2012a; Vidal *et al.*, 2013). It can be present in small quantities in several foods, including cereal-derived foods, if temperature and moisture conditions are high during grain storage. Oat grain stored at 13.5% moisture or less is recommended to prevent Ochratoxin A (Canadian Grain Commission, 2011; Health Canada, 2012a). While the risk of adverse health effects is low, to reduce public exposure Health Canada has proposed guidelines for maximum limits which range from 5 ppb for raw cereal grain to 0.5 ppb for processed infant cereals (Health Canada, 2009). These limits are similar to maximum limits introduced by the EU. Potential mycotoxin contamination in oat end products is an example of a quality concern that is of primary importance to all participants in the value chain.

3.7 Summary

Each member of the oat value chain (breeders, growers, grain handlers, millers, food manufacturers, and consumers) has their own criteria to determine quality. Oat quality to the producer means high grain yield and physical properties that determine grade and market opportunity, such as high test weight and lack of discoloration. Other physical tests, such as plumps and thins and thousand kernel weight, are also indicators of oat quality. Millers strive to source oats with high yield but from the perspective of increasing the high value portions of the kernel. Improving milling yield requires oats with a high groat to hull ratio, good hullability, and low susceptibility to groat breakage. Processing oats into products such as flakes, flour and bran for use in a variety of food products, including oatmeal, noodles, pasta, bread, and extruded snacks and cereals, requires heat and mechanical treatments. These processing treatments impart changes to a number of physical and functional properties, including granulation, water absorption, pasting properties, color, flavor and aroma, and shelf stability, all of which can impact end-product quality and consumer acceptance. Millers and food processors also must meet requirements for food safety and health claim regulations, namely low mycotoxin content and high β-glucan. The role of the oat breeder is to develop improved cultivars by taking into consideration all heritable quality traits important to each member of the value chain. Improving the quality of oats includes marker assisted breeding, which will facilitate linking genome data to phenotypes in order to help determine the molecular basis for variation in oat quality.

Acknowledgements

The authors would like to gratefully acknowledge Lindsey Boyd, Tracy Exley and Natalie Middlestead for their assistance with review and editing.

References

AACC (1989) AACC committee adopts oat bran definition. *Cereal Foods World* **34**, 1033.

AACC International (1989) Approved Methods of Analysis, 11th edn. Method 32–21.01. Insoluble and soluble dietary fiber in oat products – enzymatic-gravimetric method. Approved November 1, 1989. AACC International, St. Paul, MN.

AACC International (2007) Approved Methods of Analysis, 11th edn. Method 76-22.01. Pasting properties of oat – rapid viscosity analysis. Approved 10 October 2007. AACC International, St. Paul, MN.

AACC International (2009) Approved Methods of Analysis, 11th edn. Method 32-45.01. Total dietary fiber (Codex Alimentarius definition). Approved December 2009. AACC International, St. Paul, MN.

AACC International (2011) Approved Methods of Analysis, 11th Ed. Method 32-50.01. Insoluble, soluble, and total dietary fiber (Codex definition) by enzymatic-gravimetric method and liquid chromatography. Approved August 2011. AACC International, St. Paul, MN.

Adler, A. *et al.* (2003) Microbiological and mycotoxicological quality parameters of naked and covered oats with regard to the production of bran and flakes. *Die Bodenkultur* **54**, 41–48.

Agriculture and Agri-Food Canada (2010) Oats: Situation and outlook. *Market Outlook Report* [Online]. Available: http://www.agr.gc.ca/pol/mad-dam/index_e.php?s1=pubs& s2=rmar&s3=php&page=rmar_02_03_2010-08-03 (last accessed 22 April 2013).

Åman, P. *et al.* (2004) Molecular weight distribution of β-glucan in oat-based foods. *Cereal Chemistry* **81**, 356–360.

Ames, N.P. and Rhymer, C.R. (2003) Development of a laboratory-scale flaking machine for oat end product testing. *Cereal Chemistry* **80**, 699–702.

Andersson, A.A. M. and Börjesdotter, D. (2011) Effects of environment and variety on content and molecular weight of β-glucan in oats. *Journal of Cereal Science* **54**, 122–128.

Andersson, K.E. and Hellstrand, P. (2012) Dietary oats and modulation of atherogenic pathways. *Molecular Nutrition & Food Research* **56**, 1003–1013.

Andersson, M. *et al.* (2002) Oat bran stimulates bile acid synthesis within 8 h as measured by 7alpha-hydroxy-4-cholesten-3-one. *The American Journal of Clinical Nutrition* **76**, 1111–1116.

Angelov, A. *et al.* (2006) Development of a new oat-based probiotic drink. *International Journal of Food Microbiology* **112**, 75–80.

AOAC International (2009) AOAC Official Method 2009.01. Total dietary fiber in foods, enzymatic-gravimetric-chromatographic method. Official Methods of Analysis. AOAC International, Gaithersburg, MD.

AOAC International (2011)AOAC Official Method 2011.25. Insoluble, soluble, and total dietary fiber in foods. Official Methods of Analysis. AOAC International, Gaithersburg, MD.

Aydin, E. and Gocmen, D. (2011) Cooking quality and sensorial properties of noodle supplemented with oat flour. *Food Science and Biotechnology* **20**, 507–511.

Bayles, R.A. (1977) Poorly filled grain in the cereal crop. 1. The assessment of poor grain filling. *Journal of National Institute of Botany* **14**, 232–240.

Beck, E.J. *et al.* (2009) Oat β-glucan increases postprandial cholecystokinin levels, decreases insulin response and extends subjective satiety in overweight subjects. *Molecular Nutrition and Food Research* **53**, 1343–1351.

Beer, M.U. *et al.* (1995) Effects of oat gum on blood cholesterol levels in healthy young men. *European Journal of Clinical Nutrition* **49**, 517–522.

Beer, M.U. *et al.* (1997) Effect of cooking and storage on the amount and molecular weight of $(1{\rightarrow}3)(1{\rightarrow}4)$-β-D-glucan extracted from oat products by an in vitro digestion system. *Cereal Chemistry* **74**, 705–709.

Berg, A. *et al.* (2003) Effect of an oat bran enriched diet on the atherogenic lipid profile in patients with an increased coronary heart disease risk. A controlled randomized lifestyle intervention study. *Annals of Nutrition & Metabolism* **47**, 306–311.

Branson, C.V. and Frey, K.J. (1989) Correlated response to recurrent selection for groat-oil content in oats. *Euphytica* **43**, 21–28.

Browne, R.A. *et al.* (2002) Hullability of oat varieties and its determination using a laboratory dehuller. *Journal of Agricultural Science* **138**, 185–191.

Brummer, Y. *et al.* (2012) Glycemic response to extruded oat bran cereals processed to vary in molecular weight. *Cereal Chemistry* **89**, 255–261.

Buerstmayr, H. *et al.* (2007) Agronomic performance and quality of oat (*Avena sativa* L.) genotypes of worldwide origin produced under Central European growing conditions. *Field Crops Research* **101**, 343–351.

Bustos, M.C. *et al.* (2011) Effect of four types of dietary fiber on the technological quality of pasta. *Food Science and Technology* **17**, 213–219.

Caldwell, E.F. *et al.* (2000) Hot Cereals. In: Breakfast Cereals and How They are Made. (eds R.B. Fast and E.F. Caldwell), 2nd edn, pp. 315–342. American Association of Cereal Chemists, Inc., St. Paul, MN.

Canadian Grain Commission (2011) *Prevent ochratoxin A in stored grain* [Online]. Available: http://www.grainscanada.gc.ca/storage-entrepose/ota/ota-eng.htm#h (last accessed 22 April 2013).

Canadian Grain Commission (2012a). Oats. In: Official Grain Grading Guide, 1 August 2012, Chapter 7. Canadian Grain Commission, Winnipeg, MB.

Canadian Grain Commission (2012b) Determining test weight. In: Official Grain Grading Guide, 1 August 2012, Chapter 1. Canadian Grain Commission, Winnipeg, MB.

Cenkowski, S. *et al.* (2006) Infrared processing of oat groats in a laboratory-scale electric micronizer. *Canadian Biosystems Engineering* **48**, 3.17–3.25.

Cenkowski, S. *et al.* (2007) Decontamination of food products with superheated steam. *Journal of Food Engineering* **83**, 68–75.

Chillo, S. *et al.* (2009) Properties of quinoa and oat spaghetti loaded with carboxymethyl-cellulose sodium salt and pregelatinized starch as structuring agents. *Carbohydrate Polymers* **78**, 932–937.

Cognat, C. *et al.* (2012) Comparison of two headspace sampling techniques for the analysis of off-flavour volatiles from oat based products. *Food Chemistry* **134**, 1592–1600.

Colleoni-Sirghie, M. *et al.* (2004) Prediction of β-glucan concentration based on viscosity evaluations of raw oat flours from high β-glucan and traditional oat lines. *Cereal Chemistry* **81**, 434–443.

Doehlert, D.C. (2002) Quality improvement in oat. *Journal of Crop Production* **5**, 165–189.

Doehlert, D.C. *et al.* (1997) Influence of heat pretreatments of oat grain on the viscosity of flour slurries. *Journal of the Science of Food and Agriculture* **74**, 125–131.

Doehlert, D.C. *et al.* (1999) Factors affecting groat percentage in oat. *Crop Science* **39**, 1858–1865.

Doehlert, D. C. *et al.* (2001) Genotypic and environmental effects on grain yield and quality of oat grown in North Dakota. *Crop Science* **41**, 1066–1072.

Doehlert, D.C. *et al.*, (2006) Oat grain/groat size ratios: A physical basis for test weight. *Cereal Chemistry* **83**, 114–118.

Doehlert, D.C. *et al.* (2008) Size distributions of different orders of kernels within the oat spikelet. *Crop Science* **48**, 298–304.

Doehlert, D.C. *et al.* (2009) The green oat story: Possible mechanisms of green color formation in oat products during cooking. *Journal of Food Science* **74**, S226–S231.

Doehlert, D.C. *et al.* (2012) Extraction of β-glucan from oats for soluble dietary fiber quality analysis. *Cereal Chemistry* **89**, 230–236.

Doehlert, D.C. and McMullen, M.S. (2000) Genotypic and environmental effects on oat milling characteristics and groat hardness. *Cereal Chemistry* **77**, 148–154.

Doehlert, D.C. and McMullen, M.S. (2001) Optimizing conditions for experimental oat dehulling. *Cereal Chemistry* **78**, 675–679.

Doehlert, D.C. and McMullen, M.S. (2008) Oat grain density measurement by sand displacement and analysis of physical components of test weight. *Cereal Chemistry* **85**, 654–659.

Doehlert, D.C. and Moore, W.R. (1997) Composition of oat bran and flour prepared by three different mechanisms of dry milling. *Cereal Chemistry* **74**, 403–406.

Doehlert, D.C. and Simsek, S. (2012) Variation in β-glucan fine structure, extractability, and flour slurry viscosity in oats due to genotype and environment. *Cereal Chemistry* **89**, 242–246.

Doehlert, D.C. and Wiessenborn, D.P. (2007) Influence of physical grain characteristics on optimal rotor speed during impact dehulling on oats. *Cereal Chemistry* **84**, 294–300.

Dombrink-Kurtzman, M.A. *et al.* (2010) Determination of deoxynivalenol in infant cereal by immunoaffinity column cleanup and high-pressure liquid chromatography-UV detection. *Journal of Food Protection* **73**, 1073–1076.

Engleson, J.A. and Fulcher, R.G. (2002) Mechanical behavior of oats: specific groat characteristics and relation to groat damage during impact dehulling. *Cereal Chemistry* **79**, 790–797.

Englyst, H.N. *et al.* (1989) Dietary fibre (non-starch polysaccharides) in cereal products. *Journal of Human Nutrition and Dietetics* **2**, 253–271.

European Food Safety Authority (EFSA) (2011) Scientific Opinion on the substantiation of health claims related to resistant starch and reduction of post-prandial glycaemic responses (ID 681), "digestive health benefits" (ID 682) and "favours a normal colon metabolism" (ID 783) pursuant to Article 13(1) of Regulation (EC) No 1924/2006. *EFSA Journal* **9**(4):2024.

European Mycotoxin Awareness Network (2012) *Mycotoxin legislation worldwide* [Online]. Available: http://services.leatherheadfood.com/eman/FactSheet.aspx?ID=79 (last accessed 22 April 2013).

Fast, R.B. and Caldwell, E.F. (2000) Manufacturing technology of ready-to-eat cereals. In: Breakfast Cereals and How they are Made. (eds R.B. Fast and E.F. Caldwell), 2nd edn, pp. 17–54. American Association of Cereal Chemists, Inc., St. Paul, MN.

Fernández-Artigas, P. *et al.* (1999a) Browning indicators in model systems and baby cereals. *Journal of Agricultural and Food Chemistry* **47,** 2872–2878.

Fernandez-Artigas, P. *et al.* (1999b) Blockage of available lysine at different stages of infant cereal production. *Journal of the Science of Food and Agriculture* **79**, 851–854.

Fernández-Artigas, P. *et al.* (2001) Changes in sugar profile during infant cereal manufacture. *Food Chemistry* **74**, 499–505.

Flander, L. *et al.* (2007) Optimization of ingredients and baking process for improved wholemeal oat bread quality. *LWT – Food Science and Technology* **40**, 860–870.

Flander, L. *et al.* (2011) Effects of wheat sourdough process on the quality of mixed oat-wheat bread. *LWT – Food Science and Technology* **44**, 656–664.

Forsberg, R.A. and Reeves, D.L. (1992) Breeding oat cultivars for improved grain quality. In: Oat Science and Technology (eds H.G. Marshall and M.E. Sorrells), pp. 751–775. American Society of Agronomy, Inc. and Crop Science Society of America, Inc., Madison, WI.

Gamel, T.H. *et al.* (2012) Application of the Rapid Visco Analyzer (RVA) as an effective rheological tool for measurement of β-glucan viscosity. *Cereal Chemistry* **89**, 52–58.

Ganβmann, W. and Vorwerck, K. (1995) Oat milling, processing and storage. In: The oat crop: Production and utilization (ed. R.W. Welch), pp. 369–408. Chapman and Hall, London.

Gates, F.K. *et al.* (2004) Influence of some processing and storage conditions on the mechanical properties of oat flakes. *Transactions of the American Society of Agricultural Engineers* **47**, 223–226.

Gates, F.K. *et al.* (2008) Interaction of heat–moisture conditions and physical properties in oat processing: II. Flake quality. *Journal of Cereal Science* **48**, 288–293.

Gonzalez, J.T. and Stevenson, E.J. (2011) Glycaemic and appetitive responses to porridge made from oats differing by degree of processing. *Proceedings of the Nutrition Society* **70**, E121.

Goulet, G. *et al.* (1986) Protein nutritive value of Hinoat and Scott oat cultivars and concentrates. *Journal of Food Science* **51**, 241–242.

Groh, S. *et al.* (2001) Analysis of factors influencing milling yield and their association to other traits by QTL analysis in two hexaploid oat populations. *Theoretical and Applied Genetics* **103**, 9–18.

Guan, X. *et al.* (2007) Some functional properties of oat bran protein concentrate modified by trypsin. *Food Chemistry* **101**, 163–170.

Gujral, H.S. *et al.* (2011) Effect of sand roasting on beta glucan extractability, physicochemical and antioxidant properties of oats. *LWT – Food Science and Technology* **44**, 2223–2230.

Güler, M. (2011) Nitrogen and irrigation effects on grain β-glucan content of oats (*Avena sativa* L). *Australian Journal of Crop Science* **5**, 242–247.

Gutkoski, L.C. and El-Dash, A.A. (1999) Effect of extrusion process variables on physical and chemical properties of extruded oat products. *Plant Foods for Human Nutrition* **54**, 315–325.

Hager, A.S. *et al.* (2012a) Development of gluten-free fresh egg pasta based on oat and teff flour. *European Food Research and Technology* **235**, 861–871.

Hager, A.S. *et al.* (2012b) Investigation of product quality, sensory profile and ultrastructure of breads made from a range of commercial gluten-free flours compared to their wheat counterparts. *European Food Research and Technology* **235**, 333–344.

Hartunian-Sowa, S.M. and White, P.J. (1992) Characterization of starch isolated from oat groats with different amounts of lipid. *Cereal Chemistry* **69**, 521–527.

Head, D. *et al.* (2011) Storage stability of oat groats processed commercially and with superheated steam. *LWT – Food Science and Technology* **44**, 261–268.

Health Canada, Bureau of Chemical Safety, Food Directorate, Health Products and Food Branch (2009) *Information document on Health Canada's proposed maximum limits (standards) for the presence of the mycotoxin ochratoxin A in foods* [Online]. Available: http://www.hc-sc.gc.ca/fn-an/consult/limits-max-seuils/myco_consult_ochra-eng.php (last accessed 22 April 2013).

Health Canada, Bureau of Nutritional Sciences, Food Directorate, Health Products and Food Branch (2010) *Oat products and cholesterol lowering. Summary of assessment of a health claim about oat products and blood cholesterol lowering* [Online]. Available: http://www.hc-sc.gc.ca/fn-an/label-etiquet/claims-reclam/assess-evalu/oat-avoine-eng.php (last accessed 22 April 2013).

Health Canada, Bureau of Chemical Safety, Food Directorate, Health Products and Food Branch (2012a) *Summary of comments received as part of Health Canada's 2010 call for data on Ochratoxin A* [Online]. Available: http://www.hc-sc.gc.ca/fn-an/consult/limits-max-seuils/myco_ochra-2012-summary-resume-eng.php (last accessed 22 April 2013).

Health Canada, Bureau of Nutritional Sciences, Food Directorate, Health Products and Food Branch (2012b) *Policy for labelling and advertising of dietary fibre-containing food products* [Online]. Available: http://www.hc-sc.gc.ca/fn-an/legislation/pol/fibre-label-etiquetage-eng.php (last accessed 22 April 2013).

Health Canada (2012c) *Do Canadian adults meet their nutrient requirements through food intake alone?* [Online]. Available: http://www.hc-sc.gc.ca/fn-an/surveill/nutrition/commun/art-nutr-adult-eng.php (last accessed 22 April 2013).

Heiniö, R.L. *et al.* (2001) Effect of drying treatment conditions on sensory profile of germinated oat. *Cereal Chemistry* **78**, 707–714.

Heiniö, R.L. *et al.* (2002) Differences between sensory profiles and development of rancidity during long-term storage of native and processed oat. *Cereal Chemistry* **79**, 367–375.

Heydanek, M.G. and McGorrin, R.J. (1981) Gas chromatography-mass spectroscopy investigations on the flavor chemistry of oat groats. *Journal of Agricultural and Food Chemistry* **29**, 950–954.

Hoffenberg, E.J. *et al.* (2000) A trial of oats in children with newly diagnosed celiac disease. *Journal of Pediatrics* **137**, 361–366.

Holguín-Acuña, A.L. *et al.* (2008) Maize bran/oat flour extruded breakfast cereal: A novel source of complex polysaccharides and an antioxidant. *Food Chemistry* **111**, 654–657.

Hoover, R. and Vasanthan, T. (1992) Studies on isolation and characterization of starch from oat (*Avena nuda*) grains. *Carbohydrate Polymers* **19**, 285–297.

Howlett, J.F. *et al.* (2010) *The definition of dietary fiber – discussions at the Ninth Vahouny Fiber Symposium: building scientific agreement* [Online]. Available: http://www.ncbi.nlm.nih.gov/pmc/articles/PMC2972185/ (last accessed 22 April 2013).

Hu, X. *et al.* (2009) Relationship between kernel size and shape and lipase activity of naked oat before and after pearling treatment. *Journal of the Science of Food and Agriculture* **89**, 1424–1427.

Hu, X. *et al.* (2010) The effects of steaming and roasting treatments on β-glucan, lipid and starch in the kernels of naked oat (Avena nuda). *Journal of the Science of Food and Agriculture* **90**, 690–695.

Humphreys, D.G. *et al.* (1994) Nitrogen fertilizer and seeding date induced changes in protein, oil and beta-glucan contents of four oat cultivars. *Journal of Cereal Science* **20**, 283–290.

Hüttner, E.K. *et al.* (2010a) Rheological properties and bread making performance of commercial wholegrain oat flours. *Journal of Cereal Science* **52**, 65–71.

Hüttner, E.K. *et al.* (2010b) Fundamental study on the effect of hydrostatic pressure treatment on the bread-making performance of oat flour. *European Food Research and Technology* **230**, 827–835.

Hüttner, E.K. *et al.* (2011) Physicochemical properties of oat varieties and their potential for breadmaking. *Cereal Chemistry* **88**, 602–608.

Immerstrand, T. (2010) *Cholesterol lowering properties of oats: Effects of processing and the role of oat components.* Doctoral Thesis. Lund University, Sweden.

Janatuinen, E.K. *et al.* (2002) No harm from five year ingestion of oats in coeliac disease. *Gut* **50**, 332–335.

Jones, J.M. (2013) Dietary fiber future directions: Integrating new definitions and findings to inform nutrition research and communication. *Advances in Nutrition* **4**, 8–15.

Kälviäinen, N. *et al.* (2002) Sensory attributes and preference mapping of muesli oat flakes. *Journal of Food Science* **67**, 455–460.

Kaukonen, O. *et al.* (2011) Foaming of differently processed oats: role of nonpolar lipids and tryptophanin proteins. *Cereal Chemistry* **88**, 239–244.

Kaur, G. *et al.* (2012) Functional properties of pasta enriched with variable cereal brans. *Journal of Food Science and Technology* **49**, 467–474.

Kent, N.L. and Evers, A.D. (1994). Technology of cereals, 4th edn. Pergamon Press, Oxford.

Kerckhoffs, D.A.J.M. *et al.* (2003) Cholesterol-lowering effect of β-glucan from oat bran in mildly hypercholesterolemic subjects may decrease when β-glucan is incorporated into bread and cookies. *American Journal of Clinical Nutrition* **78**, 221–227.

Kim, H.J. and White, P.J. (2012) Interactional effects of β-glucan, starch, and protein in heated oat slurries on viscosity and in vitro bile acid binding. *Journal of Agricultural and Food Chemistry* **60**, 6217–6222.

Kirkkari, A. *et al.* (2004) Dehulling capacity and storability of naked oat. *Agricultural and Food Science* **13**, 198–211.

Klensporf, D. and Jeleń, H.H. (2008) Effect of heat treatment on the flavor of oat flakes. *Journal of Cereal Science* **48**, 656–661.

Klose, C. and Arendt, E.K. (2012). Proteins in oats; their synthesis and changes during germination: A Review. *Critical Reviews in Food Science and Nutrition* **52**, 629–639.

Lan-Pidhainy, X. *et al.* (2007) Reducing beta-glucan solubility in oat bran muffins by freeze-thaw treatment attenuates its hypoglycemic effect. *Cereal Chemistry* **84**, 512–517.

Lapveteläinen, A. *et al.* (2001) Relationships of selected physical, chemical, and sensory parameters in oat grain, rolled oats, and cooked oatmeal – A three-year study with eight cultivars. *Cereal Chemistry* **78**, 322–329.

Lapveteläinen, A. and Rannikko, H. (2000) Quantitative Sensory Profiling of Cooked Oatmeal. *LWT – Food Science and Technology* **33**, 374–379.

Lásztity, R. (ed.) (1996) The chemistry of cereal proteins, 2nd edn. CRC Press, Boca Raton, FL.

Lásztity, R. (1998) Oat grain – a wonderful reservoir of natural nutrients and biologically active substances. *Food Reviews International* **14**, 99–119.

Lehto, S. *et al.* (2003) Enzymatic oxidation of hexanal by oat. *Journal of Cereal Science* **38**, 199–203.

Liu, Y. *et al.* (2010) Individual and interactional effects of β-glucan, starch, and protein on pasting properties of oat flours. *Journal of Agricultural and Food Chemistry* **58**, 9198–9203.

Liu, S. *et al.* (2011) Extruded Moringa leaf-oat flour snacks: Physical, nutritional, and sensory properties. *International Journal of Food Properties* **14**, 854–869.

Liu, Y. and White, P.J. (2011) Molecular weight and structure of water soluble (1→3), (1→4)-β-glucans affect pasting properties of oat flours. *Journal of Food Science* **76**, C68–C74.

Majzoobi, M. *et al.* (2012) Inclusion of oat flour in the formulation of regular salted dried noodles and its effects on dough and noodle properties. *Journal of Food Processing and Preservation*. doi: 10.1111/j.1745-4549.2012.00742.x.

Manitoba Co-operator (2013) Variety selection and growers source guide. In: *Seed Manitoba* [Online]. Available: http://www.agcanada.com/pub/seed-manitoba/ (last accessed 22 April 2013).

Manthey, F.A. *et al.* (1999) Soluble and insoluble dietary fiber content and composition in oat. *Cereal Chemistry* **76**, 417–420.

Mariotti, F. *et al.* (2008) Converting nitrogen into protein – beyond 6.25 and Jones' factors. *Critical Reviews in Food Science and Nutrition* **48**, 177–184.

Mastromatteo, M. *et al.* (2012) A multistep optimization approach for the production of healthful pasta based on nonconventional flours. *Journal of Food Process Engineering* **35**, 601–621.

May, W.E. *et al.* (2004) Effect of nitrogen, seeding date and cultivar on oat quality and yield in the eastern Canadian prairies. *Canadian Journal of Plant Science* **84**, 1025–1036.

McCleary, B.V. *et al.* (2013) Measurment of total dietary fiber; which validated method to use. *Cereal Chemistry* (in press).

McMullen, M. S. *et al.* (2005) Registration of 'HiFi' oat. *Crop Science* **45**, 1664.

Miller, S.S. *et al.* (1993) Mixed linkage β-glucan, protein content, and kernel weight in *Avena* species. *Cereal Chemistry* **70**, 231–233.

Miller, S.S. and Fulcher, R.G. (2011) Microstructure and chemistry of the oat kernel. In: Oats: Chemistry and Technology. (eds H. Webster and P.J. Wood), 2nd edn, pp. 77–94. American Association of Cereal Chemists International, St. Paul, MN.

Mitchell Fetch, J.W. *et al.* (2003a) Pinnacle oat. *Canadian Journal of Plant Science* **83**, 97–99.

Mitchell Fetch, J.W. *et al.* (2003b) Ronald oat. *Canadian Journal of Plant Science* **83**, 101–104.

Mitchell Fetch, J.W. *et al.* (2006) Furlong oat. *Canadian Journal of Plant Science* **86**, 1153–1156.

Mitchell Fetch, J.W. *et al.* (2007) Leggett oat. *Canadian Journal of Plant Science* **87**, 509–512.

Mitchell Fetch, J.W. *et al.* (2009) Jordan oat. *Canadian Journal of Plant Science* **89**, 67–71.

Mitchell Fetch, J.W. *et al.* (2011a) Summit oat. *Canadian Journal of Plant Science* **91**, 787–791.

Mitchell Fetch, J.W. *et al.* (2011b) Stainless oat. *Canadian Journal of Plant Science* **91**, 357–361.

Mitra, S. *et al.* (2012) Evaluation of white salted noodles enriched with oat flour. *Cereal Chemistry* **89**, 117–125.

Molteberg, E.L. *et al.* (1996) Variation in oat groats due to variety, storage and heat treatment. II: Sensory quality. *Journal of Cereal Science* **24**, 273–282.

Newton, A.C. *et al.* (2003) Susceptibility of oat cultivars to groat discoloration: Causes and remedies. *Plant Breeding* **122**, 125–130.

Núñez, M. *et al.* (2009) Thermal characterization and phase behavior of a ready-to-eat breakfast cereal formulation and its starchy components. *Food Biophysics* **4**, 291–303.

Ohm, H.W. (1976) Response of 21 oat cultivars to nitrogen fertilization. *Agronomy Journal* **68**, 773–775.

Parker, J.K. *et al.* (2000) Sensory and instrumental analysis of volatiles generated during the extrusion cooking of oat flours. *Journal of Agricultural and Food Chemistry* **48**, 3497–3506.

Paton, D. (1977) Oat starch. Part 1. Extraction, purification and pasting properties. *Die Stärke* **5**, 149–153.

Peltonen-Sainio, P. *et al.* (2004) Impact dehulling oat grain to improve quality of on-farm produced feed. I. Hullability and associated changes in nutritive value and energy content. *Agriculture and Food Science* **13**, 18–28.

Peräaho, M. *et al.* (2004) Oats can diversify a gluten-free diet in celiac disease and dermatitis herpetiformis. *Journal of the American Dietetic Association* **104**, 1148–1150.

Peterson, D.M. (1991) Genotype and environment effects on oat beta-glucan concentration. *Crop Science* **31**, 1517–1520.

Peterson, D.M. and Brinegar, C. (1986) Oat storage proteins. In: Oats: Chemistry and Technology (ed. F. Webster), 1st edn, pp. 153–203. American Association of Cereal Chemists. St. Paul, MN.

Prairie Recommending Committee for Oat and Barley (2010). *PRCOB operating procedures*. Prairie Grain Development Committee, Canada [Online]. Available: http://www.pgdc.ca/committees_ob.html (last accessed 22 April 2013).

Prairie Recommending Committee for Oat and Barley (2012). Western cooperative oat registration test reports. Prairie Grain Development Committee, Canada [Online]. Available: http://www.pgdc.ca/committees_ob.html (last accessed 22 April 2013).

Regand, A. *et al.* (2009) Physicochemical properties of beta-glucan in differently processed oat foods influence glycemic response. *Journal of Agricultural and Food Chemistry* **57**, 8831–8838.

Ren, Q. and Tian, Y. (2012) Studies of aroma active components in naked oat by GC-MS. *Journal of Food, Agriculture and Environment* **10**, 67–71.

Renzetti, S. *et al.* (2010) Oxidative and proteolytic enzyme preparations as promising improvers for oat bread formulations: Rheological, biochemical and microstructural background. *Food Chemistry* **119**, 1465–1473.

Reungmaneepaitoon, S. *et al.* (2006) Nutritive improvement of instant fried noodles with oat bran. *Journal of Science and Technology* **28**, 89–97.

Rhymer, C. (2002) Effects of nitrogen fertilization, genotype and environment on the quality of oats (*Avena sativa* L.) grown in Manitoba. M.Sc. Thesis. University of Manitoba, Canada.

Rhymer, C. *et al.* (2005) Effects of genotype and environment on the starch properties and end-product quality of oats. *Cereal Chemistry* **82**, 197–203.

Rivera-Reyes, J.G. *et al.* (2008) Agronomic traits associated to yield and quality in oat seeds. *Asian Journal of Plant Sciences* **7**, 767–770.

Ronald, P.S. *et al.* (1999). Heritability of hull percentage in oat. *Crop Science* **39**, 52–57.

Roscoe, V. *et al.* (2008) Mycotoxins in breakfast cereals from the Canadian retail market: A 3-year survey. *Food Additives & Contaminants: Part A* **25**, 347–355.

Ruge, C. *et al.* (2012) The effects of different inactivation treatments on the storage properties and sensory quality of naked oat. *Food and Bioprocess Technology* **5**, 1853–1859.

Salmenkallio-Marttila, M. *et al.* (2004) Effects of gluten and transglutaminase on microstructure, sensory characteristics and instrumental texture of oat bread. *Agricultural and Food Science* **13**, 138–150.

Salo, T. *et al.* (2007) Reduced fertiliser use and changes in cereal grain weight, test weight and protein content in Finland in 1990–2005. *Agricultural and Food Science* **16**, 407–420.

Sandoval, A.J. *et al.* (2009) Glass transition temperatures of a ready to eat breakfast cereal formulation and its main components determined by DSC and DMTA. *Carbohydrate Polymers* **76**, 528–534.

Saskatchewan Ministry of Agriculture (2011). *Oat production and markets (factsheet)* [Online]. Available: http://www.agriculture.gov.sk.ca/Default.aspx?DN=68b33116-9944-4df6-8575-5fc379b84d3b (last accessed 22 April 2013).

Sayer, S. and White, P.J. (2011) Oat starch: physicochemical properties and function. In: Oats: Chemistry and Technology (eds. H. Webster and P.J. Wood), 2nd edn, pp. 109–122. American Association of Cereal Chemists International, St. Paul, MN.

Schuh, C. and Schieberle, P. (2005) Characterization of (E,E,Z)-2,4,6-nonatrienal as a character impact aroma compound of oat flakes. *Journal of Agricultural and Food Chemistry* **53**, 8699–8705.

Scudamore, K.A. *et al.* (2007) Occurrence and fate of Fusarium mycotoxins during commercial processing of oats in the UK. *Food Additives and Contaminants* **24**, 1374–1385.

Sgrulletta, D. *et al.* (2005) Naked oat-based pasta. Quality variability in relation to cultivar characteristics. *Tecnica Molitoria* **56**, 116–125.

Shewry, P.R. *et al.* (2008) Phytochemical and fiber components in oat varieties in the HEALTHGRAIN diversity screen. *Journal of Agricultural and Food Chemistry* **56**, 9777–9784.

Shotwell, M.A. *et al.* (1990) Analysis of seed storage protein in oats. *The Journal of Biological Chemistry* **265**, 9652–9658.

Sides, A. *et al.* (2001) Changes in the volatile profile of oats induced by processing. *Journal of Agricultural and Food Chemistry* **49**, 2125–2130.

Šliková, S. *et al.* (2010) Response of oat cultivars to *Fusarium* infection with a view to their suitability for food use. *Biologia* **65**, 609–614.

Sobrova, P. *et al.* (2010) Deoxynivalenol and its toxicity. *Interdisciplinary Toxicology* **3**, 94–99.

Šubarić, D. *et al.* (2011) Isolation and characterisation of starch from different barley and oat varieties. *Czech Journal of Food Sciences* **29**, 354–360.

Tahvonen, R. *et al.* (1998) Black currant seeds as a nutrient source in breakfast cereals produced by extrusion cooking. *European Food Research and Technology* **206**, 360–363.

Tekauz, A. *et al.* (2004) Fusarium head blight of oat – current status in western Canada. *Canadian Journal of Plant Pathology* **26**, 473–479.

Tekauz, A. *et al.* (2008) Progress in assessing the impact of fusarium head blight on oat in western Canada and screening of avena germplasm for resistance. *Cereal Research Communications* **39**, 49–56.

Tiwari, U. *et al.* (2012) A modelling approach to estimate the level and molecular weight distribution of β-glucan during the baking of an oat-based bread. *Food and Bioprocess Technology* **5**, 1990–2002.

Tiwari, U. and Cummins, E. (2009) Simulation of the factors affecting β-glucan levels during the cultivation of oats. *Journal of Cereal Science* **50**, 175–183.

Tosh, S.M. *et al.* (2010) Processing affects the physicochemical properties of beta-glucan in oat bran cereal. *Journal of Agricultural and Food Chemistry* **58**, 7723–7730.

US Department of Agriculture and US Department of Health and Human Services (2010) *Dietary Guidelines for Americans*. 7th edn [Online]. Available: http://www.health.gov/dietaryguidelines/2010.asp (last accessed 22 April 2013).

US FDA (Food and Drug Administration) (2009) *Guidance for Industry: A Food Labeling Guide*, [Online], Available: http://www.fda.gov/FoodLabelingGuide (last accessed 22 April 2013).

US FDA (Food and Drug Administration), Department of Health and Human Services (1997) Food labeling: Health claims; oats and coronary heart disease: Final rule. *Federal Register* **62**, 3584–3601.

Vidal, A. *et al.* (2013) Determination of aflatoxins, deoxynivalenol, ochratoxin A and zearalenone in wheat and oat based bran supplements sold in the Spanish market. *Food and Chemical Toxicology* **53**, 133–138.

Wang, F. *et al.* (2011) Effects of transglutaminase on the rheological and noodle-making characteristics of oat dough containing vital wheat gluten or egg albumin. *Journal of Cereal Science* **54**, 53–59.

Welch, R.W. (2006) Cereal grains. In: The encyclopedia of human nutrition (eds B. Caballero, L. Allen and A. Prentice), 2nd edn, pp. 346–357. Academic Press, New York.

Winfield, K. *et al.* (2007) *Milling oat and feed oat quality – what are the differences?* Bulletin 4703, Western Australia Department of Agriculture and Food [Online]. Available: http://www.agric.wa.gov.au/objtwr/imported_assets/content/fcp/cer/oat/oat_grain_quality.pdf (last accessed 22 April 2013).

Wieser, H. *et al.* (1983) Tryptophan content of protein fractions from different cereals. *Zeitschrift für Lebensmittel-Untersuchung und Forschung* **177**, 457–460.

Willenborg, C.J. *et al.* (2005) Effects of relative time of emergence and density of wild oat (*Avena fatua* L.) on oat quality. *Canadian Journal of Plant Science* **85**, 561–567.

Wolever, T.M.S. *et al.* (2010) Physicochemical properties of oat β-glucan influence its ability to reduce serum LDL cholesterol in humans: A randomized clinical trial. *American Journal of Clinical Nutrition* **92**, 723–732.

Wood, P.J. (2002) Relationships between solution properties of cereal β-glucans and physiological effects - A review. *Trends in Food Science and Technology* **15**, 313–320.

Wood, P.J. *et al.* (1989) Large-scale preparation and properties of oat fractions enriched in (1-3)(1-4)-beta-D-Glucan. *Cereal Chemistry* **66**, 97–103.

Wood, P.J. *et al.* (2000) Evaluation of role of concentration and molecular weight of oat β-glucan in determining effect of viscosity on plasma glucose and insulin following an oral glucose load. *British Journal of Nutrition* **84**, 19–23.

Yan, W. *et al.* (2007) Associations among oat traits and their responses to the environment. *Journal of Crop Improvement* **20**, 1–29.

Yan, W. *et al.* (2011) Genotype × location interaction patterns and texting strategies for oat in the Canadian prairies. *Crop Science* **51**, 1903–1914.

Yao, N. *et al.* (2006) Physical and sensory characteristics of extruded products made from two oat lines with different β-glucan concentrations. *Cereal Chemistry* **83**, 692–699.

Yao, N. *et al.* (2007) Molecular weight distribution of (1→3)(1→4)-β-glucan affects pasting properties of flour from oat lines with high and typical amounts of β-glucan. *Cereal Chemistry* **84**, 471–479.

Yao, N. *et al.* (2011a) Impact of β-glucan and other oat flour components on physico-chemical and sensory properties of extruded oat cereals. *International Journal of Food Science and Technology* **46**, 651–660.

Yao, N. *et al.* (2011b) Textural properties of food systems having different moisture concentrations as impacted by oat bran with different β-glucan concentrations. *Journal of Texture Studies* **42**, 359–368.

Zhang, D. *et al.* (1997) Factors affecting viscosity of slurries of oat groat flours. *Cereal Chemistry* **74**, 722–726.

Zhang, D. *et al.* (1998) Effects of oat grain hydrothermal treatments on wheat-oat flour dough properties and breadbaking quality. *Cereal Chemistry* **75**, 602–605.

Zhou, M.X. *et al.* (1998a) Structure and pasting properties of oat starch. *Cereal Chemistry* **75**, 273–281.

Zhou, M.X. *et al.* (1998b) Effects of sowing date, nitrogen application, and sowing rate on oat quality. *Australian Journal of Agricultural Research* **49**, 845–852.

Zhou, M.X. *et al.* (1999a) Effects of oat lipids on groat meal pasting properties. *Journal of the Science of Food and Agriculture* **79**, 585–592.

Zhou, M.X. *et al.* (1999b) The effect of growing sites on grain quality of oats and pasting properties of oatmeals. *Australian Journal of Agricultural Research* **50**, 1409–1416.

Zhou, M.X. *et al.* (1999c) Effects of processing and short-term storage on the pasting characteristics of slurries made from raw and rolled oats. *Food Australia* **51**, 251–258.

Zhou, M.X. *et al.* (1999d) Analysis of volatile compounds and their contribution to flavor in cereals. *Journal of Agricultural and Food Chemistry* **47**, 3941–3953.

Zhou, M.X. *et al.* (2000a) Contribution of volatiles to the flavour of oatmeal. *Journal of the Science of Food and Agriculture* **80**, 247–254.

Zhou, M.X. *et al.* (2000b) Effects of enzyme treatment and processing pasting and thermal properties of oats. *Journal of the Science of Food and Agriculture* **80**, 1486–1494.

Zhou, B.L. *et al.* (2011) Gluten enhances cooking, textural, and sensory properties of oat noodles. *Cereal Chemistry* **88**, 228–233.

Part III
Oat Nutrition and Chemistry

4

Nutritional Comparison of Oats and Other Commonly Consumed Whole Grains

Apeksha A. Gulvady[1], Robert C. Brown[1], and Jenna A. Bell[2]
[1]*Global R&D Nutrition, PepsiCo Inc., Barrington, IL, USA*
[2]*The Sports, Cardiovascular and Wellness Nutrition Dietetic Practice Group, Academy for Nutrition and Dietetics, Chicago, IL, USA*

4.1   Introduction to oats as a cereal grain

Because of the beneficial nutritional profile of whole grains, the US dietary guidelines (DGAs) recommend that at least one-half of the total grain consumption be whole grain with at least three servings of whole grain per day. Currently, United States consumption of whole grains is less than one-half of this recommendation. In addition to increasing total whole grain intake, the DGAs name dietary fiber as a nutrient of concern, since the average intake is only 15 g per day (compared to the recommended 25 g for women and 38 g for men) (USDA and HHS, 2010). Consuming oats helps individuals achieve the whole grain recommendation and also provides a unique source of water-soluble fiber that helps increase total fiber intake (Chapter 7 provides more information on oats and β-glucan) (Kumar *et al.*, 2011). To better understand the nutritional profile of oats and other commonly consumed whole grains, this chapter describes the nutritional composition of oats, compares the overall macro- and micronutrient content between oats and other grains, and delineates how oats can be distinguished from other grains.

4.1.1   Global grain production

Globally, the three major cereal grains, wheat, corn and rice, account for more than 60% of the total calories consumed. The world's total food grain production in 2012 was 1080 million tons, with rice accounting for almost 50% of the total human consumption of grains (FAO, 2013). Worldwide production of oats is less than 40 million tons, with the predominant volume going to animal feed, such that the use of oats for human consumption accounts for less than 1% of total grain intake.

4.1.2 Oat grain structure

A mature whole oat grain has an unpalatable, dry, and brittle outer layer, termed the hull, which accounts for approximately 25–36% of the total dry weight of the oat grain (Ganßmann and Vorwerck, 1995; Welch, 1995). The hull is mainly composed of cellulose and hemicellulose, with a smaller percentage of lignin (Welch, 1995). In the maturing oat grain, the hull serves as a protective coating and contributes to nutrient transport to the developing grain. Upon maturity, however, the hull hardens and becomes unfit for consumption by humans, and therefore must be removed. The removal of this outer hull leaves behind the intact "groat," with three major fractions—the bran, the starchy endosperm, and the germ. Although similar to other grains in morphology, the oat groat is usually more elongated than some common whole grains (Miller and Fulcher, 2011).

The bran comprises the outer layer of the oat groat and serves as the main source of the oat's vitamin and mineral content (Peterson *et al.*, 1975; Frølich and Nyman, 1988), phytates (Fulcher *et al.*, 1981), and phenolics (Gray *et al.*, 2000). In the mature oat, the outer pericarp, testa or seed coat, and nucellus compartments of the bran are metabolically inactive and are composed of insoluble polysaccharides and phenolic compounds (Miller and Fulcher, 2011). The aleurone layer, present just under the nucellus, contains phenolics and a small amount of soluble fiber, β-glucan, in addition to protein bodies (aleurone grains) surrounded by lipids (Bechtel and Pomeranz, 1981; Peterson *et al.*, 1985). The aleurone, together with the subaleurone layer, which contains numerous protein bodies and some starch granules, form the interface between the bran and the starchy endosperm.

The starchy endosperm, the largest tissue fraction of the grain and, constituting up to 70% of the oat groat's dry weight, serves as the storehouse of starch, protein, and lipids. The highest concentration of endosperm protein is found towards the periphery and decreases towards the interior of the kernel, whereas starch is found most concentrated at the center of the endosperm and least towards the subaleurone layer (Miller and Fulcher, 2011). The endosperm, which is also rich in lipids, accounts for up to 90% of the oat groat's total lipid content (Youngs *et al.*, 1977). An abundance of lipids are found in the subaleurone and endosperm cells in the vicinity of the germ layer (Heneen *et al.*, 2009).

The germ is made up of the embryonic axis attached to the scutellum, which in turn is composed of parenchymal and epithelial tissues that contain high levels of protein, and lipids, but little starch. The protein in the germ is present in the scutellar parenchyma and, similar to the aleurone layer, is surrounded by lipids (White *et al.*, 2006). Although rich in proteins and lipids, the germ accounts for a small proportion of the total oat groat lipids (Youngs *et al.*, 1977; Miller and Fulcher, 2011).

4.2 Overview of the nutritional composition of oats

At the level of 58.7 g carbohydrate/100 g of grain, the main constituent of oats is carbohydrate, with starch comprising the majority of this carbohydrate reserve. Very few sugars and oligosaccharides are also present and account for less than 1 g carbohydrate/100 g oats (Welch, 1995). Relatively high amounts of dietary fiber (9 g/100 g oats) and proteins, at 14 g per 100 g oats, make up a significant proportion of the grain's macronutrient content. Proteins account for 15–20% of the oat groat weight (Peterson, 1992). Oats have a relatively high lipid content, with 1.2 g saturated fat, 2.2 g monounsaturated fat, and 2.5 g polyunsaturated fat (2.4 g linoleic acid and 0.11 g alpha linolenic acid). Vitamins and minerals comprise the minor organic and inorganic (ash) components of the grain, respectively. The nutrient densities of oats macronutrients are presented in Table 4.1.

The following sections describe the macro- and micronutrient compositions of oats in detail, in addition to providing a comparison between the nutritional profiles of oats and other commonly consumed grains. Recommended intake levels of nutrients vary by age and gender. However, for the purpose of nutritional labeling, one recommended intake level for each nutrient, known as the daily value (DV), is selected. In turn, the percentage daily value (%DV) is calculated as the level of each nutrient in a standard serving of the cereal grain in relation to the requirement of the nutrient (FDA, 2009). The following sections compare the relative nutrient %DVs among cereal grains based on a 2000 calorie diet for adults and children four years of age and older, per FDA nutrition labeling guidelines (Table 4.2).

4.2.1 Fiber

The dietary fiber component of cereal grains is normally classified into soluble and nonsoluble components, referring to their capacity to dissolve in water, which has a profound impact on their physiological effects in human nutrition. The nonsoluble component of cereal grains is made up primarily of lignins, cellulose, and hemicellulose, whereas the soluble fraction primarily consists of the nonstarchy

Table 4.1 Dietary fiber, protein and fat content per 100 g grains

Grain Nutrient	Oats, dry	Wheat flour, WG	Corn meal, WG	Rice, white, long grain, raw, unenriched	Rice, brown, long grain, raw, unenriched
Fiber (g)	11	11	7	1	4
Protein (g)	17	13	8	7	8
Lipid (g)	7	3	4	1	3

Based on the value of dietary fiber, protein, and fat in grams per 100 g grain (US Department of Agriculture, 2012).

Table 4.2 Daily value for macro- and micronutrient content per 100 g grains

Nutrient	Daily Value
Total fat	65 grams (g)
Saturated fat	20 g
Cholesterol	300 milligrams (mg)
Sodium	2400 mg
Potassium	3500 mg
Total carbohydrate	300 g
Dietary fiber	25 g
Protein	50 g
Vitamin A	5000 International Units (IU)
Vitamin C	60 mg
Calcium	1000 mg
Iron	18 mg
Vitamin D	400 IU
Vitamin E	30 IU
Vitamin K	80 micrograms (μg)
Thiamin	1.5 mg
Riboflavin	1.7 mg
Niacin	20 mg
Vitamin B6	2 mg
Folate	400 μg
Vitamin B12	6 μg
Biotin	300 μg
Pantothenic acid	10 mg
Phosphorus	1000 mg
Iodine	150 μg
Magnesium	400 mg
Zinc	15 mg
Selenium	70 μg
Copper	2 mg
Manganese	2 mg
Chromium	120 μg
Molybdenum	75 μg
Chloride	3400 mg

Daily value of macro- and micronutrients based on a caloric intake of 2000 calories, for adults and children 4 years and older (FDA, 2009).

polysaccharide fraction, of which β-glucan is a major component and is especially high in oats. Nonsoluble fiber is generally more effective as a bulking agent in human health and thus provides a laxative action, whereas water-soluble fiber may have a positive impact on human health (Chapter 7 details more information on oats and β-glucan) (Kumar *et al.*, 2011).

Figure 4.1 details the fiber content of the major cereal grains as they compare to oats. Whole wheat and whole oats contain the highest concentrations of total fiber (11%) by weight compared to whole grain brown rice (4%) and whole grain corn (7%). However, the ratio of soluble to insoluble fiber is much higher in oats (58%) compared to whole wheat (22%) and whole corn (16%). Thus, oats are

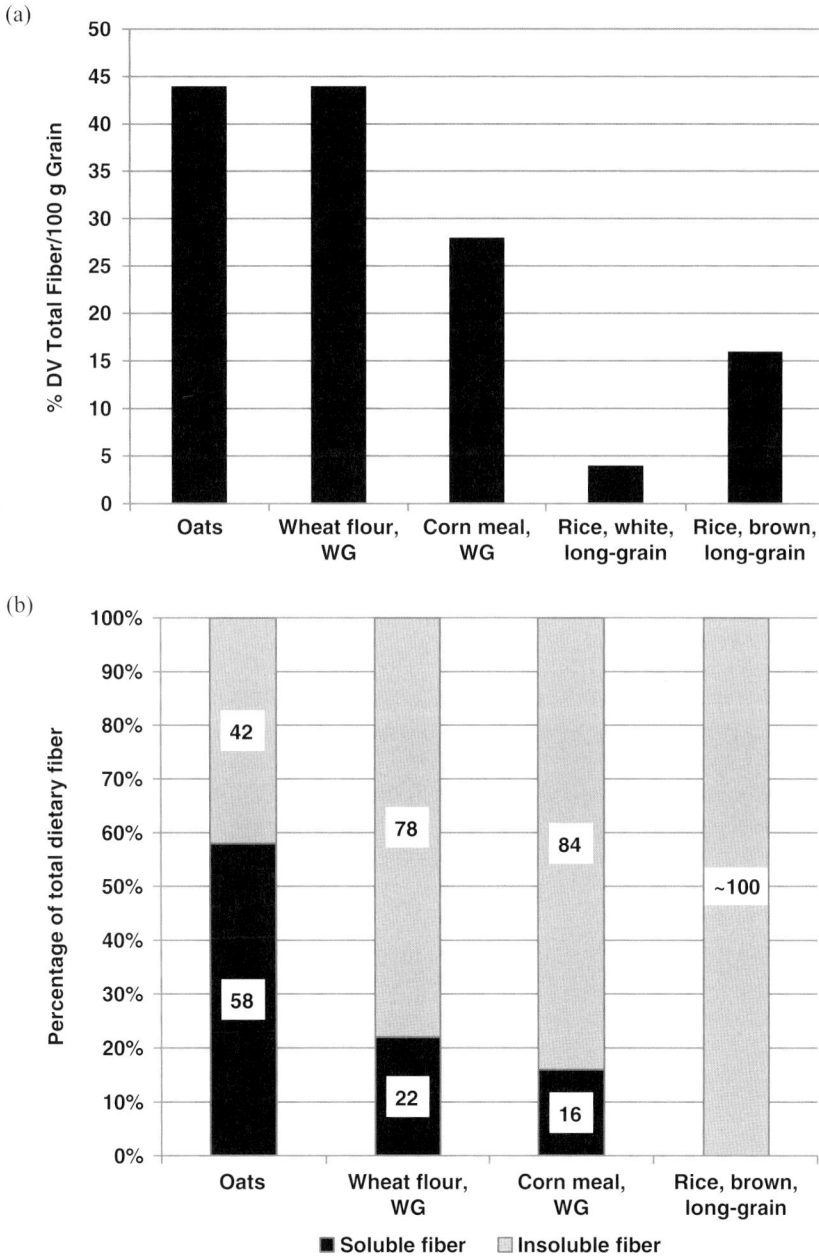

Figure 4.1 Percentage daily value and composition of fiber per 100 g grain. (a) Comparison between the percentage daily value (%DV) for total dietary fiber, calculated using the amount of dietary fiber per 100 grams cereal grain (US Department of Agriculture, 2012) and the 25 g daily value for fiber (FDA, 2009). (b) Differences in soluble and insoluble content as a percentage of total fiber.

uniquely high in soluble fiber compared to the major cereal grains and are particularly rich in β-glucan. Oats contain five times the level of β-glucan as compared to that found in whole wheat. Other minor cereal grains rich in soluble fiber and β-glucan include whole rye and barley.

The soluble fiber component of whole oats includes a higher percentage of β-glucan (69%) compared to whole wheat (36%), corn (23%), and brown rice, which contains only insoluble fiber. Research has indicated that β-glucan may have the potential to lower serum lipids and postprandial glucose and insulin levels. However, the clinical response to β-glucan may be affected by processing or other physical factors (Biorklund *et al.*, 2005) (its health impact is discussed in depth in Chapter 6). However, the higher overall ratio of soluble fiber in oats and the higher concentration of β-glucan in the soluble fraction distinguish oats from other commonly consumed whole grains.

In addition to the β-glucan content and, unlike in other grains, oats are more likely to be consumed as whole grains compared to wheat and corn. This relates to their predominate use as a milled flour to produce bread, pasta, and tortillas, or rice, which is primarily consumed with the bran removed. Therefore, recommendations to specifically increase the consumption of oats would have a positive impact on reducing the gap in whole grain intake and increasing the fiber content of the diet. However, the current gap in consumption of whole grains is so large that it will require that a greater percentage of all grains be consumed as whole grains.

4.2.2   Protein

Oats contain naturally high amounts of protein, averaging 11–15% in an oat kernel with a hull. In groats, with the high cellulose, low protein hull removed, oat protein can be as high as 12.4–24.5%, making it the highest amount of protein among commonly consumed cereal grains, including corn and rice (Lásztity, 1998). As indicated in Figure 4.2, whole grain wheat contains a high amount of protein compared to corn and rice and provides 26% of the 50 g DV of protein per 100 g serving, based on a 2000 calorie diet. However, this is less than the protein content of oats, which accounts for 34% of the DV of protein. Corn and rice provide 16% and 14–16% of the DV of protein, respectively.

In addition to the amount of protein, oat protein quality is superior to other grains due to the unique amino acid composition of the oat protein fractions— globulin (avenalin), albumins, prolamin, and glutelin (Wu *et al.*, 1972, Draper, 1973)—which, in turn, are classified based on solubility by Osborne fractionization (Klose and Arendt, 2012).

Saline-soluble globulins account for up to 50–80% of total oat proteins and are the major storage form of protein in oats, whereas the alcohol-soluble prolamin fraction constitutes a minor component and 4–15% of the total protein contained in oats. Both globulin and prolamin are found mostly in the protein bodies of the endosperm and aleurone layers of the grain; however, proportions of these two fractions can differ between grains. For instance, compared to some commonly consumed grains like wheat, which contains higher amounts of prolamin storage proteins and, to lesser extent, globulins, the high globulin:prolamin ratio of oats

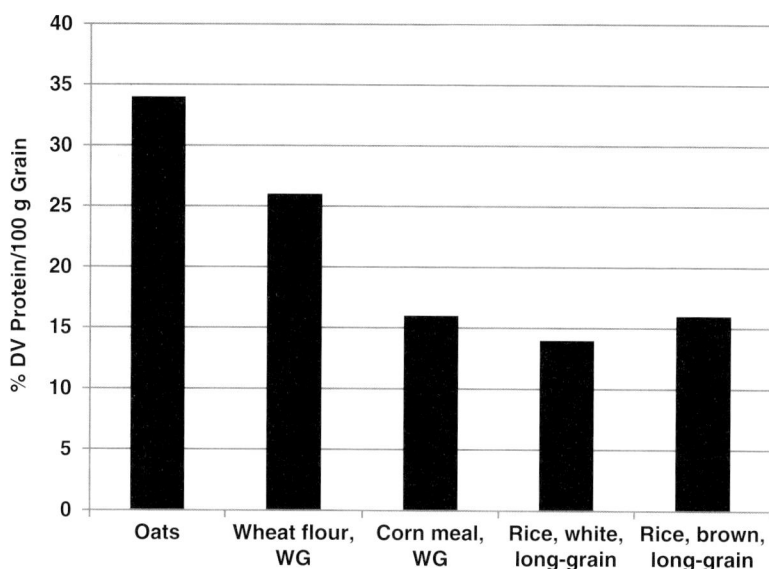

Figure 4.2 Percentage daily value protein per 100 g grain. Comparison between the percentage daily value (%DV) for protein, calculated using the amount of protein per 100 g cereal grain (US Department of Agriculture, 2012) and the 50 g DV for protein (FDA, 2009).

differentiates them from the major food grains (Shotwell *et al.*, 1990; Draper, 1973). Another minor fraction is water-soluble protein, which ranges from 1–12% of total protein, whereas glutelins make up less than 10% of total protein and are alkaline buffer-soluble (Klose and Arendt, 2012, Peterson, 2011; Lásztity, 1998). Albumins, along with some globulins, exist in the germ compartment of the groat (Draper, 1973).

With respect to the amino acid composition of the groat, the globulin fraction contains the highest amounts of basic amino acids (lysine, histidine, and arginine) and aspargine-aspartic acid, whereas the prolamin proteins are distinguishably high in glutamine-glutamic acid and proline, but lower in lysine. The amino acids that dominate the albumin fraction include lysine, aspargine-aspartic acid, and alanine. In addition, high levels of tryptophan are found in the albumin and glutelin fractions.

The protein quality of the grains can be determined by the concentration of essential amino acids in the grain in relation to their nutritional requirement. Essential amino acids, which cannot be synthesized by the body and must be provided in the diet, include histidine, isoleucine, leucine, lysine, methionine, phenylalanine, threonine, tryptophan, and valine. Nonessential amino acids, on the other hand, can be synthesized by the body, but may require indispensable amino acids as precursors. Due to the higher concentration of the globulin fraction in oats, key essential amino acids are higher in oats, specifically lysine, the limiting amino acid in wheat and other cereal grains including corn. The higher concentration of lysine in oats results in a better balance of the essential amino

Table 4.3 Amino acid composition per 100 g of commonly consumed grains

Grain / Amino acid (g)	Oats, dry	Wheat flour, WG	Corn meal, WG	Rice, white, long grain, raw, unenriched	Rice, brown, long grain, raw, unenriched
Essential amino acids					
Histidine	0.405	0.357	0.248	0.168	0.202
Isoleucine	0.694	0.443	0.291	0.308	0.336
Leucine	1.284	0.898	0.996	0.589	0.657
Lysine	0.701	0.359	0.228	0.258	0.303
Methionine	0.312	0.228	0.17	0.168	0.179
Phenylalanine	0.895	0.682	0.399	0.381	0.41
Threonine	0.575	0.367	0.305	0.255	0.291
Tryptophan	0.234	0.174	0.057	0.083	0.101
Valine	0.937	0.564	0.411	0.435	0.466
Nonessential amino acids					
Alanine	0.881	0.489	0.608	0.413	0.463
Arginine	1.192	0.648	0.405	0.594	0.602
Aspartic acid	1.448	0.722	0.565	0.67	0.743
Cysteine	0.408	0.275	0.146	0.146	0.096
Glutamic acid	3.712	4.328	1.525	1.389	1.618
Glycine	0.841	0.569	0.333	0.325	0.391
Proline	0.934	2.075	0.709	0.335	0.372
Serine	0.75	0.62	0.386	0.375	0.411
Tyrosine	0.573	0.275	0.33	0.238	0.298

Based on the value of essential and nonessential amino acids in grams per 100 g grain (US Department of Agriculture, 2012).

acids and, therefore, results in a higher amino acid score for oats (protein quality). The amino acid composition of oats and commonly consumed grains is indicated in Table 4.3. Looking at the oat protein fractions collectively, while the glutamine content is lower than in other grains, the combined glutamine-glutamic acid content, which comprises 25% of total amino acid residues, is higher. The lysine content, which averages 4.2% in oats, is greater than that of other grains apart from rice (Peterson, 2011). However, the lysine level still falls short of the recommended Food and Agriculture Organization reference standard of 5.5%/ 100 g and is also the limiting amino acid in oats. As a whole, the amino acids in oats exceed the requirement for all but two essential amino acids, lysine and threonine. However, the overall balance of essential amino acids makes oats superior to whole grain wheat and corn for overall protein quality (Pomeranz *et al.*, 1973).

4.2.3 Lipids

Whole grains are generally low in total fat on a dry weight basis; however, compared to whole wheat, whole corn, and whole grain brown rice, whole grain oats have approximately twice the level of total lipid content compared to these major cereal grains, as shown in Table 4.4 (approximately 18% of total calories).

Table 4.4 Grain fatty acid content as a percentage of calories

Grain Fatty Acid	Oats	Wheat flour WG	Corn meal WG	Brown rice
Polyunsaturated fat (%)	42	63	53	37
Monounsaturated fat (%)	37	16	30	41
Saturated fat (%)	20	21	17	22

(US Department of Agriculture, 2012).

Figure 4.3 compares the fatty acid distribution of whole oats compared to rice, wheat, and corn. It is noted that the saturated fat content of all four grains have very similar levels as a percentage of total calories, varying between 17 and 22%. The polyunsaturated fat fraction coming from both corn and wheat is the predominant type of fatty acid in these two grains, whereas rice and oats have an approximately equal distribution of polyunsaturated and monounsaturated fatty acids. Compared to US dietary recommendations, all of these grains have very favorable compositions of fatty acids. In considering the impact of dietary fat from whole grains in the United States, it has been noted that in the United States very little total grain consumption is as whole grain: approximately 1.1 ounces/day (Lin and Yen, 2007). Therefore, the total dietary fat intake coming from grains is relatively low compared to the total fat consumption in the typical US diet; thus, the favorable lipid profile of whole grains has little impact on the quality of total dietary fat intake for most individuals. Exceptions are found among some individuals eating a vegetarian diet and in countries where whole grain consumption provides a much larger percentage of total daily calories.

4.2.4 Vitamins

Vitamins are minor organic compounds that cannot be synthesized by the body and are, therefore, essential components of the diet. Based on their solubility, vitamins can be classified as water soluble—vitamin C and B vitamins—and fat soluble—Vitamins A, D, E, and K. Both water- and fat-soluble vitamins are present in oats. Table 4.5 shows the vitamin content of oats and commonly consumed cereal grains.

Among the water-soluble vitamins, neither oats nor other cereals grains naturally contain vitamin C (ascorbic acid) or vitamin B12 (cobalamin). However, other B vitamins, including thiamin, riboflavin, niacin, B6, and folate, are present in significant quantities. These B vitamins play important roles in energy and amino acid metabolism, and contribute methyl groups via their role as enzyme cofactors. Based on calculations using the US Department of Agriculture nutrient database (US Department of Agriculture, 2012) and US Food and Drug Administration nutrition labeling guidelines for a 2000 calorie diet (FDA, 2009) (Figure 4.4), 100 g oats provides 51% of the 1.5 mg DV of thiamin, and 14% of the 400 µg DV of folate, both of which are highest among commonly consumed cereals, that is, wheat, corn, and rice. At 9% of the 1.5 mg DV, oats have a similar level of riboflavin as compared to wheat and corn, which provide 10% and

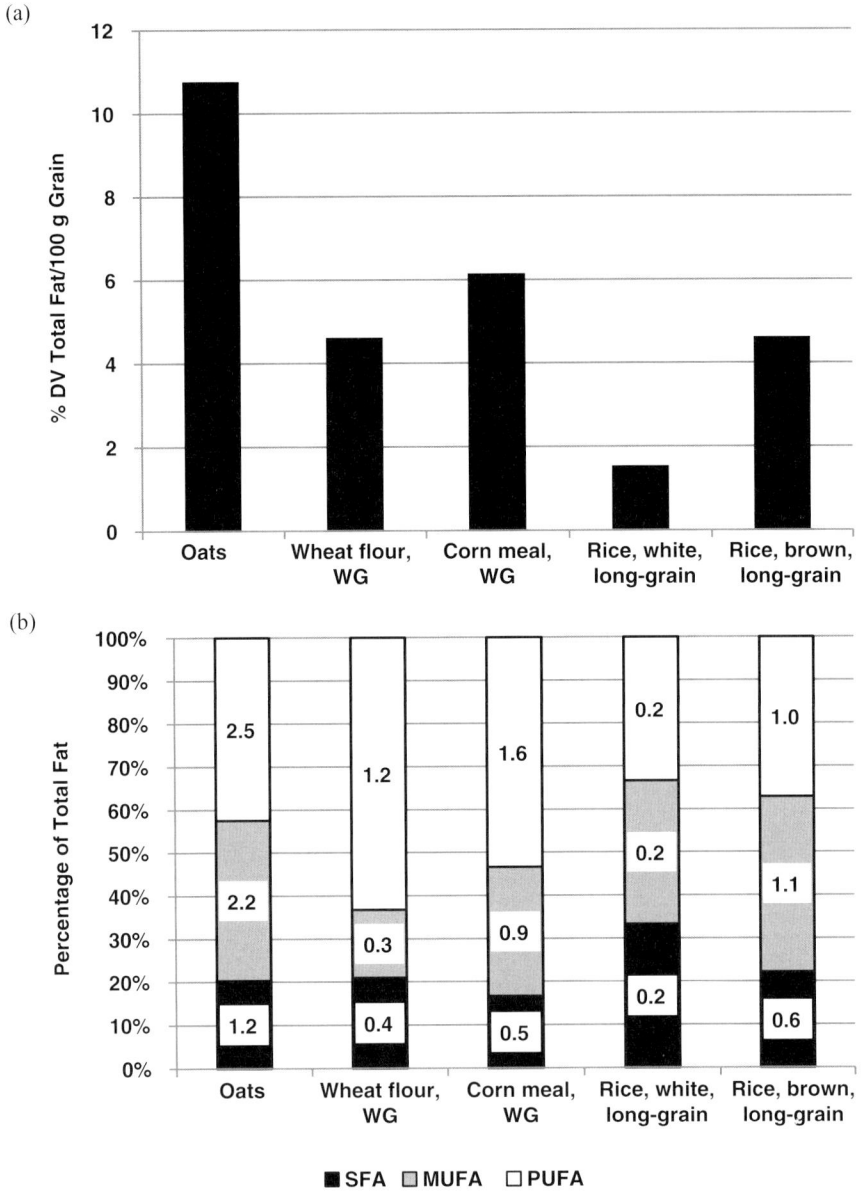

Figure 4.3 Percentage daily value and composition of fat per 100 g grain. (a) Comparison between the percentage daily value (%DV) for total fat content, calculated using the amount of dietary fiber per 100 g cereal grain (US Department of Agriculture, 2012) and the 65 g DV for fat (FDA, 2009). (b) Differences in the saturated fat (SFA), monounsaturated fatty acid (MUFA), and polyunsaturated fatty acid (PUFA) content as a percentage of total fat.

Table 4.5   Vitamin content per 100 g grains

Vitamin \ Grain	Oats, dry	Wheat flour, WG	Corn meal, WG	Rice, white, long grain, raw, unenriched	Rice, brown, long grain, raw, unenriched
Thiamin (mg)	0.763	0.502	0.385	0.07	0.401
Riboflavin (mg)	0.139	0.165	0.201	0.049	0.093
Niacin (mg)	0.961	4.957	3.632	1.6	5.091
Vitamin B6 (mg)	0.119	0.407	0.304	0.164	0.509
Folate (μg)	56	44	25	8	20

Based on the value of vitamins in mg or μg per 100 g grain (US Department of Agriculture, 2012).

12% DV, respectively. The inherent niacin content of oats is the lowest among the common grains but among wheat, corn, and rice, oats contain a comparatively high amount of the amino acid tryptophan (Figure 4.4), the precursor of niacin synthesis in the liver. Of the fat-soluble vitamins, vitamins D and A are not inherently preformed in plant-based foods. Rather, carotenoids such as β-carotene, which serve as precursors for vitamin A synthesis, are found in trace amounts in wheat and corn but not in oats. Significantly high concentrations of vitamin E are also present in oats (White *et al.*, 2006).

4.2.5   Minerals

Minerals form the inorganic component of the micronutrients in food and, depending upon the amount required in nutrition, can be classified as major or trace minerals. Major minerals are those that are required in the diet at amounts greater than 100 mg/day and include calcium, magnesium, potassium, phosphorus, and sodium. Trace minerals, on the other hand, are required in amounts less than 100 mg/day in the diet and include iron, zinc, manganese, and copper. Major and minor minerals occur in oats and commonly consumed grains but the mineral levels in oats are relatively higher than in other grains, as shown in Table 4.6 (US Department of Agriculture, 2012).

Of the major minerals, potassium, an electrolyte, and phosphorus, a component of phospholipids as well as bones and teeth, are the two major minerals that are most prominent in oats. Figure 4.5 provides a comparison of the %DV of minerals in 100 grams of the cereal grains based on a 2000 calorie diet. The potassium in oats accounts for 12% of the 3500 mg DV, followed closely by wheat, which provides 10% of the DV. The high amount of phosphorus in oats accounts for 52% of the 1000 mg DV. Wheat and brown rice supply 36% and 33% of the DV for phosphorus, respectively, followed by corn, which accounts for less than one-half the % DV provided by oats, and white rice provides the least amount of phosphorus of the grains. A very low amount (<1% DV) of sodium is provided by oats, wheat, and rice, while whole corn contains a small amount of sodium in 100 g (1.5% of the DV). Compared to the three major minerals that dominate the mineral composition, calcium and magnesium are present in lower amounts. However, their levels in oats are slightly higher than in wheat, rice, and corn.

(a)

(b)

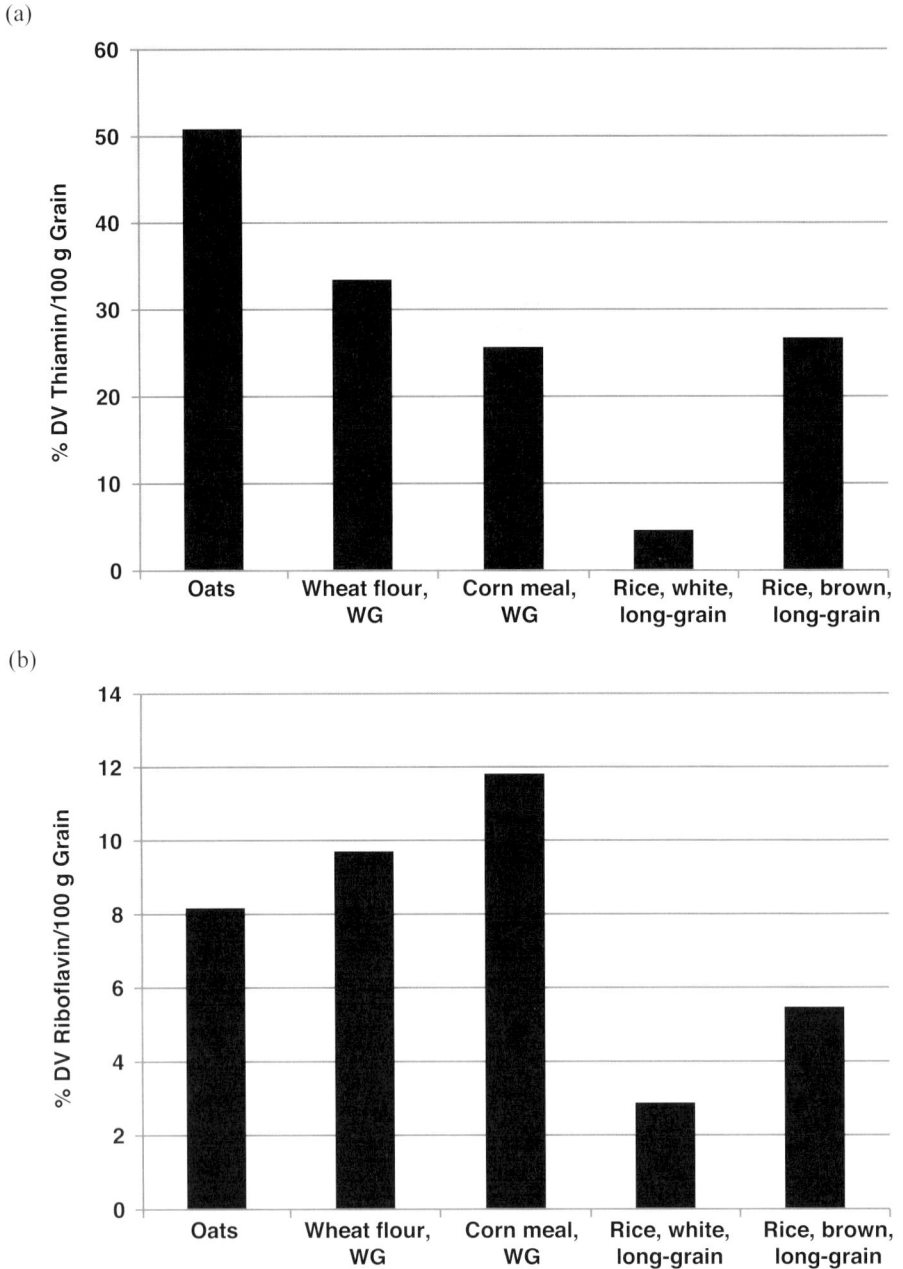

Figure 4.4 Percentage daily value vitamins per 100 g grain. Comparison between the percentage daily value (%DV) for vitamin content, calculated using the amount of vitamin per 100 g cereal grain (US Department of Agriculture, 2012) and the following DVs: (a) thiamin = 1.5 mg; (b) riboflavin = 1.7 mg; (c) niacin = 20 mg; (d) B6 = 2 mg; and (e) folate = 400 μg (FDA, 2009).

(c)

(d)

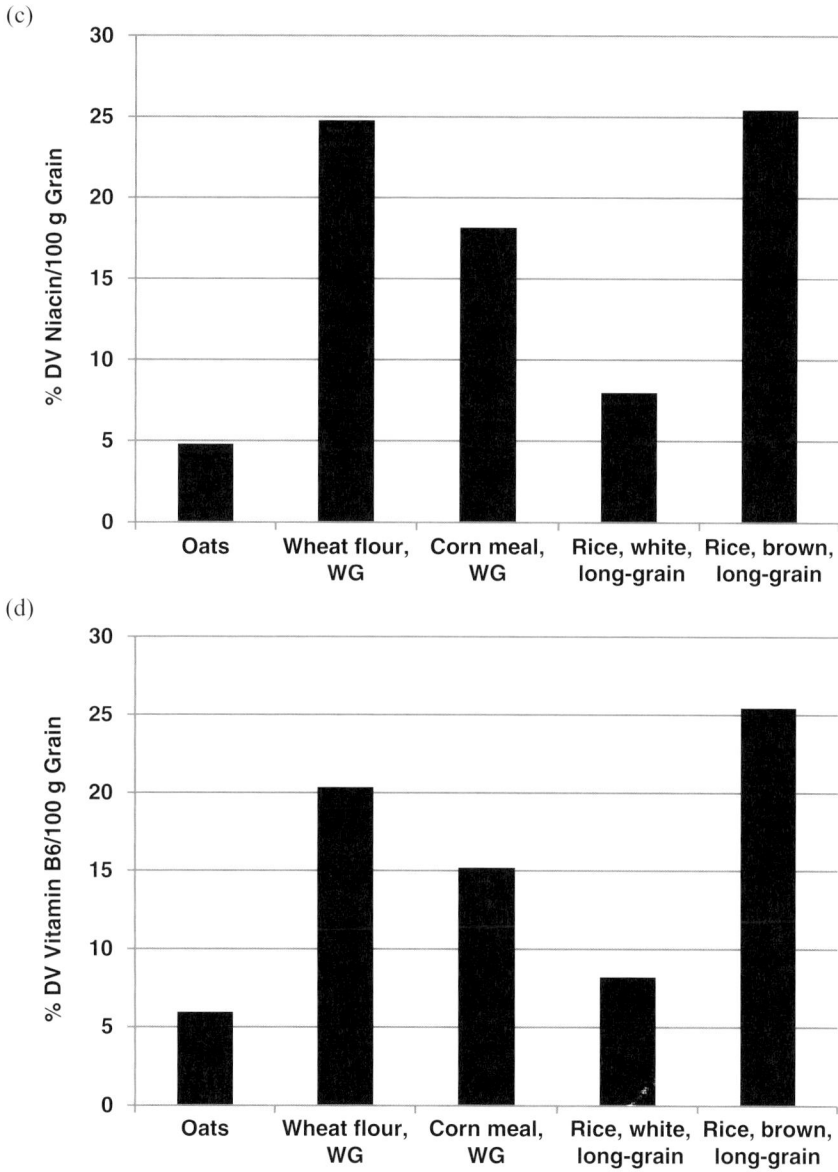

Figure 4.4 *(Continued)*

(e)

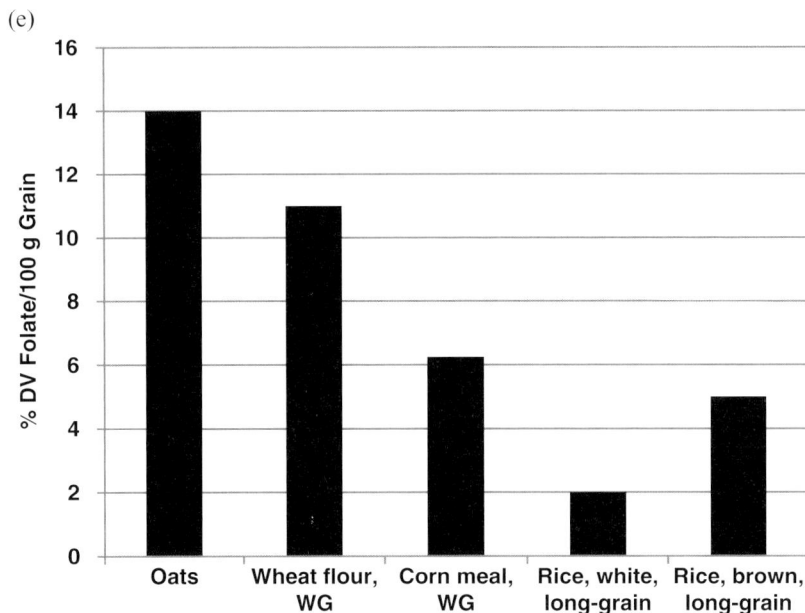

Figure 4.4 (Continued)

Calcium and magnesium are both important components of bones and teeth, and serve as enzyme cofactors in biochemical reactions, with levels in oats accounting for 5% of the 1000 mg DV and 44% of the 400 mg DV, respectively.

Similar to the major minerals, the minor minerals are also generally present in slightly greater quantities in oats than in other cereals, as indicated in Table 4.6, and, thus, provide a higher %DV per 100 g oats. The minor minerals compose 26% of the 18 mg DV of iron, an essential component of hemoglobin and an enzyme cofactor, 31% of the 2 mg DV of copper, and more than double the

Table 4.6 Mineral content per 100 g grains

Mineral (mg)	Oats, dry	Wheat flour, WG	Corn meal, WG	Rice, white, long grain, raw, unenriched	Rice, brown, long grain, raw, unenriched
Calcium	54	34	6	28	23
Iron	4.72	3.6	3.45	0.8	1.47
Magnesium	177	137	127	25	143
Phosphorus	523	357	241	115	333
Potassium	429	363	287	115	223
Sodium	2	2	35	5	7
Zinc	3.97	2.6	1.82	1.09	2.02
Copper	0.626	0.41	0.193	0.22	0.277
Manganese	4.916	4.067	0.498	1.088	3.743

Based on the value of major and trace minerals in mg per 100 g grain (US Department of Agriculture, 2012).

(a)

(b)

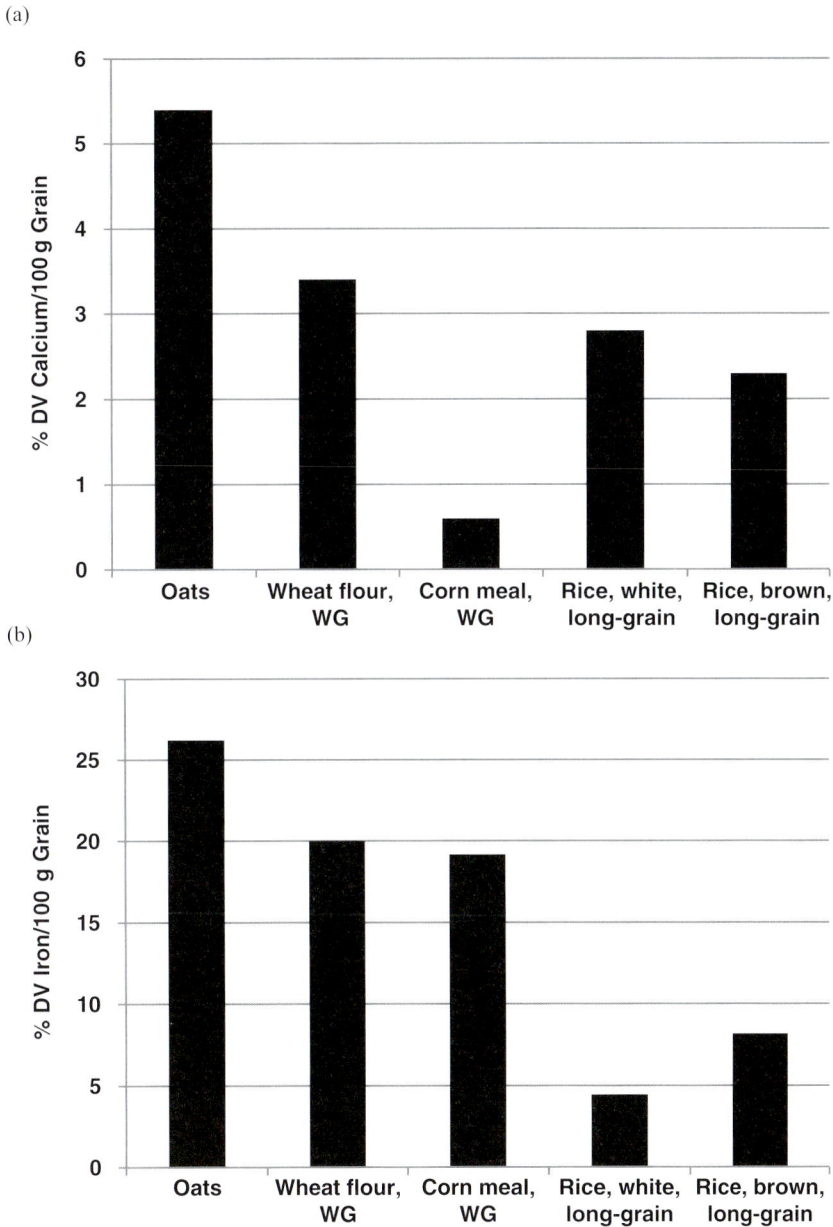

Figure 4.5 Percentage daily value minerals per 100 g grain. Comparison between the percentage daily value (%DV) for major and trace mineral content, calculated using the amount of mineral per 100 g cereal grains (US Department of Agriculture, 2012) and the following DVs: (a) calcium = 1000 mg; (b) iron = 18 mg; (c) magnesium = 400 mg; (d) phosphorus = 1000 mg; (e) potassium = 3500 mg; (f) sodium = 2400 mg; (g) zinc = 15 mg; (h) copper = 2 mg; and (i) manganese = 2 mg (FDA, 2009).

(c)

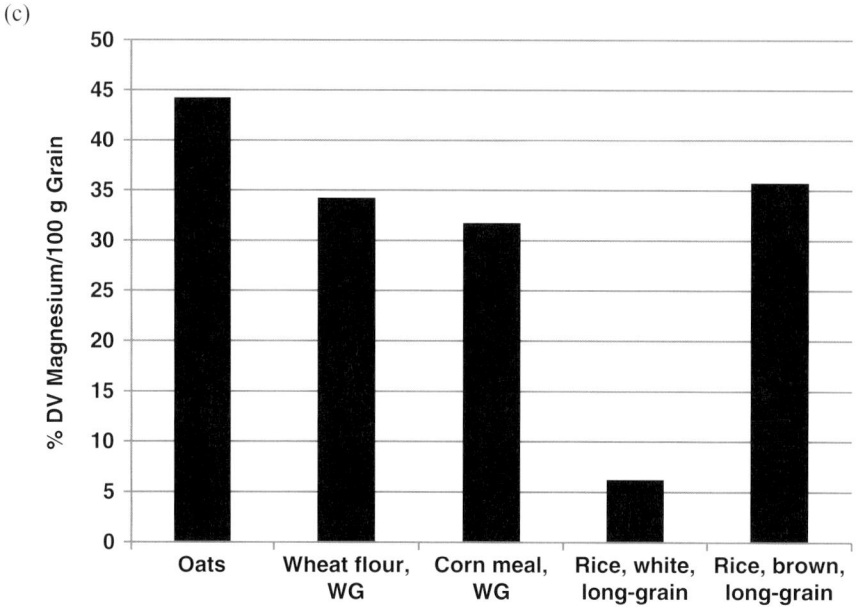

(d)

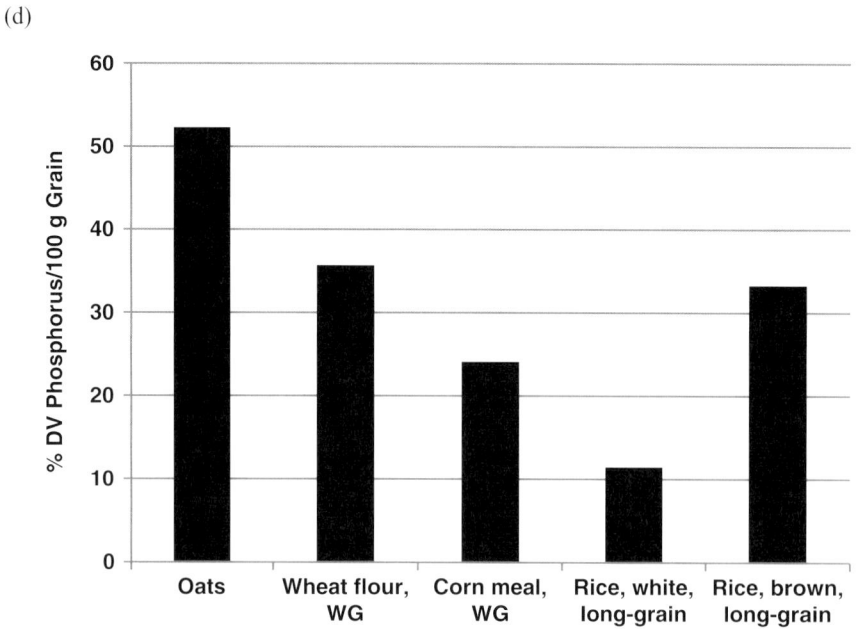

Figure 4.5   *(Continued)*

(e)

(f)

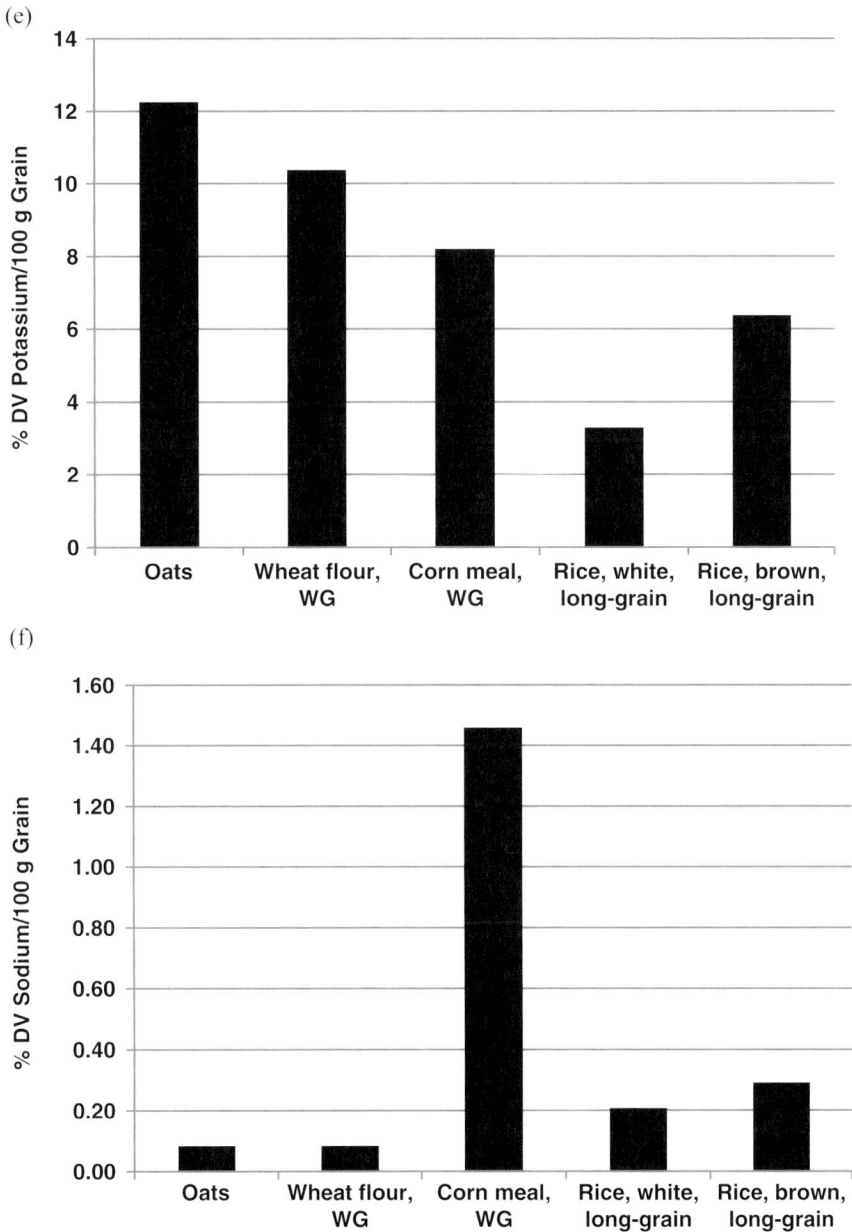

Figure 4.5 *(Continued)*

(g)

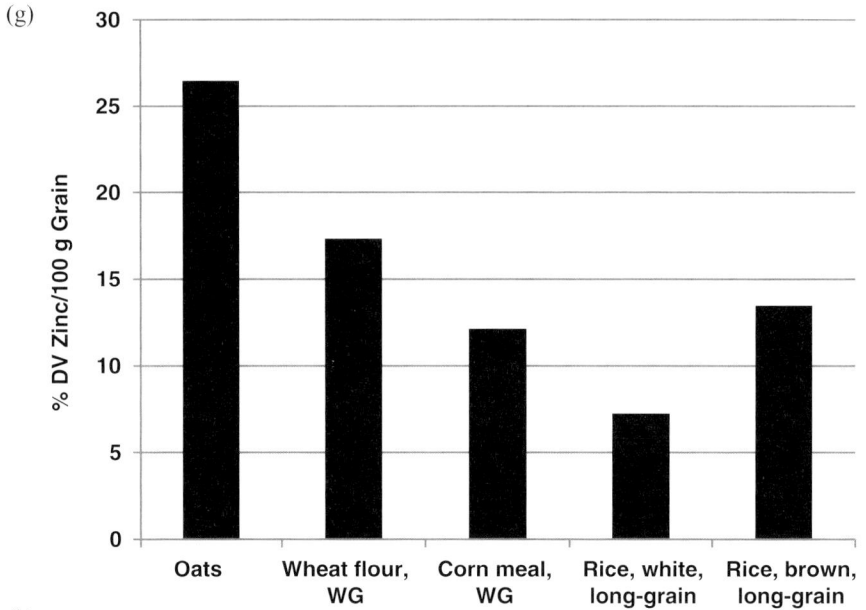

(h)

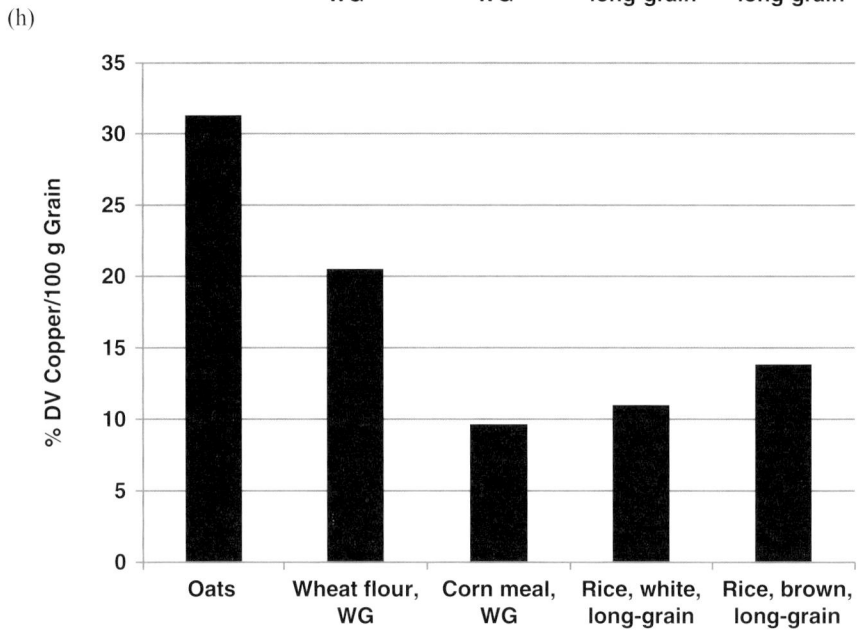

Figure 4.5 *(Continued)*

(i)

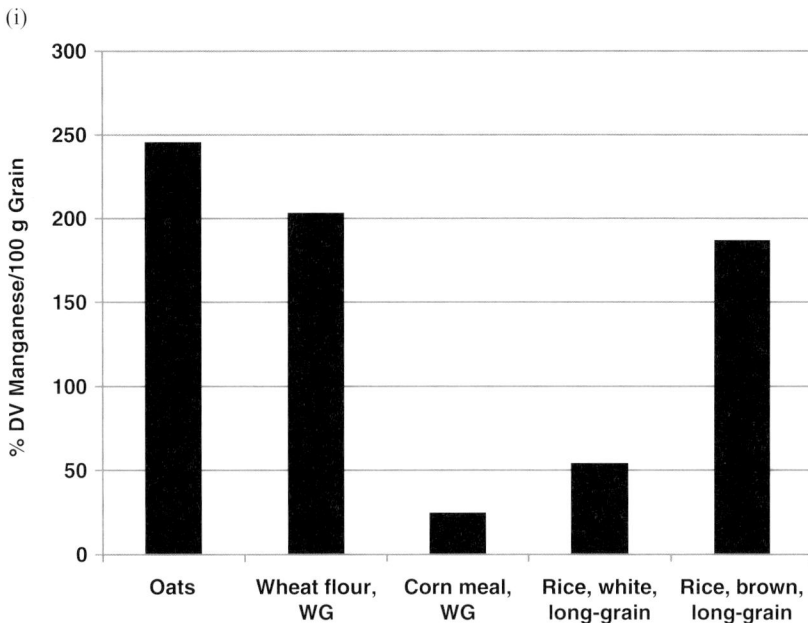

Figure 4.5 *(Continued)*

%DV of Manganese, another essential enzyme cofactor. The mineral content of the cereal grains, however, may not reflect their bioavailability due to potential interactions with the fiber component of whole grains.

4.3 Conclusion

Cereal grains contain a wide spectrum of nutrients that have potential benefits for human health. Compared to the major food grain cereals consumed globally—corn, wheat and rice—oats have a comparable nutrition profile with the exception of a higher amount of protein, some micronutrients, and, most notably, its water-soluble fiber content in the form of β-glucan. Because oats are more likely to be consumed as whole grains compared to wheat and corn, which are used in flour to produce bread, pasta, and tortillas, or rice, which is primarily consumed with the bran removed, increasing oat consumption would have a positive impact on reducing the gap in whole grain intake and increasing fiber in the diet. Therefore, recommendations that emphasize increasing oat consumption specifically, while increasing overall whole grain intake, would be expected to have a positive effect on human health.

References

Bechtel, D.B. and Pomeranz, Y. (1981) Ultrastructural and cytochemistry of mature oat (*Avena sativa* L.) endosperm. The aleurone layer and starchy endosperm. *Cereal Chemistry*, **58**, 61–69.

Biorklund, M., Van Rees, A., Mensink, R.P., and Onning, G. (2005) Changes in serum lipids and postprandial glucose and insulin concentrations after consumption of beverages with [beta]-glucans from oats or barley: a randomised dose-controlled trial. *European Journal of Clinical Nutrition*, **59**, 1272–1281.

Draper, S.R. (1973) Amino acid profiles of chemical and anatomical fractions of oat grains. *Journal of the Science of Food and Agriculture*, **24**, 1241–1250.

FAO (2013) *Higher 2012 world cereal production than was forecast in December, but stocks still expected to decline* [Online]. www.fao.org (last accessed 26 April 2013).

FDA (2009) *Guidance for Industry: A Food Labeling Guide* [Online]. US Department of Health and Human Services. http://www.fda.gov/Food/GuidanceRegulation/GuidanceDocumentsRegulatoryInformation/LabelingNutrition/ucm2006828.htm (last accessed 14 May 2013).

Frølich, W. and Nyman, M. (1988) Minerals, phytate and dietary fibre in different fractions of oat-grain. *Journal of Cereal Science*, **7**, 73–82.

Fulcher, R.G., O'Brien, T.P., and Wong, S.I. (1981) Microchemical detection of niacin, aromatic amine, and phytin reserves in cereal bran. *Cereal Chemistry*, **58**, 130–135.

Ganßmann, W. and Vorwerck, K. (1995) Oat milling, processing and storage. In: Welch, R. (ed.) *The Oat Crop*. Springer The Netherlands. 369–408

Gray, D.A., Auerbach, R.H., Hill, S., *et al.* (2000) Enrichment of oat antioxidant activity by dry milling and sieving. *Journal of Cereal Science*, **32**, 89–98.

Heneen, W.K., Banaś, A., Leonova, S., *et al.* (2009) The distribution of oil in the oat grain. *Plant Signaling & Behavior*, **4**, 55–56.

Klose, C. and Arendt, E.K. (2012) Proteins in oats; their synthesis and changes during germination: a review. *Critical Reviews in Food Science and Nutrition*, **52**, 629–639.

Kumar, V., Sinha, A.K., Makkar, H.P.S., *et al.* (2011) Dietary roles of non-starch polysaccharides in human nutrition: A review. *Critical Reviews in Food Science and Nutrition*, **52**, 899–935.

Lásztity, R. (1998) Oat grain – a wonderful reservoir of natural nutrients and biologically active substances. *Food Reviews International*, **14**, 99–119.

Lin, B.H. and Yen., S.T. (2007) The U.S. grain consumption landscape: who eats grain, in what form, where, and how much? Economic Research Report; no. 50, United States Deptartment of Agriculture. *Economic Research Service.*

Miller, S.S., and Fulcher, R.G. (2011) Microstructure and chemistry of the oat kernel. In: Webster, F.H. and Wood, P.J. (eds) *OATS: Chemistry and Technology*, 2nd edn, Chapter 5. AACC International, Inc., AACC International, St. Paul, MN.

Peterson, D.M. (1992) Composition and Nutritional Characteristics of Oat Grain and Products. *Oat Science and Technology, agronomymonogra*, 265–292.

Peterson, D.M. (2011) Storage Proteins. In: Webster, F.H. and Wood, P.J. (eds) *OATS: Chemistry and Technology*, 2nd edn, Chapter 8. AACC International, Inc., AACC International, St. Paul, MN.

Peterson, D.M., Saigo, R.H., and Holy, J. (1985) Development of oat aleurone cells and their protein bodies. *Cereal Chemistry*, **62**, 366–371.

Peterson, D.M., Senturia, J., Youngs, V.L., and Schrader, L.E. (1975) Elemental composition of oat groats. *Journal of Agricultural and Food Chemistry*, **23**, 9–13.

Pomeranz, Y., Youngs, V.L., and Robbins, G.S. (1973) Protein content and amino acid composition of oat species and tissues. *Cereal Chemistry*, **50**, 702–707.

Shotwell, M.A., Boyer, S.K., Chesnut, R.S., and Larkins, B.A. (1990) Analysis of seed storage protein genes of oats. *Journal of Biological Chemistry*, **265**, 9652–9658.

US Department Of Agriculture, ARS (2012) USDA National Nutrient Database for Standard Reference. Release 25. http://www.ars.usda.gov/main/site_main.htm?modecode=12-35-45-00 (last accessed 14 May 2013).

USDA and HHS (2010) Dietary Guidelines for Americans, 7th Edition [Online]. Available: http://www.health.gov/dietaryguidelines/2010.asp (last accessed 22 April 2013)].

Welch, R. (1995) The chemical composition of oats. In: Welch, R. (ed.) *The Oat Crop.* Springer, The Netherlands, pp. 279–320.

White, D.A., Fisk, I.D. and Gray, D.A. (2006) Characterisation of oat (*Avena sativa* L.) oil bodies and intrinsically associated E-vitamers. *Journal of Cereal Science*, **43**, 244–249.

Wu, Y.V., Sexson, K.R., Cavins, J.F. and Inglett, G.E. (1972) Oats and their dry-milled fractions: protein isolation and properties of four varieties. *Journal of Agricultural and Food Chemistry*, **20**, 757–761.

Youngs, V.L., Püskülcü, M., and Smith, R.R. (1977) Oat lipids I. Composition and distribution of lipid components in two oat cultivars. *Cereal Chemistry*, **54**, 803–812.

5

Oat Starch

Prabhakar Kasturi and Nicolas Bordenave

*Global R&D Technical Insights – Analytical Department, PepsiCo Inc.,
Barrington, IL, USA*

5.1 Introduction

Oats were the fifth most-produced cereal in the 1960s, accounting for about 5% of the world's total cereal production but their importance has been declining. Oats now rank seventh with 20 million metric tons in 2010 (compared with 844 million metric tons for maize). Hence, oat starch has not received as much attention as other major starch sources. However, some important characterization studies have been carried out since the mid-1950s, allowing for a comparison of oat starch with other sources, highlighting its specificities.

Starch accounts for 40–65% (w/w) of the weight of the groats and is largely located in the endosperm (Verhoeven *et al.*, 2004; Rhymer *et al.*, 2005). It consists of two polymers of glucose, amylose and amylopectin, packed together in a semicrystalline, granular form. Amylose is an essentially linear polymer of $(1 \rightarrow 4)$-linked α-D-glucopyranosyl units. Amylopectin is densely branched and consists of short, linear $(1 \rightarrow 4)$-linked chains branched by α-D-$(1 \rightarrow 6)$ linkages.

Starch has multiple and distinct levels of organization, from its glucose building blocks to its naturally occurring granular form, providing for its properties and characteristics. The specificities of oat starch lay in most of these levels of organization, such as the branching patterns of amylopectin, the molecular weight distribution of amylose and amylopectin, granule morphology, and the amount and characteristics of their extraneous components.

Starch granules swell and are disrupted when heated in water, leading to gels and pastes, consisting of three-dimensional networks of amylose and amylopectin. Upon cooling, the polymers undergo a reorganization phenomenon called retrogradation. Gelatinization, pasting, and retrogradation characteristics are dependent on the starch molecular characteristics and vary among starch sources. These properties allow starch to be used as an ingredient or food additive to provide moisture retention, to modify texture, to stabilize emulsions, and

---

Oats Nutrition and Technology, First Edition. Edited by YiFang Chu.

so on. From the perspective of these physicochemical properties, studies focusing on oat starch have allowed its comparison with other starches and enhanced knowledge on the most adequate uses of oat starch.

The structural properties of oat starch in comparison with starches from other sources are reviewed, as well as the consequences of these structural characteristics on the physicochemical characteristics of oat starch.

5.2   Native oat starch organization: From the molecular to the granular level

As stated previously, at the molecular level starch consists of two polymers of glucose: amylose and amylopectin.

Amylose is a linear unbranched polymer of α-(1,4)-linked D-glucopyranosyl residues, although it has been shown that amylose can exhibit a few very long branches consisting of α-D-(1,4)-glucopyranosyl chains linked to the backbone chain by an α-(1,6) linkage (BeMiller, 2007a) (Figure 5.1). However, these random and infrequent branches and the significant length of the branched chains make amylose behave as a purely linear polymer from a physicochemical standpoint (Biliaderis, 1998). The molecular weight of amylose has been reported to generally range between 2.0×10^5 and 1.2×10^6. Amylopectin is a highly branched α-glucan, consisting of short α-D-(1,4)-glucopyranosyl chains linked by α-(1,6) linkages (Figure 5.1). Branching patterns and chain lengths of branches vary among starch sources and growing conditions (Wrolstad, 2012). The general description of amylopectin has been extensively reviewed (Thompson, 2000), guided by a deeper understanding of starch biosynthesis. There is general agreement on a cluster model built around three types of linear α-D-(1,4)-glucopyranosyl chains: from the outside to the core of the amylopectin molecule,

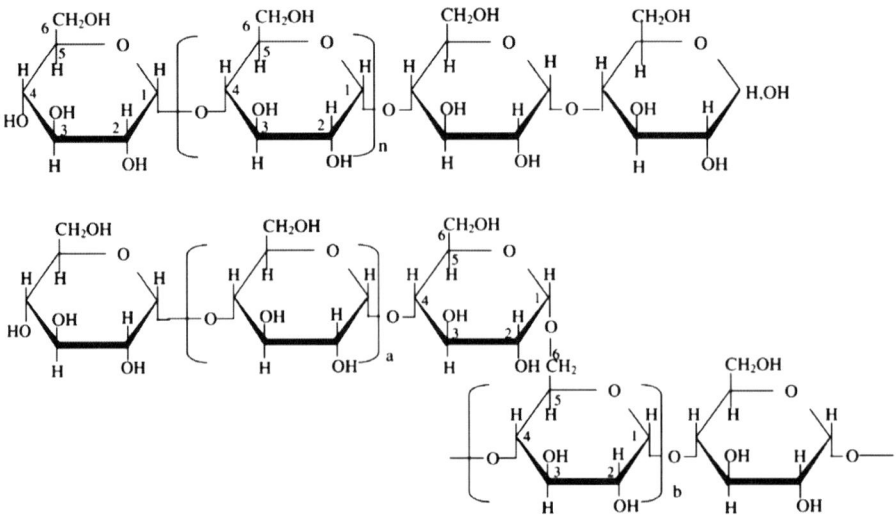

Figure 5.1   Molecular structure of amylose (top) and amylopectin (bottom).

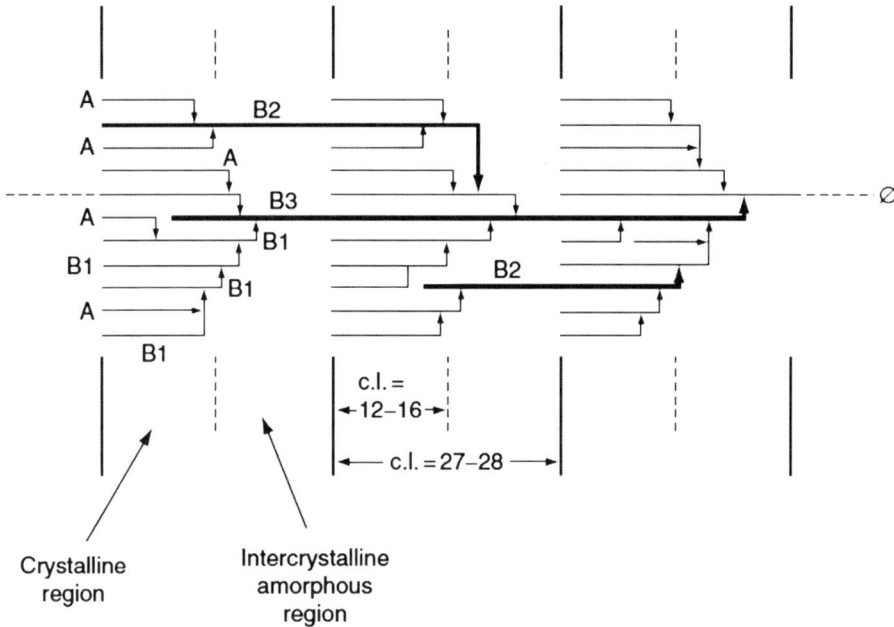

Figure 5.2 Proposed cluster model for amylopectin. A and B denote nomenclature of branch chains. Φ: reducing end; c.l.: chain length in degree of polymerization. *Source:* Adapted from Hizukuri (1986). Reproduced with permission of Pergamon.

short A-chains are unbranched and linked to B-chains, themselves linked to a single C-chain that carries the reducing group of the amylopectin molecule. Both A- and short B-chains have a chain length of 14–18 α-D-glucopyranosyl units, whereas longer B-chains have a chain length of 45–55 units (Figure 5.2).

The stereochemistry of α-(1,4) linkages yields a helicoidal shape with six units per turn to linear sequences of α-D-glucopyranosyl units (Gidley and Bociek, 1985). This provides linear glycosidic segments the opportunity for strong hydrogen bonding, that is, intermolecular and intramolecular associations, especially their association in double helices when two chains are adjacent.

5.2.1 Oat starch molecular analysis and characterization

The main molecular properties of starch are the amylose:amylopectin ratio, molecular weights of both amylose and amylopectin, and fine structures (branching patterns and branch chain lengths). The analytical methods used to investigate these characteristics are reviewed here.

5.2.1.1 Amylose:amylopectin ratio   The ratio of amylopectin and amylose in starch is generally measured via iodine-binding procedures based on the different abilities of amylose and amylopectin to form inclusion complexes with iodine (Herrero-Martínez *et al.*, 2004). The most commonly used method is spectrophotometry, with amylose and amylopectin having different maximum absorption wavelengths: approximately 620 nm for amylose and 540 nm for amylopectin. Additionally, amylose content determination can be achieved by amperometric or potentiometric measurements that rely upon the electrical properties of the iodine solution in the presence of starch. However, the inclusion complexes formed by amylopectin with iodine interfere with amperometric and potentiometric measurements, which generally lead to an overestimation of amylose content.

Other methods based on complex formation with concanavalin A (Gibson et al., 1997) or on the selective precipitation of amylose with butan-1-ol (Young, 1984) or other alcohols (Kim and Willett, 2004) have been developed. However, these latter procedures may be questionable in terms of specificity, as the underlying mechanism is not always well understood. Finally, chromatographic and capillary electrophoresis-based methods applied to debranched starches also allow an estimation of the amylose:amylopectin ratio by the attribution of long linear chains [degree of polymerization (DP) >100] to amylose and shorter chains to amylopectin, although these last statements are empirical and questionable (Chen and Bergman, 2007).

The general range of amylose content has been reported to be 19.4–29.4% for oats and approximately 28% for maize and wheat (Hoover et al., 2003). Hoover found that, among two oat cultivars grown under the same environmental conditions, the amylose content was 22.7–22.9% (Hoover and Senanayake, 1996a). A later study on six oat cultivars showed a range of 19.60–24.50% of total starch. In other studies, the range was found to be 30.3–33.6% (Hartunian-Sowa and White, 1992) and 33.6% (Stevenson et al., 2007) of total starch. Such differences were attributed to the interference of nonextracted lipids into the amylose-lipid complex, leading to a possible underestimation of amylose content. A low amylose content of 14.5% was determined on a naked Polar oat variety (Berski et al., 2011).

5.2.1.2 Starch, amylose, and amylopectin molecular weights and fine structures

One key issue in starch characterization is its molecular weight and molecular weight distribution. Indeed, molecular weight is the main driver of certain starch properties, such as viscosity in pastes or suspensions with regard to processing concerns, the rate of digestibility for nutritional concerns, film-forming properties for packaging applications, and so on (Biliaderis, 1998). While the weight-average molecular weight of starch (M_w) can easily be determined by light scattering (Wyatt, 1993; Roger et al., 1999), it is easily understandable that a single average molecular weight value can result from very different M_w distributions, resulting in different properties of the respective starches. The determination of M_w is currently one of the biggest challenges in starch characterization (Table 5.1).

These issues have been recently reviewed (Gidley et al., 2010; Gilbert et al., 2010) and are categorized as two main problems: starch dissolution and polymer separation.

Starch polymers are only metastable in close-to-neutral aqueous pH solutions and undergo phase separation over time. Starch solubilization can be achieved in either aqueous alkaline solutions or dimethyl sulfoxide (DMSO) and can be facilitated by thermal or mechanical treatments. However, all of these methods

Table 5.1 Weight-averaged molecular weights (M_w in Da) of cereal starches

Sample	High M_w amylopectin	Low M_w amylopectin	Amylose
Oats	1.36×10^6 (8.9×10^4)	3.19×10^6 (1.5×10^5)	1.68×10^5 (6.8×10^3)
Barley	6.85×10^6 (5.7×10^5)	2.72×10^6 (1.2×10^5)	1.43×10^5 (1.5×10^4)
Buckwheat	6.68×10^6 (3.9×10^5)	2.37×10^6 (6.6×10^5)	1.36×10^5 (3.9×10^4)
Corn	6.31×10^6 (3.9×10^5)	2.88×10^6 (9.3×10^5)	1.56×10^5 (5.0×10^4)
Durum wheat	7.27×10^6 (5.1×10^5)	2.94×10^6 (1.2×10^5)	1.36×10^5 (3.5×10^4)
Rice	7.85×10^6 (9.1×10^5)	2.92×10^6 (3.4×10^5)	1.63×10^5 (1.1×10^4)
Rye	7.43×10^6 (3.5×10^5)	2.87×10^6 (5.2×10^5)	1.55×10^5 (6.2×10^4)
Spring wheat	7.90×10^6 (5.6×10^5)	3.23×10^6 (3.5×10^5)	1.33×10^5 (1.8×10^4)
LSD ($p<0.05$)	7.59×10^6	1.11×10^5	2.17×10^4

Standard deviations in parentheses; LSD: least significant difference ($p < 0.05$).
Source: Adapted from Simsek *et al.*, 2012. Reproduced with permissions from Kluwer Academic Publishers.

can degrade starch, and particularly amylopectin, which is very sensitive to shear scission, due to its large size and high branching density. This is likely to result in an underestimation of starch M_w.

Several methods exist for polymer separation. The most common is size exclusion chromatography (SEC). While the needed characteristic value is molecular weight, SEC segregates polymers according to their hydrodynamic volume, V_h (their "size in space") (Jones *et al.*, 2009). However, due to the variety of branching patterns and molecular architectures of amylopectin (and amylose to a much lesser extent), there is no reciprocal relationship between V_h and molecular weight. Thus, polymers with identical V_h but different molecular weights will elute at the same time, and polymers with identical molecular weight but different V_h will elute at different times, making chromatograms very difficult to interpret (Figure 5.3). Other separation techniques, such as asymmetric-flow field flow

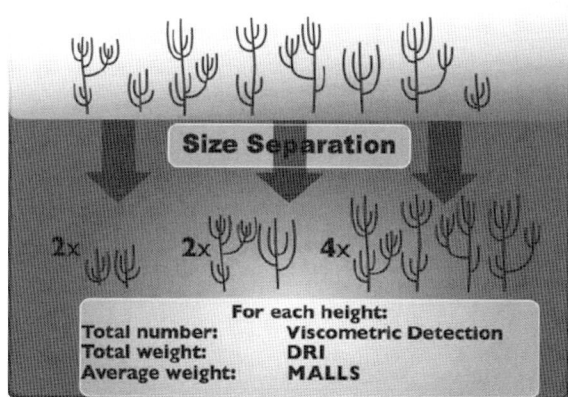

DRI: differential refractive index; MALLS: multiple-angle laser light scattering

Figure 5.3 Characterization of starch, a highly branched molecule, which is analogous to characterizing the trees in a forest by first separating by size (or height, for trees), then using three separate detectors to find, for the trees of each height, the number (distribution), the total weight (distribution), and the weight-average weight. *Source:* Gilbert *et al.* (2010). Reproduced with permissions from the American Association of Cereal Chemists (AACC).

fractionation or analytical ultracentrifugation, can also achieve polymer separation but face the same main issues. Another difficulty is the choice of a detector after polymer separation and the characteristic value it would provide: the most common and practical value measured is weight-average molecular weight M_w, which is actually a weight-average molecular weight for polymer that is separated by its V_h (Gidley *et al.*, 2010). This M_w value can be obtained primarily by light scattering techniques or through the use of linear polymer standards with known chain lengths (Gilbert *et al.*, 2010).

Similarly, SEC coupled with fluorophore-assisted carbohydrate electrophoresis is the method of choice for amylopectin fine structure determination. It is used after enzymatic treatment using debranching enzymes to elucidate chain length distribution of amylopectin (Yao *et al.*, 2005).

Nevertheless, viscosity measurements at fixed concentrations can provide quick and practical assessments and comparisons of M_w in industrial setups.

Berski and colleagues measured an overall M_w of 9.02×10^7 Da for oat starch, but amylose and amylopectin must be regarded separately (Berski *et al.*, 2011).

In the case of oat amylose, an estimate average chain length value of 3200 units has been reported (5.1×10^5 Da) with an iodine affinity of 18 g/100 g starch (Manelius and Bertoft, 1996). This value seems higher than results reported in other works. Indeed, the amylose fraction of oat cultivars E77, Dal, and L996 studied by Wang and White (1994a) exhibited a weight-average DP_w between 392 and 2920 (6.4×10^4–4.7×10^5 Da) with a peak DP_w of 939–1208 (1.5×10^5–2.0×10^5 Da) and an iodine affinity of 18.4–18.9 g iodine per 100 g amylose. This range of iodine affinity is lower than in other cereals. After debranching with isoamylase, the amylose fraction revealed a branching pattern that consisted of an average weight-average CL (CL_w) of 593–703 with a peak CL_w of 182–204 (Wang and White, 1994b, 1994c) (Table 5.2, adapted from Wang and White a, b, c). More recently, the amylose fraction of starch from commercial oat flakes was attributed a M_w of 1.68×10^5 Da (Figure 5.4) (Simsek *et al.*, 2012).

Table 5.2 Characterization of amylose, intermediate, and amylopectin fractions in E77, Dal, and L996 oat starches

Sample	DP populations after isoamylase debranching	Iodine affinity (g/100 g material)	λ_{max} (nm)	[η] (mL/g)
Amylose	392–568 2149–2920	18.4–18.9	659–662	167–173
Intermediate fraction	21.8–22.6 34.5–79.9 280–310.9	0.62–1.26	567–575	145–148
Amylopectin	16.6–20.1 30.7–31.8 181.7–204.2	0.30–0.58	557–560	124–146

DP: degree of polymerization.
Adapted from Wang and White 1994a, 1994b, 1994c.

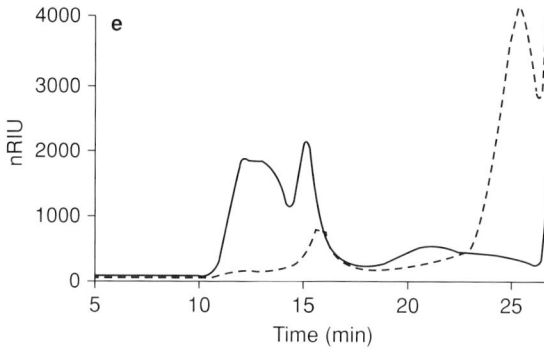

Figure 5.4 Size exclusion chromatogram of oat starch. *Source:* Simsek *et al.* (2012). Reproduced with permission from Kluwer Academic Publishers.

Oat amylopectin has been reported to have an overall average chain length of 17 (Manelius and Bertoft, 1996). In the study conducted by Simsek and colleagues (Simsek *et al.*, 2012), starch from commercial oat flakes exhibited a M_w of 1.36×10^7 Da for the amylopectin fraction. This latter study showed that while oat amylose is comparable to other cereal amyloses, oat amylopectin seems to have a significantly higher M_w than other cereal amylopectins. According to Wang and White (1994a, b, c), isoamylase debranching of oat amylopectin yielded three distinct fractions of linear chains characterized as high, intermediate, and low CL_w, which ranged between 181.7 and 204.2, 30.7 and 31.8, and 16.6 and 20.1, respectively, with iodine affinity that ranged between 0.30 and 0.58 g/100 g material.

Finally, a fraction of intermediate M_w was identified in early works (Banks and Greenwood, 1967; Paton, 1979) and it was suggested to be the result of limited complex formation of amylopectin with amylose (Tester and Karkalas, 1996). Mua and Jackson (1995) attempted to validate this theory by submitting the oat starch intermediate fraction to various solvent and harsh heat treatments. However, these results should be interpreted with caution, as the susceptibility of such branched structures to breakdown under these types of conditions was mentioned previously. Wang and White (1994a) showed the isoamylase debranching of this fraction yielded three fractions of linear branches with CL_w that ranged between 280 and 310.9, 34.5 and 79.9, and 21.8 and 22.6, respectively, with iodine affinity that ranged between 0.62 and 1.26 g/100 g material (Wang and White, 1994a). This fraction, eluting between amylopectin and amylose by SEC, was later characterized as having less branching, longer branches, and lower M_w than amylopectin on the one hand, and more branching, shorter branches, and higher M_w than amylose on the other hand, with a M_w of 3.19×10^6 Da (Simsek *et al.*, 2012).

MacArthur and D'Appolonia (1979) and Wang and White (1994a, 1994b, 1994c) measured the intrinsic viscosity of these fractions, which exhibited values ranging from 124 mL/g for the amylopectin fraction to 207 mL/g for the amylose fraction.

5.2.2   Native starch crystallinity and supramolecular organization

Starch naturally occurs in the cereal grain's endosperm in a granular form in which amylose and amylopectin are packed in a semicrystalline fashion. Starch granules exhibit growth rings that are alternatively highly ordered and amylopectin rich on the one hand, amorphous and amylose richer on the other hand. The highly order growth rings show that adjacent side chains of amylopectin are associated in double helices, which themselves are arranged according to two crystalline structures (A- and/or B-type).

5.2.2.1   Oat starch granules   Oat starch granules grow in clusters. This leads most of them to have irregular or polygonal shapes, whereas others (growing on the outer layer of the clusters) are polygonal on one side and ovoid on the other side (Figure 5.5) (Hartunian-Sowa and White, 1992; Jane *et al.*, 1994; Hoover *et al.*, 2003; Tester *et al.*, 2004). Clusters are generally 20–150 μm in diameter, 60 μm on average (Bechtel and Pomeranz, 1981; Matz, 1991). Starch extraction generally yields granules ranging from 2 to 12 μm in diameter (Gudmundsson and Eliasson, 1989; Hartunian-Sowa and White, 1992; Jane, 1994; Hoover and Senanayake, 1996a). Later work on six oat cultivars grown under the same environmental conditions yielded granules ranging from 3.8 to 10.5 μm in diameter, averaging 7.0–7.8 μm in diameter (Hoover *et al.*, 2003). Berski and colleagues measured an average granule size of 6 μm in diameter, with 10% of the sample volume less than 4 μm and another 10% of the sample volume greater than 8.5 μm in diameter (Berski *et al.*, 2011). Oat starch granules are generally smaller than observed in other common cereals, such as corn, wheat, and barley. They also have a higher specific surface area, measured at 1.224 m^2/g, compared to 0.534–0.687 m^2/g for corn, wheat, and barley (Juszczak *et al.*, 2002). This is in

Figure 5.5   Scanning electron micrograph of native oat starch. *Source:* Mirmoghtadaie *et al.*, 2009b. Reproduced with permission from Elsevier.

good agreement with the fact that oat starch granules generally reveal a smooth surface without significant porosity, by observation with noncontact atomic force and scanning electron microscopy (Juszczak *et al.*, 2003).

Starch granules are insoluble in cold water. However, they are able to absorb water, and therefore swell. This swelling is very limited in cold water and increases steadily upon moderate heating, this process being reversible and exothermic. At this stage, water absorption is believed to start amorphous growth rings, whereas intermolecular associations are less important. However, a second phase of swelling occurs when the semicrystalline order of the granule is lost (the so-called gelatinization) and starch polymers begin to leach out of the granule (mainly amylose initially), this phase being much more pronounced and irreversible (BeMiller, 2007b). This phenomenon is reviewed later under the "Gelatinization, pasting, and pastes" section.

5.2.2.2 Granule crystallinity   As mentioned previously, an important characteristic of starch granules that drives their behavior towards water and heat is their semicrystallinity. As a measure of their level of organization, crystallinity also influences starch granule mechanical properties. It is commonly measured by X-ray diffraction. Crystallinity of starch is due to the arrangement of paired, adjacent, glycosidic linear segments of amylopectin arranged in double helices, themselves arranged collectively in ordered structures. This crystallinity is at the origin of the birefringence phenomenon (appearance of a Maltese cross) that starch granules exhibit under cross-polarized light. Starch granules generally exhibit two types of X-ray diffraction patterns, the A- and B-types. In these crystalline patterns, the helical structures are essentially identical, only their arrangements differ. The A-type is compact with few water molecules in the inter-helical space, whereas the B-type is less compact and more hydrated, the water molecules being localized in the hexagonal inter-helical core (Figure 5.6). A third crystalline type,

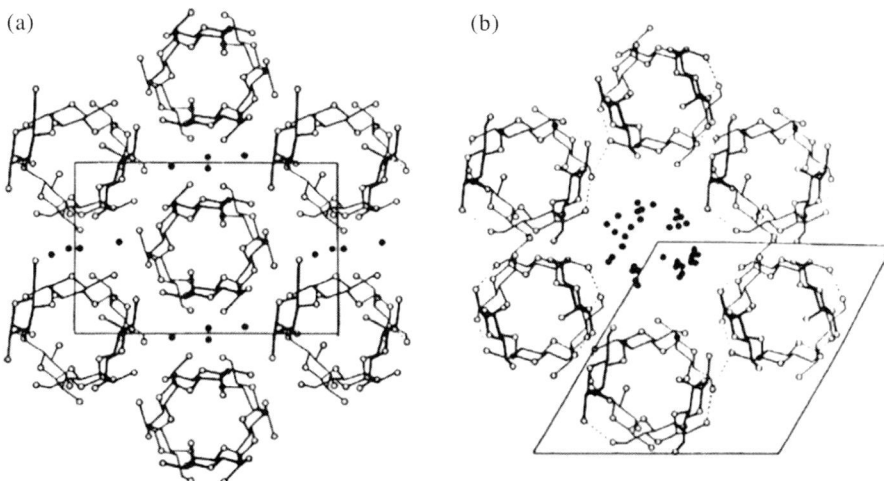

Figure 5.6   Unit cells (outlined in each diagram) and helix packing in a and b polymorphs of starch. *Source:* Hsein-Chih and Sarko (1978). Reproduced with permission from Elsevier.

Table 5.3   Proportion of crystalline material in starches of different botanical origins

Native starch	Amount of crystalline amylopectin (%)
Oat	28–37
Maize	39–45
Maize, waxy	48
Maize, 0% amylose	42
Maize, 28% amylose	30
Maize, 40% amylose	22
Maize, 56% amylose	20
Maize, 65% amylose	18
Maize, 84% amylose	17
Wheat	36–39
Rice	47–51
Barley (includes fractions)	22–27
Barley, normal	20–24
Barley, waxy	33–37

Source: Adapted from Tester *et al.* (2004). Reproduced with permission from Elsevier.

called the C-type, has been observed and is actually thought to be a mixture of A- and B-types. The A-type is most common in cereal starches, whereas the B-type is more characteristic of tubers and high-amylose starches (Verhoeven *et al.*, 2004; Wang and White, 1994c).

Oat starch clearly exhibits an A-type diffraction pattern but is weakly birefringent under cross-polarized light. This is well correlated with a degree of crystallinity ranging between 28 and 37%, generally lower than the other common cereal starches (Table 5.3) (Wang and White, 1994c; Hoover *et al.*, 2003). Nevertheless, heat-moisture treatment (annealing) increases the relative crystallinity of native oat starch (Hoover and Vasanthan, 1994).

In addition to the formation of ordered double helical structures, starch can exhibit crystals made of ordered lipid-amylose complexes. These so-called V-complexes are characterized by the inclusion of a free fatty acid aliphatic chain into the hydrophobic inner cavity of an amylose single helix. These ordered structures are detectable at a 2θ angle of approximately $20°$ on X-ray diffraction patterns. However, due to the limited complex formation of amylose with lipids or the lack of order of these complexes, no diffraction peak is detected in this zone for oat starches (Gibson *et al.*, 1997; Hoover *et al.*, 2003). Two studies showed that only 9.02–18.91% and 14.1–15.3% of total amylose was complexed with lipids (Hoover and Senanayake, 1996a; Hoover *et al.*, 2003.

5.3   Starch minor components, isolation, and extraction

According to previous remarks on the interaction between starch polymers and extraneous compounds, it appears that starch granules are not exclusively composed of amylose and amylopectin, and that extraneous compounds can drive

Table 5.4 Oat and other cereal starches composition related to minor components

Cereal variety	Lipids (%)	Proteins (%)	Ash (%)	Phosphorus (%)	Amylose (%)
Oats	0.7–2.5	0.13–0.95	0.05–0.4	0.002–0.19	19.4–33.6
Wheat	0.4–1.2	0.04–0.6	0.1–0.4	0.05–0.12	26.3–30.6
Corn	0.5–0.9	0.2–0.4	0.05–0.1	0.02–0.03	25.8–32.5
Rice	0.03–0.9	0.01–0.1	—	0.002–0.0045	12.2–28.6
Barley	0.7–1.2	0.1	—	0.0006–0.007	25.3–30.1

Adapted from Paton, 1977; MacArthur and D'Appolonia, 1979; Juliano, 1984; Morrison, 1984; Hartunian-Sowa and White, 1992; Hoover and Vasanthan, 1992; Gibinski *et al.*, 1993; Wang and White, 1994b; Hoover and Senanayake, 1996a; Tester and Karkalas, 1996; Shamekh *et al.*, 1999; Song and Jane, 2000; Hoover *et al.*, 2003; Singh *et al.*, 2003; Berski *et al.*, 2011.

some of the properties of oat starch. Moreover, both the nature and amount of these extraneous compounds are dependent on the process used to extract the starch granules from the oat endosperm.

5.3.1 Oat starch minor components

By order of decreasing content, the main minor constituents of oat starch are lipids, proteins, inorganic compounds (ashes), and phosphorus (Table 5.4).

Total lipids are generally present in oat starch at a level of 0.7–2.5% (MacArthur and D'Appolonia, 1979; Paton, 1979; Hartunian-Sowa and White, 1992; Hoover and Senanayake, 1996a; Hoover *et al.*, 2003). These values of lipid content are obtained after the acid hydrolysis of the starch. However, a sequential solvent extraction method allows the differentiation of unbound lipids (in the range 0.05–0.1% as extracted with 2:1 chloroform:methanol) and extractable bound lipids (in the range 0.75–1.2% as extracted with a 1:1 ratio of 1-propanol:water) (Hoover *et al.*, 2003). Starch lipids can also be categorized according to their localization, whether they are present in the granule or on its surface, most of them being complexed with amylose (Morrison, 1981, 1988; Zhou *et al.*, 1998). They are mainly lysophospholipids (up to 70% of total lipids) or free fatty acids (up to 30% of total lipids) (Morrison, 1984; Hoover and Vasanthan, 1992; Liukkonen and Laakso, 1992). Overall, oat starches contain more lipids than other cereals starches. This is in good agreement with the fact that oat grains have the highest lipid content of all cereals apart from maize and that, contrary to maize, 90% of oat oil is located in the groat endosperm (Morrison, 1977; Barthole *et al.*, 2012).

Although the efficiency of the defatting step of oat starch may sometimes be questionable in published studies, the physical properties of oat starches seem to correlate well with their lipid content. Indeed, formation of lipid-amylose complexes prevents amylose from being involved in amylose-amylose and amylose-amylopectin intermolecular associations that drive the physical properties of starch. Hence, the properties of starch granules, such as swelling power, and the properties of starch gels, pastes, solutions, and films, such as mechanical and optical properties, as well as amylose and amylopectin solubility, retrogradation, and precipitation, are affected by higher lipid content. With higher lipid content, the swelling power of oat starch granules is decreased (Hoover and Senanayake, 1996a; Hoover *et al.*, 2003; Rhymer *et al.*, 2005), oat starch gels are stiffer and retrogradation is decreased (Gudmundsson and Eliasson, 1989), starch films are less cohesive and more opaque, and, finally, amylose and amylopectin are less soluble (Swinkels, 1985; Wu *et al.*, 2012; Schmidt *et al.*, 2013).

Finally, it appears that the higher lipid content of oat starch and its fine structure may be associated with higher amylose content and longer average chain length of amylopectin (Wang and White, 1994c).

As mentioned previously, lysophospholipids account for approximately two-thirds of oat starch lipids. They are also the main compounds that account for oat starch phosphorus content, as suggested by the disappearance of phosphorus after complete removal of lipids from starch (Hartunian-Sowa and White, 1992). Oat starch being richer in lipids than other cereal starches, it is not surprising that phosphorus is present in higher amounts in oat starch than in other cereals (Paton, 1977; Hartunian-Sowa and White, 1992; Gibinski *et al.*, 1993).

The protein content of starch is calculated based on nitrogen content. Thus, it actually includes all the other nitrogen-containing compounds that are not proteins *per se* (e.g., peptides, amino acids, enzymes, etc.). Oat starch contains significantly more protein than other cereals, although nitrogen content varies considerably among studies, ranging from 0.02–0.09% in Alymer, Antoine, Baton, Ernie, Francis, and Gosline lines (Hoover *et al.*, 2003) to 0.001% in NO 753-2 and AC Stewart lines (Hoover and Senanayake, 1996a), for example.

5.3.2   Oat starch extraction and isolation

Within the oat grain, starch is located in the endosperm, surrounded by the β-glucan and protein-rich bran layers. Due to the adhesion of the bran and the endosperm, oat starch extraction is more challenging than for other cereals. Protein removal is a critical step that positively affects the starch extraction yield. Oat cultivars with higher protein content give the lowest starch extraction yields. Moreover, as mentioned previously, the lipid content of starch greatly affects the starch properties. Hence, special attention must be paid to lipid removal along the extraction process.

The general extraction pattern includes protein removal, multiple mechanical separations to eliminate cell wall debris (centrifugation, sieving, filtration, etc.), water washing/neutralization, and recovery by centrifugation.

Three paths have been investigated to separate efficiently proteins from starch: alkaline extraction, water extraction with high-shear homogenization, and use of proteases. These three methods have been compared elsewhere (Lim *et al.*, 1992). The highest yield was obtained with the action of proteases over 6 hours, with 78% starch on an oat flour basis and a residual protein content of 1.1%. The second highest yield was obtained with alkaline extraction using sodium hydroxide (NaOH; pH 10.5) or calcium hydroxide (Ca(OH)$_2$; pH 11.0) over 1 hour, with 71–75% starch on an at flour basis and a residual protein content of 0.3–0.4%. It must be noted that sodium hydroxide was more efficient at removing proteins than calcium hydroxide. Finally, the least-efficient method was regular water extraction at 20°C for 6 hours, followed by high-shear treatment with a tissue homogenizer, with 70% starch recovery on an oat flour basis and a residual protein content of 1.3%.

Proteolysis and alkaline extraction, in addition to being the most efficient methods for obtaining high-purity starch, do not involve high-shear treatments, which lead to substantial starch damage (up to 13%), such as granule breaking (Hoover and Vasanthan, 1992). Finally, although proteolysis appears to be more efficient than alkaline extraction at removing proteins, alkaline extraction seems to give better results regarding lipid removal, probably due to the partial saponification of the lipids and removal of the subsequent fatty acids (Paton, 1977).

At the industrial scale, processes based on dry or wet milling and the actions of various enzymes on oat groats yield starches with 94–98% purity (Biopolymer Network Ltd, New Zealand; Alko Ltd or Primalco Ltd., Rajamaki, Finland) (Autio and Eliasson, 2009). Recently, the addition of a pearling step in the dry-milling process was suggested to obtain a better separation of the bran and the endosperm (Wang *et al.*, 2007). Lipid extraction by supercritical carbon dioxide prior to milling was also suggested to improve the separation of starch from cell wall materials, possibly by breaking the starch granule clusters (Sibakov *et al.*, 2011). Steaming and roasting, which are commonly used processes for the deactivation of oat enzymes in the oat processing industry, produce the same effect on starch granule clusters as on naked oat grain (Hu *et al.*, 2010). Breakdown of granule clusters appears to be a critical step for improved starch extraction.

5.4 Beyond native starch granule: Gelatinization, pasting, retrogradation, and interactions with other polysaccharides

5.4.1 Gelatinization

Gelatinization of starch is defined as the irreversible loss of molecular order that a starch granule undergoes when heated over a certain temperature, called the gelatinization temperature T_{gel}, in the presence of excess water. This process is accompanied by absorption of water to a much greater extent than in cold water and by leaching of starch polymers (primarily amylose initially) out of the granule. T_{gel} depends on the molecular characteristics of the amylose and amylopectin within the granule, the types and amounts of extraneous compounds, and the level of molecular order of the granule. In that sense, T_{gel} is specific to each individual granule, but it occurs over a temperature range for a population of starch granules (BeMiller, 2007b; Ratnayake and Jackson, 2008). This loss of crystallinity is an endothermic process. Thus, the main techniques used to characterize starch gelatinization are the observation of birefringence loss with a hot-stage microscope (Chen *et al.*, 2007) or the measurement of heat absorption by differential scanning calorimetry (DSC) (Farkas and Mohácsi-Farkas, 1996).

Many cereal starch granules (e.g., wheat, barley, and rye) first swell along their radial axis, followed by swelling along other axes. Oat starch granules differ from wheat, barley, and rye in that respect and are similar to maize: They tend to swell simultaneously and equally along their three axes (Williams and Bowler,

1982). Their swelling power, which is a measure of their water absorption capacity, generally reflects their level of organization. Swelling power is calculated as the ratio of the volume of swollen starch granules to the volume of the dry starch. The oat starch swelling factor at between 60 and 85°C increases progressively and ranges between 7.3 and 25.3, whereas amylose leaching ranges between 0.7 and 2.5% for AC Stewart (*Avena sativa L.*) and NO 753-2 (*Avena nuda L.*). However, swelling power and amylose leaching increased dramatically around 90–95°C to reach 57.7–75.1 and 19.8–25.0%, respectively, at 95°C (Hoover and Senanayake, 1996a). The same temperature effect was observed on oat and corn starches between 85 and 95°C (Wang and White, 1994b). This swelling factor ranged between 5.6 and 20.1, whereas amylose leaching ranged between 1.1 and 3.7% for Alymer, Antoine, Baton, Ernie, Francis, and Gosline oat cultivars at 70°C (Hoover *et al.*, 2003). In this study, the starch granule-swelling factor was positively correlated with the amount of lipid complexed with the starch and negatively correlated with amylopectin chain length. Granules least prone to swelling exhibited less amylose leaching, due to amylose-amylose and amylose-amylopectin long branch associations. In comparison, wheat and maize starches exhibited swelling factors and amylose leaching values of 11.0 and 12.1, and 4.8 and 5.2%, respectively (Hoover and Vasanthan, 1992; Hoover and Manuel, 1996).

Two endothermic peaks are observed by DSC upon heating of starch (Figure 5.7a): the first one (at temperature T_{m1}) is associated with actual gelatinization and the second one occurs at higher temperatures (at temperature T_{m2}), corresponding to the melting endotherm of the amylose-lipid complexes. In lipid-free starch samples, the second melting endothermic peak does not appear (Figure 5.7b). While gelatinization is irreversible, the association–dissociation of amylose with lipids is reversible. Hence, if starch is sequentially heated above T_{m2} in a first DSC run, cooled down below T_{m1}, and heated up again above T_{m2} in a second DSC run, the first thermogram will exhibit two endotherm peaks at T_{m1} and T_{m2}, whereas the second thermogram will exhibit only one endotherm peak at T_{m2} (Figure 5.7a). Nevertheless, regarding the gelatinization endotherm peak, as gelatinization occurs over a range of temperatures, DSC provides three characteristic temperatures for starch gelatinization: onset temperature (T_o), peak temperature (T_p), and conclusion temperature (T_c). These temperatures are the temperature boundaries (T_o and T_c) and the temperature minimum (T_p) of the main endothermic peak observed on the thermogram of starch upon heating at a constant rate (Figure 5.7). For starches in general, this endotherm is actually thought to be comprised of two superimposed peaks, one corresponding to the glass transition of the starch amorphous regions, the other one corresponding to the melting of the starch's crystalline regions (BeMiller, 2007b; Ratnayake and Jackson, 2008). In the specific case of oat starch, a third component may add to this endotherm. Despite a substantial amount of lipids, amylose-lipid complexes occur on a limited basis in oat starch. However, it has been suggested that such complex formation, which is exothermic, occurs upon starch gelatinization, when amylose mobility is largely increased. Hence, there may be an exothermic component to the gelatinization endotherm of oat starch (Biliaderis, 1998; Hoover *et al.*, 2003).

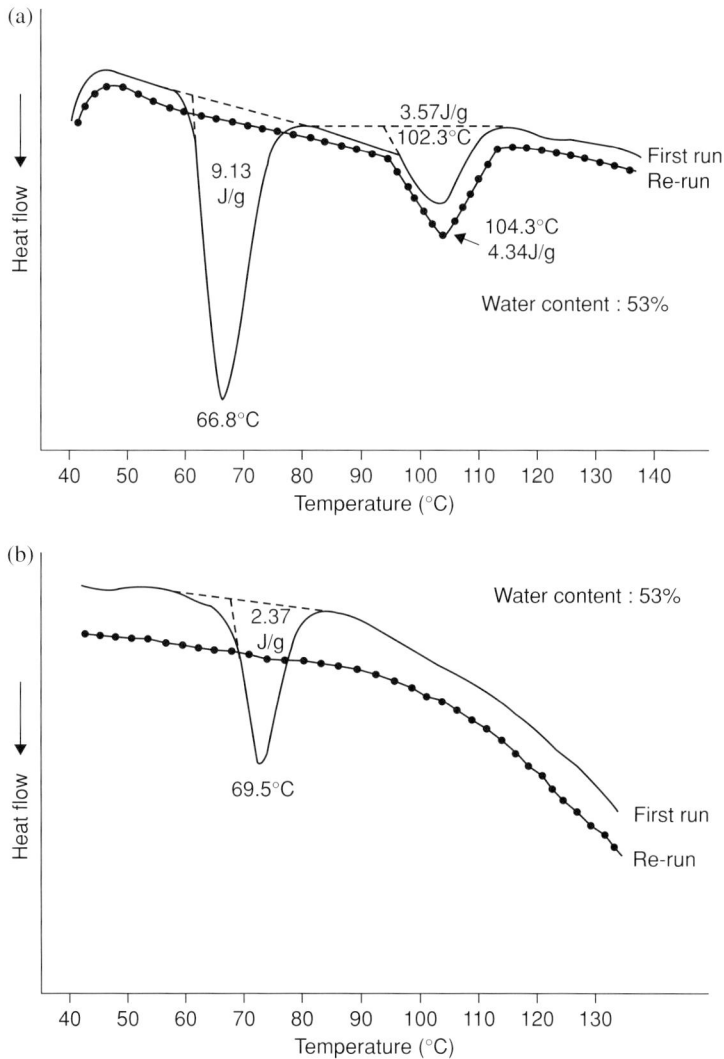

Figure 5.7   Differential scanning calorimetry thermograms of native oat starch (a) and lipid-free oat starch (b). *Source:* Doublier *et al.* (1987). Reproduced with permissions from the American Association of Cereal Chemists (AACC).

Finally, DSC also provides the enthalpy of gelatinization, ΔH, in J/g (i.e., the amount of energy required to complete the gelatinization process). The enthalpy of gelatinization is correlated with the order level of the starch sample studied, thus, with the relative crystallinity of the sample. DSC parameters of oat starches from various studies are presented in Table 5.5. Crystallinity and enthalpy data compiled in Table 5.3 (adapted from Tester *et al.*, 2004) and Table 5.6 support this relationship between gelatinization enthalpy and molecular order within the granule before gelatinization. With this perspective, oat starch seems more ordered than wheat starch (Hoover and Vasanthan, 1992) but less ordered than

Table 5.5 Swelling power and amylose leaching of oat starches at different temperatures

Oat/Corn-type starch	Lipid content (%)	Swelling power							Amylose leaching (%)					Reference
		0°C	60°C	70°C	80°C	85°C	90°C	95°C	60°C	70°C	80°C	85°C	95°C	
AC Stewart	1.67	6.3	12.9	16.7	22.3	25.3	51	75.1	0.7	1.8	2.5			(Hoover and Senanayake, 1996a, b)
NO 753-2	1.64	3.5	7.3	8.6	9.5	15.7	17.9	57.7	0.7	1.6	2.2			Hoover et al., 2003
Alymer	0.95			20.1						2.3				
Antoine	1.08			15.0						1.9				
Baton	1.29			6.0						1.2				
Francis	1.10			14.2						1.6				
Ernie	0.85			18.2						3.7				
Gosline	1.31			5.6						1.1				
E77	1.08					9.6		27.8				5.6	33.5	Wang and White. 1994b
Dal	1.16					8.7		29.0				4.1	37.4	
L996	1.18					9.1		34.8				6.0	43.3	
BXMO (corn)	0.66					13.9		28.1				10.2	29.6	
PFP (corn)	0.55					11.6		24.4				9.2	29.8	

Table 5.6 DSC parameters of oat and other cereal starches for starch gelatinization and transition of the amylose-lipid complex

Sample	Gelatinization properties				Amylose-lipid complex melting			Reference
	T_o (°C)	T_p (°C)	T_c (°C)	ΔH (J/g)	T_o (°C)	T_c (°C)	ΔH (J/g)	
Regular Corn	64.0	69.0	75.5	13.0	—	—	—	Shi and Seib, 1992; Jane et al., 1999; Cruz-Orea et al., 2002; Varavinit et al., 2003; Liu et al., 2005
High-amylose corn	68.9	80.5	106.1	11.5	—	—	—	Rhymer et al., 2005
Waxy corn	66.0	70.7	78.4	15.5	—	—	—	Hoover and Senanayake, 1996a
Wheat	57.1	61.6	66.2	10.7	—	—	—	Hoover et al., 2003
Rice	61.5	70.0	78.6	7.1	—	—	—	
Waxy rice	76.1	81.1	87.0	19.2	—	—	—	
	—	58.4–60.3	—	8.7–9.5	—	—	—	
	—	57.6–67.0	—	5.9–8.4	—	—	—	
	56.0–63.5	59.5–66.0	65.5–74.0	12.4–14.6	—	—	—	
	55.5–62.4	—	—	8.6–9.2	90.3–91.1	—	2.6–3.5	Hartunian-Sowa and White, 1992
Native oat starch	60.5	65.2	—	9.5	—	—	—	Berski et al., 2011
	56.1–69.5	61.0–73.1	—	10.0–12.8	91.9–92.4	—	1.1–3.0	(Wang and White, 1994a)
	51.1	58.1	65.2	—	—	—	—	Mua and Jackson, 1995
	60.0	66.0	70.0	11.5	—	—	—	Hoover and Vasanthan, 1994
	—	57.8–61.6	—	9.4–10.6	—	94.1–97.1	2.4–3.7	Gudmundsson and Eliasson, 1989
Lipid-free oat starch	61	66	73	10.4	—	106	—	Hoover and Vasanthan, 1992
	—	66.8	—	9.1	—	102.3	3.6	Doublier et al., 1987
	56.5	62.7	68.4	7.1	—	—	—	Mirmoghtadaie et al., 2009a
	58	64	70	12.0	—	—	—	Hoover and Vasanthan, 1994
	—	69.5	—	2.4	—	—	—	Doublier et al., 1987

corn or rice starch (Doublier *et al.*, 1987; Wang and White, 1994a, 1994d; Hoover *et al.*, 2003; Rhymer *et al.*, 2005). The presence of lipids also influences the DSC parameters of oat starch. However, conflicting data exist on this point: while Doublier and colleagues observed an increase in T_p and a decrease in ΔH upon removal of lipids from oat starch (from 66.8 to 69.5°C and from 9.13 to 2.37 J/g, respectively, Figure 5.7) (Doublier *et al.*, 1987), Hoover and Vasanthan (1994) observed the inverse phenomenon (from 66.0 to 64.0°C and from 11.5 to 12.0 J/g, respectively).

The effects of cross-linking and acetylation of oat starch on DSC parameters have been observed. On the one hand, cross-linking has minor effects on thermal transition temperatures and enthalpy: While T_o remains unchanged after varying degrees of cross-linking with phosphoryl chloride ($POCl_3$), only an increase in T_c (from 68.4 to 78.9°C) and in ΔH (from 7.1 to 7.9 J/g) were observed. On the other hand, acetylation with a degree of substitution of 0.11 considerably decreased T_o, T_p, and T_c (from 56.5, 62.7, and 68.4°C to 46.0, 53.7, and 60.3°C, respectively) and ΔH (from 7.1 to 4.9 J/g) (Mirmoghtadaie *et al.*, 2009a). It can be thought that acetyl groups act as internal plasticizer, disrupting hydrogen bonds and inter-chains interactions, hence lowering the amount of energy needed to achieve gelatinization.

5.4.2   Pasting, pastes, and retrogradation

In the presence of water and above gelatinization temperature, starch granules are swollen, all molecular order has disappeared, and amylose and amylopectin begin to leach out. At this stage, the granules are susceptible to mechanical stress: shear leads to granule disruption and dispersion of amylose and amylopectin, resulting in a paste. A starch paste is characterized by its mechanical and rheological properties.

The pasting process is usually studied with a Rapid Visco Analyzer (RVA), a Brabender Amylograph, or an Ottawa Starch Viscometer. These instruments record the change in viscosity of starch pastes/suspensions under shear as a function of temperature. Figure 5.8 shows a typical RVA pasting curve for starch. As temperature increases from room temperature, granules swell and viscosity increases. This viscosity increase is particularly dramatic above gelatinization temperature (noted pasting temperature in Figure 5.8), when granules swell most. Then, as temperature is maintained above gelatinization temperature, granules reach their maximum volume before disruption, resulting in maximum viscosity of the starch suspension (peak viscosity). Then, granules are disrupted and polymers are dispersed in water, resulting in a dramatic decrease of viscosity (breakdown). Upon cooling, an increase in viscosity is observed (setback) as starch polymers reassociate (initially, mainly amylose) (Biliaderis, 1998; Zhou *et al.*, 1998). This process, called retrogradation, is the result of the rearrangement of starch polymers into insoluble crystallites, which are ordered structures of linear glycosidic chains associated into double helices. It is crucial to note that while the crystallinity of native starch granules is primarily due to ordered structures of amylopectin chains, crystallinity of retrograded starch is primarily due to ordered

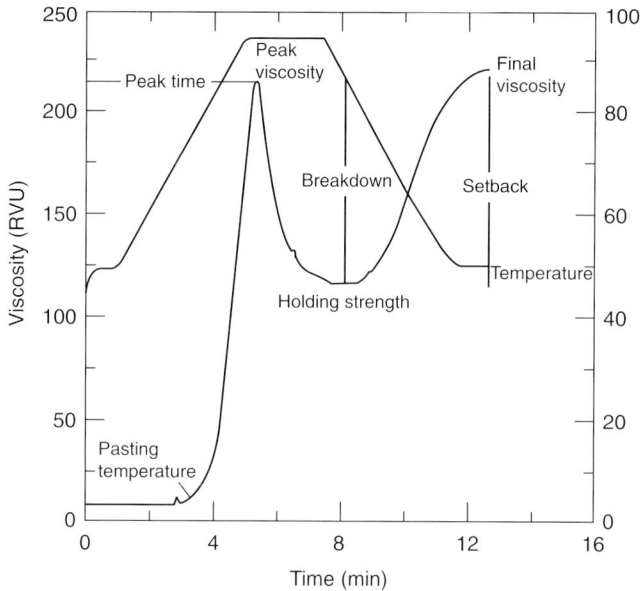

Figure 5.8 Typical Rapid Visco Analyzer pasting curve, identifying characteristic features. *Source:* Zhou *et al.* (1998). Reproduced with permissions from the American Association of Cereal Chemists (AACC).

structures of amylose chains (BeMiller, 2007b). Retrogradation of a starch paste results in a firm viscoelastic gel.

In pasting behavior, oat starch generally has a greater susceptibility to shear than other cereals starches. It also exhibits higher pasting temperature and lower peak viscosity than corn starch (Wang and White, 1994a). However, lipid removal leads to a decrease in pasting temperature, highlighting again the role of high lipid content in oat properties compared to other cereals (Hoover and Vasanthan, 1992). Pasting characteristics of oat starches from four studies are reported in Table 5.7.

The subsequent cooled gels were observed to be clearer, less firm, more elastic, and more adhesive than corn and wheat starch pastes (Paton, 1977; MacArthur and D'Appolonia, 1979; Wang and White, 1994b). These mechanical characteristics were attributed to the presence of lipids, possibly preventing amylose from participating efficiently in the three-dimensional polymeric network to form the gel.

As observed using a Ottawa Starch Viscometer, although it undergoes a very fast initial retrogradation upon the first 30 seconds of cooling and a higher initial torque upon cooling than wheat starch, oat starch pastes are overall less susceptible to retrogradation than other cereal starches: oat starch undergoes fast (initially) but limited (overall) retrogradation (Paton, 1977, 1979; MacArthur and D'Appolonia, 1979). Retrograded oat starch gels stored for seven days at 4°C were studied and their DSC parameters determined (Wang and White, 1994a). Overall, T_o and ΔH were lower for retrograded gels than for native starches (38.2–42.6°C vs 56.1–69.5°C and 0.32–1.89 J/g vs 2.40–3.07 J/g, respectively),

Table 5.7 Pasting characteristics of oat starches

Sample	Pasting temperature (°C)	Peak height (BU)	Hold (BU)	Final viscosity (BU)	Setback (BU)	Pasting conditions	Reference
Three oat cultivars	81–83.5	760–855	—	870–1130	—	95°C for 30 min	MacArthur and D'Appolonia, 1979
Hinoat	—	455	85	245	—	97°C for 30 min and cooled to 25°C	Paton, 1981
Three US oat cultivars	83.6–93	155–310	145–295	285–470	105–240	95°C for 30 min and cooled to 50°C	Wang and White, 1994a
Commercial oat cultivars	83	390	280	790	—	95°C for 30 min and cooled to 85°C	Mua and Jackson, 1995

whereas the range of phase transition was larger (17.7–$25.6°$C vs 6.9–$12.1°$C), suggesting a much lower degree of order and a broader range of crystalline types for retrograded starches than for native starches. Moreover, the degree of retrogradation (28.2–60.9%), calculated as $\Delta H_{retrograded\ starch}/\Delta H_{native\ starch}$, was positively correlated with the breadth of phase transition, supporting again the idea of a broader range of crystalline types. Lipid removal leads to faster retrogradation and an increase in degree of retrogradation, although it remained more limited than in other cereals, confirming that the presence of lipids interferes with retrogradation: 50% after 10 days of storage for defatted oat starch versus 32–40% after 28 days of storage for nondefatted oat starch (Gudmundsson and Eliasson, 1989; Hartunian-Sowa and White, 1992; Hoover and Vasanthan, 1994).

The characteristics of acid-hydrolyzed oat starches were also investigated (Virtanen *et al.*, 1993). Acid hydrolysis results in reduced paste viscosity. While oat starch pastes were observed to be strongly thixotropic (Doublier *et al.*, 1987), acid-hydrolyzed oat starch had a viscoelasticity behavior below $40°$C. In addition to that effect, acid-hydrolyzed gels had their rigidity and elasticity reduced and underwent enhanced phase separation between amylose and amylopectin upon cooling. Amylose was thought to be the main contributor for gel and paste properties of acid-hydrolyzed oat starch (Virtanen *et al.*, 1993).

5.5 Industrial uses

While oat flour has been intensively studied, oat starch has not received much attention for industrial uses, probably due to its availability and cost, in comparison with other cereals. It has noticeable properties for specific applications, however. Its high lipid content and small granule size make it relevant to the pulp and paper industry, for coating and sizing of papers. Alpine Gloves Inc. has patented latex gloves powdered with oat starch, claiming that it reduces the risk of latex allergies because, unlike cornstarch, oat starch does not adhere to latex protein. It has been sparsely tested as a food ingredient: as a texture agent in sauces (Gibiński *et al.*, 2006), as an water-in-oil emulsion stabilizer (Bodor *et al.*, 1986), and as an ingredient in bread dough (Toufeili *et al.*, 1999) and cake batter (Mirmoghtadaie *et al.*, 2009b). A noticeable industrial use of oat starch is Oatrim, a fat substitute patented by the US Department of Agriculture and made by the conversion of oat starch into amylodextrin (Inglett, 1991, 1993; Inglett *et al.*, 1994). Despite their limited use, oats and oats starches are increasingly scrutinized as potential candidates for wheat-based products in order to avoid allergy risks linked to gluten (Hüttner *et al.*, 2010; Flander *et al.*, 2011; Rezvani *et al.*, 2011).

Finally, oat starch raises interest from a nutritional perspective. Starch is usually divided into three categories regarding its digestion: rapidly digestible starch (which is digested in less than 20 min), slowly digestible starch (SDS, which is digested in 20–120 min) and resistant starch (RS, which is not digested after 120 min) (Lehmann and Robin, 2007). In addition to oat β-glucans and their health benefits, oat starch exhibits interesting nutritional properties. Indeed, oat starch contains a significant amount of SDS and RS (Mishra and Monro, 2009), possibly due to its relatively high lipid content and the possibility of creating amylose–lipid complexes, which are generally slowly digested or resistant to digestion (Kawai *et al.*, 2012). In a later study (Kim and White, 2012), oat starches isolated from four oat lines were tested *in vitro* for their digestion kinetics. These starches were tested as uncooked or cooked slurries in water, over 180 min. Upon heating the slurry, the extent of starch digestion increased from 31–39% to 52–64%, leading to as increase of their predictive Glycemic Index from 61–67 to 77–86 (Goñi *et al.*, 1997). However, within the uncooked and cooked slurries, no significant differences were found among the four oat starches regarding extent and kinetics of *in vitro* digestion. According to these observations and Glycemic Index values of other starches (Foster-Powell *et al.*, 2002), oat starch itself (not oat flour) does not seem to be particularly nutritionally advantageous. Nevertheless, a chapter of this book covering oat β-glucans gives more insights on oats nutritional characteristics related to these dietary fibers (Chapter 6).

5.6 Conclusion and perspectives

Oat starch appears in the form of granules with 2–12 μm diameter and can contain a broad range of amylose (20–34 %). Although oat starch extraction is difficult due to a low separability with the bran, amylopectin and amylose have been characterized with M_w of $1–2 \times 10^6$ and $1–2 \times 10^5$ Da, respectively. While oat amylose M_w is not significantly different from other cereals amyloses, oat amylopectin seems to have a lower M_w than other cereals amylopectins. Native oat starch exhibits an A-type semicrystalline pattern with a degree of crystallinity ranging from 28 to 37%. It contains 0.7–2.5% lipids, a level that is fairly high in comparison with other cereals starches. The amount of lipids seems to be positively correlated with amylose content and complexation between these two compounds is common. Oat starch thermal properties do not differ significantly from other cereals starches: the only noticeable difference here is that upon cooling after pasting, oat starch retrogrades faster but to a lower extent, as compared to other cereals starches.

Despite these slightly differentiating properties (e.g., lipid content, granule size, crystallinity, etc.), oat starch is not different enough from the starches of other more available cereals to be able to compete with them commercially. Due to the limited production of oats, oat starch remains a specialty product with a relatively high price and limited varieties (by chemical/enzymatic modification or by breeding) compared to the other main cereal starches.

Nevertheless, oat starch may focus more attention on its unique nutritional properties. Indeed, in addition to its exceptionally high soluble dietary fiber content (β-glucans), oat starch shows a promising profile regarding its slowly digestible to resistant to digestibility fractions. These characteristics are increasingly being examined for the development of food and beverage products intended to help treat cholesterol and diabetes, or to manage obesity. Beyond whole oat grains, oat starch may be able to play a significant role in this area.

References

Autio, K. and Eliasson, A.-C. (2009) Oat Starch. In: *Starch: Chemistry and Technology* (eds J.N. BeMiller and R.L. Whistler), 3rd edn, pp. 589–599. Academic Press, New York.

Banks, W. and Greenwood, C.T. (1967) The fractionation of laboratory-isolated cereal starches using dimethyl sulphoxide. *Starch/Staerke* **19**, 394–398.

Barthole, G. *et al.* (2012) Controlling lipid accumulation in cereal grains. *Plant Science* **185–186**, 33–39.

Bechtel, D.B. and Pomeranz, Y. (1981) Ultrastructure and cytochemistry of mature oat (*Avena sativa* L.) endosperm. The aleurone layer and starchy endosperm. *Cereal Chemistry* **58**, 61–69.

BeMiller, J.N. (2007a) *Carbohydrate Chemistry for Food Scientists*, 2nd edn. AACC International Press, St. Paul, MN.

BeMiller, J.N. (2007b) Starches, modified starches, and other starch products. In: *Carbohydrate Chemistry for Food Scientists* (ed. J.N. BeMiller), pp. 173–223. AACC International Press, St. Paul, MN.

Berski, W. *et al.* (2011) Pasting and rheological properties of oat starch and its derivatives. *Carbohydrate Polymers* **83**, 665–671.

Biliaderis, C.G. (1998) Structures and phase transitions of starch polymers. In: *Food Science and Technology* (eds S.R. Tannenbaum and P. Walstra), pp. 57–168. Marcel Dekker, New York.

Bodor, J. *et al.* (1986) *Edible water-in-oil emulsion spreads containing hydrated starch particles dispersed in the aqueous phase*. Lever Brothers Company.

Chen, M.H. and Bergman, C.J. (2007) Method for determining the amylose content, molecular weights, and weight- and molar-based distributions of degree of polymerization of amylose and fine-structure of amylopectin. *Carbohydrate Polymers* **69**, 562–578.

Chen, P. *et al.* (2007) Phase transition of starch granules observed by microscope under shearless and shear conditions. *Carbohydrate Polymers* **68**, 495–501.

Cruz-Orea, A. *et al.* (2002) Phase transitions in the starch-water system studied by adiabatic scanning calorimetry. *Journal of Agricultural and Food Chemistry* **50**, 1335–1344.

Doublier, J.-L. *et al.* (1987) A rheological investigation of oat starch pastes. *Cereal Chemistry* **64**, 21–26.

Farkas, J. and Mohácsi-Farkas, C. (1996) Application of differential scanning calorimetry in food research and food quality assurance. *Journal of Thermal Analysis* **47**, 1787–1803.

Flander, L. *et al.* (2011) Effects of tyrosinase and laccase on oat proteins and quality parameters of gluten-free oat breads. *Journal of Agricultural and Food Chemistry* **59**, 8385–8390.

Foster-Powell, K. *et al.* (2002). International table of glycemic index and glycemic load values: 2002. *American Journal of Clinical Nutrition* **76**(1), 5–56

Gibinski, M. *et al.* (1993) Physicochemical properties of defatted oat starch. *Starch/Staerke* **45**, 354–357.

Gibiński, M. *et al.* (2006) Thickening of sweet and sour sauces with various polysaccharide combinations. *Journal of Food Engineering* **75**, 407–414.

Gibson, T.S. *et al.* (1997) A procedure to measure amylose in cereal starches and flours with concanavalin A. *Journal of Cereal Science* **25**, 111–119.

Gidley, M.J. and Bociek, S.M. (1985) Molecular organization in starches: A 13C CP/MAS NMR study. *Journal of the American Chemical Society* **107**, 7040–7044.

Gidley, M.J. *et al.* (2010) Reliable measurements of the size distributions of starch molecules in solution: Current dilemmas and recommendations. *Carbohydrate Polymers* **79**, 255–261.

Gilbert, R.G. *et al.* (2010) Characterizing the size and molecular weight distribution of starch: Why it is important and why it is hard. *Cereal Foods World* **55**, 139–143.

Goñi, I. I. (1997). A starch hydrolysis procedure to estimate glycemic index. *Nutrition Research* **17**(3), 427–437.

Gudmundsson, M. and Eliasson, A.-C. (1989) Some physico-chemical properties of oat starches extracted from varieties with different oil content. *Acta Agriculturae Scandinavica* **39**, 101–111.

Hartunian-Sowa, S.M. and White, P.J. (1992) Characterization of starch isolated from oat groats with different amounts of lipid. *Cereal Chemistry* **69**, 521–527.

Herrero-Martínez, J.M. *et al.* (2004) Determination of the amylose-amylopectin ratio of starches by iodine-affinity capillary electrophoresis. *Journal of Chromatography A* **1053**, 227–234.

Hizukuri, S. (1986) Polymodal distribution of the chain lengths of amylopectins, and its significance. *Carbohydrate Research* **147**, 342–347.

Hoover, R. and Manuel, H. (1996) The effect of heat-moisture treatment on the structure and physicochemical properties of normal maize, waxy maize, dull waxy maize and amylomaize V starches. *Journal of Cereal Science* **23**, 153–162.

Hoover, R. and Senanayake, S.P.J.N. (1996a) Composition and physicochemical properties of oat starches. *Food Research International* **29**, 15–26.

Hoover, R. and Senanayake, S.P.J.N. (1996b) Effect of sugars on the thermal and retrogradation properties of oat starches. *Journal of Food Biochemistry* **20**, 65–83.

Hoover, R. and Vasanthan, T. (1992) Studies on isolation and characterization of starch from oat (Avena nuda) grains. *Carbohydrate Polymers* **19**, 285–297.

Hoover, R. and Vasanthan, T. (1994) Effect of heat-moisture treatment on the structure and physicochemical properties of cereal, legume, and tuber starches. *Carbohydrate Research* **252**, 33–53.

Hoover, R. *et al.* (2003) Physicochemical properties of Canadian oat starches. *Carbohydrate Polymers* **52**, 253–261.

Hsein-Chih, H.W. and Sarko, A. (1978) The double-helical molecular structure of crystalline A-amylose. *Carbohydrate Research* **60**, 27–40.

Hu, G. *et al.* (2010) Efficient measurement of amylose content in cereal grains. *Journal of Cereal Science* **51**, 35–40.

Hüttner, E.K. *et al.* (2010) Fundamental study on the effect of hydrostatic pressure treatment on the bread-making performance of oat flour. *European Food Research and Technology* **230**, 827–835.

Inglett, G.E. (1991) A method of making a soluble dietary fiber composition from oats. The United States of America, as represented by the Secretary of Agriculture. US Patent 4,996,063.

Inglett, G.E. (1993) Amylodextrins containing β-glucan from oat flours and bran. *Food Chemistry* **47**, 133–136.

Inglett, G.E. *et al.* (1994) Sensory and nutritional evaluation of Oatrim. *Cereal Foods World* **39**, 755–759.

Jane, J. *et al.* (1994) Anthology of starch granule morphology by scanning electron microscopy. *Starch/Stärke* **46**, 121–129.

Jane, J. *et al.* (1999) Effects of amylopectin branch chain length and amylose content on the gelatinization and pasting properties of starch. *Cereal Chemistry* **76**, 629–637.

Jones, R.G. *et al.* (2009) Compendium of polymer terminology and nomenclature. In: *IUPAC Recommendations 2008*. Royal Society of Chemistry, Cambridge.

Juliano, B.O. (1984) Rice starch: production, properties, and uses. In: *Starch: Chemistry and Technology*. (eds R.L. Whistler and J.N. BeMiller), 2nd edn, pp. 507–528. Academic Press, New York.

Juszczak, L. *et al.* (2002) Characteristics of cereal starch granules surface using nitrogen adsorption. *Journal of Food Engineering* **54**, 103–110.

Juszczak, L. *et al.* (2003) Non-contact atomic force microscopy of starch granules surface – Part II. Selected cereal starches. *Starch/Staerke* **55**, 8–16.

Kawai, K. *et al.* (2012) Complex formation, thermal properties, and in-vitro digestibility of gelatinized potato starch-fatty acid mixtures. *Food Hydrocolloids* **27**, 228–234.

Kim, H. J. and White, P. J. (2012). In vitro digestion rate and estimated glycemic index of oat flours from typical and high β-glucan oat lines. *Journal of Agricultural and Food Chemistry* **60**(20), 5237–5242.

Kim, S. and Willett, J.L. (2004) *Isolation* of amylose from starch solutions by phase separation. *Starch/Staerke* **56**, 29–36.

Lehmann, U. and Robin, F. (2007) Slowly digestible starch – its structure and health implications: A review. *Trends in Food Science and Technology* **18**, 346–355.

Lim, W.J. *et al.* (1992) Isolation of oat starch from oat flour. *Cereal Chemistry* **69**, 233–236.

Liu, H. *et al.* (2005) Thermal behaviour of high amylose cornstarch studied by DSC. *International Journal of Food Engineering* **1**, 1–6.

Liukkonen, K. and Laakso, S. (1992) Characterization of internal and surface lipids of oat starches from two isolation processes. *Starch/Staerke* **44**, 128–132.

MacArthur, L.A. and D'Appolonia, B.L. (1979) Comparison of oat and wheat carbohydrates. II. starch. *Cereal Chemistry* **56**, 458–461.

Manelius, R. and Bertoft, E. (1996) The effect of Ca^{2+} ions on the α-amylolysis of granular starches from oats and waxy-maize. *Journal of Cereal Science* **24**, 139–150.

Matz, S.A. (1991) Oats. In: *The Chemistry and Technology of Cereals as Food and Feed* (ed S.A. Matz), pp. 107–134. Springer, New York.

Mirmoghtadaie, L. *et al.* (2009a) Effects of cross-linking and acetylation on oat starch properties. *Food Chemistry* **116**, 709–713.

Mirmoghtadaie, L. *et al.* (2009b) Effect of modified oat starch and protein on batter properties and quality of cake. *Cereal Chemistry* **86**, 685–691

Mishra, S. and Monro, J.A. (2009) Digestibility of starch fractions in wholegrain rolled oats. *Journal of Cereal Science* **50**, 61–66.

Morrison, W.R. (1977) Cereal lipids. *Proceedings of the Nutrition Society* **36**, 143–148.

Morrison, W.R. (1981) Starch lipids: A reappraisal. *Starch/Staerke* **33**, 408–410.

Morrison, W.R. (1984) A relationship between the amylose and lipid contents of starches from diploid cereals. *Journal of Cereal Science* **2**, 257–271.

Morrison, W.R. (1988) Lipids in cereal starches: A review. *Journal of Cereal Science* **8**, 1–15.

Mua, J.-P. and Jackson, D.S. (1995) Gelatinization and solubility properties of commercial oat starch. *Starch/Staerke* **47**, 2–7.

Paton, D. (1977) Oat starch Part 1. Extraction, purification and pasting properties. *Starch/Staerke* **29**, 149–153.

Paton, D. (1979) Oat starch: Some recent developments. *Starch/Staerke* **31**, 184–187.

Paton, D. (1981) Behavior of Hinoat oat starch in sucrose, salt, and acid. *Cereal Chemistry*, **58**, 35–39.

Ratnayake, W.S. and Jackson, D.S. (2008) Starch Gelatinization. In: *Advances in Food and Nutrition Research* (ed. S. Taylor), pp. 221–268. Elsevier, Amsterdam, The Netherlands.

Rezvani, V. *et al.* (2011) The effect of "real oat bread" compared with "barley bread offered in Tehran" on serum glucose and lipid profiles in dislipidemic and type 2 diabetic subjects. *Iranian Journal of Endocrinology and Metabolism* **13**, 233–242.

Rhymer, C. *et al.* (2005) Effects of genotype and environment on the starch properties and end-product quality of oats. *Cereal Chemistry* **82**(2), 197–203.

Roger, P. *et al.* (1999) Contribution of amylose and amylopectin to the light scattering behaviour of starches in aqueous solution." *Polymer* **40**(25), 6897–6909.

Schmidt, V.C.R. *et al.* (2013) Water vapor barrier and mechanical properties of starch films containing stearic acid. *Industrial Crops and Products* **41**(1), 227–234.

Shamekh, S. *et al.* (1999) Fragmentation of oat and barley starch granules during heating. *Journal of Cereal Science* **30**(2), 173–182.

Shi, Y.C. and Seib, P.A. (1992) The structure of four waxy starches related to gelatinization and retrogradation. *Carbohydrate Research* **227**, 131–145.

Sibakov, J. *et al.* (2012) Minireview: β-Glucan extraction methods from oats. *Agro Food Industry Hi-Tech* **23**(1), 10–12.

Simsek, S. *et al.* (2012) Analysis of cereal starches by high-performance size exclusion chromatography. *Food Analytical Methods* **6**(1), 181–190.

Singh, N. *et al.* (2003) Morphological, thermal and rheological properties of starches from different botanical sources. *Food Chemistry* **81**(2), 219–231.

Song, Y. and Jane, J. (2000) Characterization of barley starches of waxy, normal, and high amylose varieties. *Carbohydrate Polymers* **41**(4), 365–377.

Swinkels, J.J.M. (1985) Composition and properties of commercial native starches. *Starch/Staerke* **37**(1), 1–5.

Tester, R.F. and Karkalas, J. (1996) Swelling and gelatinization of oat starches. *Cereal Chemistry* **73**(2), 271–277.

Tester, R.F. *et al.* (2004) Starch – Composition, fine structure and architecture. *Journal of Cereal Science* **39**(2), 151–165.

Thompson, D.B. (2000) On the non-random nature of amylopectin branching. *Carbohydrate Polymers* **43**(3), 223–239.

Toufeili, I. *et al.* (1999) Substitution of wheat starch with non-wheat starches and cross-linked waxy barley starch affects sensory properties and staling of Arabic bread. *Journal of the Science of Food and Agriculture* **79**(13), 1855–1860.

Varavinit, S. *et al.* (2003) Effect of amylose content on gelatinization, retrogradation and pasting properties of flours from different cultivars of Thai rice. *Starch/Staerke* **55**(9), 410–415.

Verhoeven, T. *et al.* (2004) Isolation and characterisation of novel starch mutants of oats. *Journal of Cereal Science* **40**(1), 69–79.

Virtanen, T. *et al.* (1993) Heat-induced changes in native and acid-modified oat starch pastes. *Journal of Cereal Science* **17**(2), 137–145.

Wang, L.Z. and White, P.J. (1994a) Functional properties of oat starches and relationships among functional and structural characteristics. *Cereal Chemistry* **71**(5), 451–458.

Wang, L.Z. and White, P.J. (1994b) Structure and properties of amylose, amylopectin, and intermediate materials of oat starches. *Cereal Chemistry* **71**(3), 263–268.

Wang, L.Z. and White, P.J. (1994c) Structure and physicochemical properties of starches from oats with different lipid contents. *Cereal Chemistry* **71**(5c), 443–450.

Wang, R. *et al.* (2007) Dry processing of oats – Application of dry milling. *Journal of Food Engineering* **82**(4), 559–567.

Williams, M. R. and Bowler, P. (1982) Starch gelatinization: A morphological study of triticeae and other starches. *Starch/Staerke* **34**(7), 221–223.

Wrolstad, R.E. (2012) *Food carbohydrate chemistry*. John Wiley & Sons, Inc., Hoboken, NJ.

Wu, X. *et al.* (2012) Effect of stearic acid and sodium stearate on cast cornstarch films. *Journal of Applied Polymer Science* **124**(5), 3782–3791.

Wyatt, P.J. (1993) Light scattering and the absolute characterization of macromolecules. *Analytica Chimica Acta* **272**(1), 1–40.

Yao, Y. *et al.* (2005) High-performance size-exclusion chromatography (HPSEC) and fluorophore-assisted carbohydrate electrophoresis (FACE) to describe the chain-length distribution of debranched starch. *Carbohydrate Research* **340**(4), 701–710.

Zhou, M. *et al.* (1998) Structure and pasting properties of oat starch. *Cereal Chemistry* **75**(3), 273–281.

6

Oat β-Glucans: Physicochemistry and Nutritional Properties

Madhuvanti Kale[1], Bruce Hamaker[1], and Nicolas Bordenave[2]

[1]*Whistler Center for Carbohydrate Research, Purdue University, West Lafayette, IN, USA*
[2]*Global R&D Technical Insights – Analytical Department, PepsiCo Inc., Barrington, IL, USA*

6.1   Introduction

Oat β-glucans are part of a large family of mixed-linkage β-glucans. Among the various glucans found in plants, fungi, or of microbial origin, with $(1\rightarrow2)$, $(1\rightarrow3)$, $(1\rightarrow4)$, or $(1\rightarrow6)$ β-linkages, cereals and lichens exhibit specific mixed $(1\rightarrow3);(1\rightarrow4)$ β-linked glucans.

Historically discovered in lichens and then in barley (because of the issues they may create during the beer brewing process), they were also identified in oats. Over several decades of study and characterization, β-linked glucans have been accepted as bioactive components of cereal grains, and ultimately as bioactive ingredients, supporting the consumption of products such as oatmeal and other oat-based products, including oat-based ready-to-eat cereals. Indeed, oats, through the activity of their soluble dietary fiber, namely β-glucans, have earned a US Food and Drug Administration (FDA) claim that they "may lower the risk of heart disease," provided an intake of 3 g β-glucans per day at a dose of 0.75 g per serving be consumed (Anonymous, 1997).

However, multiple varieties of oats exist and growing conditions vary among crop seasons and growing regions. Hence, oat β-glucan characteristics may expected to be variable along with their functionalities (physicochemical as well as nutritional). With this perspective, oat β-glucans (as well as β-glucans from other sources, such as barley and wheat) have been intensely studied over the past three decades (Wood, 2011).

Oats Nutrition and Technology, First Edition. Edited by YiFang Chu.
© 2014 John Wiley & Sons, Ltd. Published 2014 by John Wiley & Sons, Ltd.

This chapter aims to review the physicochemical properties of oat β-glucans from the molecular to the macroscopic scale, in solution or in gels, with the tentative establishment of their structure-function relationships. Subsequently, the nutritional properties of oat β-glucans are reviewed with respect to their physicochemical characteristics.

6.2   Molecular structures and characteristics

Oat β-glucans are linear polymers of β-D-anhydroglucopyranosyl units connected by a mixture of (1→3) and (1→4) linkages. Molecular structures and characteristics of oat β-glucans include the relative abundances of these linkages, their distribution along the polymeric β-glucan chain, and the molecular weight distribution of the β-glucan chains.

(1→3) and (1→4) linkages are not distributed randomly along β-glucan polymers. According to Burton and colleagues (Burton *et al.*, 2010), this distribution can be considered semirandom and reveals the two levels of organization of β-glucans.

Indeed, enzymatic hydrolysis by lichenase reveals that a large portion of β-glucans, about 90%, is made of cellotriosyl and cellotetraosyl units joined by (1→3) linkages. These cellotriosyl and cellotetraosyl blocks are cellulose-like trimers and tetramers: three or four β-(1→4) linked D-glucopyranosyl units. The remainder of the polymer is composed of cellulose-like oligomers with Degree of Polymerization (DP) greater than or equal to five and up to 13–16 (Wang *et al.*, 2003). Thus, the building blocks of β-glucans are nonrandom sequences of glucopyranosyl units (Figure 6.1).

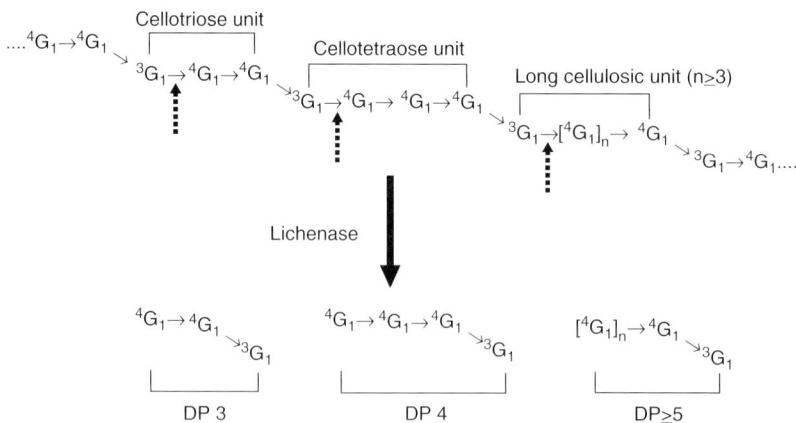

G: β-D-glucopyranosyl unit; DP3: 3-*O*-β-cellobiosyl-D-glucose; DP4: 3- *O*-β-cellotriosyl-D-glucose; DP ≥ 5: cellodextrin-like oligosaccharides containing more than three consecutive 4-*O*-linked glucose residues.

Figure 6.1   Generalized structure of cereal β-glucans and their debranching with lichenase; dotted arrows indicate the lichenase hydrolysis sites on the polysaccharide chain. *Source:* Lazaridou and Biliaderis (2007). Reproduced with permission from Academic Press.

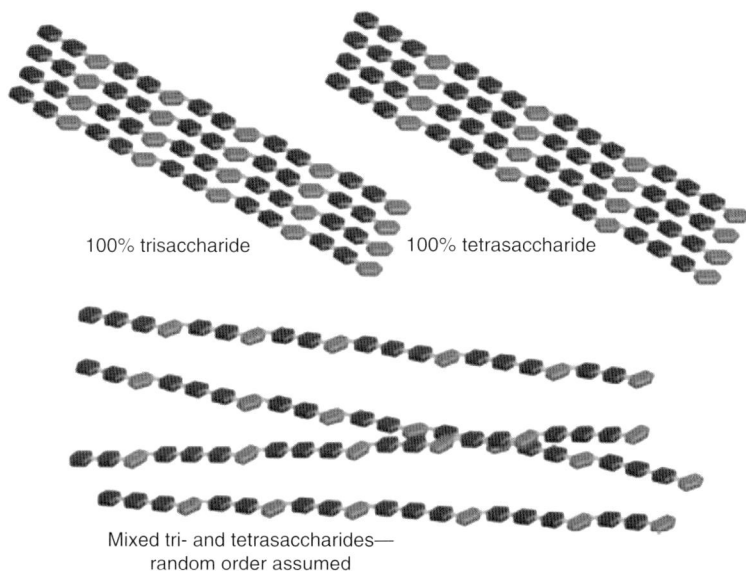

100% trisaccharide   100% tetrasaccharide

Mixed tri- and tetrasaccharides—
random order assumed

Figure 6.2 The effects of (1,3;1,4)-β-glucan fine structure on aggregation and solubility. *Source:* Fincher (2009). Reproduced with permission from the Nature Publishing Group.

However, Staudte and colleagues (Staudte *et al.*, 1983) showed that these building blocks are arranged according to random sequences along the polymeric chain. This random distribution of cellotriosyl and cellotetraosyl units gives β-glucans their solubility properties. Indeed, long cellotriosyl (or cellotetraosyl) sequences allow extensive aggregation through interchain cooperative hydrogen bonding as it exists in cellulose fibrils, making them insoluble in water (Figure 6.2). The random distribution of β-glucan building blocks prevents this cellulose-like aggregation (Figure 6.3). It is thought that this structural feature comes from the biosynthetic pathway to β-glucans. It has been suggested that cellotriosyl and cellotetraosyl blocks are first synthesized in the Golgi apparatus of the plant cell in two steps. Cellobiosyl units are synthesized (and possibly assembled into higher even-numbered DP blocks), then a glycosyl transferase adds a glucose unit onto these blocks to make odd-numbered DP blocks (Buckeridge *et al.*, 2004). These oligomers may then be transported to the plasma membrane where they may be randomly assembled in polymeric chains (Peng *et al.*, 2002; Fincher, 2009).

Relative amounts of the different DP blocks composing β-glucans have been measured in several studies. The general procedure consists of the selective hydrolysis of (1→3) linkages by lichenase, followed by the quantitation of the resulting oligomers by various chromatographic methods, including high-performance anion exchange chromatography with pulsed amperometric detection (HPAEC-PAD), reverse phase high-performance liquid chromatography, capillary electrophoresis, and mass spectrometry (Jiang and Vasanthan, 2000; Johansson *et al.*, 2000; Colleoni-Sirghie *et al.*, 2003a; Mikkelsen *et al.*, 2013).

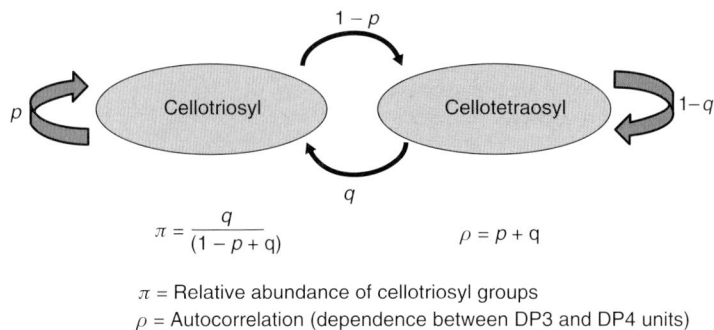

$$\pi = \frac{q}{(1-p+q)} \qquad\qquad \rho = p + q$$

π = Relative abundance of cellotriosyl groups
ρ = Autocorrelation (dependence between DP3 and DP4 units)

Figure 6.3   Fine structural analysis of the (1,3;1,4)-β-glucan from barley. The structure of a water-soluble barley (1,3;1,4)-β-glucan was investigated by treating it as a Markov chain. The polysaccharide consists mostly of cellotriosyl and cellotetraosyl units linke by (1,3)-β-linkages. This analysis was used to explore whether the addition of a particular unit oligosaccharide structure to the elongating chain depended on the adjacent unit oligosaccharide in the chain. Through controlled enzymatic hydrolysis such that most of the products represented two units of the polysaccharidic chain, it was possible to quantitate the amount of hexasaccharide (which consisted of two cellotriosyl units), heptasaccharide (which consisted of one cellotriosyl unit and one cellotetraosyl unit), and octasaccharide (which consisted of two cellotetraosyl units). Applying this quantitative measure of near-neighbor analysis in the Markov chain would give a value for ρ of about 1.0 for blocks of adjacent cellotriosyl units and adjacent cellotetraosyl units, with values near -1.0 if the polysaccharide consisted of alternating cellotriosyl and cellotetraosyl units and values close to zero if the cellotriosyl and cellotetraosyl units were randomly arranged. The experimentally determined values for ρ for two different but related (1,3;1,4)-β-glucan preparations were -0.003 and 0.050. The data therefore indicated that the (1,3)-β-linked cellotriosyl and cellotetraosyl units were arranged essentially at random along the polysaccharide chain. *Source:* Fincher (2009). Reproduced with permission from the Nature Publishing Group.

In oats, DP3, DP4, and DP $\geq$ 5 represent 53.4–66.1%, 29.1–41.4%, and 3.6–9.7% (w/w) of the polymer, respectively, with a DP3:DP4 ratio of 1.4–2.3. In comparison, barley β-glucans possess more DP3 and DP $\geq$ 5 blocks and fewer DP4 blocks at levels of 51.8–69.3%, 24.8–32.9%, and 4.5–17.5%, respectively. Thus, the DP3:DP4 ratio in barley is generally higher than that in oats, ranging from 1.6 to 3.5. The DP3:DP4 ratio is even higher in wheat, ranging from 3.7 to 4.5 (Table 6.1). However, the total abundance of combined DP3 and DP4 is similar in these cereals. Differences in DP3:DP4 among oats were attributed to genotypic and environmental differences linked to β-glucan synthase activities (Miller *et al.*, 1993b; Buckeridge *et al.*, 1999, 2001; Johansson, 2006). Drier environments seems to lead to lower DP3:DP4 ratios (Doehlert and Simsek, 2012). Furthermore, differences have also been found according to the location of the β-glucans within the grain, with DP3:DP4 being higher in β-glucans extracted from oat bran compared to β-glucans extracted from grain endosperm (Wood *et al.*, 1994b).

Quantification of higher DP blocks can be challenging due to decreasing solubility with increasing DP (Doublier and Wood, 1995). Nevertheless, it appears that their amount decreases with increasing DP, with the noticeable exception of DP9, which is relatively more abundant than other DP blocks and can make up to 1.6% of the total material (Wood *et al.*, 1994b; Izydorczyk *et al.*, 1998; Lazaridou *et al.*, 2004).

Table 6.1 Molecular structures of cereal β-glucans

Soure	DP3[a]	DP4[a]	DP≥5[a]	Molar ratio DP3/DP4	$(1\rightarrow4)/$ $(1\rightarrow3)$	Molecular weight (10^{-3})	References
Oat	55.0–58.1	34.2–36.0	7.7–8.9	2.1–2.3	2.3–2.6	—	Dais and Perlin (1982)
	—	—	—	—	2.4	360–3100	Doublier and Wood (1995) and Wood et al. (1991a–c)
	—	—	—	—	—	1500	Autio et al. (1992)
	—	—	—	—	—	1100–1500	Malkki et al. (1992)
	—	—	—	1.5–2.3	2.5	—	Miller and Fulcher (1995)
	—	—	—	—	—	—	Westerlund et al. (1993)
	—	—	—	—	—	600–840	Jaskari et al. (1995)
	57.6	34.1	8.2	1.7	—	1200–2500	Beer et al. {1997a, b)
	—	—	—	—	—	—	Izydorczyk et al. (1998)
	58.3	33.5	8.1	2.2	—	120–2400	Zhang et al. (1998)
	—	—	—	—	—	1160	Cui et al. (2000)
	—	40.4–41.4	—	1.7–1.8	—	1100–1600	Johansson et al. (2000)
	—	—	—	—	2.4	214–257	Roubrocks et al. (2000a, 2001)
	56.7	34.6	8.7	2.2	—	611–1700	Wang et al. (2002, 2003)
	55.6–55.9	33.6–34.4	7.1–7.5	1.6–1.7	2.4	—	Colleoni-Sirghie et al. (2003a)
	54.2–60.9	33.8–36.7	3.6–9.7	2.0–2.3	2.4–2.8	65–250	Lazaridou et al. (2003, 2004)
	54.6–56.8	35.3–36.3	7.7–9.2	2.0–2.1	2.3–2.6	180–850	Skendi et al. (2003)
	—	—	—	—	—	2060–2300	Aman et al. (2004)

Table 6.1 (*Continued*)

Source	DP3[a]	DP4[a]	DP≥5[a]	Motar ratio DP3/DP4	$(1\to4)/(1\to3)$	Molecular weight (10^{-3})	References
Barley	—	—	—	—	1.9–2.3	—	Balance and Manners (1978)
	—	—	—	—	2.3–2.6	—	Dais and Perlin (1982)
	56–61	28–32	6–13	2.3–2.9	2.2–2.6	150–290	Woodward et al. (1983b, 1988)
	62.1	29.4	8.4	2.8–3.4	2.4	1700–2700	Wood et al. (1991a–c) and Wood (1994)
	59.2–64.9	25.3–30.4	9.4–10.2	2.6–3.4	2.4	80–150	Saulnier et al. (1994)
	—	—	—	—	2.4	—	Henriksson et al. (1995)
	—	—	—	—	—	1300–1500	Beer et al. (1997a)
	—	—	—	—	—	200–600	Gomez et al. (1997a)
	—	—	—	—	—	570–2340	Knuckles et al. (1997b)
	56.8–61.6	26.1–32.3	10.6–11.2	1.8–2.4	—	—	Izydorczyk et al. (1998a,c)
	—	—	—	—	—	31–560	Morgan and Ofman (1998)
	—	—	—	—	—	100–375	Bohm and Kulicke (1999a)
	63.7	28.5	7.8	3.3	—	—	Cui et al. (2000)
	51.8–61.9	28.1–32.1	6.3–12.5	2.3–2.8	—	708	Jiang and Vasanthan (2000)
	66.0	25.7	8.2	3.4	—	693	Wang et al. (2003)
	61.5–64.3	27.9–30.1	7.8–8.6	2.7–3.0	—	—	Wood et al. (2003)
	59.4–64.3	24.8–31.0	8.2–17.5	2.5–3.2	—	—	Storsley et al. (2003)
	57.7–62.4	29.4–32.9	7.7–9.5	2.3–2.8	1.9–2.2	—	Irakli et al. (2004)
	62.0–63.3	27.5–29.2	8.8–9.1	2.8–3.0	2.2–2.7	1320–450	Lazaridou et al. (2004)
	62.0–69.3	26.2–29.1	4.5–8.9	2.8–3.5	2.1–2.8	213	Vaikousi et al. (2004)
Rye	—	—	—	2.7–3.0	—	250	Wood et al. (1991a–c)
	—	—	—	1.9–2.3	2.3	1100	Roubroeks et al. (2000b)
	—	—	—	3.0–3.8	—	21	Wood et al. (1991a)
Wheat	72.3	21.0	6.7	4.5	—	267–487	Cui et al. (2000) and Li et al. (2006)
	67.1	24.2	8.7	3.7	—	209	Lazaridou et al. (2004)

[a] Hydrolysis products of cereal β-glucans by lichenase: DP3 is 3-O-β-cellobiosyl-D-glucose, DP4 is 3-O-β-cellotriosyl-D-glucose and DP ≥ 5 is cellodextrin-like oligosaccharides containing more than three consecutive 4-0-linked glucose residues.

Source: Lazaridou and Biliaderis (2007). Reproduced with permission from Academic Press.

The weight-average molecular weight (M_w) of β-glucans has been reported to range between 6.5×10^4 and 3.1×10^6 Da. Such a range can be partially explained by genotypic and environmental factors. In a study on four oat varieties grown in 11 different environments, Andersson (Andersson and Börjesdotter, 2011) found a more restricted range of molecular weights ($1.73 - 2.02 \times 10^6$ Da) for β-glucans; molecular weights of several hundreds of kDa are the most common (Johansson, 2006; Mikkelsen, 2013). Nevertheless, they found that molecular weight was influenced to a greater extent by the environment than by genotype. Moreover, they found a positive correlation between β-glucan content and molecular weight. A similar observation was made by Colleoni-Sirghie and colleagues (Colleoni-Sirghie et al., 2003b) through the higher viscosity (at identical concentration) of β-glucans from oat lines with higher β-glucan content. However, extraction and purification conditions, which are reviewed in a later section, can considerably affect molecular weight, either through bacterial degradation, extraction efficiency, depolymerization, and/or aggregation.

Historically, sample-average M_w could be determined using viscosimetry and the relationship between the intrinsic viscosity of the polymer [η] and its M_w, namely the Mark–Houwink–Sakurada law:

$$[\eta] = \lim_{c \to 0} \frac{\eta_r - 1}{c} = K M_w^\alpha.$$

where $\eta_r \frac{\eta}{\eta_0}$ is the relative viscosity of the polymer solution at concentration c (with η_0 as the viscosity of the pure solvent and η as the viscosity of a solution using that solvent) and K and α are constants that depend on the polymer-solvent system and the polymer conformation. [η] can be determined for β-glucans of differents M_w by their Huggins plot ($\frac{\eta_r - 1}{c}$ as a function of c, extrapolated to $c \to 0$). Thus, α and K can be calculated as well as M_w for any new sample of measured intrinsic viscosity. α has been found to range between 0.57–0.71 for oat β-glucans (Gómez et al., 1997a, b, c; Wang et al., 2001; Li et al., 2006) (Figure 6.4). It is an indicator of the polymer's conformation in solution; this particular aspect of oat β-glucans is reviewed in a later section.

Nevertheless, whereas an overall sample-average M_w can be useful and related to a polymer's physical properties, M_w distribution gives a more insightful picture of its structure. A method of choice for the determination of β-glucan M_w distribution is size exclusion chromatography (SEC), where polymers are separated according to their hydrodynamic volume V_h, that is, their "size in space." After separation, β-glucan fractions must be detected; traditionally, Calcofluor fluorescence has been used for this purpose. Indeed, Calcofluor has been known since the early 1980s to bind specifically to and precipitate β-glucans from alkaline extracts of oats and barley (Wood, 1980, 1982; Jensen and Aastrup, 1981; Jørgensen 1983, 1988; Jørgensen and Aastrup, 1988) and β-glucans-Calcofluor complex thermodynamic characteristics are now well understood (Wu et al., 2008). Thus, it has been used as a post-size exclusion column dye to detect and quantify β-glucan fractions (Wood et al., 1991a; Rimsten et al., 2003). With the help of β-glucan standards with known M_w, the M_w distribution of a β-glucan sample can then be obtained (Figure 6.5). However, the binding of Calcofluor

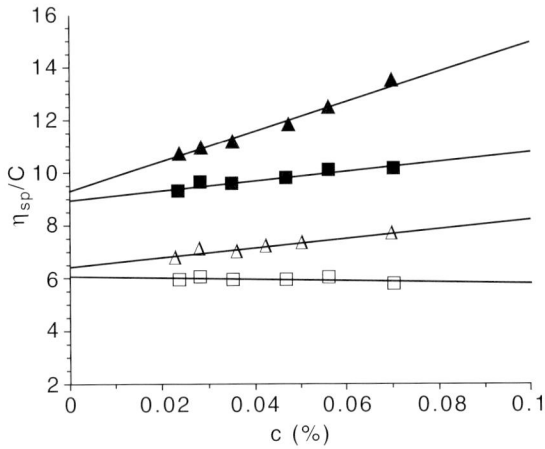

▲ = before autoclaving in H_2O, Δ = after autoclaving in H_2O, ■ = before autoclaving in 0.5 (V_{cad}) Cadoxen, □ = after autoclaving in 0.5 (V_{cad}) Cadoxen, c = concentration (% w/v), η_{sp} = specific viscosity, V_{cad} = volume fraction of Cadoxen.

Figure 6.4   Huggins plots of oat β-glucan solutions before and after autoclaving. *Source:* Wang *et al.* (2001). Reproduced with permission from Pergamon.

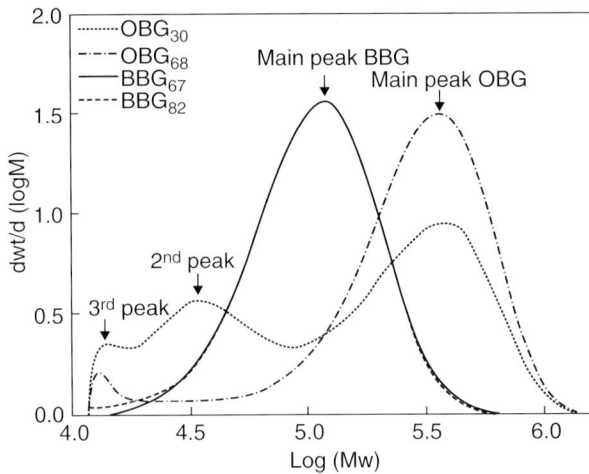

BG$_{no}$ refers to the percentage purity of the sample.
Dwt/d (logM) expresses the different weight
fractions with the area under the curve expressing
100% of the sample.

Figure 6.5   High performance size exclusion chromatograms of crude and purified barley (BBG) and oat (OBG) β-glucan samples. *Source:* Mikkelsen *et al.* (2010). Reproduced with permission from Pergamon.

with β-glucans decreases dramatically for M_w below 10 000–20 000 Da, rendering this method less reliable for low M_w fractions (Gómez *et al.*, 2000).

Multi-angle light scattering coupled with differential refractive index (MALS-dRI) and viscosimetric detection can also be used to determine this M_w distribution (Li *et al.*, 2006; Kim *et al.*, 2008). This technique is based on the relationship between M_w and the angular dependence of the light scattered by a polymer in solution at a certain concentration (Wyatt, 1993). Interestingly, as is reviewed further in another section, this technique allowed the discovery of the aggregative behavior of β-glucans in aqueous solution (Gómez *et al.*, 2000; Håkansson *et al.*, 2012), possibly leading to overestimation of β-glucan M_w. MALS-dRI also proved useful when coupled with asymmetric field-flow fractionation (Håkansson *et al.*, 2012).

6.3 Extraction

6.3.1 Occurrence and location

β-glucans are primarily located in the starchy endosperm and bran of the oat kernel (Wood and Fulcher, 1978; Wood *et al.*, 1983; Miller *et al.*, 1995). Small amounts have been found in the germ, but no evidence of β-glucan presence has been found in the hull. These occurrences of β-glucans in oats have been detected with Calcofluor and Congo red staining. It is interesting to note that relative occurences of β-glucans within oat grain depend on the oat line (Sikora *et al.*, 2013).

The bran is composed of tissues surrounding the starchy endosperm of the groat. From the outside to the endosperm, the bran is composed of several cell layers: the pericarp, the seed coat, the aleurone layer, and the subaleurone layer. A minority of an oat kernel's β-glucans is found in the inner layer of the aleurone cell wall.

The starchy endosperm is the primary location for β-glucans in oat kernels. The wall of the endosperm cells can be modeled as a gel made of β-glucans and other soluble polysaccharides reinforced with insoluble polysaccharides, primarily cellulose (Miller and Fulcher, 1994; Miller *et al.*, 1995; Somerville *et al.*, 2004). β-glucan content varies within the endosperm with cell wall thickness: From the core to the outer layers of the endosperm, cell walls are increasingly thicker and contain increasing amounts of β-glucans (Welch *et al.*, 1991).

β-glucan content also varies considerably among oat cultivars (Saastamoinen *et al.*, 1992; Miller *et al.*, 1993a; Miller and Fulcher, 1994; Genç *et al.*, 2001; Demirbas, 2005; Havrlentová and Kraic, 2006) and according to environmental growing conditions. Cho and White (1993) found that the vast majority of oat varietals exhibit β-glucan contents ranging between 4.5 and 5.5%, although β-glucan content can be as low as 1.8% and as high as 8.5% (Saastamoinen *et al.*, 1992; Sikora *et al.*, 2013) in native kernels. In four oats varietals grown in 11 different environments, Andersson (Andersson and Börjesdotter, 2011) found that the β-glucan content of oats (ranging between 2.3 and 3.2%) was impacted by oat genotype to a greater extent than by growing conditions. The importance of the oat cultivar

Table 6.2   Different simulated scenarios compared with a baseline model

Model/ scenarios	Input summary of scenario analysis	β-Glucan level in harvested HO (g/100 g)	β-Glucan level in harvested NO (g/100 g)
Baseline	Baseline model	3.50	4.25
Scenario 1	No fertiliser application	3.40 (−3.0)[a]	4.12 (−3.0)
Scenario 2	Harvesting on physiological maturity	3.73 (6.4)	4.52 (6.4)
Scenario 3	No storage	3.88 (11)	4.71 (11)

[a]Value in parentheses denotes the percentage change over baseline model.
Source: Tiwari and Cummins (2009). Reproduced with permissions from Academic Press.

was emphasized by a Monte Carlo simulation study on the factors influencing the level of β-glucans in oats (Tiwari and Cummins, 2009), where agronomic practices and environmental conditions seemed to play minor roles in oat β-glucan content (Table 6.2). However, these levels can be improved in processed oat products, such as oat flakes or oat flours, via defatting or partial bran removal by milling and sieving (Wood *et al.*, 1991b; Vasanthan and Temelli, 2008).

6.3.2   Quantification in oats

Rapid methods for β-glucan quantification based on Calcofluor staining can be used for β-glucans in solution (Beer *et al.*, 1997a). However, these methods are limited as: Calcofluor response varies according to the ionic strength of the solution and decreases at molecular weight below 10 000–20 000 Da (Figure 6.6); β-glucans are usually not fully soluble (hence, leading to an underestimation of β-glucan content) (Cui and Wood, 2000); Calcofluor can possibly interfere with hemicelluloses and some proteins (Takenaka and Shibata, 1969; Wood, 1982).

The standard accepted methods for quantification of β-glucans (AACC Method 32–23, AOAC Method 995–16, EBC Methods 3.11.1, 4.16.1, and 8.11.1, and ICC Standard Method number 166) are based on work carried out by McCleary and colleagues (McCleary and Glennie-Holmes, 1985; McCleary and Nurthen, 1986; McCleary and Codd, 1991; McCleary and Mugford 1992, 1997).

The general principle is the following. β-glucans are hydrolyzed by lichenase into cello-oligosaccharides as described previously (approximately 90% of DP3 and DP4, approximately 10% of DP $\geq$ 5), which are in turn converted into glucose by β-glucosidase. Then, the amount of glucose released is measured by UV absorbance with glucose oxidase/peroxidase.

However, the standard method can be time consuming and has not been adapted for high-throughput screening of the β-glucan content of oats products. Thus, near-infrared spectroscopy (Mikkelsen *et al.*, 2010; Bellato *et al.*, 2011) and an immunoassay (Rampitsch *et al.*, 2006) were developed for this purpose. A method based on viscosity of raw oat flour was also developed in an attempt to quickly predict β-glucan content of oats (Colleoni-Sirghie *et al.*, 2004).

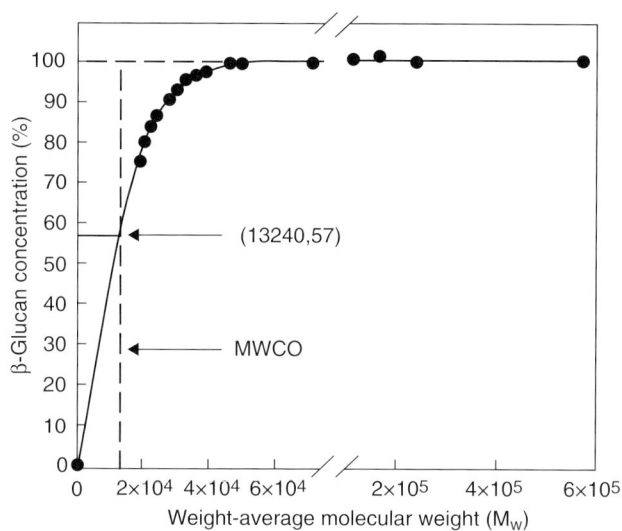

Figure 6.6 Yield response curve of the detector (%) as a function of the average molecular weights of fractions of commercial β-glucan depolymerized to different extents using β-glucanase. Initial concentration of β-glucan, 2 mg/mL. Experimental points were fitted to a four-parameter sigmoid. Adjusted values were $y_0 = -47.87$; $a = 147.94$; $x_0 = 6003.25$, and $b = 8140.22$ ($r = 0.996$). *Source:* Gomez et al. (2000). Reproduced with permissions from Academic Press.

6.3.3 Extraction and purification

As suggested above, β-glucans may not be fully soluble in water, despite their classification as water-soluble gums (BeMiller, 2007a), and their extraction from oats can be a challenge. This challenge is due to two main difficulties. On one hand, the extraction process must be mild and sufficiently controlled to avoid depolymerization of the polymer (thus, loss or false measurement of its properties). On the other hand, it must be sufficiently harsh to avoid extracting only a nonrepresentative portion of β-glucans from the sample.

Depolymerization can occur through alkaline or acidic hydrolysis, chemically or thermally induced oxidative cleavage, or enzymatic hydrolysis due to nondeactivated enzymes remaining in the analyzed sample.

While alkaline or acidic hydrolyses can be avoided by staying within the limits of pH 1.5 to 13, oxidative cleavage can be caused by various compounds present in the sample (e.g., phenolics, ascorbic acid, ferrous ions, free radicals generated *in situ*, etc.) and cannot readily be controlled (Kivelä *et al.*, 2009, 2011, 2012; Kivelä 2011). However, Makinen and colleagues (Mäkinen *et al.*, 2012) stated that oxidative degradation of β-glucans was mainly due to endogenous H_2O_2 and that it could be overcome by the addition of cadmium ethylenediamine (Cadoxen): following the viscosity of β-glucan extracts in solution, while β-glucans in water lost more than 50% of their initial viscosity, β-glucans in solution with Cadoxen retained approximately 90% of their initial viscosity.

Regarding enzymatic depolymerization, it is important to make certain that enzymes are deactivated in the sample, but this is usually the case during milling, where the kilning phase is aimed at deactivating lipases to avoid the development of rancidity (Hutchinson *et al.*, 1951; Kazi and Cahill, 1969). The enzymes most likely to affect β-glucan structure are β-glucanases or cellulases, which are inherent to the sample or produced by microorganisms growing in the sample. If oats were not milled or subjected to processes specifically designed for enzyme deactivation, enzyme deactivation could be achieved by hot aqueous ethanol treatment (typically under reflux for several hours, with 50–85% aqueous ethanol) (Papageorgiou *et al.*, 2005).

Thus, β-glucan extraction is usually achieved at neutral to alkaline pH and at high temperatures (generally 60–100°C) (Wood *et al.*, 1978), although acidic extraction is occasionally used (Bhatty, 1992). Due to the location of β-glucans in the oat starch-rich endosperm, this process is most likely to solubilize and co-extract starch along with β-glucans. This starch can be eliminated by hydrolysis with α-amylase. However, most commercial α-amylases exhibit β-glucanase activity (McCleary, 2000; Doehlert *et al.*, 2012). Thus, it is interesting to use thermostable α-amylase at elevated temperatures to ensure that starch hydrolysis and deactivation of β-glucanases are achieved simultaneously. Then, α-limit dextrins produced by the action of α-amylase on the starch can then be dialyzed from the sample extract containing β-glucans. Ahmad and colleagues (Ahmad *et al.*, 2010) also suggested treating the extract with protease to increase the yield of extraction.

Benito-Roman and colleagues (Benito-Román *et al.*, 2013) also suggested the assistance of ultrasound for β-glucan extraction, but as the energy output was increased to increase extraction efficacy, the molecular weight of the recovered β-glucans dropped significantly.

After dialysis, further purification must be carried out to ensure the β-glucan extract is free from co-extracted products, such as water-soluble hemicelluloses or proteins (Ahmad *et al.*, 2010; Mikkelsen *et al.*, 2010). This step can be achieved by selective precipitation with 20–30% ammonium sulfate, 50–70% ethanol, or dyes such as Calcofluor or Congo Red (Wood *et al.*, 1989, 1994b; Colleoni-Sirghie *et al.*, 2003a; Wang *et al.*, 2003).

Purity can be confirmed by nuclear magnetic resonance (NMR). A full ^{13}C-NMR spectrum of purified oat β-glucan is shown in Figure 6.7.

Overall, β-glucan extractability may not be expected to be 100% in oats. Yields of β-glucan extracted from oats can be as low as 30% of total β-glucans in the case of hot water extraction and as high as 90% in hot alkali. However, Ahmad and colleagues (Ahmad *et al.*, 2010) found lower yields in hot alkali (< 80% recovery) compared to enzyme-assisted extraction (87% recovery). Nevertheless, Doehlert and coworkers (Doehlert *et al.*, 2012) showed that, if extractions are given sufficient time and enough repeats, pH and temperature extraction conditions have little effect on final yields, although it can be argued that prolonged extraction processes increase the chances of β-glucans degradation.

In comparison, oat β-glucans are more extractable than barley β-glucans (~80 vs. 20% in water at 40°C, respectively), which are in turn more extractable than wheat β-glucans (~0% in water at 40°C) (Fincher, 2011). The DP3:DP4 ratio

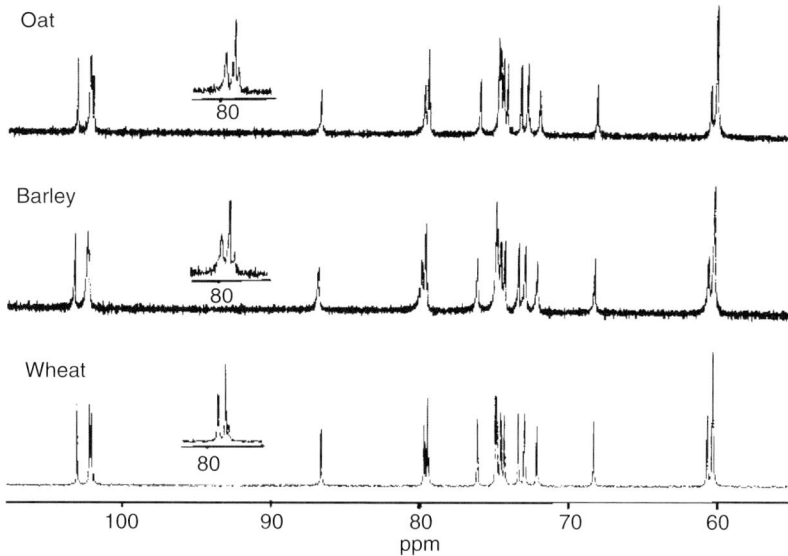

Figure 6.7 ^{13}C NMR spectra of cereal β-glucans. *Source:* Cui and Wang (2009). Reproduced with permissions from Plenum Publishers.

for these cereals are 1.5–2.3:1, 1.8–3.5:1, and 3.7–4.5:1, respectively, suggesting a relationship between a higher DP3:DP4 ratio and a lower solubility. This has been supported by work from Mikkelsen and colleagues (Mikkelson *et al.*, 2013), where β-glucans with higher DP3:DP4 ratios, hence higher chances of block structures within chains and higher chances of interchain associations, exhibited lower solubility.

A general scheme for oat β-glucan extraction is shown in Figure 6.8 (reprinted from Lazaridou *et al.*, 2004). This scheme is complemented by a general protocol published by Pettolino and others (Pettolino *et al.*, 2012) aimed at extracting and characterizing plant cell walls, including oat β-glucans.

The same general principles are used for industrial-scale extraction of β-glucans from oats. Numerous patents have been filed relative to these processes, claiming β-glucan extracts containing up to 90% β-glucans by weight (Fox, 1998; Morgan, 2003; Potter *et al.*, 2003; Van Lengerich *et al.*, 2004; Kvist and Lawther, 2005; Vasanthan *et al.*, 2010; Hellweg *et al.*, 2011; Redmond and Fielder, 2011; Sibakov *et al.*, 2012).

6.4 Solution properties

6.4.1 Conformation

HPSEC and viscosimetry were mentioned earlier, regarding the molecular weight determination of β-glucans. These can also provide useful information regarding β-glucan conformation in solution, which is also an important driver for its rheological properties.

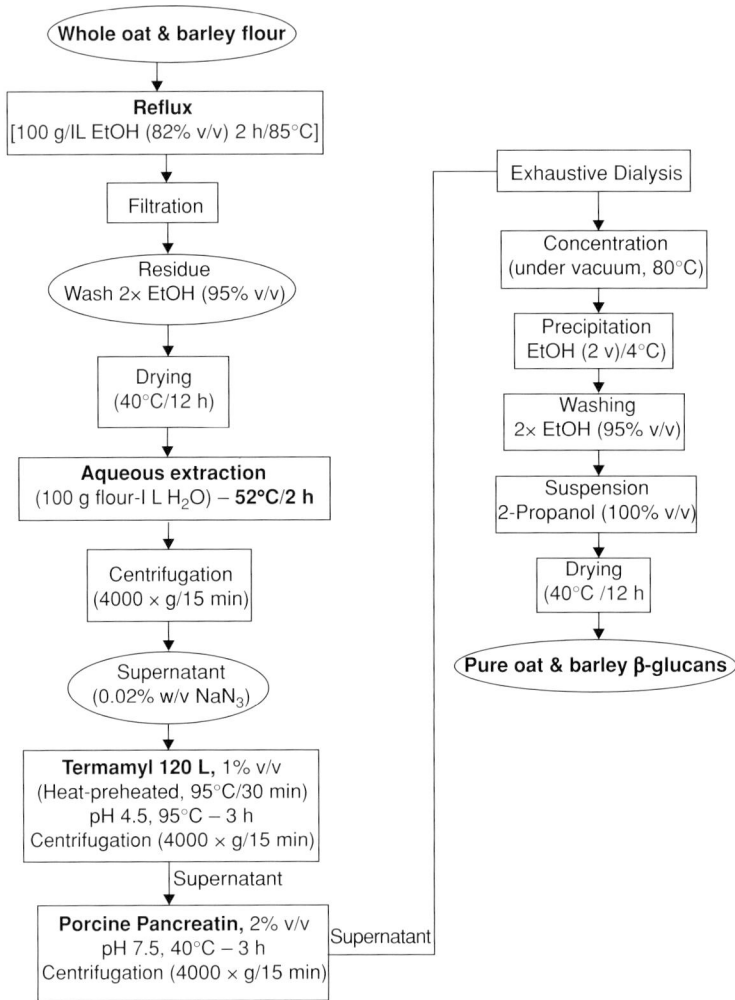

Figure 6.8   Extraction–purification scheme of oat and barley β-glucans from whole flours of oat and barley Greek cultivars. *Source:* Lazaridou *et al.* (2004). Reproduced with permission from Elsevier.

$<R_g>$, the radius of gyration of polymers, is the root mean square of the distance of each monomer of the polymer to its center of mass. $<R_g>$ is an approach of the "size of the polymer in space." HPSEC coupled with multi-angle light scattering (MALS) or right angle laser light scattering (RALLS) detection can determine $<R_g>$. Indeed, a polymeric sample in the measurement cell of a MALS detector scatters light according to the following law (Wyatt, 1993):

$$\frac{K^*c}{R(\theta, c)} = \frac{1}{M_w P(\theta)} + 2A_2c$$

where K^* is a constant dependent on the polymer-solvent and light wavelength system; c is the concentration of polymer in the measurement cell; M_w is the weight-average molecular weight of the polymeric sample in the measurement cell; A_2 is the second virial coefficient, a constant dependent on the specific polymer-solvent system; $R(\theta, c)$ is the excess Rayleigh ratio of the solution, dependent on angle θ at which the intensity of light scattered is measured and the concentration c of the polymer in the measurement cell, and directly proportional to the intensity excess of light scattered by the sample as compared to the pure solvent; and $P(\theta)$ is the factor of angular dependence of the scattered light. This relationship allows the calculation of M_w for each fraction of the eluted polymer sample.

The $P(\theta)$ factor can be calculated as $P(\theta) = 1 - \frac{16\pi^2 n_0^2}{3\lambda_0^2} <R_g^2> sin^2 \frac{\theta}{2} + O\left(sin^4 \frac{\theta}{2}\right)$, where n_0 is the index of refraction of the solvent and λ_0 is the vacuum wavelength of the laser. This relationship, which establishes a relation between the intensity of light scattered by the sample in the measurement cell and $<R_g>$, allows the calculation of $<R_g>$ for each fraction of the eluted polymer sample.

Additionally, a plot of $\log(<R_g>)$ against $\log(M_w)$ usually gives a linear relationship whose slope is a conformational factor of the polymer: The slope is approximately 0.33 for a spherical conformation, 0.5–0.6 for a random coil conformation, and approximately 1.0 for a rigid rod conformation.

Similar conformational information can be obtained from Mark–Houwink–Sakurada plots, where $\log([\eta])$, the intrinsic viscosity of the polymer, is plotted against $\log(M_w)$. The slope of the linear relationship once again reflects the conformation of the polymer in solution: 0 for a sphere, 0.5–0.8 for a random coil conformation, and approximately 1.8–2.0 for a rigid rod conformation.

Data gathered from Wang and colleagues (Wang et al., 2003) allow a $\log(<R_g>)$ versus $\log(M_w)$ conformational plot, yielding a slope of 0.51 (from linear regression, $R^2 = 0.95$), which suggests a random coil conformation (Figure 6.9). In the same study, the Mark–Houwink–Sakurada plots showed a slope of 0.62, confirming this result (Figure 6.10, Table 6.3). Similarly, Varum and colleagues (1992) reported a slope of 0.59 for the Mark–Houwink–Sakurada plot (Vårum et al., 1992).

In work reporting shape factors (from MALS and RALLS) of β-glucans hydrolyzed to different extents, Roubroeks and colleagues (2000, 2001) reported a shape factor from 0.78 to 1.07, with a tendency to increase, that is, give more

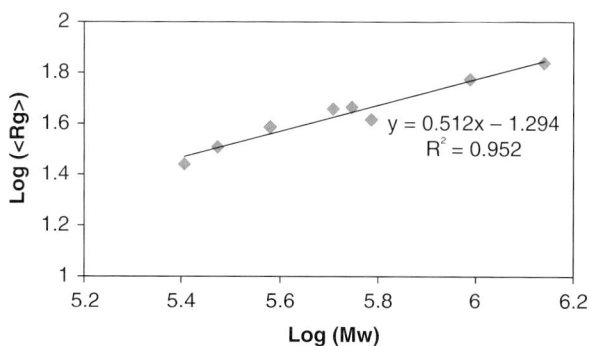

Figure 6.9   Conformational plot Log($<R_g>$) versus Log(M_w) of β-glucan from oat. *Source:* Adapted from Wang et al. (2003). Reproduced with permission from Elsevier.

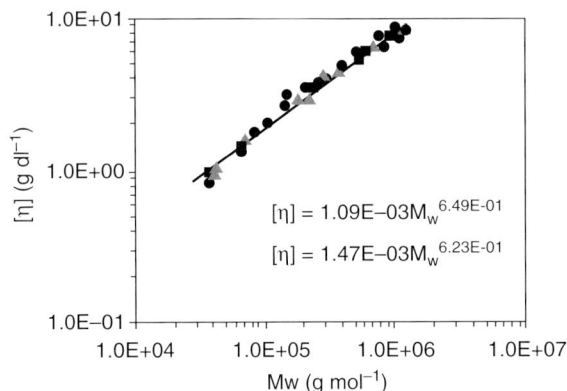

Figure 6.10 Mark–Houwink–Sakurada plots of β-glucan from (▲) oat and (●) barley in aqueous solutions. *Source:* Adapted from Wang *et al.* (2003). Reproduced with permission from Elsevier.

elongated stiffer chains, with the degree of hydrolysis (Table 6.4). However, this generally random coil conformation is also supported by the random distribution of building blocks of various DP units along the polymer's chains, as discussed previously. This irregular distribution may be responsible for the nonordered conformation of β-glucans. Li and colleagues (Li *et al.*, 2012) performed molecular modeling studies of cereal β-glucans that confirmed "moderately extended sinuous chain conformation" with increasing stiffness correlated with an increasing DP3:DP4 ratio.

Nevertheless, these measurement and characterization studies at the molecular level must be taken with caution because of the ability of β-glucans to form aggregates in solution. Indeed, fractions of β-glucans sometimes scatter light at an abnormally high intensity, considering their concentration (as measured by refractive index) (Håkansson *et al.*, 2012). This is a sign of aggregate formation as shown previously for bovine serum albumin (Ye, 2006), for example. However, these potential aggregates show the conformational attributes of single β-glucan chains, with a slope of 0.59, typical of random coils, in the conformational plots

Table 6.3 Weight-average molecular weight (M_w), radius of gyration (R_g), and intrinsic viscosity ($[\eta]$) of unfractionated oat β-glucan (F0) and its seven fractions (F1–F7) obtained by stepwise precipitation with ammonium sulfate. Last column lists the yield (% of original material, F0) of each fraction

	$M_w \times 10^5$ (g/mol)	R_g (nm)	$[\eta]$(dl/g)	Yield (%)
F0	6.11	41.3	6.3	
F1	13.8	69.0	7.8	18.8
F2	9.73	59.4	7.0	14.0
F3	5.58	46.3	6.0	8.8
F4	5.10	45.5	5.3	8.8
F5	3.80	38.6	4.5	10.2
F6	2.98	32.3	3.7	12.6
F7	2.55	27.6	3.2	12.6

Source: Adapted from Wang *et al.* (2003). Reproduced with permission from Elsevier.

Table 6.4   Conformational parameters, exponents, and persistence lengths of fractions with increasing hydrolysis time

Hydrolysis time (h)	a (SEC–RI–RALLS–Vise)	a (SEC–RI–MALLS–Vise)	v^a	l_p^b (nm)
0	0.67	0.78	0.56	2.44
1	0.73	0.85	0.57	2.22
2	0.81	0.87	0.60	2.19
4	0.79	0.86	0.60	2.07
8	0.80	0.91	0.60	2.07
10	0.82	0.94	0.61	1.83
12	0.88	1.02	0.63	1.85
18	0.85	1.07	0.63	1.80
24	0.93	1.05	0.68	1.51
30	0.98	1.15	0.68	1.54
48	0.82	0.99	0.62	1.27
70	1.00	0.86	0.70	1.27

Source: Roubroeks, Mastromauro *et al.* (2000). Reproduced with permission from the American Chemical Society.

(Vårum *et al.*, 1992). This latter observation, in addition to the reversibility of aggregation between neutral (aggregates formed) and alkaline pH (aggregates dissociated), suggests that these aggregates are labile and subject to exchange. One explanation that has been suggested to explain the aggregation of β-glucans is the possibility of hydrogen bonding between cellulose-like segments of high DP units, as occurs in cellulose (Vårum and Smidsrød, 1988; Cavallero *et al.*, 2002). This point is discussed further in the next section.

6.4.2   Rheology

In solution, oat β-glucans exhibit nonNewtonian pseudoplastic (shear thinning) behaviors above a critical concentration and Newtonian behavior below this concentration. Conceptually, random coils do not interact at low concentrations and applying higher shear rate does not impact this noninteracting state. However, these random coils "physically" overlap above a critical concentration. Thus, when the shear rate increases, overlapping polymer chains in random coil conformations are stretched and aligned, leading to a viscosity decrease and a more rapid flow, which is typical of linear polymers. This is revealed in rheological data as a constant viscosity plateau at low shear rates, followed by a viscosity drop as shear rate increases (Figure 6.11). In this respect, viscosity of β-glucan solutions as a function of shear rate follows a power law as described by Morris (1989):

$$\eta = \frac{\eta_0}{1 + \left(\frac{\dot{\gamma}}{\dot{\gamma}_{1/2}}\right)^{0.76}}$$

where η is the viscosity of the solution, η_0 is the viscosity of the Newtonian plateau, $\dot{\gamma}$ is the shear rate, and $\dot{\gamma}_{1/2}$ the shear rate at which $\eta = \frac{\eta_0}{2}$. The degree of pseudoplasticity increases with the polymer's molecular weight and

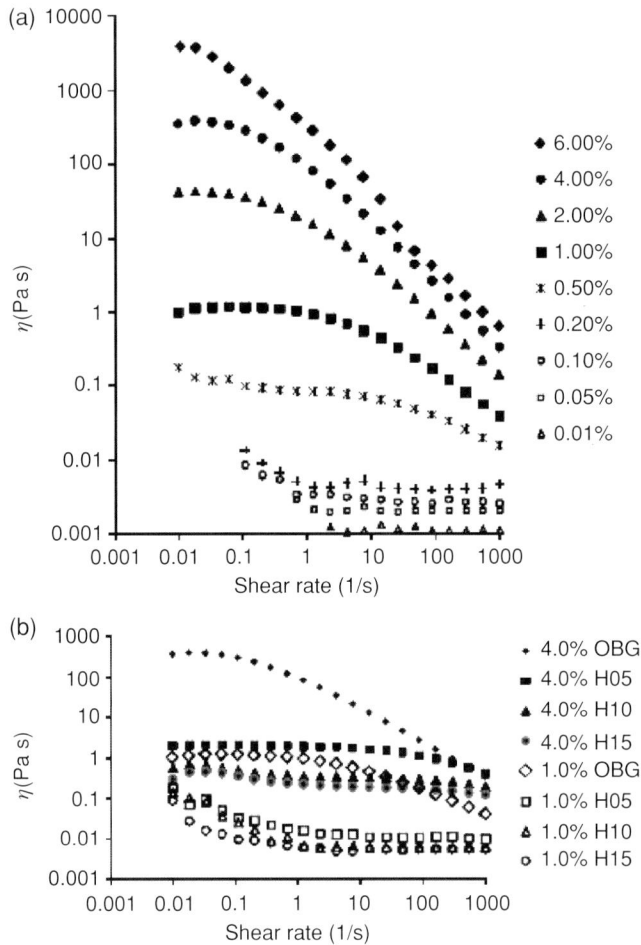

Figure 6.11   Viscosity dependence on shear rate for oat β-glucan dispersions differing in (a) concentration and (b) different molecular weights at 1% and 4% (w/v). *Source:* Agbenorhevi *et al.* (2011). Reproduced with permission from IPC Business Press.

concentration (Xu *et al.*, 2013). For high molecular weight oat β-glucans, the critical concentration where the Newtonian/nonNewtonian transition occurs is around 0.2% (Ren *et al.*, 2003) (Figure 6.12).

Low molecular weight β-glucans show the same behavior (Skendi *et al.*, 2003): the shear rate span of this plateau decreases as β-glucan concentration increases at constant molecular weight, and vice versa. Interestingly, Kivelä and colleagues (Kivelä *et al.*, 2010) confirmed these observations by studying the rheological properties of high molecular weight β-glucans broken down by high-pressure homogenization, which lost their initial viscosity and pseudoplastic behavior. The impact of the DP3:DP4 ratio is less clear, although Ryu and coworkers (Ryu *et al.*, 2012) found that at equivalent molecular weight and $<R_g>$, β-glucans with the lowest DP3:DP4 ratios tended to have the highest viscosities in solution.

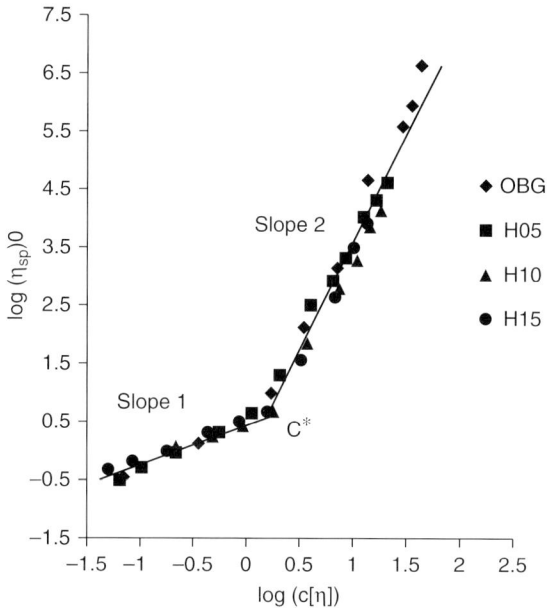

Figure 6.12 Zero shear-specific viscosity $(\eta_{sp})_0$ versus the reduced concentration $c[\eta]$ for β-glucan isolates. The intercept between the two linear slopes indicates the critical concentration c^* that demarcates the transition from dilute to concentrated solution behavior. *Source:* Agbenorhevi *et al.* (2011). Reproduced with permission from IPC Business Press.

Although β-glucans mainly form viscous solutions, they also have the ability to form gels, especially after prolonged storage time. The formation of these gels can be attributed to physical cross-linking of the polymer chains through interchain hydrogen bonding, as described for aggregation, leading to three-dimensional gel networks. Gel behavior is measured by rheometers that characterize the viscoelastic response of a material subjected to oscillatory shear (BeMiller, 2007b). For a certain oscillatory frequency, ω, the complex shear modulus, G^* (obtained from complex viscosity η^* divided by ω), of the material provides two measurements: the real part of G^*, G', is its elastic component (elastic or storage modulus); the imaginary part of G^*, G'', is its viscous component (viscous or loss modulus). A viscoelastic gel has the characteristics of both a solid and a viscous liquid: G' describes the solid component of the gel and G'' describes the liquid component of the gel. Thus, by definition of these two components, $G'' > G'$ in a viscous solution and $G' > G''$ in a gel. Thus, gelation time (G_t) upon storage can be determined by the measurement of G^*: G_t is the time at which G' crosses G''. Gelation rate (also known as elasticity increment I_E) is defined as the maximum slope of the G' plot versus time.

At comparable molecular weight and concentrations, oat β-glucans have longer gelation times than barley and wheat, the latter having the shortest gelation time (Bohm and Kulicke, 1999; Cui and Wood, 2000; Lazaridou *et al.*, 2004; Tosh *et al.*, 2004a). Within oat β-glucans, gelation kinetics appears to be mainly correlated with the DP3:DP4 ratio: Gelation occurs more rapidly (higher rate and shorter time) when interchain interactions increase, that is, when the DP3:DP4 ratio increases, leading to more ordered structures. Increased DP3:DP4 ratio not only leads rapidly to a gel, but also leads to more "solid-like" gels with higher G' (elastic modulus) for comparable molecular weights and concentrations (Bohm and Kulicke, 1999). Moreover, G_t and I_E are negatively correlated with molecular weight of β-glucans: high molecular weight β-glucans (250 000 Da) showed no gelation behavior (Lazaridou *et al.*, 2003; Skendi *et al.*, 2003) (Figure 6.13). Gelation characteristics G_t and I_E are also positively correlated with β-glucan concentration and are dependent upon temperature: G_t and I_E

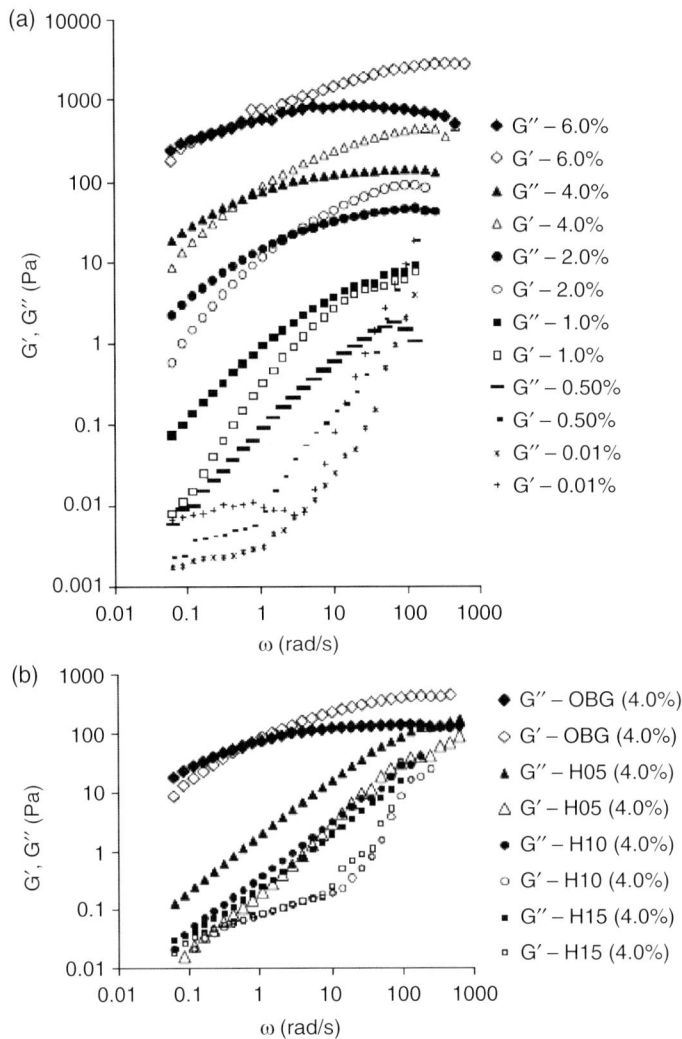

Figure 6.13 Frequency dependence of storage (G′) and loss (G) of oat β-glucan dispersions at (a) different concentrations (OBG) and (b) with different molecular weights at 4% (w/v). *Source:* Agbenorhevi *et al.* (2011). Reproduced with permission from IPC Business Press.

also have a maximum around temperatures between 25 and 35°C (Agbenorhevi *et al.*, 2011).

Once a gel is obtained, increasing the temperature leads to the melting of the gel. This melting process can be studied by differential scanning calorimetry, which shows that melting occurs over a temperature range, generally between 55 and 70°C (Lazaridou *et al.*, 2004; Tosh *et al.*, 2004). However, the peak temperature associated with the melting of β-glucan gels tends to decrease with decreasing molecular weight of the polymer, whereas the temperature range of melting is broadened. As observed with the thermal characteristics of retrograded starch, the broader the melting peak, the more various are the intermolecular associations leading to gelation. This highlights once again the importance of the fine structure of β-glucans in terms of their physicochemical properties in general and their gelling properties in particular. Surprisingly, the occurrence of cellulose-like sequences (DP $\geq$5) does not correlate with β-glucan gelling properties (gelation time, mechanical strength of the gel, melting temperature, etc.), despite their apparent ability to be involved in intermolecular hydrogen bonding as occurs in cellulose. In fact, gelling properties appear to be highly correlated with the occurrence of DP3 units, which fundamentally means a higher probability of cellotriosyl sequences along the polysaccharide chain and cooperative interchain hydrogen bonding to form junction zones (Lazaridou *et al.*, 2003; Tosh *et al.*, 2003, 2004a, 2004b). This was clearly shown by the comparison between lichenan and oat, rye, and barley β-glucans, relative to their DP3 unit content and storage modulus G' of the gels they formed (Figure 6.14).

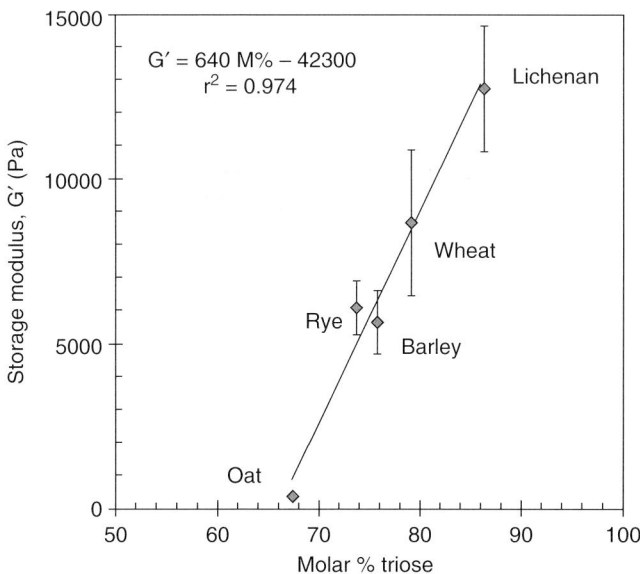

Figure 6.14   Correlation of storage modulus, G'; with mole % cellotriosyl units in each of the (1,3;1,4)-β-D-glucans. Gels were aged 7 days at 5°C. *Source:* Tosh (2004). Reproduced with permission from Elsevier.

6.5   Oat β-glucan nutritional properties

The nutritional implications of cereal β-glucans are well documented in the literature. In particular, the effects of oat and barley β-glucans on blood glucose regulation in diabetic subjects and serum cholesterol reduction in hypercholesterolemic subjects have been extensively studied (Annapurna, 2011; Othman *et al.*, 2011; Tiwari and Cummins, 2011; Cloetens *et al.*, 2012; Daou and Zhang, 2012; Kumar *et al.*, 2012). The latter effects have been recognized by FDA, which approved the health claim that "oats may lower the risk of heart disease," in 1997 (Anonymous, 1997). Recent studies have also shown immunomodulatory effects for these polymers, which may affect cancer prevention and treatment. The physiological effects of β-glucans are significantly affected by their physicochemical properties, such as flow viscosity, gelation, molecular weight, and chemical structure. Discussed in this section are the effects of β-glucans at every stage during transit through the gastrointestinal (GI) track and their effects on glucose and cholesterol metabolism, immune function, satiety and energy intake, with emphasis on the roles of physicochemical properties, processing history, and the food matrix on these effects. GI transit is considered in terms of the effects of β-glucans on the properties of the digesta as well as the changes in β-glucan properties throughout transit. The potential relationships between molecular weight, structure, solution viscosity, and solubility of β-glucans, as well as their hypoglycemic and hypocholesterolemic effects, is explored, along with the need for adequate reporting of their chemical and rheological properties in the context of clinical trials.

6.5.1   Gastrointestinal transit

The effects of β-glucans on the transit of food material through the GI tract are mainly attributed to swelling, water binding capacity, and solution viscosity. The molecular weight and degree of solubilization of the β-glucans, which in turn are largely affected by the food matrix, are important determinants of viscosity and, therefore, of behavior during transit. In the colon, β-glucans are fermented by colonic microbiot, and the resulting production of short-chain fatty acids (SCFAs) and changes in microbial populations have significant physiological effects. It is important to note that transit through the upper GI tract also affects the β-glucans themselves, causing reactions such as depolymerization and structural changes. Mälkki and Virtanen (2001) reviewed the GI effects of oat β-glucans in terms of their transit through specific regions of the GI tract. A similar approach is taken to this discussion, while at the same time emphasizing the changes in the β-glucan molecules as they move through the GI tract.

6.5.1.1   Oat β-glucans in the stomach   In liquid foods such as beverages, complete hydration of β-glucans occurs *ex vivo* in the food matrix. β-glucans incorporated into solid or semisolid food matrices are hydrated within the GI tract. The hydration process begins in the mouth and continues as the material moves through the GI tract until complete hydration is achieved. The hydration and swelling rates depend on particle size, processing history, and other constituents in the food matrix. Since hydration is the key to viscosity development (which is highly significant for the physiological effects of β-glucans), the factors affecting hydration will greatly influence the nutritional properties of β-glucans. Swelling of β-glucans in the stomach leads to gastric distension, which is associated with satiation (Woods, 2004). Thus, increased stomach distension due to swelling of β-glucans may lead to a reduction in meal size by promoting satiation. The molecular properties of β-glucans are largely unchanged in the stomach. *In vitro* studies on the incubation of oat β-glucans with gastric fluid at pH 1.5 (Johansen *et al.*, 1993) and pepsin (Wood *et al.*, 1991c) did not show significant changes in molecular weight after incubation.

6.5.1.2 Transit through the small intestine As a meal containing β-glucans exits the stomach, its viscosity begins to have effects on digestion and absorption of nutrients in the small intestine. The high viscosity caused by β-glucans is largely maintained in the small intestine (via increased mucin production), as the β-glucans are not digested there (although some depolymerization does occur) (Malkki and Virtanen, 2001). As the digesta move through the small intestine, the high viscosity probably alters the flow pattern, leading to lower nutrient absorption by limiting diffusion of nutrients to the intestinal wall. Notably, starch digestibility and glucose uptake may be affected by reduced transport of substrates to the intestinal wall, which possesses mucosal α-glucosidases. This has been suggested as one of the mechanisms for a lower postprandial glucose response in the presence of β-glucans (Dunaif and Schneeman, 1981; Regand *et al.*, 2011). Some authors have suggested that oat β-glucans may reduce the activities of enzymes, such as amylase, lipase, and chymotrypsin (Dunaif and Schneeman, 1981; Jenkins *et al.*, 1982). However, such assays have been performed *in vitro* and conclusions based on such experiments may not be transferrable to *in vivo* systems (Malkki and Virtanen, 2001). Lund and coworkers (Lund *et al.*, 1989) also suggested that increased intestinal viscosity leads to an increase in thickness of the unstirred water layer at the intestinal wall, further limiting diffusion of nutrients to intestinal epithelial cells for absorption. Lipid and cholesterol absorption is reduced, since emulsification of fats is impaired by high viscosity, leading to larger fat droplet sizes (Lazaridou and Biliaderis, 2007).

The human ileostomy model is commonly used to study the effects of β-glucans on digestion and absorption of nutrients in the upper GI tract, and to study changes in the β-glucans caused by transit. Increased excretion of bile acids (Lazaridou and Biliaderis, 2007) and fats (Lia *et al.*, 1997) has been observed in ileostomy models in the presence of oat β-glucans. In another study, Lia and colleagues (Lia *et al.*, 1996) observed increased protein recovery in ileal effluents in the presence of oat bran, suggesting decreased absorption of protein. They did not observe any difference in starch recovery in the ileal effluent after oat bran-containing or wheat bread control meals were consumed.

Transit through the upper GI tract may also causes changes in the physic-ochemical properties of the β-glucans. Molecular weight can be affected by depolymerization, which is regarded as an effect of the enzymatic activity of microorganisms in the gut. In ileostomy subjects, 88.5% of the ingested β-glucans were recovered in the ileal effluent and some degradation of the polymers was observed (Sundberg *et al.*, 1996). It is noteworthy that the small intestine in ileostomy patients is known to have relatively high microbial loads, which may be the reason behind the observed degradation of β-glucans (Malkki and Vir-tanen, 2001). The decrease in molecular weight is associated with a decrease in viscosity *in vitro*, but the system may be more complex *in vivo*. Concentration of β-glucans in solution is another major factor determining viscosity; this may change during transit through the GI tract. Proteolysis of proteins from the food matrix in the small intestine was shown to lead to greater solubilization of β-glucans (Robertson *et al.*, 1997). Increased degradation of β-glucans was also found to occur, along with solubilization. The degradation of β-glucans in the upper GI tract is also dependent on the molecular weight of the original substrate, with high molecular weight β-glucans possibly degraded to a greater extent than low molecular weight β-glucans (Lazaridou and Biliaderis, 2007).

6.5.1.3 Fermentation in the large intestine

Oat β-glucans are highly fer-mentable by microorganisms in the colon. The major fermentation products are SCFAs, such as acetate, propionate, and butyrate. Each SCFA has different func-tions and metabolic fates in the body. Acetate is used in peripheral tissues as an energy source (Kim and White, 2009), whereas propionate influences glucose and lipid metabolism in the liver (Anderson *et al.*, 1990). Butyrate serves as an energy source for colonic epithelial cells. SCFA production in the colon may also play a role in the cholesterol-lowering effects of β-glucans, although such evi-dence is not very clear. Butyrate is produced in high amounts by oat β-glucans compared with other dietary fibers (Malkki and Virtanen, 2001). In addition to serving as an energy source, butyrate may also retard the growth of carcinoma cells and induce apoptosis, thus mitigating colorectal cancers (Hague *et al.*, 1993, 1995). Some studies have focused on the mechanisms of this action of butyrate, and although epidemiological evidence for the reduction of colorectal cancer by β-glucans is weak at best, a generally beneficial role of butyrate production has been acknowledged.

Oat β-glucans are also known to have a prebiotic effect, which implies that they selectively stimulate the growth of certain microbial strains in the colon, thus pro-viding a health benefit (Malkki and Virtanen, 2001). *In vitro* human fecal fermen-tation studies have demonstrated the selectivity of oat bran and flour (Kim and White, 2009) and purified β-glucans from oats and barley (Hughes *et al.*, 2008) for certain *Bifidobacteria* and *Lactobacillus* species, which are considered to be bene-ficial. The prebiotic effect is most pronounced with β-glucan oligosaccharides. In an *in vitro* human fecal fermentation study with barley and oat β-glucans of dif-ferent molecular weights, Hughes and coworkers (Hughes *et al.*, 2008) found that the higher molecular weight polymers stimulated the growth of *Lactobacillus* and *Enterococcus*, but not *Bifidobacterium*. β-glucan oligosaccharides, obtained by hydrolyzing the native polymers, stimulated the growth of *Bifidobacteria* as well. It is noteworthy that the prebiotic effect and SCFA production of β-glucans and β-glucan oligosaccharides have been studied mainly in *in vitro* models (Drzikova *et al.*, 2005; Kim and White, 2009; Cloetens *et al.*, 2012); more *in vivo* studies are required to establish the prebiotic effect in more detail and to elucidate structure-function relationships.

Oat bran is also known to increase stool weight, decrease transit time through the colon, and relieve constipation (Cloetens *et al.*, 2012). This effect is mainly attributed to insoluble fiber and would be significant in the case of β-glucan-enriched foods where the degree of solubilization is low. For soluble β-glucans, the increase in stool dry weight is mainly attributed to an increase in microbial cells (Chen *et al.*, 1998). Oat β-glucans, due to their butyrogenic effects, are also considered to be advantageous for the treatment of diarrhea, since absorption of butyrate by colonic epithelial cells promotes absorption of water and sodium, thus assisting rehydration therapy (Malkki and Virtanen, 2001).

Thus, the transit of oat β-glucans through the GI tract greatly affects nutrient absorption and metabolism. The physicochemical properties (molecular weight and cellotriosyl:cellotetraosyl ratio, for example) and the matrix effects (solubility of β-glucans, for example) are important determinants of the effects of the polymers throughout such transit, while these properties are altered by conditions in the GI tract (pH and ionic strength, for example): *in vivo* effects of oat β-glucans on transit are then hardly predictable and only very partially linked to strict physicochemical properties of β-glucans. Some of the specific physiological consequences of dietary oat β-glucans are now considered, including the cholesterol-lowering effect, regulation of glucose and insulin responses, effects on appetite and energy intake, and immunomodulatory effects.

6.5.2 Cholesterol-lowering effect

The cholesterol-lowering effect of oat β-glucans was first reported by Anderson and coworkers in the mid-1980s (Anderson *et al.*, 1984; Anderson and Tietyen-Clark, 1986). Since then, several clinical studies have confirmed this result. In 1997, the FDA, after reviewing 33 clinical studies, approved a health claim for oats for the reduction of the risk of coronary heart disease (Anonymous, 1997). A daily dosage of 3 g β-glucan was deemed to be effective and food products must contain 0.75 g β-glucans per serving to qualify for this claim. The claim was later expanded to include barley β-glucans as well. The hypocholesterolemic effect of β-glucans is generally considered to be higher in subjects who have higher than normal levels of total cholesterol and low-density lipoprotein (LDL) cholesterol in their blood (Lazaridou and Biliaderis, 2007). The actual values of cholesterol reduction seen in clinical trials vary, depending on the dosage, type, and physicochemical properties of the β-glucans, as well as the food matrix.

Tiwari and Cummins (2011) used a meta-analysis approach to establish a dose response of plasma total cholesterol levels to daily consumption of β-glucans. They found a decrease in cholesterol levels with increasing β-glucan consumption up to 3 g/day, with no further decrease beyond that value. This correlates well with the FDA recommended daily dosage. A recent review by Othman and others (Othman *et al.*, 2011) summarizes some clinical studies in terms of study size, β-glucan dosage, duration, diet, and changes in total, high-density lipoprotein, and LDL cholesterol levels. Based on this review, the reduction in total cholesterol ranged from 0 to 13%, whereas a reduction in LDL cholesterol ranged from 0 to 16.5%. In view of this broad range due to the factors mentioned above, the focus here is on the mechanisms of the hypocholesterolemic effect and the roles of the physicochemical properties and food matrices, rather than reporting the actual values observed in various studies. For summaries of the actual values of cholesterol reduction, the reader is referred to the meta-analysis by Tiwari and Cummins (2011) and reviews by Othman and colleagues (Othman *et al.*, 2011) and Kelly and coworkers (Kelly *et al.*, 2007).

6.5.2.1 Mechanism of cholesterol reduction by dietary β-glucans Othman
and colleagues (Othman *et al.*, 2011) recently reviewed the cholesterol-lowering effect of β-glucans, including their various mechanisms of action. The most widely recognized mechanism is the increase in viscosity of digesta in the small intestine. Enhanced intestinal viscosity leads to reduced uptake of dietary cholesterol and impairs reabsorption of bile acids (Lund *et al.*, 1989; Nauman *et al.*, 2006; Lazaridou and Biliaderis, 2007; Othman *et al.*, 2011). The former effect could be the result of a change in fat emulsification due to increased viscosity and the resulting large fat droplet size (Lazaridou and Biliaderis, 2007). The latter effect causes increased synthesis of bile acids in the liver, utilizing LDL cholesterol and lowering serum LDL and total cholesterol levels in the process. This mechanism has been discussed with more details by Daou and Zhang (2012). Although there is no direct evidence of the reduced absorption and reabsorption of cholesterol and bile acids in the small intestine in the presence of β-glucans, an *in vitro* study of cholesterol and D-galactose absorption by rat small intestines (Lund *et al.*, 1989), as well as the observation of increased fecal excretion of bile acids and increased fat excretion in ileostomy patients (Lia *et al.*, 1995), are considered as evidence in support of these mechanisms. It is clear that viscosity plays a central role in the mechanism of cholesterol reduction by β-glucans.

Bile acid binding by β-glucans has also been suggested as a possible mechanism for the hypocholesterolemic effect (Drzikova *et al.*, 2005; Dongowski, 2007). However, in a ^{13}C-NMR study of the interaction between glycocholic acid (a bile acid) and barley β-glucans, Bowles and others (Bowles *et al.*, 1996) found no evidence of any specific molecular interactions. They suggest physical entrapment of bile acids in the digesta, due to increased viscosity, as a more likely mechanism for the increased bile acid excretion observed in the presence of β-glucans. Bile acid binding also seems to be affected by non-β-glucan components of oat flour. A recent study by Kim and White (2010) showed that, based on β-glucan amounts, oat flour bound greater amounts of bile acids *in vitro* compared to extracted β-glucans. Sayar and colleagues (Sayar *et al.*, 2004) made a similar suggestion based on the observation that the amount of β-glucan in oat flour did not correlate well with bile acid binding ability, whereas there was a significant correlation with the amount of insoluble fiber. However, previous *in vivo* studies have shown increased bile acid excretion that was dependent on the properties of the β-glucans (Bae *et al.*, 2010; Othman *et al.*, 2011). Thus, although β-glucans are not the only significant component involved in bile acid binding, it is clear that the properties of the β-glucans themselves and their relation with their carrying matrix do affect the binding and excretion of bile acids.

Another possible mechanism of cholesterol reduction by β-glucans involves the fermentation of these polymers by microorganisms in the colon and the resulting production of SCFAs, such as acetate, propionate, and butyrate (Hughes *et al.*, 2008; Barsanti *et al.*, 2011; Othman *et al.*, 2011). These SCFAs, particularly the propionate:acetate ratio, are known to affect lipid metabolism. Propionic acid was shown to inhibit cholesterol synthesis in rat hepatocytes at concentrations of 1–2.5 mM (Anderson *et al.*, 1990). However, the significance of this effect in humans is unclear, as the concentration of propionate in the hepatic portal vein may be lower than this level (Othman *et al.*, 2011). Additionally, a study by Battilana and coworkers (Battilana *et al.*, 2001) showed that the administration of β-glucan-containing meals at small intervals of one hour over a nine-hour period did not cause any significant change in *de novo* lipogenesis in the liver (or in glucose metabolism). The authors interpreted this as evidence that the primary mechanism by which β-glucans affect carbohydrate and lipid metabolism was via delayed or reduced nutrient absorption, rather than the production of SCFAs in the colon.

The effect of β-glucans on carbohydrate metabolism may also play an indirect role in their hypocholesterolemic effect (Lazaridou and Biliaderis, 2007). Insulin has profound effects on lipid metabolism in the liver and is known to be associated with increased hepatic fatty acid and lipoprotein synthesis (Tobin *et al.*, 2002). Increased insulin levels and insulin resistance are known to increase hepatic cholesterol synthesis (Pihlajamaki *et al.*, 2004) and lower insulin responses have been associated with lower serum cholesterol levels (Jenkins *et al.*, 1989). As a result, the lowered insulin response that is caused by dietary β-glucans may be of significance not only to the regulation of carbohydrate metabolism but also to their hypocholesterolemic effect.

Thus, the mechanisms of cholesterol reduction by dietary β-glucans are manifold. For further details about the cholesterol-lowering effect, the reader is referred to reviews by Lazaridou and Biliaderis (2007), Barsanti and coworkers (Barsanti *et al.*, 2011), Othman and colleagues (Othman *et al.*, 2011), and Daou and Zhang (2012).

6.5.2.2 Effects of the physicochemical properties of β-glucans and the food matrix on the cholesterol-lowering effect

As mentioned earlier, the physicochemical properties of β-glucans have profound effects on their cholesterol-lowering ability. In particular, molecular weight, structure, and solution viscosity play significant roles in determining bile acid binding and prevention of cholesterol absorption. In addition, the food matrix in which these polymers are incorporated affects their effectiveness through interactions with other matrix components and changes in solubility.

The molecular weight of β-glucans affects their physiological consequences in an indirect manner by affecting viscosity. Barsanti and colleagues (Barsanti *et al.*,

2011) suggested that a M_w between 26.8 and 3000 kDa is required to provide significant intestinal viscosity. It is important to note, however, that higher molecular weight does not seem to be a direct predictor of a heightened cholesterol-lowering effect. Indeed, intestinal viscosity build up due to β-glucans is mainly a consequence of both their molecular weight and their concentration, this latter being partly dependent on the molecular weight itself. It is important to note that although at equal concentration higher molecular weight β-glucans provide higher viscosity, higher molecular weight β-glucans tend to be less soluble than the lower molecular weight fraction. These effects may explain unclear or conflicting conclusions reported thereafter, especially *in vivo*, regarding the effect of β-glucans molecular weight on their nutritional properties.

Wolever and colleagues (Wolever *et al.*, 2010) found in a recent clinical trial that the administration of 3 g of β-glucans per day in the form of an extruded breakfast cereal (two preparations with high and medium M_w, 2210 and 530 kDa, respectively) led to similar reductions in serum cholesterol of about 5%, whereas a low M_w preparation (210 kDa) led to a much lower reduction. Bile acid excretion is also affected by the molecular weight of the β-glucan preparation. In a study with ileostomy patients, bile acid excretion was decreased by more than 50% in subjects fed β-glucans degraded by β-glucanase compared to subjects that were given native β-glucans (Lazaridou and Biliaderis, 2007). Thus, extensive degradation of the β-glucans seems to negatively affect bile acid excretion, whereas slight degradation does not have much effect. The molecular weight of β-glucans can also be reduced during storage of the food, due to endogenous β-glucanase activity in flour and during GI transit. Thus, it is likely that both factors affect the cholesterol-lowering effect of β-glucans.

Kim and White (2010) reported the effect of the molecular weight of β-glucans on the *in vitro* bile acid binding capacity. They reported an inverse relationship between bile acid binding and molecular weight, in the M_w range of 156–687 kDa. In contrast, Sayar and others (Sayar *et al.*, 2004) report no significant effect of lichenase degradation (to reduce molecular weight) on bile acid binding capacity of oat β-glucans. In a more recent *in vitro* study with oat flour muffins, Kim and White (2011) reported higher bile acid binding by high M_w (319 kDa) compared to lower M_w (114 and 40 kDa) β-glucans. The inconsistency of the relationship between molecular weight and *in vitro* bile acid binding ability suggests that some other factor or combination of factors, perhaps including the fine structure of the polymer molecules, temperature, and pH of the medium, are important determinants of bile acid binding. Gel formation, which in the case of oat β-glucans is not normally observed at high molecular weight but may occur with low molecular weight polymers (Wood, 2002), may also have been a confounding factor in the above studies. Dongowski (2007) suggested that the variability of methods used for these studies may also be a major factor contributing to the inconsistency in observations.

Although the importance of solution viscosity in the cholesterol-lowering effect of β-glucans is clear, there is a striking lack of clinical data relating viscosity of β-glucan preparations with serum cholesterol levels. Wood (2002) commented on this problem, and there have been very few reports since then that have addressed it. Wolever and colleagues (Wolever *et al.*, 2010) presented data correlating the viscosity of β-glucans, solubilized from test food (cereal containing oat bran) using *in vitro* digestion methods, and serum LDL cholesterol reductions observed in subjects consuming the cereal. They found a clear inverse relationship between log(viscosity) and serum LDL cholesterol levels after four weeks of dietary intervention.

Currently, the link between viscosity and cholesterol lowering is largely one that has been inferred using data on molecular weight and solubility of β-glucans, both of which affect solution viscosity. Viscosity is related to the product of molecular weight and concentration in solution as an exponential function (Wood, 2001). The solubilization of β-glucans from the food matrix in the GI tract is dependent on processing history and the food matrix itself. Beer and coworkers (Beer *et al.*, 1997b) reported solubilization of oat β-glucans from different matrices, such as bran, porridge, and muffins. The percentage of β-glucan solubilized during *in vitro* digestion ranged from 13% for the bran to 85% for the muffins. Freezing of the muffins for 8 weeks led to a significant decrease (about 50%) in subsequent solubilization. This may be a significant source of variability in clinical trials, as the test foods are often stored frozen for different times before offering them to the subjects.

The type of food matrix is also a significant determinant of the cholesterol-lowering effect. In general, liquid food matrices show greater effects than solid food matrices, probably due to the complete hydration and viscosity development in the former (Nauman *et al.*, 2006; Lazaridou and Biliaderis, 2007; Othman *et al.*, 2011). Kerkchoffs and colleagues (Kerckhoffs *et al.*, 2003) reported that the cholesterol-lowering effect of oat β-glucans is less when they are incorporated into breads or cookies, as compared to incorporation into orange juice. Processing history, including extraction methods and storage conditions, is also a factor affecting the solution properties of β-glucans (Malkki and Virtanen, 2001; Wood, 2002; Lazaridou and Biliaderis, 2007). For instance, endogenous β-glucanases in flour may not be inactivated under mild extraction conditions (50–60°C) (Keogh *et al.*, 2003). This may cause depolymerization of the β-glucans in the food matrix during processing and storage, and thus affect molecular weight and viscosity development. Notably, depolymerization of β-glucans occurs during bread dough fermentation, as a result of endogenous β-glucanase activity in flour, which is favored by the warm, moist conditions (Tosh, 2007). Freeze–thaw cycles can also lead to lowering of solubility by strengthening polymer–polymer interactions. As mentioned before, frozen storage of prepared products can decrease solubilization of β-glucans (Wood, 2002, 2010).

Therefore, molecular weight, structure, solubility, solution viscosity, processing history, and type of food matrix are all interrelated factors that determine the effectiveness of a β-glucan preparation in decreasing serum cholesterol levels. In a notable example of a combination of all of these factors, Torronen and others (Torronen *et al.*, 1992) found in a clinical study that daily intake of 11.2 g β-glucans did not lead to significant lowering of serum cholesterol levels. This was in contrast to several other studies that observed significant cholesterol reduction at lower dosages. The authors suggested that the lack of effect observed was likely to be due to the low molecular weight and/or solubility of the β-glucans, which would lead to lower intestinal viscosity. The study had used an oat bran concentrate, high in β-glucans, incorporated into a bread matrix. Although the authors did not actually measure the solubilization of β-glucans from the bread under physiological conditions, it is possible that only a fraction of the β-glucans was actually solubilized in the GI tract. It is also possible that endogenous β-glucanase activity in the oat bran concentrate (which was extracted using a cold water milling process) and/or the wheat flour degraded the β-glucans, thus decreasing the molecular weight. Because of the low molecular weight and low solubility, it is likely that the intestinal viscosity of the digesta in the test subjects was not very high. Thus, in spite of the large amount of β-glucans provided in the diet, there was no effect on the serum cholesterol levels.

Many such apparent inconsistencies in the scientific literature may be explained by differences in physicochemical properties. Thus, it should be highlighted once again that: (1) *in vitro* studies do not translate very well *in vivo*, due to the lack of practical insights that can be obtained on phenomena occuring in the body, and (2) *in vivo* effects of oat β-glucans on cholesterol levels are then hardly predictable and only very partially linked to strict physicochemical properties of β-glucans. It is, therefore, important that clinical trials investigating the physiological effects of β-glucans must report relevant physicochemical properties of the β-glucan preparations administered.

6.5.3   Attenuation of glucose and insulin response

One of the most widely recognized effects of dietary β-glucans is their ability to suppress postprandial peak plasma glucose and insulin concentrations. Along with the obvious benefit to patients with type 2 diabetes, this attenuated response is also of some benefit in reducing the risk of the disease and insulin insensitivity (Wood, 2002, 2007). Several clinical studies have investigated the mechanisms of the attenuation of the glycemic response (Braaten *et al.* 1991; Tappy *et al.*, 1996; Cavallero *et al.*, 2002; Biorklund *et al.*, 2005; Panahi *et al.*, 2007; Regand *et al.*, 2011). Different levels of attenuation have been reported, depending on the dose and physicochemical properties of the β-glucans and the type of food matrix. For summaries of the actual values of glucose and insulin response attenuation, the reader is referred to reviews by Würsch and Pi-Sunyer (1997), Wood (2010), Tiwari and Cummins (2011), and Cloeten and colleagues (Cloetens *et al.*, 2012). This discussion once again focuses on the various mechanisms considered to be involved in the hypoglycemic effects of dietary β-glucans, followed by a detailed consideration of how physicochemical properties affect these mechanisms.

6.5.3.1   Mechanism of postprandial glucose and insulin level attenuation by β-glucans   Blood glucose response curves after consumption of a meal containing viscous, soluble β-glucans tend to be flatter (lower peak blood glucose values) than the response curves after meals with comparable glycemic carbohydrate content and no β-glucans (Barsanti *et al.*, 2011) (Figure 6.15).

Increased gut viscosity due to hydration of β-glucans is generally considered the major contributor to this effect. Several researchers have commented on the importance of viscosity in this regard (Wood *et al.*, 1994a; Malkki and Virtanen, 2001; Wood, 2002, 2007, 2010; Lazaridou and Biliaderis, 2007; Tosh, 2007; Daou

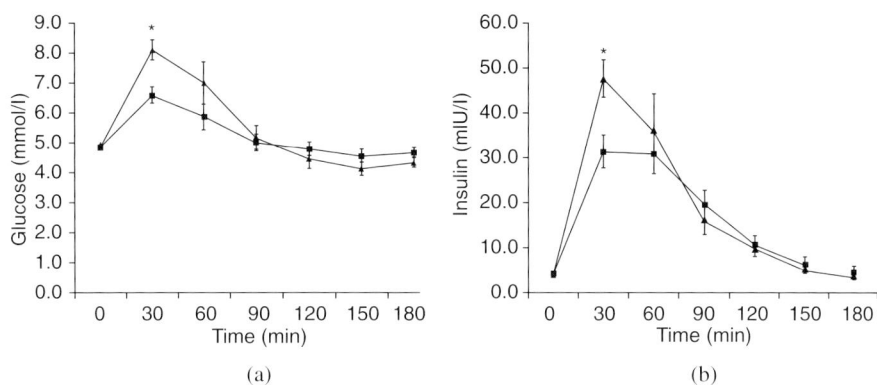

Figure 6.15   Postprandial glucose and insulin response curves of type 2 diabetic subjects after consumption of a beverage containing 5 g oat β-glucans (circles) and a control beverage (triangles). *Source:* Biorklund *et al.* (2005). Reproduced with permission from the Nature Publishing Group.

and Zhang, 2012). The increase in viscosity of the chyme delays gastric emptying, lengthens transit time through the small intestine, and affects nutrient absorption by impairing diffusion through the unstirred water layer in the small intestine (Barsanti *et al.*, 2011). The delayed gastric emptying due to β-glucans, although mentioned by many researchers as a consequence of viscosity, has not been established. Hlebowicz and others (Hlebowicz *et al.*, 2008) found that a muesli containing 4 g oat bran did not delay gastric emptying, although it did lower postprandial blood glucose concentrations. This observation suggests that delaying of gastric emptying is not a significant mechanism involved in glucose response attenuation by β-glucans. Increased gastrointestinal viscosity also affects starch digestion by modulating the activity of pancreatic α-amylase and by limiting access of the substrates to mucosal α-glucosidases (Dunaif and Schneeman, 1981). The net decrease in the rate of glucose absorption into the blood affects pancreatic insulin response, effectively reducing postprandial insulin concentrations in the blood (Daou and Zhang, 2012). The lower insulin response improves insulin sensitivity over time, thus improving glucose metabolism (Behall *et al.*, 2006).

While some studies have attempted to relate consumption of β-glucans to the long-term maintenance of normal blood glucose levels, a recent report by the European Food Safety Authority concluded that there is insufficient evidence currently to support such a claim (Tiwari and Cummins, 2011). Thus, β-glucans, by increasing the viscosity of chyme, have multiple effects on the transit of material through the GI tract and, therefore, on blood glucose and insulin response. Swelling and gelation of β-glucans, leading to entrapment of other food components, can also affect digestibility and influence postprandial glucose and insulin response (Tappy *et al.*, 1996). The effects of dose, molecular weight, and viscosity of β-glucans on the attenuation of blood glucose and insulin response have been studied extensively by many researchers, and will be discussed in detail in a later section.

β-glucans also modify starch digestibility by affecting the properties of the food matrix and by limiting water availability for starch gelatinization (Cleary and

Brennan, 2006). The modification of starch digestibility is dependent not only on the properties of the β-glucans but also on the processing history and food matrix. Regand and coworkers (Regand *et al.*, 2011) studied the *in vitro* starch digestibility of a baked test food containing different amounts and molecular weight of β-glucans at two different starch levels. The high and medium M_w β-glucans (2133 kDa and 435 kDa, respectively) significantly decreased the percentage of rapidly digestible starch and increased the percentage of slowly digestible starch. They also measured the postprandial peak blood glucose response (PBGR) and the area under the glucose response curve (AUC) after consumption of the test foods. The high and medium molecular weight β-glucans significantly decreased both the PBGR and AUC.

6.5.3.2 Effect of physicochemical properties of β-glucans on their hypoglycemic effect   The role of viscosity of β-glucans in their hypoglycemic effect can hardly be overemphasized. Since viscosity is a function of molecular weight and solubility (concentration in solution), these two factors are intimately connected with the hypoglycemic effect. Unlike the hypocholesterolemic effect, there are clinical data available that correlate viscosity (or the product of molecular weight and concentration in solution, as a predictor of viscosity) and the regulation of postprandial glucose and insulin response. Some studies have used different dosage levels in order to study the effect of viscosity. For instance, Tappy and colleagues (Tappy *et al.*, 1996) used 4 g, 6 g, and 8.4 g of oat β-glucans from oat concentrate in an extruded cereal to study plasma glucose and insulin responses in noninsulin-dependent diabetic subjects. They found a significant decrease in glucose response at all three dosage levels but the response in the case of 6 g and 8.4 g β-glucans was significantly lower than that for the 4-g dosage. However, in attributing the differences observed in this study to the different GI viscosities developed in each of the test foods, the authors did not comment on the possible dose-response factor, which is likely to confound the results in such a study.

Biorklund and colleagues (Biorklund *et al.*, 2005) used a combination of different dosage levels and different molecular weight of β-glucans in a beverage model to compare postprandial glucose and insulin responses. They observed a significant decrease in the response when the beverage contained 5 g β-glucan with a M_w of 70 kDa but no significant difference from the fiber-free control with the beverage containing 4 g of a 40 kDa β-glucan. While the most likely explanation, as commented by Wood (2010), is that the viscosity of the latter formulation was much lower (viscosity was not directly measured by the authors), it is also noteworthy that the 40-kDa β-glucan was from barley, whereas the 70-kDa β-glucan was from oats. It is unclear how much effect the difference in chemical structure of these polymers is likely to have on the glucose and insulin response.

Wood and coworkers (Wood *et al.*, 1994a) used different partially hydrolyzed oat β-glucans at different concentrations in a drink to correlate the measured viscosity with the postprandial blood glucose response, while avoiding the confounding effects of dose response. They reported an inverse relationship between viscosity and PBGR, a finding that has been reported since by many other authors using other food systems (Brummer *et al.*, 2012). More recently, Panahi and

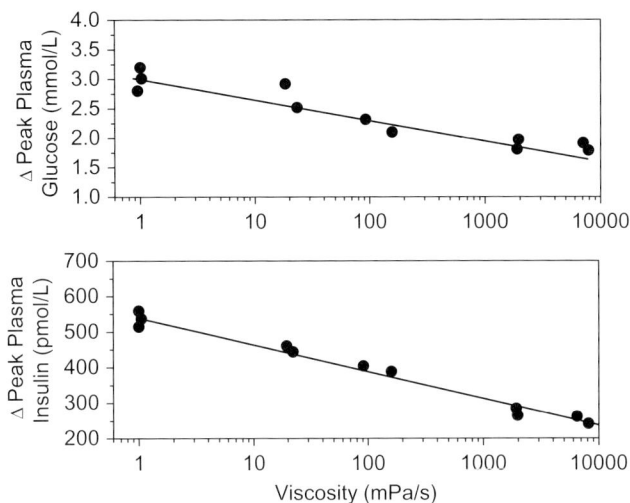

Figure 6.16   Postprandial glucose (a) and insulin (b) responses after an oral glucose load of 50 g in the presence of 7.2 g β-glucan isolate as a function of viscosity. Adapted from Wood *et al.*, 1991c; Wood 1994, 2002.

colleagues (Panahi *et al.*, 2007), using a drink system similar to Wood (1994b) found that while a high molecular weight, high-viscosity β-glucan significantly reduced PBGR, a low-viscosity sample had no significant effect.

Processing history also affects glucose response attenuation by β-glucans. A recent study by Lan-Pidhainy and others (Lan-Pidhainy *et al.*, 2007) found that subjecting oat bran muffins to freeze–thaw cycles reduced β-glucan solubility and significantly altered the hypoglycemic effect. Processing history can also affect the molecular weight of β-glucans as discussed before. Thus, any factor that changes the physicochemical properties in such a manner to influence solution viscosity has an important bearing on the hypoglycemic effect. Figure 6.16 hows the relationship between glucose and insulin responses, after a glucose load of 50 g with a 7.2 g dose of oat β-glucan isolate, and solution viscosity.

It is clear from Figure 6.16 that glucose and insulin response attenuation is an inversely proportional, linear function of log(viscosity). The establishment of such a regression is of great significance to the understanding of the physiological effects of viscous dietary fibers, as discussed by Wood and coworkers (Wood *et al.*, 1991c). The significance to product development is also clear. The desirable hypoglycemic (and hypocholesterolemic) effects of β-glucans can only be obtained if high viscosity is developed *in vivo* in the GI tract. Thus, product development strategies aimed towards this ultimate goal are necessary to obtain a health benefit. However, while such strategies may apply very well *in vitro, in vitro* experiments may not translate very well *in vivo* and the number of factors actually involved in viscosity development *in vivo* (β-glucans physicochemical properties, solubility, matrix environment, chemical environment factors such as pH, temperature and ionic strength, etc.) make the hypolglycemic effect of given β-glucans unlikely to be accurately predicted.

6.5.3.3 Satiety and energy intake   The effect of oat β-glucans on appetite and energy intake is a subject of debate. High gut viscosity, which has been correlated with longer gastric emptying times, gelation, and production of SCFAs in the colon, which stimulates the secretion of glucagon-like peptide-1 (GLP-1, a satiety hormone), are considered the mechanisms by which β-glucans can reduce appetite and energy intake. However, none of these effects has been observed consistently across different human studies. While some studies (Rytter *et al.*, 1996; Urooj *et al.*, 1998; Bourdon *et al.*, 1999; Beck *et al.*, 2009a, b, 2010; Vitaglione *et al.*, 2009; Lyly *et al.*, 2010) have seen positive effects of β-glucans on subjective satiety scores, long-term appetite control, weight loss, and gut hormone levels, other studies (Kim *et al.*, 2006; Hlebowicz *et al.*, 2008; Peters *et al.*, 2009) found no significant effects. Animal studies have indicated that β-glucans delay gastric emptying in a dose-dependent manner (Begin *et al.*, 1989; Johansen *et al.*, 1996, 1997. However, Lia and Andersson (1994) compared gastric emptying rates of porridge containing oat β-glucans (as rolled oats or oat brans) in humans and found no significant difference compared to a wheat semolina porridge control. Mälkki and Virtanen (2001) speculated that a measurement artifact may have contributed to this finding, as the authors used a radioactive tracer that could have separated from the food bolus and emptied faster. In a later study with muesli enriched with 4 g oat bran, Hlebowicz and colleagues (Hlebowicz *et al.*, 2008) found no significant difference in gastric emptying rates of the test meal compared to a control with corn flakes.

For a more detailed review of these and other studies, the reader is referred to reviews by Mälkki and Virtanen (2001) and Cloetens and others (Cloetens *et al.*, 2012). It is interesting to note that these studies have used widely different types of food matrices, including solid and liquid foods, different doses, cereal sources, and molecular weight of β-glucans, and different measures of satiety. No clear trend relating β-glucan dose, physicochemical properties, or food matrix type with the satiety effect was noticed. There is a need for standardization in the reporting of such studies, which would allow more rigorous comparison between them to evaluate whether β-glucans do, in fact, affect appetite and energy intake.

6.5.4   Immunological effects

Recent studies have shown that cereal β-glucans can stimulate the mammalian immune system, indicating a host of possible physiological effects for these polymers, which were not fully recognized previously. Fungal β-glucans, which, unlike cereal β-glucans, consist of a β-(1, 3) glucopyranosyl backbone, have been known to have antitumor effects (Wasser, 2002; Adams *et al.*, 2008; Chan *et al.*, 2009). Barsanti and colleagues (Barsanti *et al.*, 2011) explained the mechanism of modulation of immune responses by fungal β-glucans. A study also found that a gelling barley β-glucan could have effects on the immune system (Lazaridou and Biliaderis, 2007), although these effects have not been understood as specifically as those for fungal β-glucans. A recent cell culture study by Rieder and colleagues (Rieder *et al.*, 2011) investigated cytokine secretion in human intestinal epithelial cell lines as a result of exposure to different concentrations and molecular weights of cereal β-glucans (the exact cereal sources were not mentioned). Estrada and coworkers (Estrada *et al.*, 1997) reported the immunomodulatory activities of oat β-glucans *in vitro* and *in vivo* in a mouse model. In both studies, as well as some others, some dose-dependent effects on cytokine secretion were observed, indicating that cereal β-glucans do have some immunomodulatory activity. Wood (2010) recently reviewed the available evidence for immunomodulatory activities of cereal β-glucans and concluded that while animal and *in vitro* evidence of such activity are emerging, there is currently a lack of human studies pointing to a specific effect of these polymers on the immune system. More studies designed to investigate the effect of dietary cereal β-glucans on immune function will be necessary before such an effect can be established and recognized.

6.5.5 Reporting chemical and rheological data in clinical trials

While current scientific evidence is certainly conclusive in general terms of the positive effects of dietary β-glucans on carbohydrate and lipid metabolism, much of the difficulty in comparing observed effects in various studies stems from a lack of adequate reporting of physicochemical and rheological properties of the β-glucans used. Additionally, dose responses for certain physiological effects of β-glucans have not been clearly established, probably due to the use of different amounts and sources (chemical structures) in the studies, leading to different rheological properties and physiological effects. Tiwari and Cummins (2011) attempted to establish a dose response for blood glucose levels but found too much variability, which led to unpredictable responses (although they were able to establish a dose-response effect for total cholesterol levels).

One of the most obvious omissions in the current literature on this topic is data correlating solution viscosity with serum cholesterol reductions in clinical trials (such data are available for glucose and insulin response attenuation, as seen in Figure 6.16). This is surprising, considering the generally recognized role of viscosity in the hypocholesterolemic effect. Clinical trials often use different dosages, food matrices, and even sources of β-glucans, and whereas viscosity has been considered to be the most important factor affecting physiological response, the actual viscosity values for the test foods are rarely reported. Ren and colleagues (Ren *et al.*, 2003) reported the solution viscosity properties of oat endosperm β-glucans as a function of concentration and specific viscosity and Wood (2010) has used equations to calculate viscosity data for previous studies. Such approaches are valuable for the analysis of the currently available literature, although the need for adequate reporting in the future can hardly be overemphasized. Additionally, since physiological response is a holistic sum of all the processes that occur when a test food containing β-glucans is ingested, including upper GI transit and fermentation in the colon, it is important to consider the effects of fine structure on these properties.

Fermentation in the colon is considered to be one of the mechanisms for the hypocholesterolemic effect of β-glucans. Structural attributes, such as the ratio of (1, 3) to (1, 4) linkages and cellotriosyl to cellotetraosyl residues, can affect fermentation properties. Thus, the chemical structure of the β-glucan administered in the test food will affect the physiological response to it, as observed in clinical trials. This is especially important in trials where β-glucans from different cereal (and fungal) sources may be compared, and adequate reporting of structural features in these cases would be necessary for the complete analysis of the data.

Another significant source of variation among studies stems from differences in the food matrix. While many studies focus on the relationship between β-glucans molecular weight and their physiological effects, β-glucans solubility tends to be overlooked. It has been established that the food matrix greatly affects the amount of β-glucan solubilized in the GI tract. *In vitro* methods simulating upper GI conditions should be used to evaluate solubility in an accurate manner, leading to the consideration of GI viscosity as a result of β-glucans solubility, molecular weight and chemical environment, ratherthan these individual isolated factors. This will provide a more accurate value of the dose of soluble β-glucans and will enable the evaluation of rheological properties *in vitro* as estimates of the likely effects *in vivo*.

6.6   Conclusion and perspectives

Oat soluble dietary fibers are mainly composed of mixed (1,3;1,4)-β-D-glucans, as found in barley and wheat. However, oat β-glucans are unique in terms of their structural features. They are composed of approximately 90% cellotriosyl (DP3 cellulose-like block) and cellotetraosyl (DP4 cellulose-like block) units, in a ratio ranging from 1.5 to 2.3, that are randomly arranged along the polysaccharide chain. Their abundance, generally ranging between 4.5 and 5.5% of the oat groat weight, and the molecular weight, ranging from 6.5×10^4 to 3.1×10^6 Da, are mainly characteristic of oat cultivars and are impacted to a lesser extent by environmental conditions and agronomic practices. β-glucans are primarily located in the starchy endosperm of oat groats and to a slightly lesser extent in the bran. Historically, their detection has been made possible by their unique interaction with dyes such as Calcofluor. This precise interaction also allows their specific precipitation for purification after extraction from the grain. Extraction is probably the most problematic aspect of oat β-glucan study; it is mainly carried out with hot alkali and can be enzymatically assisted. However, oat β-glucans are not fully soluble and the extraction method often leads to degradation and depolymerization, thus affecting further characterization and study of their functional properties. A recovery of 60–80% of the total β-glucan from grain is common.

At the molecular level, oat β-glucans have a random coil conformation and tend to form aggregates in solution. These two features drive their main physical properties: they have a Newtonian flow behavior at low concentrations and a nonNewtonian pseudoplastic behavior above a certain critical concentration. They can form gels upon storage. This gelling ability has been shown to be highly correlated with the abundance of DP3 units along the polysaccharide chain.

All these physicochemical properties comprise the underlying basis for oat β-glucan physiological functions. They influence GI transit through their swelling and water binding capacities, their viscosity in solution and their fermentability. Thus, they slow or lower the digestion and absorption of nutrients and may promote satiety. Their viscosity, bile acid binding capacity, fermentability, and ability to regulate blood insulin levels provide them with the capacity to lower total and LDL cholesterol levels in blood. Oats have thus gained recognition by the FDA for their role in lowering the risk of heart disease. Oat β-glucans also regulate postprandial blood glucose and blood insulin levels via similar mechanisms. Finally, they exhibit immunomodulatory properties. However, more work is needed to fully understand the intimate mechanisms underlying oat β-glucan physiological properties and immunomodulatory capacities. Additionally, given the complexity of the physiological mechanisms of action of β-glucans, the state of science is so that it is not currently possible to establish direct correlations between these mechanisms, their effects *in vivo* (on cholesterol and blood glucose levels, for example) and β-glucans properties (physicochemical properties and solubility as they relate to their physicochemical environment, interactions with colonic microbiota, for example).

On the way to a better understanding of oat β-glucans, adequate reporting of physicochemical properties in clinical trials is necessary for such relationships to be established and recognized, and to allow easier comparisons between studies for conclusions to be drawn. It is important for these relationships to be recognized by researchers as well as product developers, so that the health benefits of β-glucans may be preserved in food products. This perspective may lead to new concepts for oat β-glucans as functional ingredients relative to their targeted health benefits.

Finally, we wish to recognize here the work of a pioneer and major contributor in the current knowledge on cereal β-glucans, Peter J. Wood, former researcher at Agriculture and Agrifood Canada, who passed away in 2011 and whose legacy on β-glucans knowledge remains tremendous.

References

Adams, E. L. *et al.* (2008) Differential high-affinity interaction of Dectin-1 with natural or synthetic glucans is dependent upon primary structure and is influenced by polymer chain length and side-chain branching. *Journal of Pharmacology and Experimental Therapeutics* **325**(1): 115–123.

Agbenorhevi, J. K. *et al.* (2011) Rheological and microstructural investigation of oat β-glucan isolates varying in molecular weight. *International Journal of Biological Macromolecules* **49**(3): 369–377.

Ahmad, A. *et al.* (2010) Extraction and characterization of β-d-glucan from oat for industrial utilization. *International Journal of Biological Macromolecules* **46**(3): 304–309.

Anderson, J. W. and J. Tietyen-Clark (1986) Dietary fiber: Hyperlipidemia, hypertension, and coronary heart disease. *American Journal of Gastroenterology* **81**: 907–919.

Anderson, J. W. *et al.* (1984) Hypocholesterolemic effects of oat bran or bean intake for hypercholesterolemic men. *American Journal of Clinical Nutrition* **40**: 1146–1155.

Anderson, J. W. *et al.* (1990) Dietary fiber and coronary heart disease. *Critical Reviews in Food Science and Nutrition* **29**: 95–147.

Andersson, A. A. M. and D. Börjesdotter (2011) Effects of environment and variety on content and molecular weight of β-glucan in oats. *Journal of Cereal Science* **54**(1): 122–128.

Annapurna, A. (2011) Health benefits of barley. *Journal of Pharmaceutical Research and Health Care* **3**(2): 22.

Anonymous (1997) Food labeling: Soluble dietary fiber from certain foods and coronary heart disease. *Federal Register* **67**: 61773–61783.

Bae, I. Y. *et al.* (2010) Effect of enzymatic hydrolysis on cholesterol-lowering activity of oat β-glucan. *New Biotechnology* **27**(1): 85–88.

Barsanti, L. *et al.* (2011) Chemistry, physico-chemistry and applications linked to biological activities of beta-glucans. *Natural Product Reports* **28**: 457–466.

Battilana, P. *et al.* (2001) Mechanisms of action of beta-glucan in postprandial glucose metabolism in healthy men. *European Journal of Clinical Nutrition* **55**: 327–333.

Beck, E. J. *et al.* (2009a) Increases in peptide Y-Y levels following oat b-glucan ingestion are dose-dependent in overweight adults. *Nutrition Research* **29**: 705–709.

Beck, E. J. *et al.* (2009b) Oat beta-glucan increases postprandial cholecystokinin levels, decreases insulin response and extends subjective satiety in overweight subjects. *Molecular Nutrition and Food Research* **53**: 1343–1351.

Beck, E. J. *et al.* (2010) Oat beta-glucan supplementation does not enhance the effectiveness of an energy-restricted diet in overweight women. *British Journal of Nutrition* **103**: 1212–1222.

Beer, M. U., P. J. Wood, *et al.*, (1997a) Molecular Weight Distribution and (1→3)(1→4)-β-D-Glucan Content of Consecutive Extracts of Various Oat and Barley Cultivars. *Cereal Chem.* **74**(4): 476–480.

Beer, M. U. *et al.* (1997b) Effect of cooking and storage on the amount and molecular weight of (1,3)(1,4)-beta-D-glucan extracted from oat products by an in vitro digestion system. *Cereal Chemistry* **74**: 705–709.

Begin, F. *et al.* (1989) Effect of dietary fibres on glycemia and insulinemia and on gastrointestinal function in rats. *Canadian Journal of Physiology and Pharmacology* **67**: 1265–1271.

Behall, K. M. *et al.* (2006) Barley beta-glucan reduces plasma glucose and insulin responses compared with resistant starch in men. *Nutrition Research* **26**: 644–650.

Bellato, S. *et al.* (2011) Use of near infrared reflectance and transmittance coupled to robust calibration for the evaluation of nutritional value in naked oats. *Journal of Agricultural and Food Chemistry* **59**(9): 4349–4360.

BeMiller, J. N. (2007a) Carbohydrate nutrition, dietary fiber, bulking agents, and fat mimetics. In: *Carbohydrate Chemistry for Food Scientists*, 2nd edn (ed. J. N. BeMiller). AACC International Press, St. Paul, MN, pp. 321–346.

BeMiller, J. N. (2007b) Polysaccharides: Properties. In: *Carbohydrate Chemistry for Food Scientists*, 2nd edn (ed. J. N. BeMiller). AACC International Press, St. Paul, MN, pp. 119–172.

Benito-Román, Ó. *et al.* (2013) Ultrasound-assisted extraction of β-glucans from barley. *LWT – Food Science and Technology* **50**(1): 57–63.

Bhatty, R. S. (1992) Total and extractable β-glucan contents of oats and their relationship to viscosity. *Journal of Cereal Science* **15**(2): 185–192.

Biorklund, M. *et al.* (2005) Changes in serum lipids and postprandial glucose and insulin concentrations after consumption of beverages with beta-glucans from oats or barley: A randomized dose-controlled trial. *European Journal of Clinical Nutrition* **59**: 1272–1281.

Bohm, N. and W. M. Kulicke (1999) Rheological studies of barley (1→3)(1→4)-β-glucan in concentrated solution: Mechanistic and kinetic investigation of the gel formation. *Carbohydrate Research* **315**(3–4): 302–311.

Bourdon, I. *et al.* (1999) Postprandial lipid, glucose, insulin, and cholecystokinin responses in men fed barley pasta enriched with beta-glucan. *American Journal of Clinical Nutrition* **69**: 55–63.

Bowles, R. K. *et al.* (1996) 13C CP-MAS NMR study of the interaction of bile acids with barley beta-D-glucan. *Carbohydrate Polymers* **29**: 7–10.

Braaten, J. T. *et al.* (1991) Oat gum lowers glucose and insulin after an oral glucose load. *American Journal of Clinical Nutrition* **53**: 1425–1430.

Brummer, Y. *et al.* (2012) Glycemic response to extruded oat bran cereals processed to vary in molecular weight. *Cereal Chemistry* **89**(5): 255–261.

Buckeridge, M. S. *et al.* (1999) The mechanism of synthesis of a mixed-linkage (1→3),(1→4)β-D-glucan in maize. Evidence for multiple sites of glucosyl transfer in the synthase complex. *Plant Physiology* **120**(4): 1105–1116.

Buckeridge, M. S. *et al.* (2001) Insight into multi-site mechanisms of glycosyl transfer in (1→4) β-d-glycans provided by the cereal mixed-linkage (1→3),(1→4)β-d-glucan synthase. *Phytochemistry* **57**(7): 1045–1053.

Buckeridge, M. S. *et al.* (2004) Mixed Linkage (1→3),(1→4)-β-D-Glucans of Grasses. *Cereal Chemistry* **81**(1): 115–127.

Burton, R. A. *et al.* (2010) Heterogeneity in the chemistry, structure and function of plant cell walls. *Nature Chemical Biology* **6**(10): 724–732.

Cavallero, A. *et al.* (2002) High (1,3 1,4)-beta-glucan barley fractions in bread making and their effects on human glycemic response. *Journal of Cereal Science* **36**: 59–66.

Chan, G. C. F. *et al.* (2009) The effects of β-glucan on human immune and cancer cells. *Journal of Hematology and Oncology* **2**: 25–36.

Chen, H.-L. *et al.* (1998) Mechanisms by which wheat bran and oat bran increase stool weight in humans. *American Journal of Clinical Nutrition* **68**: 711–719.

Cho, K. C. and P. J. White (1993) Enzymatic analysis of beta-glucan content in different oat genotypes. *Cereal Chemistry* **70**(5): 539–542.

Cleary, L. and C. Brennan (2006) The influence of a (1–3)(1–4)-beta-D-glucan rich fraction from barley on the physico-chemical properties and in vitro reducing sugars release of durum wheat pasta. *International Journal of Food Science and Technology* **41**: 910–918.

Cloetens, L. *et al.* (2012) Role of dietary beta-glucans in the prevention of metabolic syndrome. *Nutrition Reviews* **70**: 444–458.

Colleoni-Sirghie, M. *et al.* (2003a) Structural features of water soluble (1,3) (1,4)-β-D-glucans from high-β-glucan and traditional oat lines. *Carbohydrate Polymers* **54**(2): 237–249.

Colleoni-Sirghie, M. *et al.* (2003b) Rheological and molecular properties of water soluble (1,3) (1,4)-β-D-glucans from high-β-glucan and traditional oat lines. *Carbohydrate Polymers* **52**(4): 439–447.

Colleoni-Sirghie, M. *et al.* (2004) Prediction of β-glucan concentration based on viscosity evaluations of raw oat flours from high β-glucan and traditional oat lines. *Cereal Chemistry* **81**(4): 434–443.

Cui, S. W. and Q. Wang (2009) Cell wall polysaccharides in cereals: Chemical structures and functional properties. *Structural Chemistry* **20**(2): 291–297.

Cui, W. and P. J. Wood (2000) Relationships between structural features, molecular weight and rheological properties of cereal β-D-glucans. In: *Hydrocolloids Vol. 1. Physical Chemistry and Industrial Applications of Gels, Polysaccharides and Proteins* (ed. N. Katsuyoshi). Elsevier Science, Amsterdam, The Netherlands, pp. 159–168.

Daou, C. and H. Zhang (2012) Oat beta-glucan: its role in health promotion and prevention of diseases. *Comprehensive Reviews in Food Science and Food Safety* **11**(4): 355–365.

Demirbas, A. (2005) β-Glucan and mineral nutrient contents of cereals grown in Turkey. *Food Chemistry* **90**(4): 773–777.

Doehlert, D. C. and S. Simsek (2012) Variation in β-glucan fine structure, extractability, and flour slurry viscosity in oats due to genotype and environment. *Cereal Chemistry* **89**(5): 242–246.

Doehlert, D. C. *et al.* (2012) Extraction of β-glucan from oats for soluble dietary fiber quality analysis. *Cereal Chemistry* **89**(5): 230–236.

Dongowski, G. (2007) Interactions between dietary fiber-rich preparations and glycoconjugated bile acids in vitro. *Food Chemistry* **104**: 390–397.

Doublier, J. L. and P. J. Wood (1995) Rheological properties of aqueous solutions of (1,3)(1,4)-beta-D-glucan from oats (*Avena sativa* L.). *Cereal Chemistry* **72**(4): 335–340.

Drzikova, B. *et al.* (2005) The composition of dietary fiber-rich extrudates from oat affects bile acid binding and fermentation in vitro. *Food Chemistry* **90**: 181–192.

Dunaif, G. and B. O. Schneeman (1981) The effect of dietary fiber on human pancreatic enzyme activity in vitro. *American Journal of Clinical Nutrition* **34**: 1034–1035.

Estrada, A. *et al.* (1997) Immunomodulatory activities of oat beta-glucan *in vitro* and *in vivo*. *Microbiology and Immunology* **41**: 991–998.

Fincher, G. B. (2009) Revolutionary times in our understanding of cell wall biosynthesis and remodeling in the grasses. *Plant Physiology* **149**(1): 27–37.

Fincher, G. B. (2011) *Fine Structure of Polysaccharides from Plant Cell Walls: From Human Health to Biofuel Production.* Belfort Lecture 2011, Whistler Center for Carbohydrate Research, Purdue University, West Lafayette, IN.

Fox, G. J. (1998) Long chained beta glucan isolates derived from viscous barley grain. Barkley Seeds, Inc., Yuma, AZ.

Genç, H. *et al.* (2001) Analysis of mixed-linked (1→3), (1→4)-β-D-glucans in cereal grains from Turkey. *Food Chemistry* **73**(2): 221–224.

Gómez, C. *et al.* (1997a) Physical and structural properties of barley (1→3),(1→4)-β-D-glucan. Part I. Determination of molecular weight and macromolecular radius by light scattering. *Carbohydrate Polymers* **32**(1): 7–15.

Gómez, C. *et al.* (1997b) Physical and structural properties of barley (1→3),(1→4)-β-D-glucan. Part II. Viscosity, chain stiffness and macromolecular dimensions. *Carbohydrate Polymers* **32**(1): 17–22.

Gómez, C. *et al.* (1997c) Physical and structural properties of barley (1→3),(1→4)-β-D-glucan – III. Formation of aggregates analysed through its viscoelastic and flow behaviour. *Carbohydrate Polymers* **34**(3): 141–148.

Gómez, C. *et al.* (2000) Determination of the apparent molecular weight cut-off for the fluorimetric calcofluor-FIA method when detecting (1→3),(1→4)-β-D-glucan using a high ionic strength eluant. *Journal of Cereal Science* **31**(2): 155–157.

Hague, A. *et al.* (1993) Sodium butyrate induces apoptosis in human colonic tumour cell lines in a p53-independent pathway: Implications for the possible role of dietary fibre in the prevention of large-bowel cancer. *International Journal of Cancer* **55**(3): 498–505.

Hague, A. *et al.* (1995) Apoptosis in colorectal tumour cells: Induction by the short chain fatty acids butyrate, propionate and acetate and by the bile salt deoxycholate. *International Journal of Cancer* **60**(3): 400–406.

Håkansson, A. *et al.* (2012) Asymmetrical flow field-flow fractionation enables the characterization of molecular and supramolecular properties of cereal β-glucan dispersions. *Carbohydrate Polymers* **87**(1): 518–523.

Havrlentová, M. and J. Kraic (2006) Content of β-D-glucan in cereal grains. *Journal of Food and Nutrition Research* **45**(3): 97–103.

Hellweg, J. H. *et al.* (2011) Methods for preparing oat bran enriched in beta-glucan and oat products prepared therefrom. General Mills IP Holdings II, LLC, Minneapolis, MN.

Hlebowicz, J. *et al.* (2008) Effect of muesli with 4 g oat beta-glucan on postprandial blood glucose, gastric emptying and satiety in healthy subjects: A randomized crossover trial. *Journal of the American College of Nutrition* **27**: 470–475.

Hughes, S. A. *et al.* (2008) *In vitro* fermentation of oat and barley derived beta-glucans by human fecal microbiota. *FEMS Microbiology Ecology* **64**: 482–493.

Hutchinson, J. B. *et al.* (1951) Location and destruction of lipase in oats. *Nature* **167**(4254): 758–759.

Izydorczyk, M. S. *et al.* (1998) Fractionation of oat (1→3), (1→4)-β-d-glucans and characterisation of the fractions 1. *Journal of Cereal Science* **27**(3): 321–325.

Jenkins, D. J. A. *et al.* (1982) Relationship between rate of digestion of foods and postprandial glycemia. *Diabetologia* **22**: 450–455.

Jenkins, D. J. A. *et al.* (1989) Nibbling versus gorging: Metabolic advantages of increased meal frequency. *New England Journal of Medicine* **321**: 929–934.

Jensen, S. Å. and S. Aastrup (1981) A fluorimetric method for measuring 1,3:1,4-β-glucan in beer, wort, malt and barley by use of Calcofluor. *Carlsberg Research Communications* **46**(1–2): 87–95.

Jiang, G. and T. Vasanthan (2000) MALDI-MS and HPLC quantification of oligosaccharides of lichenase-hydrolyzed water-soluble β-glucan from ten barley varieties. *Journal of Agricultural and Food Chemistry* **48**(8): 3305–3310.

Johansen, H. N. *et al.* (1993) Molecular weight changes in the mixed linkage-beta-D-glucan of oats incurred by the digestive processes in the upper gastrointestinal tract of pigs. *Journal of Agricultural and Food Chemistry* **41**: 2347–2352.

Johansen, H. N. *et al.* (1996) Effects of varying content of soluble dietary fiber from wheat flour and oat milling fractions on gastric emptying in pigs. *British Journal of Nutrition* **75**: 339–351.

Johansen, H. N. *et al.* (1997) Physicochemical properties and the degradation of oat bran polysaccharides in the gut of pigs. *Journal of the Science of Food and Agriculture* **73**: 81–92.

Johansson, L. (2006) Structural analyses of (1→3),(1→4)-β-D-glucan of oats and barley. Academic Dissertation, University of Helsinki, Finland.

Johansson, L. *et al.* (2000) Structural characterization of water soluble β-glucan of oat bran. *Carbohydrate Polymers* **42**(2): 143–148.

Jørgensen, K. G. (1983) An improved method for determining β-glucan in wort and beer by use of Calcofluor. *Carlsberg Research Communications* **48**(5): 505–516.

Jørgensen, K. G. (1988) Quantification of high molecular weight (1→3)(1→4)-β-d-glucan using Calcofluor complex formation and flow injection analysis. I. analytical principle and its standardization. *Carlsberg Research Communications* **53**(5): 277–285.

Jørgensen, K. G. and S. Aastrup (1988) Quantification of high molecular weight (1→3)(1→4)-β-d-glucan using Calcofluor complex formation and flow injection analysis. II. determination of total β-glucan content of barley and malt. *Carlsberg Research Communications* **53**(5): 287–296.

Kazi, T. and T. J. Cahill (1969) A rapid method for the detection of residual lipase activity in oat products. *The Analyst* **94**(1118): 417.

Kelly, S. A. *et al.* (2007) Whole grain cereals for coronary heart disease. *Cochrane Database of Systematic Reviews* 2 (Art. No.: CD005051). doi: 10.1002/14651858 .CD005051.pub2.

Keogh, G. F. *et al.* (2003) Randomized controlled cross-over study of the effect of a highly beta-glucan-enriched barley on cardiovascular disease risk factors in mildly hypercholesterolemic men. *American Journal of Clinical Nutrition* **78**: 711–718.

Kerckhoffs, D. A. J. M. *et al.* (2003) Cholesterol-lowering effect of beta-glucan from oat bran in mildly hypercholesterolemic subjects may decrease when incorporated into bread and cookies. *American Journal of Clinical Nutrition* **78**: 221–227.

Kim, H. J. and P. J. White (2009) In vitro fermentation of oat flours from typical and high beta-glucan oat lines. *Journal of Agricultural and Food Chemistry* **57**: 7529–7536.

Kim, H. J. and P. J. White (2010) *In vitro* bile acid binding and fermentation of high, medium and low molecular weight beta-glucan. *Journal of Agricultural and Food Chemistry* **58**: 628–634.

Kim, H. J. and P. J. White (2011) Molecular weight of β-glucan affects physical characteristics, *in vitro* bile acid binding, and fermentation of muffins. *Cereal Chemistry* **88**: 64–71.

Kim, H. *et al.* (2006) Short-term satiety and glycemic response after consumption of whole grains with various amounts of beta-glucans. *Cereal Food World* **51**: 29–33.

Kim, S. *et al.* (2008) Content and molecular weight distribution of oat β-glucan in oatrim, nutrim, and C-trim products. *Cereal Chemistry* **85**(5): 701–705.

Kivelä, R. (2011) Non-Enzymatic Degradation of (1→3)(1→4)-β-D-Glucan in Aqueous Pocessing of Oats. Academic Dissertation, University of Helsinki, Finland.

Kivelä, R. *et al.* (2009) Degradation of cereal beta-glucan by ascorbic acid induced oxygen radicals. *Journal of Cereal Science* **49**(1): 1–3.

Kivelä, R. *et al.* (2010) Influence of homogenisation on the solution properties of oat β-glucan. *Food Hydrocolloids* **24**(6–7): 611–618.

Kivelä, R. *et al.* (2011) Oxidative and radical mediated cleavage of β-glucan in thermal treatments. *Carbohydrate Polymers* **85**(3): 645–652.

Kivelä, R. *et al.* (2012) Oxidation of oat β-glucan in aqueous solutions during processing. *Carbohydrate Polymers* **87**(1): 589–597.

Kumar, V., A. K. Sinha, *et al.* (2012) Dietary Roles of Non-Starch Polysachharides in Human Nutrition: A Review. *Critical Reviews in Food Science and Nutrition* **52**(10): 899–935.

Kvist, S. and J. M. Lawther (2005) Concentration of Beta-Glucans. Biovelop International B.V., Amstelveen, The Netherlands.

Lan-Pidhainy, X. *et al.* (2007) Reducing beta-glucan solubility in oat bran muffins by freeze-thaw treatment attenuates its hypoglycemic effect. *Cereal Chemistry* **84**: 512–517.

Lazaridou, A. and C. G. Biliaderis (2007) Molecular aspects of cereal beta-glucan functionality: Physical properties, technological applications and physiological effects. *Journal of Cereal Science* **46**: 101–118.

Lazaridou, A. *et al.* (2003) Molecular size effects on rheological properties of oat β-glucans in solution and gels. *Food Hydrocolloids* **17**(5): 693–712.

Lazaridou, A. *et al.* (2004) A comparative study on structure-function relations of mixed-linkage (1→3), (1→4) linear β-D-glucans. *Food Hydrocolloids* **18**(5): 837–855.

Li, W. *et al.* (2006) Solution and conformational properties of wheat β-D-glucans studied by light scattering and viscometry. *Biomacromolecules* **7**(2): 446–452.

Li, W. *et al.* (2012) Study of conformational properties of cereal β-glucans by computer modeling. *Food Hydrocolloids* **26**(2): 377–382.

Lia, A. and H. Andersson (1994) Glycemic response and gastric emptying rate of oat bran and semolina porridge meals in diabetic subjects. *Scandinavian Journal of Nutrition* **38**: 154–158.

Lia, A. *et al.* (1995) Oat beta-glucan increases bile acid excretion and a fibre-rich barley fraction increases cholesterol excretion in ileostomy subjects. *American Journal of Clinical Nutrition* **62**: 1245–1251.

Lia, A. *et al.* (1996) Substrates available for colonic fermentation from oat, barley and wheat bread diets: A study in ileostomy subjects. *British Journal of Nutrition* **76**: 797–808.

Lia, A. *et al.* (1997) Postprandial lipemia in relation to sterol and fat excretion in ileostomy subjects given oat-bran and wheat test meals. *American Journal of Clinical Nutrition* **66**: 357–365.

Lund, E. K. *et al.* (1989) Effect of oat gum on the physical properties of the gastrointestinal contents and on the uptake of D-galactose and cholesterol by rat small intestine *in vitro*. *British Journal of Nutrition* **62**: 92–101.

Lyly, M. *et al.* (2010) The effect of fiber amount, energy level, and viscosity of beverages containing oat fibre supplement on perceived satiety. *Food and Nutrition Research* **54**: 2149.

Mäkinen, O. E. *et al.* (2012) Formation of oxidising species and their role in the viscosity loss of cereal beta-glucan extracts. *Food Chemistry* **132**(4): 2007–2013.

Malkki, Y. and E. Virtanen (2001) Gastrointestinal effects of oat bran and oat gum: A review. *Lebensmittel Wissenschaft und Technologie* **34**: 337–347.

McCleary, B. V. (2000) Importance of enzyme purity and activity in the measurement of total dietary fiber and dietary fiber components. *Journal of AOAC International* **83**(4): 997–1005.

McCleary, B. V. and R. Codd (1991) Measurement of $(1\rightarrow3),(1\rightarrow4)$-β-D-glucan in barley and oats: A streamlined enzymic procedure. *Journal of the Science of Food and Agriculture* **55**(2): 303–312.

McCleary, B. V. and M. Glennie-Holmes (1985) Enzymic quantification of (1–3)(1–4)-β-D-glucan in barley and malt. *Journal of the Institute of Brewing* **91**: 285–295.

McCleary, B. V. and D. C. Mugford (1992) Interlaboratory evaluation of β-glucan analysis methods. Fourth International Oat Conference, Adelaide, Australia.

McCleary, B. V. and D. C. Mugford (1997) Determination of β-glucan in barley and oats by streamlined enzymatic method: summary of collaborative study. *Journal of AOAC International* **80**(3): 580–583.

McCleary, B. V. and E. J. Nurthen (1986) Measurement of (1–3)(1–4)-β-D-glucan in malt, wort and beer. *Journal of the Institute of Brewing* **92**: 168–173.

Mikkelsen, M. S. *et al.* (2010) Comparative spectroscopic and rheological studies on crude and purified soluble barley and oat β-glucan preparations. *Food Research International* **43**(10): 2417–2424.

Mikkelsen, M. S. *et al.* (2013) Molecular structure of large-scale extracted β-glucan from barley and oat: Identification of a significantly changed block structure in a high β-glucan barley mutant. *Food Chemistry* **136**(1): 130–138.

Miller, S. S. and R. G. Fulcher (1994) Distribution of (1–3),(1–4)-beta-D-glucan in kernels of oats and barley using microspectrofluorometry. *Cereal Chemistry* **71**(1): 64–68.

Miller, S. S. *et al.* (1993a) Oat β-glucans: An evaluation of eastern Canadian cultivars and unregistered lines. *Canadian Journal of Plant Science* **73**(2): 429–436.

Miller, S. S. *et al.* (1993b) Mixed linkage beta-glucan, protein content, and kernel weight in avena species. *Cereal Chemistry* **70**(2): 231–233.

Miller, S. S. *et al.* (1995) Oat Endosperm Cell Walls: I. Isolation, Composition, and Comparison with Other Tissues. *Cereal Chem.* **72**(5): 421–425.

Morgan, K. R. (2003) Beta-Glucan Products and Extraction Processes from Cereals. Gracelinc Limited, Lower Hutt, NZ.

Morris, E. R. (1989) Polysaccharide solution properties: Origin, rheological characterization and implications for food systems. In: *Frontiers in Carbohydrate Research. 1: Food Applications* (eds R.P. Millane, J.N. BeMiller, and R. Chandrasekaran). Elsevier, London, pp. 132–163.

Nauman, E. *et al.* (2006) Beta-glucan incorporated into a fruit drink effectively lowers serum LDL-cholesterol concentrations. *American Journal of Clinical Nutrition* **83**: 601–605.

Othman, R. A. *et al.* (2011) Cholesterol-lowering effects of oat beta-glucan. *Nutrition Reviews* **69**: 299–309.

Panahi, S. *et al.* (2007) Beta-glucan from two sources of oat concentrates affect postprandial glycemia in relation to the level of viscosity. *Journal of the American College of Nutrition* **26**: 639–644.

Papageorgiou, M. *et al.* (2005) Water extractable (1→3,1→4)-β-D-glucans from barley and oats: An intervarietal study on their structural features and rheological behaviour. *Journal of Cereal Science* **42**(2): 213–224.

Peng, L. *et al.* (2002) Sitosterol-β-glucoside as primer for cellulose synthesis in plants. *Science* **295**(5552): 147–150.

Peters, H. P. F. *et al.* (2009) No effect of added beta-glucan or of fructooligosaccharide on appetite or energy intake. *American Journal of Clinical Nutrition* **89**: 58–63.

Pettolino, F. A. *et al.* (2012) Determining the polysaccharide composition of plant cell walls. *Nature Protocols* **7**(9): 1590–1607.

Pihlajamaki, J. *et al.* (2004) Insulin resistance is associated with increased cholesterol synthesis and decreased cholesterol absorption in normoglycemic men. *Journal of Lipid Research* **45**: 507–512.

Potter, R. C. *et al.* (2003) Method for concentrating beta-glucan film. Nurture, Inc., Missoula, MT.

Rampitsch, C. *et al.* (2006) Early generation β-glucan selection in oat using a monoclonal antibody-based enzyme-linked immunosorbent assay. *Cereal Chemistry* **83**(5): 510–512.

Redmond, M. J. and D. A. Fielder (2011) Oat extracts: refining, compositions and methods of use. CEAPRO, Inc., Edmonton, AB.

Regand, A. *et al.* (2011) The molecular weight, solubility and viscosity of oat beta-glucan affect human glycemic response by modifying starch digestibility. *Food Chemistry* **129**: 297–304.

Ren, Y. *et al.* (2003) Dilute and semi-dilute solution properties of (1→3), (1→4)-β-D-glucan, the endosperm cell wall polysaccharide of oats (*Avena sativa* L.). *Carbohydrate Polymers* **53**(4): 401–408.

Rieder, A. *et al.* (2011) Cereal beta-glucan preparations of different molecular weights induce variable cytokine secretion in human intestinal epithelial cell lines. *Food Chemistry* **128**: 1037–1043.

Rimsten, L. *et al.* (2003) Determination of α-glucan molecular weight using SEC with calcofluor detection in cereal extracts. *Cereal Chemistry* **80**(4): 485–490.

Robertson, J. A. *et al.* (1997) Solubilization of mixed linkage (1–3)(1–4)-beta-D-glucans from barley: Effects of cooking and digestion. *Journal of Cereal Science* **25**: 275–283.

Roubroeks, J. P. *et al.* (2000) Molecular weight, structure, and shape of oat (1→3),(1→4)-β-D-glucan fractions obtained by enzymatic degradation with lichenase. *Biomacromolecules* **1**(4): 584–591.

Roubroeks, J. P. *et al.* (2001) Molecular weight, structure and shape of oat (1→3),(1→4)-β-D-glucan fractions obtained by enzymatic degradation with (1→4)-β-D-glucan 4-glucanohydrolase from *Trichoderma reesei*. *Carbohydrate Polymers* **46**(3), 275–285.

Rytter, E. *et al.* (1996) Changes in plasma insulin, enterostatin, and lipoprotein levels during an energy-restricted dietary regimen including a new oat-based liquid food. *Annals of Nutrition and Metabolism* **40**: 212–220.

Ryu, J. H. *et al.* (2012) Effects of barley and oat β-glucan structures on their rheological and thermal characteristics. *Carbohydrate Polymers* **89**(4): 1238–1243.

Saastamoinen, M. *et al.* (1992) Genetic and environmental variation in β-glucan content of oats cultivated or tested in Finland. *Journal of Cereal Science* **16**(3): 279–290.

Sayar, S. *et al.* (2004) *In vitro* bile acid binding of flours from oat lines varying in percentage and molecular weight distribution of beta-glucan. *Journal of Agricultural and Food Chemistry* **53**: 8797–8803.

Sibakov, J. *et al.* (2012) Minireview: β-Glucan extraction methods from oats. *Agro Food Industry Hi-Tech* **23**(1): 10–12.

Sikora, P. *et al.* (2013) Identification of high β-glucan oat lines and localization and chemical characterization of their seed kernel β-glucans. *Food Chemistry* **137**(1–4): 83–91.

Skendi, A. *et al.* (2003) Structure and rheological properties of water soluble β-glucans from oat cultivars of Avena sativa and Avena bysantina. *Journal of Cereal Science* **38**(1): 15–31.

Somerville, C. *et al.* (2004) Toward a systems approach to understanding plant cell walls. *Science* **306**(5705): 2206–2211.

Staudte, R. G. *et al.* (1983) Water-soluble (1→3), (1→4)-β-d-glucans from barley (Hordeum vulgare) endosperm. III. Distribution of cellotriosyl and cellotetraosyl residues. *Carbohydrate Polymers* **3**(4): 299–312.

Sundberg, B. *et al.* (1996) Mixed-linked beta-glucan from breads of different cereals is partly degraded in the human ileostomy model. *American Journal of Clinical Nutrition* **64**: 878–885.

Takenaka, O. and K. Shibata (1969) States of amino acid residues in proteins: XX. fluorescence of stilbene dyes adsorbed on hydrophobic regions of protein molecules. *Journal of Biochemistry* **66**(6): 805–814.

Tappy, L. *et al.* (1996) Effects of breakfast cereals containing various amounts of beta-glucan fibers on plasma glucose and insulin response in NIDDM subjects. *Diabetes Care* **19**: 831–834.

Tiwari, U. and E. Cummins (2009) Simulation of the factors affecting β-glucan levels during the cultivation of oats. *Journal of Cereal Science* **50**(2): 175–183.

Tiwari, U. and E. Cummins (2011) Meta-analysis of the effect of beta-glucan intake on blood cholesterol and glucose levels. *Nutrition* **27**: 1008–1016.

Tobin, K. A. R. *et al.* (2002) Liver X receptors as insulin-mediating factors in fatty acid and cholesterol biosynthesis. *Journal of Biological Chemistry* **277**: 10691–10697.

Torronen, R. *et al.* (1992) Effects of oat bran concentrate on serum lipids in free-living men with mild to moderate hypercholesterolemia. *European Journal of Clinical Nutrition* **46**: 621–627.

Tosh, S. M. (2007) Factors affecting bioactivity of cereal beta-glucans. In: *Dietary Fiber Components and Functions* (eds H. Salovaara, F. Gates, and M. Tenkanen). Wageningen Academic Publishers Wageningen, The Netherlands, pp. 75–89.

Tosh, S. M. *et al.* (2003) Gelation characteristics of acid-hydrolyzed oat beta-glucan solutions solubilized at a range of temperatures. *Food Hydrocolloids* **17**(4): 523–527.

Tosh, S. M. *et al.* (2004a) Evaluation of structure in the formation of gels by structurally diverse (1→3)(1→4)-β-D-glucans from four cereal and one lichen species. *Carbohydrate Polymers* **57**(3): 249–259.

Tosh, S. M. *et al.* (2004b) Structural characteristics and rheological properties of partially hydrolyzed oat β-glucan: The effects of molecular weight and hydrolysis method. *Carbohydrate Polymers* **55**(4): 425–436.

Urooj, A. *et al.* (1998) Effect of barley incorporation in bread on its quality and glycemic responses in diabetics. *International Journal of Food Science and Nutrition* **49**: 265–270.

Van Lengerich, B. H. *et al.* (2004) Beta-glucan compositions and process therefore. General Mills, Inc., Minneapolis, MN.

Vårum, K. M. and O. Smidsrød (1988) Partial chemical and physical characterisation of (1→3),(1→4)-β-d-glucans from oat (*Avena sativa* L.) aleurone. *Carbohydrate Polymers* **9**(2): 103–117.

Vårum, K. M. *et al.* (1992) Light scattering reveals micelle-like aggregation in the (1→3),(1→4)-β-D-glucans from oat aleurone. *Food Hydrocolloids* **5**(6): 497–511.

Vasanthan, T. and F. Temelli (2008) Grain fractionation technologies for cereal beta-glucan concentration. *Food Research International* **41**(9): 876–881.

Vasanthan, T. *et al.* (2010) Preparation of high viscosity beta-glucan concentrates. United States, The Governors of the University of Alberta (Edmonton, AB). **US 7,662,418 B2**.

Vitaglione, P. *et al.* (2009) Beta-glucan-enriched bread reduces energy intake and modifies plasma ghrelin and peptide YY concentrations in the short term. *Appetite* **53**: 338–344.

Wang, Q., *et al.* (2001) The effect of autoclaving on the dispersibility and stability of three neutral polysaccharides in dilute aqueous solutions. *Carbohydrate Polymers* **45**(4): 355–362.

Wang, Q. *et al.* (2003) Preparation and characterization of molecular weight standards of low polydispersity from oat and barley (1→3)(1→4)-β-D-glucan. *Food Hydrocolloids* **17**(6): 845–853.

Wasser, S. P. (2002) Medicinal mushrooms as a source of antitumor and immunomodulating polysaccharides. *Applied Microbiology and Biotechnology* **60**: 258–274.

Welch, R. W. *et al.* (1991) Variation in the kernel (1→3) (1→4)-β-D-Glucan content of oat cultivars and wild Avena species and its relationship to other characteristics. *Journal of Cereal Science* **13**(2): 173–178.

Wolever, T. M. S. *et al.* (2010) Physicochemical properties of oat beta-glucan influence its ability to reduce serum LDL cholesterol in humans: A randomized clinical trial. *American Journal of Clinical Nutrition* **92**: 723–732.

Wood, P. J. (1980) The interaction of direct dyes with water soluble substituted celluloses and cereal β-glucans. *Industrial and Engineering Chemistry Product Research and Development* **19**(1): 19–23.

Wood, P. J. (1982) Factors affecting precipitation and spectral changes associated with complex-formation between dyes and β-d-glucans. *Carbohydrate Research* **102**(1): 283–293.

Wood, P. J. (1994) Evaluation of oat bran as a soluble fibre source. Characterization of oat β-glucan and its effects on glycaemic response. *Carbohydrate Polymers* **25**(4): 331–336.

Wood, P. J. (2001) Cereal beta-glucans: Structure, properties and health claims. In: *Advanced Dietary Fiber Technology* (eds B. V. McCleary and L. Prosky). Blackwell Science Co., Oxford, pp. 315–328.

Wood, P. J. (2002) Relationships between solution properties of cereal beta-glucans and physiological effects: A review. *Trends in Food Science and Technology* **13**: 313–320.

Wood, P. J. (2007) Cereal beta-glucans in diet and health. *Journal of Cereal Science* **46**: 230–238.

Wood, P. J. (2010) Oat and rye beta glucan: Properties and function. *Cereal Chemistry* **87**: 315–330.

Wood, P. J. (2011) Oat β-Glucan: Properties and Function. *OATS: Chemistry and Technology*, 2nd edn (eds F. H. Webster and P. J. Wood). AACC Intl. Press St. Paul, MN, pp. 219–254.

Wood, P. J. and R. G. Fulcher (1978) Interaction of some dyes with cereal beta-glucans. *Cereal Chemistry* **55**(6): 952–966.

Wood, P. J. *et al.* (1978) Extraction of high-viscosity gums from oats. *Cereal Chemistry* **55**(6): 1038–1049.

Wood, P. J. *et al.* (1983) Studies on the specificity of interaction of cereal cell wall components with Congo Red and Calcofluor. Specific detection and histochemistry of (1→3),(1→4),-β-D-glucan. *Journal of Cereal Science* **1**(2): 95–110.

Wood, P. J. *et al.* (1989) Large-scale preparation and properties of oat fractions enriched in (1–3)(1–4)-beta-D-glucan. *Cereal Chemistry* **66**(2): 97–103.

Wood, P. J. *et al.* (1991a) Molecular characterization of cereal beta-D-glucans. Structural analysis of oat beta-D-glucan and rapid structural evaluation of beta-D-glucans from different sources by high-performance liquid chromatography of oligosaccharides released by lichenase. *Cereal Chemistry* **68**(1): 31–39.

Wood, P. J. *et al.* (1991b) Potential for /3-glucan enrichment in brans derived from oat (*Avena sativa* L.) cultivars of different (1-a3),(1 -4)-f3-D-glucan concentrations. *Cereal Chemistry* **68**(1): 48–51.

Wood, P. J. *et al.* (1991c) Molecular characterization of cereal beta-glucans II. Size exclusion chromatography for comparison of molecular weight. *Cereal Chemistry* **68**: 530–536.

Wood, P. J. *et al.* (1994a) Effect of dose and modification of viscous properties of oat gum on plasma glucose and insulin following an oral glucose load. *British Journal of Nutrition* **72**: 731–743.

Wood, P. J. *et al.* (1994b) Structural studies of (1–3)(1–4)-beta-D-glucans by ^{13}C-nuclear magnetic resonance spectroscopy and by rapid analysis of cellulose- like regions using high-performance anion-exchange chromatography of oligosaccharides released by lichenase. *Cereal Chemistry* **71**(3): 301–307.

Woods, S. C. (2004) Gastrointestinal satiety signals I. An overview of gastrointestinal signals that influence food intake. *American Journal of Physiology – Gastrointestinal and Liver Physiology* **286**: G7-G13.

Wu, J. *et al.* (2008) Interactions between oat β-glucan and calcofluor characterized by spectroscopic method. *Journal of Agricultural and Food Chemistry* **56**(3): 1131–1137.

Wursch, P. and F. X. Pi-Sunyer (1997) The role of viscous soluble fiber in the metabolic control of diabetes: A review with special emphasis on cereals rich in beta-glucan. *Diabetes Care* **20**: 1774–1780.

Wyatt, P. J. (1993) Light scattering and the absolute characterization of macromolecules. *Analytica Chimica Acta* **272**(1): 1–40.

Xu, J. *et al.* (2013) Viscoelastic properties of oat β-glucan-rich aqueous dispersions. *Food Chemistry* **138**(1): 186–191.

Ye, H. (2006) Simultaneous determination of protein aggregation, degradation, and absolute molecular weight by size exclusion chromatography-multiangle laser light scattering. *Analytical Biochemistry* **356**(1): 76–85.

7
Health Benefits of Oat Phytochemicals

Shaowei Cui and Rui Hai Liu
Department of Food Science, Cornell University, Ithaca, New York, USA

7.1 Introduction

Epidemiological studies have consistently shown that consumption of whole grains and whole grain products is associated with a reduced risk of various types of chronic diseases, such as cardiovascular disease (Thompson, 1994; Anderson *et al.*, 2000; Mellen *et al.*, 2008), hypertension (Keenan *et al.*, 2003), type 2 diabetes (Meyer *et al.*, 2000; Liu *et al.*, 2000; Montonen *et al.*, 2003; Priebe *et al.*, 2008), obesity (Ripsin *et al.*, 1992), certain cancers (Jacobs *et al.*, 1998; Kasum *et al.*, 2002; Egeberg *et al.*, 2010), and all-cause mortality (Jacobs *et al.*, 1999; Steffen *et al.*, 2003).

Whole grains consist of the intact, ground, cracked, or flaked caryopsis, whose principal anatomical components—the starchy endosperm, germ, and bran—are present in the same relative proportions as they exist in the intact caryopsis (AACCI, 1999). The 2010 dietary guidelines for Americans defined whole grains as grains and grain products made from the entire grain seed, usually called the kernel, which consists of the bran, germ, and endosperm. If the kernel has been cracked, crushed, or flaked, it must retain nearly the same relative proportions of bran, germ, and endosperm as the original grain in order to be called whole grain (USDA, 2010). There are a variety of grain products that are consumed around the world, such as wheat, corn, rice, oats, millet, barley, spelt, and rye, and they are usually produced into flour, cereal, breads, and other products (Table 7.1).

The 2010 dietary guidelines for Americans recommend the consumption of at least 3-ounce equivalents of whole grain products per day based on a 2000-calorie level and at least one-half of the recommended total grain intake should be whole grains (USDA, 2010). However, fewer than 5% of Americans consume the minimum recommended amount of whole grains, and the average intake of whole grains in the US is less than 1-ounce equivalent of whole grains per day.

Oats Nutrition and Technology, First Edition. Edited by YiFang Chu.
© 2014 John Wiley & Sons, Ltd. Published 2014 by John Wiley & Sons, Ltd.

Table 7.1   Common whole grain and food products

Species	Common Name	Common food products
Triticum aestivum	Wheat	Breads, flours, pastas, baked goods
Zea mays	Corn	Corn cakes, tortillas, popcorn, hominy
Oryza sativa	Rice	White rice, brown rice, parboiled rice
Avena sativa	Oats	Oatmeal, flours
Panicum miliaceum	Millet	Bird food, porridge, millet
Hordeum vulgare	Barley	Hulled barley
Triticum aestivum spelta	Spelt	Breads, baked goods
Secale cereale	Rye	Breads

Source: Adapted from Okarter and Liu (2010). Reproduced with permission from Taylor and Francis.

Whole grains are rich sources of fibers, vitamins, minerals, and phytochemicals, including phenolics, carotenoids, lignans, γ-oryzanol, β-glucan, inulin, and sterols (Liu, 2007; Okarter and Liu, 2010). Phytochemicals are defined as bioactive compounds of plant origin that when ingested provide certain functional benefits in reducing the risk of chronic diseases beyond basic nutrition (Liu, 2004). Previous reviews have focused on phytochemicals of whole grains and their health benefits (Liu, 2007; Okarter and Liu, 2010), including lowering the risk of developing cardiovascular disease, type 2 diabetes, and certain cancers. Phytochemicals in whole grains and their antioxidant activities are receiving more and more attention with increasing knowledge of both free and bound forms as the major dietary sources of bioactive compounds in the prevention of diseases (Adom and Liu, 2002; Liu, 2007).

Oats (*Avena sativa*) are one of major important whole grains in the human diet. Oats are good sources of fibers, minerals, vitamins, and phytochemicals, including phenolics, β-glucan, avenanthramides, vitamin E, lignans, and phytostanols. Oats are mainly consumed as oatmeal, breakfast cereals, and oat flour. They are also used in a variety of baked products, including oatcakes, oat bread, and oat cookies, and are processed to brew beer. The beneficial effects associated with oat consumption are, in part, due to the existence of the unique phytochemicals of oats. The objective of this chapter is to review the current literature on oat phytochemicals and the health benefits of oat consumption.

7.2   Oat phytochemicals

The health benefits of oats have been attributed to their unique phytochemicals. Oat phytochemicals are present in free and bound forms (Adom and Liu, 2002), which are responsible for their health benefits. The most important groups of oat phytochemicals are phenolics, β-glucans, lignans, avenanthramides, carotenoids, vitamin E, and phytosterols.

7.2.1   Phenolics

Phenolics are compounds possessing one or more aromatic rings with one or more hydroxyl groups and are generally categorized as phenolic acids, flavonoids, stilbenes, coumarins, and tannins (Liu, 2004). Adom and Liu (2002) first reported both free and bound phenolics in grains; before that, phenolic compounds in grains had been commonly underestimated, as insoluble, bound forms were not included. Total phenolic content of oats was reported to be 6.53 ± 0.19 μmol gallic acid equiv/g grain, the free form accounting for about 25% of total phenolics and the bound form accounting for 75%. The most common phenolic compounds found in oats are phenolic acids and flavonoids.

7.2.1.1 Phenolic acids   Phenolic acids are compounds containing a phenolic ring and a carboxylic group, which are derived from benzoic or cinnamic acid. Therefore, phenolic acids can be subdivided into two major groups: hydroxybenzoic acid and hydroxycinnamic acid derivatives (Figure 7.1). Hydroxybenzoic acid derivatives include p-hydroxybenzoic, protocatechuic, vannilic, syringic, and gallic acids. They are commonly present in the bound form and are typically components of complex structures such as lignins and hydrolyzable tannins. They can also be found as derivatives of sugars and organic acids in plant foods. Hydroxycinnamic acid derivatives include p-coumaric, caffeic, ferulic, and sinapic acids

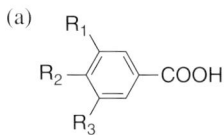

(a)

R_2—(ring with R_1, R_3)—COOH

	Substitutions		
Benzoic acid derivatives	R_1	R_2	R_3
p-Hydroxybenzoic acid	H	OH	H
Protocatechuic acid	H	OH	OH
Vanillic acid	CH_3O	OH	H
Syringic acid	CH_3O	OH	CH_3O

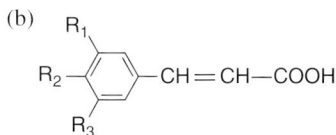

(b)

R_2—(ring with R_1, R_3)—CH=CH—COOH

	Substitutions		
Cinnamic acid derivatives	R_1	R_2	R_3
p-Coumaric acid	H	OH	H
Caffeic acid	OH	OH	H
Ferulic acid	CH_3O	OH	H
Sinapic acid	CH_3O	OH	CH_3O

Figure 7.1   Structures of common phenolic acids in oats: (a) benzoic acid derivatives and (b) cinnamic acid derivatives. *Source:* Adapted from Liu (2007). Reproduced with permission from Elsevier.

(Figure 7.1). They are mainly present in the bound form, linked to cell wall structural components, such as cellulose, lignin, and proteins, through ester bonds.

The common phenolic acids found in oats are ferulic acid, syringic acid, chlorogenic acid, vanillic acid, caffeic acid, and *p*-coumaric acid; others include *p*-hydroxybenzoic acid, protocatechuic acid, and trans-sinapic acid (Sosulski *et al.*, 1982). Sosulski and colleagues (1982) reported that phenolic acids in oats were present in different forms (free, soluble-conjugated, and insoluble). Ferulic acid was the dominant phenolic acid in oat flour; the content was 66.3 mg/kg, which accounted for 76.2% of the total phenolic acids in oat flour, and *trans*-ferulic acid was significantly higher (>96% of total) than *cis*-ferulic acid. It was also observed that ferulic acid mainly existed in the bound (insoluble) form, and the ratio of free, soluble-conjugated, and bound ferulic acid in oat flour is about 1:3.7:23. The order of the different amounts of phenolic acids in oat flour is ferulic acid > syringic acid > chlorogenic acid ≈ vanillic acid > caffeic acid > *p*-coumaric acid > *p*-hydroxybenzoic acid > protocatechuic acid.

Adom and Liu (2002) investigated the ferulic acid content of oats and reported that the total ferulic acid content in oats was 185±5 μmol ferulic acid/100 g grain, more than 97% of the total ferulic acid was found in the bound form, and the ratio of free, soluble-conjugated, and bound ferulic acid in oat was about 1:5.2:278.

Mattila and colleagues (2005) reported the phenolic acid content in oat products, finding that the ferulic acid content was 250±18 mg/kg and the sinapic acid content was 55±2.4 mg/kg. Oat flakes (whole grain) were reported to have fewer phenolic acids (472 mg/kg) than oat bran (651 mg/kg), which contains partial endosperm as well. Ferulic acid was the dominant phenolic acid in oat flakes and oat bran, accounting for more than 76% and 72% of total phenolic acids, respectively.

7.2.1.2 Flavonoids   Flavonoids are a group of phenolic compounds that commonly have a generic structure consisting of two aromatic rings (A and B rings) linked by three carbons that are usually in an oxygenated heterocyclic ring, or C ring (Figure 7.2; Liu, 2004). Structural differences in the heterocyclic C ring classify flavonoids into six groups: flavonols (e.g., quercetin, kaempferol, myricetin, galangin, and fisetin), flavones (e.g., apigenin, chrysin, and luteolin), flavanols (e.g., catechin, epicatechin, epigallocatechin, epicatechingallate, and epigallocatechingallate), flavanones (e.g., eriodictyol, hesperitin, and naringenin), anthocyanidins (e.g., cyanidin, pelargonidin, delphinidin, peonidin, and malvidin), and isoflavonoids (e.g., genistein, daidzein, glycitein, and formononetin) (Figure 7.3).

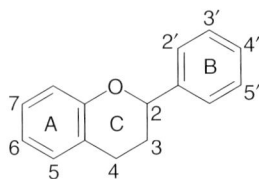

Figure 7.2   The generic structure of flavonoids.

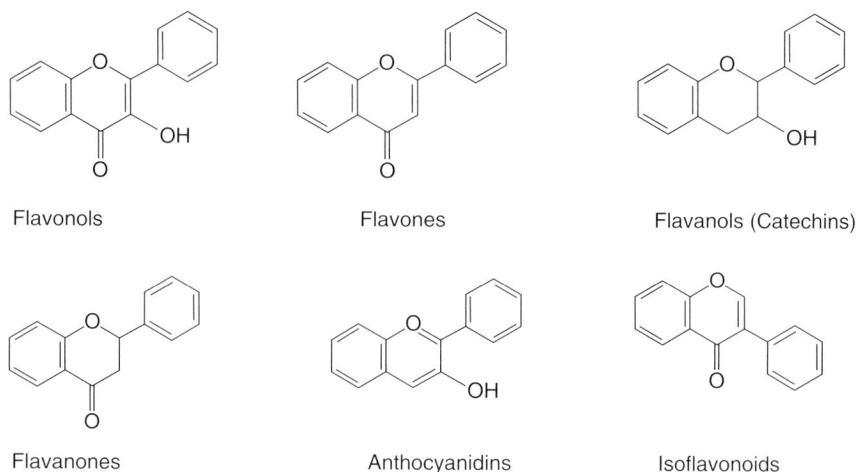

Flavonols   Flavones   Flavanols (Catechins)

Flavanones   Anthocyanidins   Isoflavonoids

Figure 7.3   Structure of main classes of dietary flavonoids. *Source:* Liu, 2004. Reproduced with permission from the American Society for Nutrition.

More than 5000 distinct flavonoids have been identified. They are most frequently found in nature as conjugates in glycosylated or esterified forms. Flavonoids cannot be synthesized by humans or animals; their food sources include fruits, vegetables, legumes, tea products, and other plant foods. Consumption of total flavonoids in the United States was estimated from 20 mg/day (Beecher, 2003) to 189.7 mg/day (Chun *et al.*, 2007), of which flavan-3-ols account for 83.5%, followed by flananones (7.6%), flavonols (6.8%), and others (Chun *et al.*, 2007).

Flavonoid content in some cereal grains has been reported by USDA. Buckwheat was reported to have a high content of quercetin (15.38 mg/100 g), and anthocyanidin content in purple wheat was high (USDA, 2011), but there are few published studies on flavonoid content in oats. Some of the detected flavonoids include apigenin, luteolin, tricin, kaempferol, and quercetin (Peterson, 2001; USDA, 2011) (Figure 7.4).

Health benefits of flavonoids as antioxidants mainly include prevention of cancer and cardiovascular disease. Other health benefits with anti-inflammatory, anticarcinogenic, and gastroprotective properties were also investigated (Kim *et al.*, 2004; Zayachkivska *et al.*, 2005; Kyle *et al.*, 2010). Mink and coworkers (2007) investigated the association between flavonoid intake and cardiovascular disease mortality by conducting a prospective study from three USDA databases, including 34 489 postmenopausal women. A significant inverse association between flavonoid intake and risk of coronary heart disease (CHD), cardiovascular disease (CVD), and total mortality was observed: Relative risks of CVD, CHD, and total mortality were reported to be 0.91 (95% CI $= 0.83$–0.99), 0.88 (95% CI $= 0.78$–0.99), and 0.90 (95% CI $= 0.86$–0.95), respectively, with anthocyanidin intake. Relative risk of CHD was 0.78 (95% CI $= 0.65$–0.94) for the highest quintile of flavanone intake versus lowest intake; relative risk of total

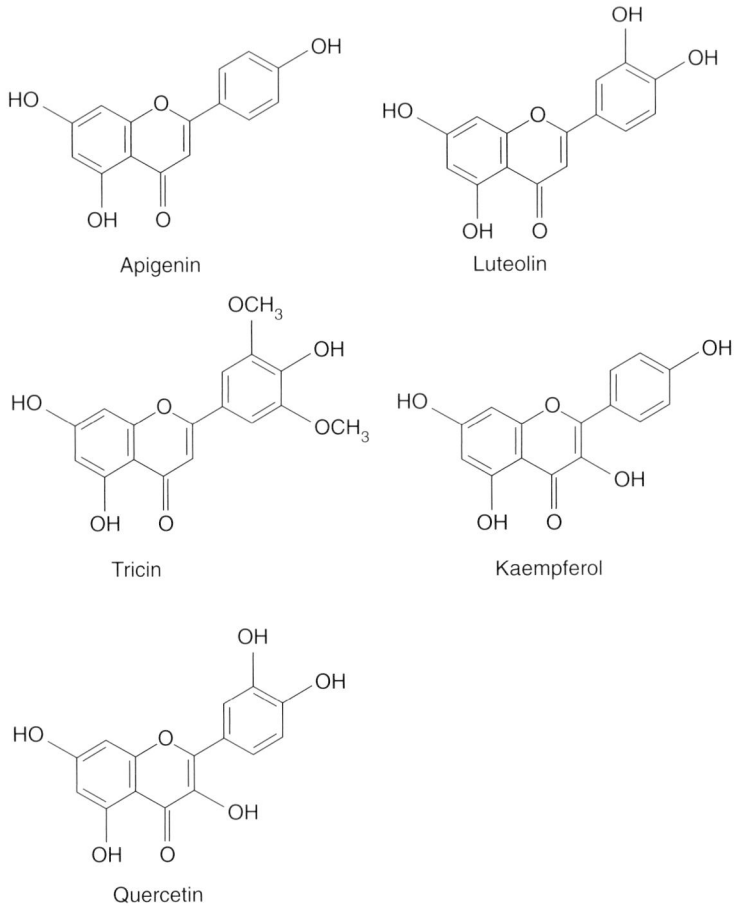

Figure 7.4   Chemical structures of common dietary flavonoids in oats.

mortality was reduced to 0.88 (95% CI = 0.82–0.96), comparing the highest quintile of flavone intake with lowest intake. Hooper and coworkers (2008) performed a meta-analysis to investigate the association between flavonoid-rich foods and CVD; they found that consumption of flavonoid-rich foods such as soy proteins and green tea correlated with reduced levels of LDL cholesterol.

7.2.2   β-Glucan

β-glucans are polysaccharides of D-glucose monomers linked by β-glycosidic bonds with side branches. They are most commonly found in the cell walls of cereal grains, baker's yeast, fungus, and bacteria. β-glucans in fungus generally consist of linear glucose chains linked by 1–3 β-glycosidic bonds and branches linked by 1–6 glycosidic bonds, whereas those in bacteria have 1–4 side branches. Those β-glucans were reported to enhance the immune system and have anticancer effects (Ooi and Liu, 2000; Chan *et al.*, 2009).

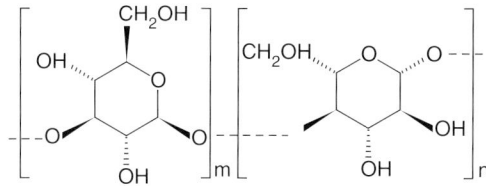

Figure 7.5   Structure of oat β-(1−3)(1−4)-glucan.

Oat β-glucan is unique with a group of linear polymers of glucose molecules connected by about 70% β-(1−4) and 30% β-(1−3)-linkages (Figure 7.5); it is classified as a soluble dietary fiber. Oat β-glucan is more flexible, soluble, and viscous than cellulose, due to the β-(1−3)-linkages, which provides its unique viscosity and other beneficial physicochemical characteristics in health promotion. Common human dietary sources of cereal β-glucan include oats, barley, rye, and wheat; oat and barley have the highest amounts of β-glucan. The FDA has approved the health claims of β-glucan in that consumption of about 3 g/day of β-glucan-soluble fiber lowers blood cholesterol levels (FDA, 1997).

The soluble β-glucan content in oats depends on cultivars and varies with isolation, purification, and detection methods. Wood (1994) reported that oats with hulls generally had 2.2–4.2% dry weight β-glucan and oats without hulls (oat groats) generally have 2.7–6.8% dry weight β-glucan. For commercial oat bran, high quality normally requires 7–10% β-glucan content. Johansson and colleagues (2000) reported that β-glucan content in oat bran was 9.5% using an AOAC method. For different oat cultivars, Wood (1994) reported the β-glucan content of oat groats and bran in 11 oat cultivars, which ranged from 3.9 to 6.8% dry weight in oat groats and from 5.8 to 8.9% dry weight in oat bran. Andersson and Börjesdotter (2011) investigated the effects of the environment and oat variety on β-glucan content and molecular weight by conducting a field experiment with four oat varieties grown in 11 different environments. The β-glucan content was reported to vary between 2.3 and 3.2% (of whole grain) in those oats.

The major health benefits of oat β-glucan include lowering blood cholesterol levels, controlling blood sugar, and enhancing the immune system (Liu, 2007). Wolever and colleagues (2010) investigated the LDL cholesterol-lowering effect of oat β-glucan with different molecular weights and observed that a high molecular weight β-glucan diet could reduce LDL cholesterol more efficiently when compared to the same amount of low molecular weight β-glucan. Tiwari and Cummins (2011) performed a meta-analysis of studies that investigated the relationship between β-glucan consumption from oats and barley, and blood cholesterol and glucose levels. Of the 126 clinical studies, 20 showed that there was a significant decrease in total cholesterol (0.60 mmol/L; 95% CI = 0.34–0.85) and LDL (0.66 mmol/L; 95% CI = 0.36–0.96) when β-glucan was consumed. Additionally, 49 clinical studies showed a significant change in blood glucose level (−2.58 mmol/L; 95% CI = −3.22 to −1.84). The dose-response model demonstrated a 0.30 mmol/L decrease in total cholesterol when 3 g β-glucans were consumed per day, which is in agreement with FDA recommendations. Othman and colleagues

(2011) also reviewed studies conducted between 1997 and 2000 that investigated the cholesterol-lowering effect of oat β-glucans, concluding that those scientific results are consistent with FDA recommendations of oat consumption. Daou and Zhang (2012) reviewed health benefits of oat β-glucans in the prevention of CVD, control of diabetes, and stimulation of immune functions. The health benefits of oat β-glucans might be explained by their physicochemical properties, such as viscosity and molecular weight.

7.2.3   Lignans

Lignans are a group of dietary phytoestrogen compounds that comprise two coupled C_6C_3 units. The common dietary plant lignans in the human diet include secoisolariciresinol, matairesinol, lariciresinol, pinoresinol, and syringaresinol (Figure 7.6).

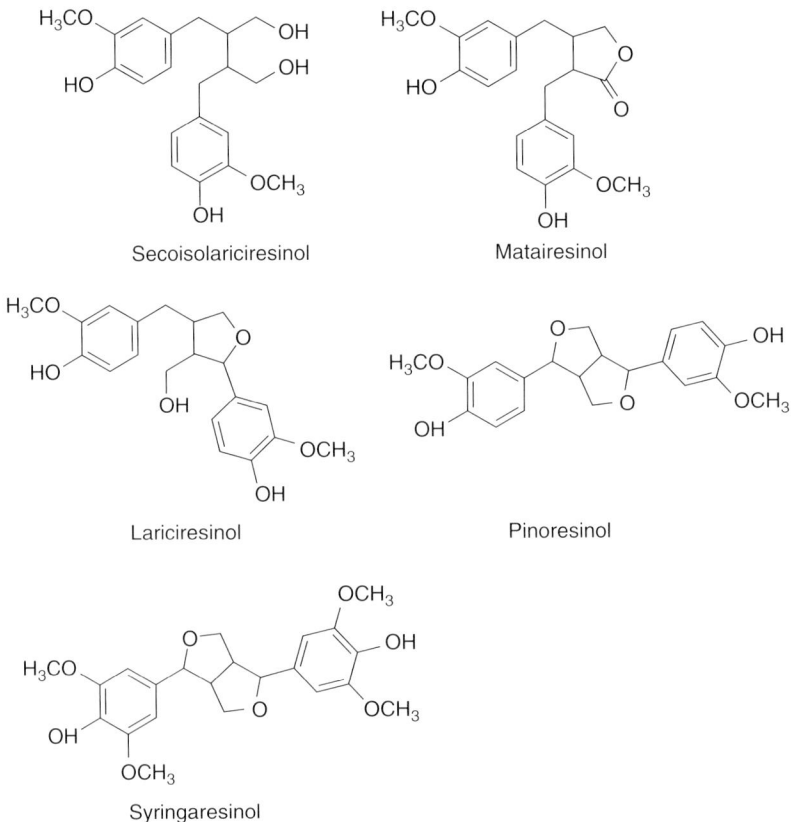

Figure 7.6   Chemical structures of common plant lignans.

Lignans are found in a wide variety of whole grains (rye, buckwheat, and oats), legumes, vegetables (asparagus and eggplant), and fruits (lemon, pineapple, kiwi, grape, and orange) (Thompson *et al.*, 1991; Penalvo *et al.*, 2005). Flax seeds are the richest dietary source of plant lignans (Thompson *et al.*, 1991). Lignan intake among postmenopausal women in the United States was estimated to be 578 μg (416–796 μg); among all lignans, matairesinol intake was 19 μg and secoisolariciresinol intake was 560 μg (de Kleijn *et al.*, 2001). Individual lignan intake in the United States was reported to range between 106 and 579 μg/day (Peterson *et al.*, 2010).

Penalvo and coworkers (2005) quantified lignan content in whole grains, vegetables, and fruits. Syringaresinol was found to be the predominant lignan in oats (352 μg/100 g), followed by pinoresinol (194 μg/100 g) and lariciresinol (183 μg/100 g). Total lignan content in oats (859 μg/100 g) was found to be higher than in wheat, barley, and millet, but lower than flax seed (335 mg/100 g), rye, and buckwheat. Smeds and colleagues (2009) reported that total lignan content of five different spring oat cultivars varied from 820 to 2550 μg/100 g. Syringaresinol, lariciresinol, and pinoresinol are the main lignans present in oats, and other lignans include medioresinol, secoisolariciresinol, and matairesinol.

Plant lignans can be converted into mammalian lignans (e.g., enterodiol and enterolactone) by intestinal microorganisms. Health benefits of mammalian lignans are related to their strong antioxidant activity and anti-estrogenic characteristics (Thompson *et al.*, 1991; Landete, 2012). Mammalian lignans, enterodiol and enterolactone, have been reported to prevent cancer, reduce the risk of CVD, and possess hepatoprotective effects. Johnsen and coworkers (2010) investigated the relationship between plasma enterolactone and risk of colon and rectal cancer by conducting a case-cohort study among 57,053 participants aged 50–64 years. Lower risk of colon cancer among women was reported (IRR = 0.76; 95% CI = 0.60–0.96) with doubled plasma enterolactone concentrations, but the same risk-reduction effect was not reported in men. Lin and coworkers (2012) conducted a case-control study investigating the association between dietary intake of lignans and risk of adenocarcinoma of the esophagus and gastroesophageal junction. The odds ratio for esophageal adenocarcinoma between the highest lignin exposure quartile and the lowest quartile was reported to be 0.65 (95% CI = 0.38–1.12), which indicated a 35% reduced risk. A 63% decreased risk of gastroesophageal junction adenocarcinoma was also observed among those with the highest lignin consumption quartile compared with those in the lowest.

Peterson and coworkers (2010) reviewed epidemiological studies on the association between dietary lignan intake and CVD risk. Out of the 11 human epidemiological studies reviewed, five showed decreased risk when the intake of lignans increased or the serum enterolactone level increased. Milder and coworkers (2006) reported an inverse association between the intake of matairesinol and risk of CHD (RR = 0.72, 95% CI = 0.53–0.98), risk of CVD (RR = 0.83, 95% CI = 0.69–1.00), and risk of all-cause mortality (RR = 0.86, 95% CI = 0.76–0.97). This study is supported by another study reporting increased flow-mediated dilation from 4.1% to 8.1% with increased matairesinol intake (Pellegrini *et al.*, 2010; Peterson *et al.*, 2010).

7.2.4   Avenanthramides

Avenanthramides are a group of alkaloids first identified and characterized in oat groats and hulls by Collins (1989). They consist of an anthranilic acid derivative linked to a hydroxycinnamic acid derivative by N-containing bonds. Oat is the only cereal grain that contain avenanthramides and the most abundant avenanthramides in oats are avenanthramide A (Bp), avenanthramide B (Bf), and avenanthramide C (Bc) (Figure 7.7).

Avenanthramide content in oats varies depending on cultivar and processing (Dimberg *et al.*, 1996). Dimberg and colleagues (1996) analyzed avenanthramide content in three different oat cultivars: Kapp, Mustang, and Svea. Avenanthramide B (Bf) content varied from 21 to 43 mg/kg oat grain, avenanthramide C (Bc) varied from 28 to 62 mg/kg oat grain, and avenanthramide A (Bp) varied from 25 to 47 mg/kg oat grain. Heat treatment (steam at $100°C$ for 10 min and dry at $100°C$ for 4 h) generally reduced the content of avenanthramide B by 18.2% and avenanthramide C by 18.8%, but the effect on avenanthramide A was more obvious with a 44.0% decrease in content. These facts suggest that avenanthramides B and C are relatively more heat stable than is avenanthramide A.

Health benefits of the avenanthramides are mainly related to their antioxidant activity (Emmons *et al.*, 1999; Peterson *et al.*, 2002; Bratt *et al.*, 2003). Potential mechanisms such as antiproliferation, anti-inflammation, anti-itch, cytoprotection, and vasodilation were also reviewed (Meydani, 2009): oat avenanthramides were reported to suppress adhesion molecules such as ICAM-1, thus inhibiting

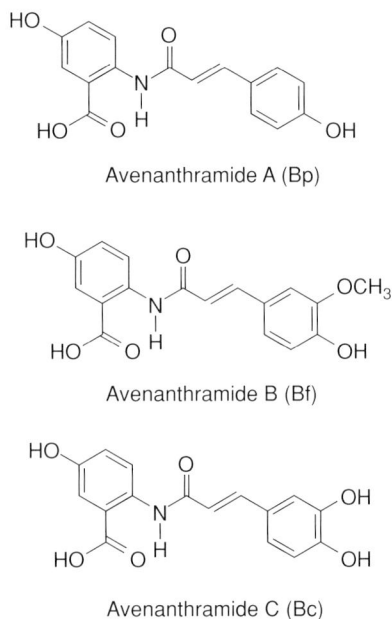

Avenanthramide A (Bp)

Avenanthramide B (Bf)

Avenanthramide C (Bc)

Figure 7.7   Chemical structures of common oat avenanthramides.

monocyte adhesion to human aortic endothelial cell monolayers and reducing inflammatory cytokine production. Several studies examined the antiprolifera- tive effects of avenanthramides on vascular smooth muscle cells and showed that avenanthramides could inhibit cell cycle signaling at the G1 to S phase tran- sition by modulating cell cycle regulatory proteins p53, p21cip1, p27kip1, and pRb (Nie *et al.*, 2006a, 2006b; Meydani 2009). Chen and colleagues (2007) tested the bioavailability of avenanthramides by collecting plasma avenanthramide concentration data from adults after consumption of 0.5 or 1.0 g of an oat avenanthramide-enriched mixture (AEM). Bioavailability was compared using the ratio of area under the curve (AUC) of plasma avenanthramide concentrate versus the time curve to the amount of avenanthramides in each AEM dose. Avenanthramide A had the highest bioavailability in both the 0.5- and 1.0 g AEM dose (77 μmol and 154 μmol avenanthramide A in AEM, respectively). Avenan- thramide B had the lowest bioavailability in the 0.5 g AEM dose but reached the same level with avenanthramide C at 1.0 g AEM dose. For the study of antiox- idant activity, plasma-reduced glutathione was reported to increase by 12% in the 1.0 g AEM dose compared with placebo, as avenanthramides enhanced some antioxidant defenses *in vivo*.

7.2.5 Carotenoids

Carotenoids are classified into hydrocarbons (e.g., α-carotene and β-carotene) and their oxygenated derivatives (e.g., β-cryptoxanthin, lutein, and zeaxanthin) (Figure 7.8). They have a 40-carbon skeleton and may be cyclized at one or both ends. The central part of the molecule is formed by a long series of conjugated double bonds; this feature plays an important role in their chemical reactivity and light-absorbance properties. Carotenoids are commonly found in fruits, veg- etables, and whole grains as yellow, orange, and red colors, and are present in all-trans forms.

Carotenoids commonly found in oats are lutein, zeacanthin, and α- and β- carotenes (Panfili *et al.*, 2004) (Figure 7.8). The average contents of lutein, zea- canthin, and the carotenes in oats are 0.23, 0.12, and 0.01 mg/kg dry weight.

Carotenoids are natural antioxidants. Their major health benefits include inhi- bition of cancer, enhancement of the immune system, prevention of macular degeneration, reduction in risk of cataracts, and prevention of CVD (Dutta *et al.*, 2005). α-Carotene, β-carotene, and β-cryptoxanthin have provitamin A activity and β-carotene is the principal precursor of vitamin A. Zeaxanthin and lutein can absorb near-UV light to protect the macula of the retina. Männistö and cowork- ers (2007) investigated the relationship between dietary carotenoids and risk of colorectal cancer. When pooling data from 11 cohort studies, there was no associ- ation between intake of each carotenoid and multivariate relative risk of colorec- tal cancer, regardless of whether multivitamin supplements were consumed. The relative risk of colorectal cancer was 1.06 (95% CI = 0.95–1.17) for α-carotene, 1.00 (95% CI = 0.90–1.12) for β-carotene, and 1.03 (95% CI = 0.93–1.14) for β-cryptoxanthin (Männistö *et al.*, 2007). In studies that included tomato sauce consumption as the major source of lycopene, relative risk of colorectal cancer

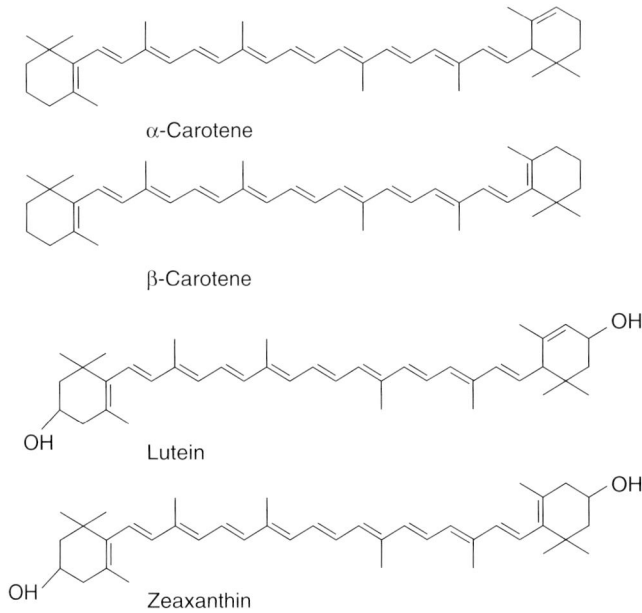

Figure 7.8 Chemical structure of common dietary carotenoids in oat.

was reported to be 1.08 (95% CI = 0.98–1.20). When compared to lutein plus zeacanthin intake of less than 1000 µg/day, the pooled multivariate relative risk of colorectal cancer was 0.87 (95% CI = 0.78–0.98) for the intake of greater than or equal to 4000 µg/day, an amount equal to approximately 200 g broccoli.

7.2.6   Vitamin E

Vitamin E is a generic name for eight lipid-soluble antioxidants that can be divided into two types: tocopherols (α-, β-, γ-, and δ-) and tocotrienols (α-, β-, γ-, and δ-) (Figure 7.9). Their basic structures consist of a six-hydroxychroman group and a phytol tail. Tocopherols contain saturated phytol side chains, whereas tocotrienols have three nonconjugated double bonds in the phytol side chain. Vitamin E compounds are commonly found in fruits, vegetables, and whole grains, especially wheat germ.

Panfili and colleagues (2003) determined the total vitamin E content and specific vitamers in oats. α-tocotrienol was the predominant vitamer in oats (56.4 mg/kg dry weight) and accounted for more than 78% of the total vitamin E of oats. The second predominant vitamer in oats was α-tocopherol (14.9 mg/kg dry weight). The other vitamers presented in oats included β-tocopherol, γ-tocopherol, and β-tocotrienol. Compared with other whole grains, oats had the highest α-tocotrienol content and vitamin E activity (33.6 mg tocopherol equivalents/kg dry weight). Oats, corn, and barley were the only grains that contain γ-tocopherol, which is a good source of vitamin E.

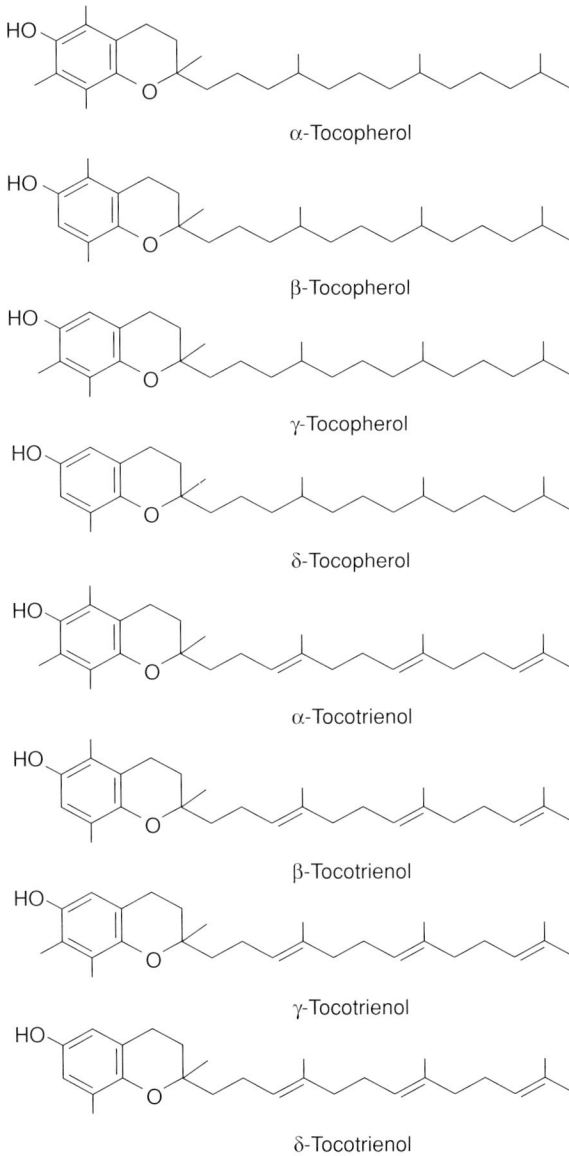

α-Tocopherol

β-Tocopherol

γ-Tocopherol

δ-Tocopherol

α-Tocotrienol

β-Tocotrienol

γ-Tocotrienol

δ-Tocotrienol

Figure 7.9 Chemical structures of tocopherols and tocotrienols. *Source:* Adapted from Liu (2007). Reproduced with permission from Elsevier.

The most important functions of vitamin E in the body are antioxidant activity and maintenance of membrane integrity. Health benefits of vitamin E include protection against photo-induced inflammation and reducing risk of type 2 diabetes (Konger, 2006; Liu *et al.*, 2006). In a recent study investigating the relationship between vitamin E consumption and prevention of CVD and cancer in the Women's Health Study, there was a significant reduction in risk of cardiovascular death (RR = 0.76, 95% CI = 0.59–0.98), but there was no significant effect on the risk of total cancer (RR = 1.01, 95% CI = 0.94–1.08), or on breast, lung, or colon cancers (Lee *et al.*, 2005). Another recent study evaluating the effects of long-term consumption of vitamins E and C in prevention of CVD in men reported that vitamin E consumption did not have a significant effect on total cardiovascular mortality (Sesso *et al.*, 2008).

7.2.7   Phytosterols

Phytosterols is a collective term for plant sterols and stanols, which are similar in structures to cholesterol. Plant stanols are the hydrogenated counterparts of their corresponding sterols, which have a double bond in the sterol ring. The most common plant sterols are sitosterol, campesterol, and stigmasterol, and their respective plant stanols are sitostanol, campestano, and stigmastanol (Figure 7.10). Plant sterols and stanols are found in oilseeds, unrefined vegetable oils, whole grains, nuts, and legumes.

The average daily intake of plant sterols in the Western diet of adults is estimated to range from 150 to 400 mg/day (Ntanios, 2001). This is lower than the estimated effective dose of 1.5–3.0 g/day, which leads to an 8–15% reduction in LDL cholesterol (Quílez *et al.*, 2003). Currently, esterified plant stanols are the major forms used in human clinical trials and in food fortification, since esterification makes plant sterols and stanols more lipid soluble, so they can be easily incorporated into fat-containing foods such as margarines and salad dressings.

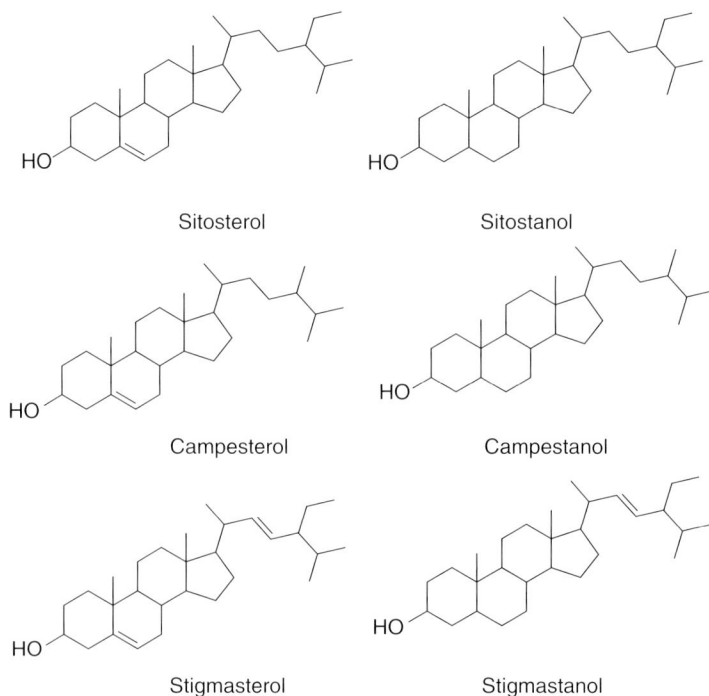

Figure 7.10   Chemical structures of common plant sterols and stanols.

Upon intake, the ester is cleaved by lipases in the small intestine and the plant stanol residues are released.

Phytosterols found in oats include β-sitosterol, sitostanol, campesterol, campestanol, $\triangle^5$-avenasterol, $\triangle^7$-avenasterol, and stigmasterol (Knights and Laurie, 1967; Määttä et al., 1999; Jiang and Wang, 2005). Määttä and colleagues (1999) analyzed cultivar and environmental effects on the phytosterol content of oats. Total sterol content in oat kernels varied from 350 to 491 μg/g dry weight. β-sitosterol was found to be the predominant phytosterol in oats (237–321 μg/g dry weight), accounting for more than 53% of total sterols in oat kernels. In Jiang and Wang's study (2005) of oat phytosterol content, they reported sitosterol was the predominant phytosterol in both oat bran and hull, accounting for 45.7% of total phytosterols in oat bran and 50% of total phytosterols in oat hull.

Health benefits of plant sterols and stanols include lowering serum total and LDL cholesterol levels, as phytosterols compete with cholesterol for micelle formation in the intestinal lumen and inhibit cholesterol absorption (Nissinen et al., 2002; Quílez et al., 2003). Demonty and coworkers (2009) performed a meta-analysis of randomized controlled trials to investigate LDL cholesterol-lowering effects of phytosterols. For an average daily phytosterol intake of 2.15 g, the pooled absolute LDL-cholesterol reduction was 0.34 mmol/L (95% CI = 0.31–0.36).

7.3 Health benefits of oat phytochemicals: Epidemiological evidence

7.3.1 Cardiovascular disease (CVD)

A variety of epidemiological studies have consistently shown that consumption of whole grain, including oats, is associated with reduced risk of CVD. Ripsin and coworkers (1992) performed a meta-analysis of studies published up to 1991 that investigated the relationship between oat consumption and blood total cholesterol levels. From the 10 trials that met the inclusion criteria, it was found that oat consumption reduced blood total cholesterol levels by 0.13 mmol/L (95% CI = 0.017–0.19 mmol/L). The reduction effect was stronger for subjects with initially higher blood cholesterol levels, especially when more than 3 g of soluble fiber were consumed per day.

Liu and colleagues (1999) reported results from the Nurses' Health Study that showed that consumption of cooked oatmeal (2–4 servings per week) reduced risk of CHD (RR = 0.70; 95% CI = 0.49–0.98) for female US nurses aged 38–63 years after adjustment for body mass index, cigarette smoking, alcohol intake, total energy intake, and other possible confounding factors. The risk-reduction effect of oatmeal when consumed 2–4 times per week was better when combined with the consumption of bread, rice, or wheat germ 2–4 times per week. Wolk and colleagues (1999) showed that consumption of cold breakfast oatmeal five or more times per week in comparison with no consumption of cold breakfast oatmeal was associated with a 29% lower risk of CHD (RR = 0.71; 95% CI = 0.38–1.34). Ruxton and Derbyshire

(2008) reviewed studies (1990–2008) of the association between cardiovascular risk factors and regular consumption of oats. Of the 21 randomized controlled trials reviewed, ten focused on American adults, two on Australian adults, two on New Zealanders, five on Europeans, one on Canadians, and one on Mexicans. Thirteen studies reported a significant reduction in total plasma cholesterol, and 14 studies reported a significant reduction in LDL cholesterol, when oat products were consumed.

7.3.2 Hypertension

Oats are reported to have antihypertensive effects. Pins and colleagues (2002) investigated the antihypertensive effects of oats by conducting a 12-week randomized, controlled, parallel group trial among 88 adults with mild or moderate hypertension [blood pressure (BP) 120/80 to 160/100 mm Hg]. Seventy-three percent of subjects in the oat cereal treatment group experienced a reduction in their requirement for BP medication compared with 42% in the low-fiber cereal control group. For those without a reduction in medication, subjects in the oat cereal treatment group had a higher decrease in both systolic blood pressure (SBP) and diastolic blood pressure (DBP) than the control group.

Keenan and colleagues (2003) investigated the antihypertensive effects of whole wheat cereals when added to a standard American diet by conducting a 6-week randomized, controlled, parallel-group pilot study among 22 adults with mild or borderline hypertension (SBP 130–160 mm Hg, DBP 85–100 mm Hg). Eighteen adults completed the trial and the results revealed that SBP decreased 7.5 mm Hg and DBP decreased 5.5 mm Hg in the oat cereal treatment group, whereas in the low-fiber cereal control group, neither SBP nor DBP significantly changed.

Maki and colleagues (2007) conducted a randomized, controlled, clinical trial among adults with elevated SBP and/or DBP to assess the effects of an oat β-glucan-containing diet on BP. The oat β-glucan diet group was treated with ready-to-eat (RTE) oat bran, oatmeal, and β-glucan powder over a 12-week period. Significant reductions in both SBP (8.3 mm Hg reduction, p = 0.008) and DBP (3.9 mm Hg reduction, p = 0.018) were observed in subjects with body mass index above 31.5 kg/m^2 compared with the control group.

Kochar and colleagues (2012) investigated the association between breakfast cereal consumption and the risk of hypertension in the Physicians' Health Study I by analyzing data from 13,368 male participants; whole grain cereal was specified as breakfast cereals that contain at least 25% oat or bran in the study. After adjustments for age, smoking, alcohol consumption, fruit and vegetable consumption, physical activity, and history of diabetes, the relative risk of hypertension was reported to decrease with higher whole grain cereal consumption. For whole grain cereal intake of more than seven servings per week, the relative risk of hypertension was 0.81 (95% CI = 0.75–0.89) when the body mass index was less than 25 kg/m^2 and 0.80 (95% CI = 0.72–0.90) when the body mass index was more than 25 kg/m^2, which indicated the hypertension risk-lowering effect worked for both lean and overweight participants.

7.3.3   Type 2 diabetes/blood sugar maintenance

Several epidemiological studies have linked oat product consumption with reduced risk of type 2 diabetes and maintenance of blood sugar. Liu and coworkers (2000) investigated the association between cooked oatmeal consumption and type 2 diabetes among 75, 521 US female nurses aged 38–63 years. After adjusting for age, body mass index, physical activity, total energy intake, and other possible confounding factors, it was found that there was a significant inverse association between cooked oatmeal intake and risk of type 2 diabetes. Consumption of cooked oatmeal five to six times per week reduced the risk of type 2 diabetes by 39% (RR = 0.61; 95% CI = 0.32–1.15).

Jenkins and coworkers (2008) investigated oats as part of a low-glycemic index diet on type 2 diabetes by conducting a randomized, parallel study among 210 participants with type 2 diabetes. The low-glycemic index diet emphasized large flake oatmeal, oat bran, beans, nuts, bulgur, flax, and other low-glycemic index foods. Glycated hemoglobin A_{1c} (HbA_{1c}) was measured as an indicator of plasma glucose concentration. Among 155 participants who completed the trial, HbA_{1c} decreased by 0.50% (absolute value, 95% CI = 0.39%–0.61%) in the low-glycemic index group, compared with a 0.18% absolute HbA_{1c} decrease in the high-cereal fiber group. The results indicated that oats were one of many low-glycemic index products that can help lower blood sugar content.

Post and coworkers (2012) performed a meta-analysis of studies published between 1980 and 2010 that involved effects of increased dietary fiber intake on HbA_{1c} and fasting blood glucose levels among participants with known type 2 diabetes mellitus. Fifteen studies were included and the analysis showed that HbA_{1c} decreased by 0.26% (95% CI = 0.02–0.51) and fasting blood sugar decreased by 0.85 mmol/L (95% CI = 0.46–1.25) over placebo.

7.3.4   Obesity and weight control

Oats were reported to help weight control among obese populations. Saltzman and colleagues (2001) conducted an 8-week trial among 43 healthy adults to investigate the effects of hypocaloric diets with and without oats on body weight, BP, and blood lipids. It was observed that both the oat feeding and control groups lost weight (oat-feeding group –3.9 ± 1.6 kg, control group –4.0 ± 1.1 kg) but the decrease in SBP of the oat-feeding group (–6 ± 7 mm Hg) was more obvious than that of the control group (–1 ± 10 mm Hg) (Saltzman *et al.*, 2001). In addition, total cholesterol and LDL cholesterol levels were significantly decreased in the oat-feeding group.

Maki and colleagues (2010) investigated the effects of the consumption of whole-grain RTE oat cereal as part of a dietary program on overweight and obese adults. A randomized, controlled trial was conducted among 204 obese adults whose baseline LDL cholesterol levels were between 130 and 200 mg/dL. After a 12-week treatment, it was observed that both LDL cholesterol and total cholesterol levels in the whole-grain RTE oat cereal group decreased more than those of the energy-matched, low-fiber food control group. Weight loss and waist circumference were both observed to decrease in the oat cereal and control groups, but waist circumference decreased to a greater extent in the whole-grain RTE oat cereal group (−3.3 ± 0.4 *vs.* −1.9 ± 0.4 cm).

7.3.5 Digestive health/colon health

Oats are unique with β-glucan promoting digestive and colon health. Janatu-inen and colleagues (1995) investigated the effects of diets containing oats on the health of the small intestine of patients with celiac disease. A randomized trial was conducted among patients with celiac disease, comparing the effects of gluten-free diets with and without oats. It was observed that patients in remission did not have worsening architecture of the duodenal villi, with or without oats. Therefore, moderate amounts of oats were recommended for patients with celiac disease on gluten-free diets, because oats did not have adverse effects on the small intestine.

Mälkki and Virtanen (2001) reviewed the effects on gastrointestinal health of oat bran and oat gum. Dietary fiber in oats could help reduce the rate of gastric emptying to affect satiety and remained nearly intact in the small intestines of humans.

7.3.6 Cancer

Many epidemiological studies have investigated the association between dietary fiber consumption and relative risk of cancer (Cummings *et al.*, 1992; Fuchs *et al.*, 1999; Terry *et al.*, 2001) but few studies have investigated oat fiber and risk of cancers Generally, oat products are included in whole grains or dietary fibers in the literature to investigate their potential health benefit in cancer prevention. Park and colleagues (2009) investigated the relationship between dietary fiber intake and risk of breast cancer among 185,598 postmenopausal women in the National Institutes of Health-AARP Diet and Health Study over a 7-year period. Dietary fiber intake was reported to be inversely associated with breast cancer risk (RR = 0.87; 95% CI = 0.77–0.98).

Suzuki and colleagues (2009) conducted a prospective analysis of the association between dietary fiber intake and prostate cancer risk among 142,590 men in the European Prospective Investigation into Cancer and Nutrition study. Incidence rate ratios of prostate cancer were not significantly reduced across increased quintiles of cereal fiber intake, suggesting that dietary fiber intake is not associated with prostate cancer risk.

Dahm and colleagues (2010) investigated the association between dietary fiber and colorectal cancer risk by conducting a prospective case-control study within seven UK cohort studies. Intake of dietary fiber was reported to be inversely associated with the risk of colorectal and colon cancers in adjusted models that took age, alcohol, energy, and other factors into consideration.

Aune and colleagues (2011) reviewed dietary fiber and whole grain intake and their relationship with colorectal cancer. Six studies were included in a whole grain intake dose-response meta-analysis. The relative risk for colorectal cancer with an increment of 90 g/day of whole grains was 0.83 (95% CI = 0.78–0.89), relative risk for colon cancer was 0.86 (95% CI = 0.79–0.94), and relative risk for rectal cancer was 0.80 (95% CI = 0.56–1.14) (2011). Eight cohort studies were investigated in the dose-response analysis of cereal fiber and risk of colorectal cancer, and the pooled relative risk was 0.90 (95% CI = 0.83–0.97) at a 10 g/day intake of cereal fiber.

7.4 Summary

Epidemiological studies have consistently shown that regular consumption of whole grains is associated with reduced risk of developing chronic diseases, such as CVD, type 2 diabetes, and certain cancers. As a unique whole grain, oats and oat products provide multiple health benefits with their distinct phytochemicals, such as phenolics, carotenoids, β-glucans, avenanthramides, lignans, vitamin E, and phytosterols. Health benefits of oat phytochemicals were also reported in lowering the risk of hypertension, weight control, blood sugar maintenance, and improvement of digestive health. Future research on health benefits of oat phytochemicals is thus warranted.

References

AACCI (1999) AACC members agree on difinition of whole grain. American Association of Cereal Chemists International, St. Paul, MN.

Adom, K.K. and Liu, R.H. (2002) Antioxidant activity of grains. *Journal of Agricultural and Food Chemistry* **50**, 6182–6187.

Anderson, J.W., Hanna, T.J., Peng, X., and Kryscio, R.J. (2000) Whole grain foods and heart disease risk. *Journal of the American College of Nutrition* **19**, 291S–299S.

Andersson, A.A.M. and Börjesdotter, D. (2011) Effects of environment and variety on content and molecular weight of β-glucan in oats. *Journal of Cereal Science* **54**, 122–128.

Aune, D., Chan, D.S., Lau, R., *et al.* (2011) Dietary fibre, whole grains, and risk of colorectal cancer: Systematic review and dose-response meta-analysis of prospective studies. *British Medical Journal* **343**, d6617.

Beecher, G.R. (2003) Overview of dietary flavonoids: Nomenclature, occurrence and intake. *Journal of Nutrition* **133**, 3248S–3254S.

Bratt, K., Sunnerheim, K., Bryngelsson, S., *et al.* (2003) Avenanthramides in oats (*Avena sativa* L.) and structure−antioxidant activity relationships. *Journal of Agricultural and Food Chemistry* **51**, 594–600.

Chan, G.C., Chan, W.K., and Sze, D.M. (2009) The effects of beta-glucan on human immune and cancer cells. *Journal of Hematologyand Oncology* **2**, 25.

Chen, C.-Y.O., Milbury, P.E., Collins, F.W., and Blumberg, J.B. (2007) Avenanthramides are bioavailable and have antioxidant activity in humans after acute consumption of an enriched mixture from oats. *Journal of Nutrition* **137**, 1375–1382.

Chun, O.K., Chung, S.J., and Song, W.O. (2007) Estimated dietary flavonoid intake and major food sources of U.S. adults. *Journal of Nutrition* **137**, 1244–1252.

Collins, F.W. (1989) Oat phenolics: avenanthramides, novel substituted N-cinnamoylanthranilate alkaloids from oat groats and hulls. *Journal of Agricultural and Food Chemistry* **37**, 60–66.

Cummings, J.H., Bingham, S.A., Heaton, K.W., and Eastwood, M.A. (1992) Fecal weight, colon cancer risk, and dietary intake of nonstarch polysaccharides (dietary fiber). *Gastroenterology* **103**, 1783–1789.

Dahm, C.C., Keogh, R.H., Spencer, E.A., *et al.* (2010) Dietary fiber and colorectal cancer risk: A nested case–control study using food diaries. *Journal of the National Cancer Institute* **102**, 614–626.

Daou, C. and Zhang, H. (2012) Oat beta-glucan: Its role in health promotion and prevention of diseases. *Comprehensive Reviews in Food Science and Food Safety* **11**, 355–365.

de Kleijn, M.J.J., van der Schouw, Y.T., Wilson, P.W.F., *et al.* (2001) Intake of dietary phytoestrogens is low in postmenopausal women in the United States: The Framingham Study1–4. *Journal of Nutrition* **131**, 1826–1832.

Demonty, I., Ras, R.T., van der Knaap, H.C.M., *et al.* (2009) Continuous dose-response relationship of the LDL-cholesterol-lowering effect of phytosterol intake. *Journal of Nutrition* **139**, 271–284.

Dimberg, L.H., Molteberg, E.L., Solheim, R., and Frølich, W. (1996) Variation in oat groats due to variety, storage and heat treatment. I: Phenolic compounds. *Journal of Cereal Science* **24**, 263–272.

Dutta, D., Chaudhuri, U.R., and Chakraborty, R. (2005) Structure, health benefits, antioxidant property and processing and storage of carotenoids. *African Journal of Biotechonology* **4**, 1510–1520.

Egeberg, R., Olsen, A., Loft, S., *et al.* (2010) Intake of wholegrain products and risk of colorectal cancers in the diet, cancer and health cohort study. *British Journal of Cancer* **103**, 730–734.

Emmons, C.L., Peterson, D.M., and Paul, G.L. (1999) Antioxidant capacity of oat (*Avena sativa* L.) extracts. 2. *In vitro* antioxidant activity and contents of phenolic and tocol antioxidants. *Journal of Agricultural and Food Chemistry* **47**, 4894–4898.

FDA (1997) Food labeling: Health claims; Oats and coronary heart disease; Rules and regulations. *Federal Register* **62**, 3584–3601.

Fuchs, C.S., Giovannucci, E.L., Colditz, G.A., *et al.* (1999) Dietary fiber and the risk of colorectal cancer and adenoma in women. *New England Journal of Medicine* **340**, 169–176.

Hooper, L., Kroon, P.A., Rimm, E.B., *et al.* (2008) Flavonoids, flavonoid-rich foods, and cardiovascular risk: A meta-analysis of randomized controlled trials. *American Journal of Clinical Nutrition* **88**, 38–50.

Jacobs, D.R., Marquart, L., Slavin, J., and Kushi, L.H. (1998) Whole-grain intake and cancer: An expanded review and meta-analysis. *Nutrition and Cancer* **30**, 85–96.

Jacobs, D.R., Meyer, K.A., Kushi, L.H., and Folsom, A.R. (1999) Is whole grain intake associated with reduced total and cause-specific death rates in older women? The Iowa Women's Health Study. *American Journal of Public Health* **89**, 322–329.

Janatuinen, E.K., Pikkarainen, P.H., Kemppainen, T.A., *et al.* (1995) A comparison of diets with and without oats in adults with celiac disease. *New England Journal of Medicine* **333**, 1033–1037.

Jenkins, D.J.A., Kendall, C.W.C., McKeown-Eyssen, G., *et al.* (2008) Effect of a low-glycemic index or a high-cereal fiber diet on type 2 diabetes: A randomized trial. *Journal of the American Medical Association* **300**, 2742–2753.

Jiang, Y. and Wang, T. (2005) Phytosterols in cereal by-products. *Journal of the American Oil Chemists' Society* **82**, 439–444.

Johansson, L., Virkki, L., Maunu, S., *et al.* (2000) Structural characterization of water soluble β-glucan of oat bran. *Carbohydrate Polymers* **42**, 143–148.

Johnsen, N., Olsen, A., Thomsen, B., *et al.* (2010) Plasma enterolactone and risk of colon and rectal cancer in a case-cohort study of Danish men and women. *Cancer Causes and Control* **21**, 153–162.

Kasum, C.M., Jacobs, D.R., Nicodemus, K., and Folsom, A.R. (2002) Dietary risk factors for upper aerodigestive tract cancers. *International Journal of Cancer* **99**, 267–272.

Keenan, J.M., Pins, J.J., Frazel, C., *et al.* (2003) Oat ingestion reduces systolic and diastolic blood pressure in patients with mild or borderline hypertension: A pilot trial. *Journal of Family Practice* **51**, 369–374.

Kim, H.P., Son, K.H., Chang, H.W., and Kang, S.S. (2004) Anti-inflammatory plant flavonoids and cellular action mechanisms. *Journal of Pharmacological Sciences* **96**, 229–245.

Knights, B.A. and Laurie, W. (1967) Application of combined gas-liquid chromatography-mass spectrometry to the identification of sterols in oat seed. *Phytochemistry* **6**, 407–416.

Kochar, J., Gaziano, J.M., and Djoussé, L. (2012) Breakfast cereals and risk of hypertension in the Physicians' Health Study I. *Clinical Nutrition* **31**, 89–92.

Konger, R.L. (2006) A new wrinkle on topical vitamin E and photo-inflammation: Mechanistic studies of a hydrophilic gamma-tocopherol derivative compared with alpha-tocopherol. *Journal of Investigative Dermatology* **126**, 1447–1449.

Kyle, J.A., Sharp, L., Little, J., *et al.* (2010) Dietary flavonoid intake and colorectal cancer: A case-control study. *British Journal of Nutrition* **103**, 429–436.

Landete, J.M. (2012) Plant and mammalian lignans: A review of source, intake, metabolism, intestinal bacteria and health. *Food Research International* **46**, 410–424.

Lee, I.-M., Cook, N.R., Gaziano, J.M., *et al.* (2005) Vitamin E in the primary prevention of cardiovascular disease and cancer: The women's health study: A randomized controlled trial. *Journal of the American Medical Association* **294**, 56–65.

Lin, Y., Yngve, A., Lagergren, J., and Lu, Y. (2012) Dietary intake of lignans and risk of adenocarcinoma of the esophagus and gastroesophageal junction. *Cancer Causes and Control* **23**, 837–844.

Liu, R.H. (2004) Potential synergy of phytochemicals in cancer prevention: Mechanism of action. *Journal of Nutrition* **134**, 3479S–3485S.

Liu, R.H. (2007) Whole grain phytochemicals and health. *Journal of Cereal Science* **46**, 207–219.

Liu, S., Stampfer, M.J., Hu, F.B., *et al.* (1999) Whole-grain consumption and risk of coronary heart disease: Results from the Nurses' Health Study. *American Journal of Clinical Nutrition* **70**, 412–419.

Liu, S., Manson, J.E., Stampfer, M.J., *et al.* (2000) A prospective study of whole-grain intake and risk of type 2 diabetes mellitus in US women. *American Journal of Public Health* **90**, 1409–1415.

Liu, S., Lee, I.-M., Song, Y., *et al.* (2006) Vitamin E and risk of type 2 diabetes in the Women's Health Study randomized controlled trial. *Diabetes* **55**, 2856–2862.

Määttä, K., Lampi, A.-M., Petterson, J., *et al.* (1999) Phytosterol content in seven oat cultivars grown at three locations in Sweden. *Journal of the Science of Food and Agriculture* **79**, 1021–1027.

Maki, K.C., Beiseigel, J.M., Jonnalagadda, S.S., *et al.* (2010) Whole-grain ready-to-eat oat cereal, as part of a dietary program for weight loss, reduces low-density lipoprotein cholesterol in adults with overweight and obesity more than a dietary program including low-fiber control foods. *Journal of the American Dietetic Association* **110**, 205–214.

Maki, K.C., Galant, R., Samuel, P., *et al.* (2007) Effects of consuming foods containing oat beta-glucan on blood pressure, carbohydrate metabolism and biomarkers of oxidative stress in men and women with elevated blood pressure. *European Journal of Clinical Nutrition* **61**, 786–795.

Mälkki, Y. and Virtanen, E. (2001) Gastrointestinal effects of oat bran and oat gum: A review. *Food Science and Technology* **34**, 337–347.

Männistö, S., Yaun, S.-S., Hunter, D.J., *et al.* (2007) Dietary carotenoids and risk of col-orectal cancer in a pooled analysis of 11 cohort studies. *American Journal of Epidemi-ology* **165**, 246–255.

Mattila, P., Pihlava, J.-m., and Hellstrom, J. (2005) Contents of phenolic acids, alkyl- and alkenylresorcinols, and avenanthramides in commercial grain products. *Journal of Agri-cultural and Food Chemistry* **53**, 8290–8295.

Mellen, P.B., Walsh, T.F., and Herrington, D.M. (2008) Whole grain intake and cardio-vascular disease: A meta-analysis. *Nutrition, Metabolism and Cardiovascular Diseases* **18**, 283–290.

Meydani, M. (2009) Potential health benefits of avenanthramides of oats. *Nutrition Reviews* **67**, 731–735.

Meyer, K.A., Kushi, L.H., Jacobs, D.R., *et al.*, 2000. Carbohydrates, dietary fiber, and incident type 2 diabetes in older women. *American Journal of Clinical Nutrition* **71**, 921–930.

Milder, I.E., Feskens, E.J., Arts, I.C., *et al.*, 2006. Intakes of 4 dietary lignans and cause-specific and all-cause mortality in the Zutphen Elderly Study. *American Journal of Clin-ical Nutrition* **84**, 400–405.

Mink, P.J., Scrafford, C.G., Barraj, L.M., *et al.* (2007) Flavonoid intake and cardiovascular disease mortality: A prospective study in postmenopausal women. *American Journal of Clinical Nutrition* **85**, 895–909.

Montonen, J., Knekt, P., Järvinen, R., *et al.* (2003) Whole-grain and fiber intake and the incidence of type 2 diabetes. *American Journal of Clinical Nutrition* **77**, 622–629.

Nie, L., Wise, M., Peterson, D., and Meydani, M. (2006a) Mechanism by which avenanthramide-c, a polyphenol of oats, blocks cell cycle progression in vascular smooth muscle cells. *Free Radical Biology and Medicine* **41**, 702–708.

Nie, L., Wise, M.L., Peterson, D.M., and Meydani, M. (2006b) Avenanthramide, a polyphenol from oats, inhibits vascular smooth muscle cell proliferation and enhances nitric oxide production. *Atherosclerosis* **186**, 260–266.

Nissinen, M., Gylling, H., Vuoristo, M., and Miettinen, T.A. (2002) Micellar distribution of cholesterol and phytosterols after duodenal plant stanol ester infusion. *American Journal of Physiology – Gastrointestinal and Liver Physiology* **282**, G1009–G1015.

Ntanios, F. (2001) Plant sterol-ester-enriched spreads as an example of a new functional food. *European Journal of Lipid Science and Technology* **103**, 102–106.

Okarter, N. and Liu, R.H. (2010) Health benefits of whole grain phytochemicals. *Critical Reviews in Food Science and Nutrition* **50**, 193–208.

Ooi, V.E. and Liu, F. (2000) Immunomodulation and anti-cancer activity of polysaccharide-protein complexes. *Current Medicinal Chemistry* **7**, 715–729.

Othman, R.A., Moghadasian, M.H., and Jones, P.J.H. (2011) Cholesterol-lowering effects of oat β-glucan. *Nutrition Reviews* **69**, 299–309.

Panfili, G., Fratianni, A., and Irano, M. (2003) Normal phase high-performance liquid chromatography method for the determination of tocopherols and tocotrienols in cere-als. *Journal of Agricultural and Food Chemistry* **51**, 3940–3944.

Panfili, G., Fratianni, A., and Irano, M. (2004) Improved normal-phase high-performance liquid chromatography procedure for the determination of carotenoids in cereals. *Jour-nal of Agricultural and Food Chemistry* **52**, 6373–6377.

Park, Y., Brinton, L.A., Subar, A.F., *et al.* (2009) Dietary fiber intake and risk of breast cancer in postmenopausal women: the National Institutes of Health–AARP Diet and Health Study. *American Journal of Clinical Nutrition* **90**, 664–671.

Pellegrini, N., Valtueña, S., Ardigò, D., *et al.*, 2010. Intake of the plant lignans matairesinol, secoisolariciresinol, pinoresinol, and lariciresinol in relation to vascular inflamma-tion and endothelial dysfunction in middle age-elderly men and post-menopausal women living in Northern Italy. *Nutrition, Metabolism, and Cardiovascular Diseases* **20**, 64–71.

Penalvo, J.L., Haajanen, K.M., Botting, N., and Adlercreutz, H. (2005) Quantification of lignans in food using isotope dilution gas chromatography/mass spectrometry. *Journal of Agricultural and Food Chemistry* **53**, 9342–9347.

Peterson, D.M. (2001) Oat antioxidants. *Journal of Cereal Science* **33**, 115–129.

Peterson, D.M., Hahn, M.J., and Emmons, C.L. (2002) Oat avenanthramides exhibit antioxidant activities in vitro. *Food Chemistry* **79**, 473–478.

Peterson, J., Dwyer, J., Adlercreutz, H., *et al.* (2010) Dietary lignans: Physiology and potential for cardiovascular disease risk reduction. *Nutrition Reviews* **68**, 571–603.

Pins, J.J., Geleva, D., Keenan, J.M., *et al.* (2002) Do whole-grain oat cereals reduce the need for antihypertensive medications and improve blood pressure control? *Journal of Family Practice* **51**, 353–359.

Post, R.E., Mainous, A.G., King, D.E., and Simpson, K.N.,(2012) Dietary fiber for the treatment of type 2 diabetes mellitus: A meta-analysis. *Journal of the American Board of Family Medicine* **25**, 16–23.

Priebe, M., Binsbergen, J.v., Vos, R.d., and Vonk, R.J. (2008) Whole grain foods for the prevention of type 2 diabetes mellitus. *Cochrane Database of Systematic Reviews* **1** (Art. No.: CD006061). doi: 10.1002/14651858.CD006061.pub2.

Quílez, J., García-Lorda, P., and Salas-Salvadó, J. (2003) Potential uses and benefits of phytosterols in diet: Present situation and future directions. *Clinical Nutrition* **22**, 343–351.

Ripsin, C.M., Keenan, J.M., Jacobs, J., *et al.* (1992) Oat products and lipid lowering: A meta-analysis. *Journal of the American Medical Association* **267**, 3317–3325.

Ruxton, C.H.S. and Derbyshire, E. (2008) A systematic review of the association between cardiovascular risk factors and regular consumption of oats. *British Food Journal* **110**, 1119–1132.

Saltzman, E., Das, S.K., Lichtenstein, A.H., *et al.* (2001) An oat-containing hypocaloric diet reduces systolic blood pressure and improves lipid profile beyond effects of weight loss in men and women. *Journal of Nutrition* **131**, 1465–1470.

Sesso, H.D., Buring, J.E., Christen, W.G., *et al.* (2008) Vitamins E and C in the prevention of cardiovascular disease in men: The physicians' health study II randomized controlled trial. *The Journal of the American Medical Association* **300**, 2123–2133.

Smeds, A.I., Jauhiainen, L., Tuomola, E., and Peltonen-Sainio, P. (2009) Characterization of variation in the lignan content and composition of winter rye, spring wheat, and spring oat. *Journal of Agricultural and Food Chemistry* **57**, 5837–5842.

Sosulski, F., Krygier, K., and Hogge, L. (1982) Free, esterified, and insoluble-bound phenolic acids. 3. Composition of phenolic acids in cereal and potato flours. *Journal of Agricultural and Food Chemistry* **30**, 337–340.

Steffen, L.M., Jacobs, D.R., Stevens, J., *et al.* (2003) Associations of whole-grain, refined-grain, and fruit and vegetable consumption with risks of all-cause mortality and incident coronary artery disease and ischemic stroke: The Atherosclerosis Risk in Communities (ARIC) Study. *American Journal of Clinical Nutrition* **78**, 383–390.

Suzuki, R., Allen, N.E., Key, T.J., *et al.* (2009) A prospective analysis of the association between dietary fiber intake and prostate cancer risk in EPIC. *International Journal of Cancer* **124**, 245–249.

Terry, P., Giovannucci, E., Michels, K.B., *et al.* (2001) Fruit, vegetables, dietary fiber, and risk of colorectal cancer. *Journal of the National Cancer Institute* **93**, 525–533.

Thompson, L.U. (1994) Antioxidants and hormone-mediated health benefits of whole grains. *Critical Reviews in Food Science and Nutrition* **34**, 473–497.

Thompson, L.U., Robb, P., Serraino, M., and Cheung, F. (1991) Mammalian lignan production from various foods. *Nutrition and Cancer* **16**, 43–52.

Tiwari, U. and Cummins, E. (2011) Meta-analysis of the effect of β-glucan intake on blood cholesterol and glucose levels. *Nutrition* **27**, 1008–1016.

USDA (2010) *Dietary Guidelines for Americans, 2010*, 7th edn. US Department of Agriculture, Department of Health and Human Services, Washington, DC.

USDA Agricultural Research Service (2011) USDA Database for the Flavonoid Content of Selected Foods, Release 3.0. http://www.ars.usda.gov/SP2UserFiles/Place/12354500/Data/Flav/Flav_R03.pdf (last accessed 16 May 2013).

Wolever, T.M., Tosh, S.M., Gibbs, A.L., *et al.* (2010) Physicochemical properties of oat β-glucan influence its ability to reduce serum LDL cholesterol in humans: A randomized clinical trial. *American Journal of Clinical Nutrition* **92**, 723–732.

Wolk, A., Manson, J.E., Stampfer, M.J., *et al.* (1999) Long-term intake of dietary fiber and decreased risk of coronary heart disease among women. *JAMA: Journal of the American Medical Association* **281**, 1998–2004.

Wood, P.J. 1994. Evaluation of oat bran as a soluble fibre source. Characterization of oat β-glucan and its effects on glycemic response. *Carbohydrate Polymers* **25**, 331–336.

Zayachkivska, O.S., Konturek, S.J., Drozdowicz, D., *et al.* (2005) Gastroprotective effects of flavonoids in plant extracts. *Journal of Physiology and Pharmacology* **56**, 219–231

8

Avenanthramides: Chemistry and Biosynthesis

Mitchell L. Wise

United States Department of Agriculture, Agricultural Research Service, Cereal Crops Research, Madison, WI, USA

8.1   Introduction

Avenanthramides are a group of phenolic alkaloids initially discovered in oats (*Avena sativa*) that function as phytoalexins (i.e., antimicrobial compounds produced in response to infection). These natural products are conjugates of one of three phenylpropanoids (*p*-coumaric, ferulic, or caffeic acid) and anthranilic acid (or a hydroxylated and/or methoxylated derivative of anthranilic acid). Avenanthramides and related compounds are also found in the eggs of white cabbage butterflies (*Pieris brassicae* and *P. rapae*) (Blaakmeer *et al.*, 1994) and have been isolated from fungus-infected carnation (*Dianthus caryophyllus*), indicating that both monocots and dicots produce these metabolites (Ponchet *et al.*, 1988). Unlike oat, carnations typically use benzoic acid rather than cinnamic acid derivatives as the acyl donor in the biosynthesis of these phytoalexins.

Shigeyuki Mayama investigated the possibility that oat produces phytoalexins in response to Crown Rust (*Puccinia coronata*) infection and found that methanolic extracts of leaves from oat cultivar Shokan-1 infected with an incompatible race of *P. coronata* (race 226) yielded a fraction of UV-absorbing compounds by LH-20 chromatography. These compounds were not found in healthy leaves or leaves infected with a compatible race. Further examination of this fraction revealed the presence of three novel metabolites, each with a benzoxazine-4-one functionality which he termed avenalumins (Figure 8.1) (Mayama *et al.*, 1981a). It was further demonstrated that the purified avenalumins inhibited fungal germination *in vitro*. These investigators synthesized the metabolites and determined that the synthetic compounds were toxic to both compatible and

Oats Nutrition and Technology, First Edition. Edited by YiFang Chu.
© 2014 John Wiley & Sons, Ltd. Published 2014 by John Wiley & Sons, Ltd.

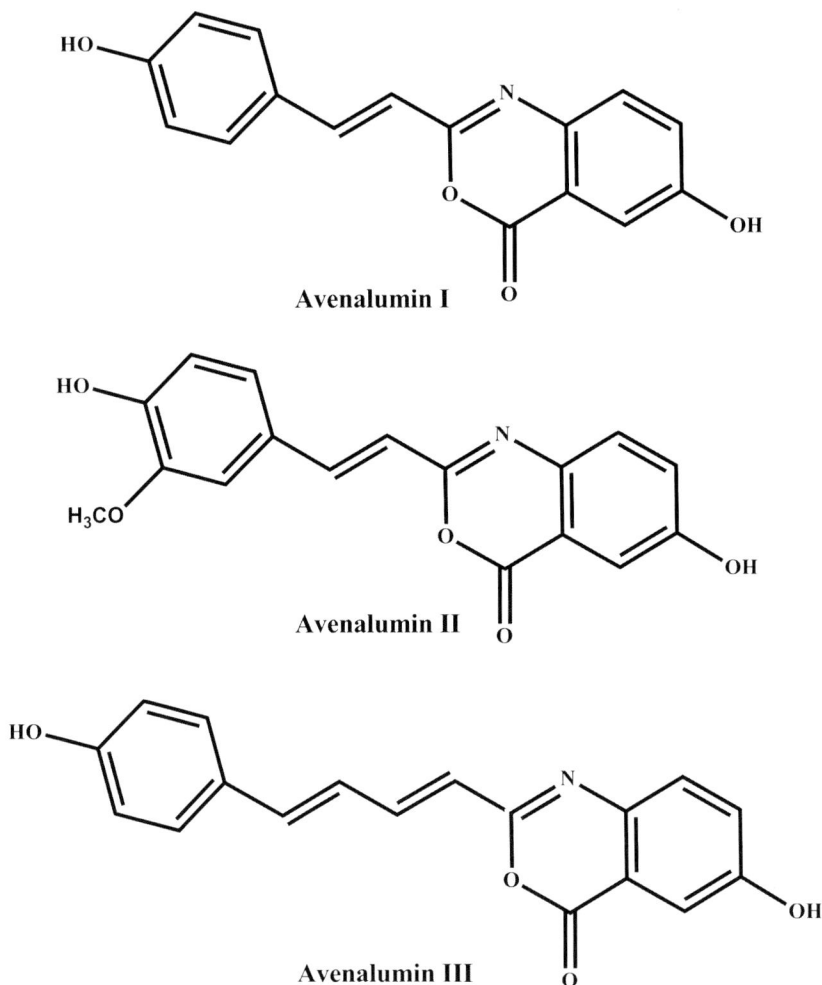

Figure 8.1   Structures of the avenalumins as originally described by Mayama (Mayama *et al.*, 1981b). Courtesy of the USDA.

incompatible races of *P. coronata*, demonstrating that they were phytoalexins, the first described in monocots. In addition, avenanthramides are also found in the oat grain. However, in contrast to leaf tissue, where they are usually either non-detectable or present at very low concentrations in the absence of fungal inoculation, grain avenanthramides appear to be constitutively expressed but in highly variable concentrations.

8.2   Nomenclature

Several nomenclatures describing the avenanthramides have been presented. Originally termed avenalumin I, II, and III (Mayama *et al.*, 1981a), Collins found multiple compounds in oat grain extracts with structures similar to the avenalumins, which he called avenanthramides (Collins, 1989). Collins assigned an alphabetic descriptor to each avenanthramide congener (Table 8.1). A more systematic nomenclature was subsequently developed by Dimberg, whereby the anthranilate derivative was assigned a letter (e.g., A = anthranilate, B = 5-hydroxy anthranilate) and the accompanying phenylpropanoid was designated by c (caffeic acid), f (ferulic acid), or p (*p*-coumaric acid). Later this nomenclature was modified to use a numeric descriptor for the anthranilic acid moiety. Thus Collins' avenanthramide C = Dimberg's Bc or 2c. Dimberg's alphanumeric nomenclature is used in this chapter. It was subsequently determined that Mayama's avenalumins I, II and III were, in fact, the open ring amides corresponding to avenanthramide 2p, 2f, and 2pd respectively (Crombie and Mistry, 1990; Miyagawa *et al.*, 1995). Although the avenalumin term has fallen out of use, the other three nomenclatures are all found in the current literature.

8.3   Synthesis

In a comprehensive report in 1989, Collins described the structure of several of these metabolites based on meticulous structural analysis, substantiated by synthesis of several of the avenanthramides (Collins, 1989). To synthesize the avenanthramides Collins used a modification of the method described by Bain and Smalley (1968). Commercially available cinnamic acids (e.g., *p*-coumaric acid) were first reacted with acetic anhydride to protect the hydroxyl group, thus yielding the acetoxy derivative. The acetoxy cinnamic acid was then reacted with thionyl chloride to form the highly reactive acid chloride, which was then reacted with either anthranilic acid or 5-hydroxy anthranilic acid (both commercially available, although in the original procedure Collins synthesized 5-hydroxyanthranilate from 5-chloro-2-nitrobenzoic acid) yielding the acetoxy avenanthramide. Hydrolysis of the protecting group was accomplished by reflux in a 50% methanolic solution with 10% ammonium hydroxide, which typically resulted in 50–60% yield (Collins, 1989). Another approach to avenanthramide synthesis uses peptide coupling reagents such as dicyclohexylcarbodiimide (Ishihara *et al.*, 1998) or benzotriazol-1-yloxy-tris (dimethylamino) phosphonium hexafluorophosphate (BOP) (Wise, 2011) to activate the carboxylate group on the phenylpropanoid moiety before reaction with anthranilic acid or 5-hydroxyanthranilate. Because avenanthramides, particularly the caffeic acid derivatives, are somewhat sensitive to high alkalinity, this author has found that acetylation before reaction with BOP and use of an organic base (e.g., pyrollidine diluted in dichloromethane) to deprotect the hydroxyl groups considerably improves yield. It should also be noted that the acid chloride intermediate in the Collins' method is highly sensitive to water. The synthetic avenanthramides can be purified from reaction byproducts by LH-20 chromatography (Collins, 1989).

8.4   Chemical stability

Because avenanthramides exhibit certain nutritional attributes (Meydani, 2009), their stability is important as it relates to the retention of these phytonutrients in processed foods. Oat processing usually involves hydrothermal treatment as an early step to inactivate certain enzymes (e.g., lipases and lipoxygenases)

Table 8.1 General structure of avenanthramides with the nomenclatures used by Collins and Dimberg. Compounds marked with an asterisk (*) are not known to occur naturally and (**) these compounds have been termed "L" in some reports

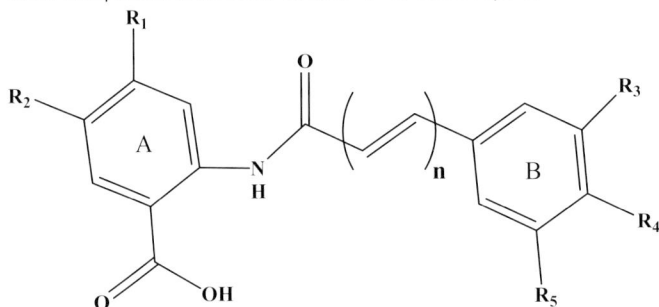

		Nomenclature					
Dimberg	Collins	n	R_1	R_2	R_3	R_4	R_5
1p	D	1	H	H	H	OH	H
$1p_d$	L	2	H	H	H	OH	H
1f	E	1	H	H	OCH_3	OH	H
$1f_d$	M	2	H	H	OCH_3	OH	H
1c	F	1	H	H	OH	OH	H
$1c_d$	N	2	H	H	OH	OH	H
*1a		1	H	H	H	H	H
*1s		1	H	H	OCH_3	OH	OCH_3
2p	A	1	H	OH	H	OH	H
$2p_d$	**O	2	H	OH	H	OH	H
2f	B	1	H	OH	OCH_3	OH	H
$2f_d$	P	2	H	OH	OCH_3	OH	H
2c	C	1	H	OH	OH	OH	H
$2c_d$	Q	2	H	OH	OH	OH	H
*2a		1	H	OH	H	H	H
*2s		1	H	OH	OCH_3	OH	OCH_3
3p	X	1	OCH_3	OH	H	OH	H
$3p_d$	U	2	OCH_3	OH	H	OH	H
3f	Y	1	OCH_3	OH	OCH_3	OH	H
$3f_d$	V	2	OCH_3	OH	OCH_3	OH	H
3c	Z	1	OCH_3	OH	OH	OH	H
$3c_d$	W	2	OCH_3	OH	OH	OH	H
*3a		1	OCH_3	OH	H	H	H
*3s		1	OCH_3	OH	OCH_3	OH	OCH3
4p	G	1	OH	H	H	OH	H
$4p_d$	**R	2	OH	H	H	OH	H
4f	H	1	OH	H	OCH_3	OH	H
$4f_d$	S	2	OH	H	OCH_3	OH	H
4c	K	1	OH	H	OH	OH	H
$4c_d$	T	2	OH	H	OH	OH	H
5p	AA	1	OH	OH	H	OH	H
$5p_d$	OO	2	OH	OH	H	OH	H
5f	BB	1	OH	OH	OCH_3	OH	H
$5f_d$	PP	2	OH	OH	OCH_3	OH	H
5c	CC	1	OH	OH	OH	OH	H
$5c_d$	QQ	2	OH	OH	OH	OH	H

responsible for rancidity (Girardet and Webster, 2011). In laboratory experiments avenanthramides have proven to be relatively heat stable. For example, in a study to determine the pH and temperature stability of 2p, 2f, and 2c, Dimberg and colleagues found that the 2p concentration was essentially unchanged in sodium phosphate buffer after three hours at room temperature or in a 95°C water bath; at pH 7 and 12, there was a slight loss at 95°C. Avenanthramide 2f appeared to be more sensitive to the higher temperatures at pH 7 and 12. Avenanthramide 2c, on the other hand, was completely degraded at pH 12 at both temperatures and diminished by more than 80% at 95°C, even at pH 7 (Dimberg *et al.*, 2001). The effect of UV light on avenanthramide *trans-cis* isomerization was also assessed in this report. Although cinnamic acid and its derivatives are known to isomerize under UV light (Kort *et al.*, 1996), Dimberg found that the three avenanthramides tested remained exclusively in the *trans* conformation after 18 hours exposure to UV light at 254 nm. In contrast, Collins reported that the avenanthramides isomerize upon exposure to daylight or UV light (Collins and Mullin, 1988).

Using samples obtained from commercial oat processing plants in Sweden, Bryngelsson and colleagues (Bryngelsson *et al.*, 2002) investigated the effect of commercial heat processing on avenanthramide content, namely steaming (two sequential treatments at 100°C for one hour and then 20 min), autoclaving (2.4 bar, 120°C, 16 min and dried at 100°C), and drum drying (rolled flakes or ground whole meal). They found that 2f and 2c levels were essentially unaffected by steaming, although 2p levels were lowered by approximately 30%. Autoclaving similarly reduced 2c and 2p levels (approximately 30%), whereas 2f levels were unchanged by autoclaving alone but reduced slightly after drying. All three of the tested avenanthramides were substantially reduced in drum-dried rolled oats, but avenanthramide levels were unchanged in whole-meal oats. The authors speculated that avenanthramides in the drum-dried whole meal were relatively stable because prior autoclaving reduced their level to a point beyond which they did not further decrease, in contrast to the drum-dried rolled oats, which were previously subject to steam inactivation. These findings indicate that the avenanthramides are relatively heat stable but somewhat labile to alkaline conditions, especially 2c. Indeed, cooking certain oat-based products actually increased avenanthramide levels in some cases, possibly by releasing bound forms of the metabolites or promoting *de novo* biosynthesis (Dimberg *et al.*, 2001).

8.5   Antioxidant properties

Oat contains numerous compounds with antioxidant properties (Emmons and Peterson, 1999; Peterson, 2001). In fact, in the early twentieth century food products were commonly protected from oxidative spoiling by packaging in oat flour-coated paper or by incorporating oat flour into the product itself (Peters, 1937). Like other phenolic natural products, avenanthramides possess antioxidant properties. Several studies have demonstrated that avenanthramides contribute substantially to the overall antioxidant capacity of oats (Peterson, 2001; Bratt *et al.*, 2003; Fagerlund *et al.*, 2009). However, measuring antioxidant activity *in vitro* is a somewhat contentious subject; none of the numerous assays available can reliably claim to measure total antioxidant capacity (Peterson, 2001; Huang *et al.*, 2005). Nevertheless, results of several assays show that avenanthramides are potent antioxidants, with 2c usually showing the greatest antioxidant activity followed by 2f and 2p, in that order (Peterson *et al.*, 2002).

In a detailed investigation of structural factors contributing to the antioxidant functions of avenanthramides, Fagerlund and colleagues (Fagerlund *et al.*, 2009) synthesized a series of naturally occurring avenanthramides as well as certain nonbiological analogs. These compounds were cinnamic (a), ferulic (f), *p*-coumaric (p), caffeic (c), and synaptic (s) acid conjugates of anthranilic (1), 5-hydroxy (2), and 5-hydroxy-4-methoxy (3) anthranilic acids (Table 8.1). These synthetic avenanthramides were assayed for radical scavenging activity using the 2,2-diphenyl-1-picrylhydrazyl assay and for antioxidant activity using an azo-initiated linoleic acid hydroperoxide formation assay (via spectrophotometric measurement of diene formation). As expected, 1a showed virtually no inhibition of diene formation or radical scavenging activity, whereas 3s, which possesses an hydroxyl and ortho methoxy group on the A (anthranilate) ring and methoxy groups on either side of the 4-hydroxyl group on the B ring, was the most potent inhibitor of linoleic acid oxidation. The catechol-containing 1c, 2c, and 3c compounds were the strongest radical scavengers but only marginally more so than their s-series counterparts. Compound 3f, which possesses ortho hydroxyl, methoxy groups showed stronger radical scavenging activity than any of the 1c,s or 2c,s congeners. Thus, substitution of a methoxy group ortho to the phenolic hydroxyl appears more important to antioxidant activity than which ring contains these substitutions. This is probably due to stabilization of the resulting radical through hyperconjugation (Fagerlund *et al.*, 2009). The anthranilate hydroxyl group also appeared to be important in inhibiting linoleic acid oxidation in that 2a, which lacks B ring hydroxyls, was nearly as effective as 3c or 3p. Because amide groups can exhibit double bond characteristics, the authors originally hypothesized that extended conjugation between ring structures might contribute to the antioxidant action. However, *ab initio* calculations did not support this hypothesis (Fagerlund *et al.*, 2009). Although the antioxidant capacity of oat avenanthramides may have some nutritional value and almost certainly reduces rancidity during storage, many of their purported health benefits are not directly attributable to their antioxidant properties (Meydani, 2009; Chapter 11).

8.6   Solubility of avenanthramides

Avenanthramides are soluble in ethyl acetate, diethyl ether, aqueous acetone, and methanol but relatively insoluble in chloroform and benzene (Collins, 1989). Their solubility in water is highly dependent on pH. Ionization of the A ring carboxyl group at higher pH promotes solubility. This ionization also results in a bathochromic shift (longer wavelength) in the band I absorption maxima, resulting in a distinct green color in some of the avenanthramides (Collins, 1989). Indeed, there have been anecdotal reports of oat products turning an unappealing grayish-green when cooked (Doehlert *et al.*, 2009). Doehlert and coworkers found that high pH or ferrous iron (Fe^{2+}) in solutions used to cook oatmeal resulted in a distinct greenish hue. Other divalent cations such as Ca^{2+} or Mg^{2+}

did not produce this effect. Although the ferrous iron is rapidly oxidized to the essentially insoluble ferric (Fe^{3+}) state in the presence of atmospheric oxygen, the authors speculated that under certain conditions, such as using well-drawn water, sufficient ferrous iron might be retained to elicit the grayish-green coloration and that phenolic compounds, avenanthramides in particular, were likely a major contributing source of the chromophores (Doehlert *et al.*, 2009).

8.7   Analysis of avenanthramides

Analysis of avenanthramide content in oat grain and oat products is typically accomplished by solvent extraction followed by high-performance liquid chromatography with UV absorption or liquid chromatography/mass spectrometry. Identification of individual avenanthramides by simple UV absorbance is problematic in that sources of authentic standards are limited. Only recently has a commercial source of any of these metabolites become available from a specialty chemical company in Switzerland, which offers avenanthramides 2c, 2f, and 2p. Previously, investigators were compelled to synthesize these metabolites in their laboratory or obtain them from others who did so. Nevertheless, avenanthramides have relatively high extinction coefficients (23–28 000 L/mol-cm) at 340 nm (Collins, 1989; Wise, unpublished). Hence, UV absorbance is a fairly sensitive method for detection. Even so, definitive identification is possible only for the avenanthramides for which authentic standards are available to determine retention time identity.

   Mass spectral detection offers a more powerful means of identifying these metabolites. Liquid chromatography interfaced with detection by mass spectrometry is easily accomplished with avenanthramides (Jastrebova *et al.*, 2006). They are readily ionized by electrospray ionization in both positive and negative modes. Appropriate ionization conditions result in partial fragmentation of the compounds at the amide bond, thus providing additional structural information. Tandem mass spectrometry or an ion trap can also provide fragmentation data. Fragmentation in the positive mode yields the acylium carbocation of the phenylpropanoid moiety, whereas fragmentation in the negative mode yields the isocyanate derivative of the anthranilate moiety (Figure 8.2). Thus, application of these ionization modes makes it possible to discern some of the isomeric avenanthramides such as 2c and 5p (Wise, 2011).

8.8   Biosynthesis of avenanthramides

The biosynthesis of avenanthramides results from the acylation of anthranilic acid and derivatives by the CoA thioester of *p*-coumaric, ferulic, or caffeic acid, catalyzed by hydroxycinnamoyl CoA: hydroxyanthranilate *N*-hydroxycinnamoyl transferase (HHT, EC 2.3.1). All of these substrates originate from carbohydrate metabolism funneled through the shikimate pathway (Mann, 1987). Thus, erythrose-4-phosphate, originating from pentose phosphate metabolism, and phosphoenolpyruvate, from glycolysis, are condensed to 3-deoxy-D-*arabio*-heptulosonate-7-phoshate (DAHP) by DAHP synthase in the first committed step of the shikimate pathway (Figure 8.3). Most lines of research indicate that

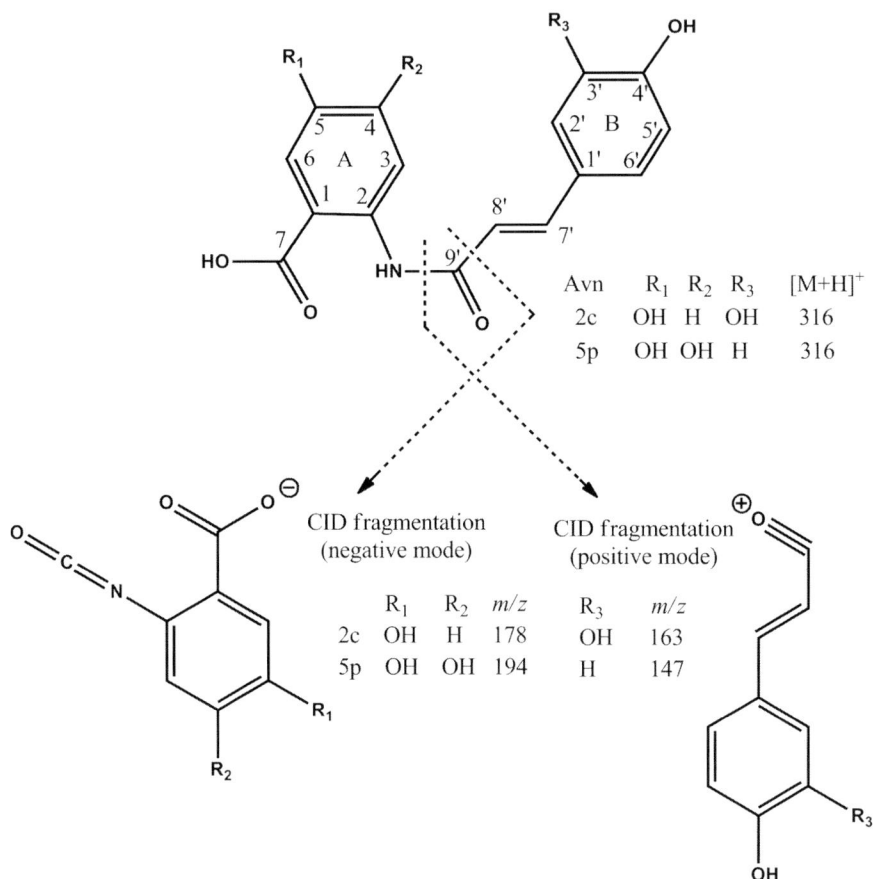

Figure 8.2   Fragmentation of avenanthramides subsequent to electrospray ionization. Fragmentation can be effected with appropriate settings in the spray chamber or by tandem mass spectrometry. Ionization in positive or negative mode results in differential bond cleavage. The protonated molecular ion and typical fragments observed for the isomeric avenanthramides 2c and 5p are provided to illustrate how these metabolites can be differentiated. CID = collision-induced dissociation.

this metabolism is compartmentalized within plastids (Tzin and Galili, 2010). However, there is some evidence that the necessary enzymes can be found in the cytosol (Maeda and Dudareva, 2012). Five additional enzymatic reactions yield chorismate, which can be transformed into either anthranilate by anthranilate synthase, or prephenate by chorismate mutase. Anthranilate is subsequently converted to tryptophan, whereas prephenate is the precursor to tyrosine or phenylalanine.

   This important biosynthetic pathway provides not only the aromatic amino acids but also the precursors to lignin and a multitude of natural products involved in biotic and abiotic stress responses. Thus, regulation of carbon flux through this pathway is important in plant physiology but presently not well

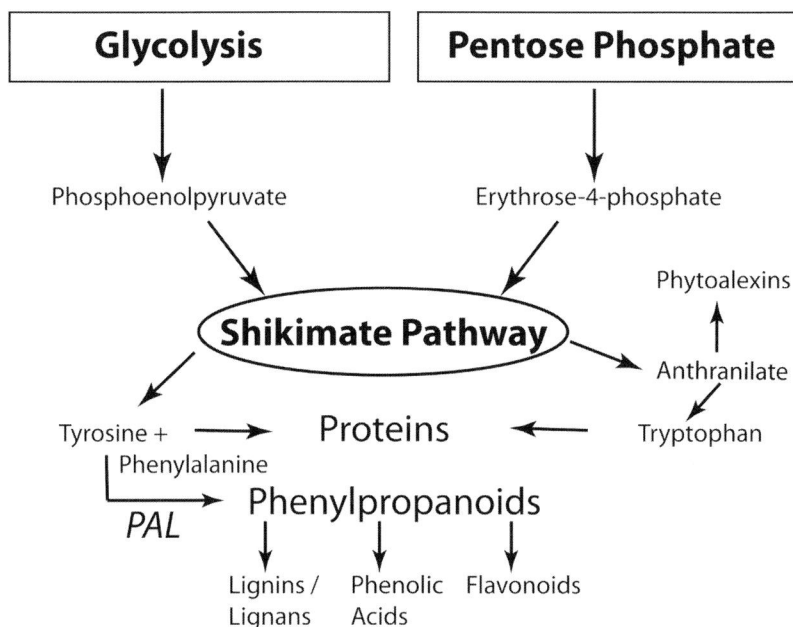

Figure 8.3   Generalized flow of carbohydrate metabolism into and out of the shikimate pathway.

understood (Tzin and Galili, 2010), although increased expression of several of the enzymes involved is influenced by pathogen infection and other environmental cues (Tzin and Galili, 2010; Maeda and Dudareva, 2012). The penultimate biosynthetic step leading to chorismate is catalyzed by 5-enolpyruvylshikimate-3-phosphate synthase (EC 2.5.1.19), the target of glyphosate, one of the most widely used herbicides in the world, and has, therefore, been subject to extensive research efforts.

   Conversion of phenylalanine (or tyrosine) into the phenylpropanoids is the next stage in avenanthramide biosynthesis. Because of the importance of lignins in wood chemistry and their propensity to inhibit cellulosic biofuel production, these critical steps in lignin biosynthesis have received substantial attention for many years. Phenylalanine ammonia lyase (PAL, EC 4.3.1.24) mediates transformation of the amino acid to *trans*-cinnamic acid as the first committed step in phenylpropanoid biosynthesis (Figure 8.4). Like many enzymes involved in phenylpropanoid metabolism, PAL occurs as an enzyme family, with at least four known isoforms in *Arabidopsis*. The various isoforms of PAL appear to be either constitutively expressed or induced by environmental factors, including UV light, pathogen attack, wounding, and plant growth regulators (Jones, 1984; Dixon and Paiva, 1995). Differential expression of PAL isoforms appears to regulate the flux of compounds through the shikimate pathway to their final metabolic fate (i.e., lignin biosynthesis or secondary metabolism) (Cochrane *et al.*, 2004; Rohde *et al.*, 2004).

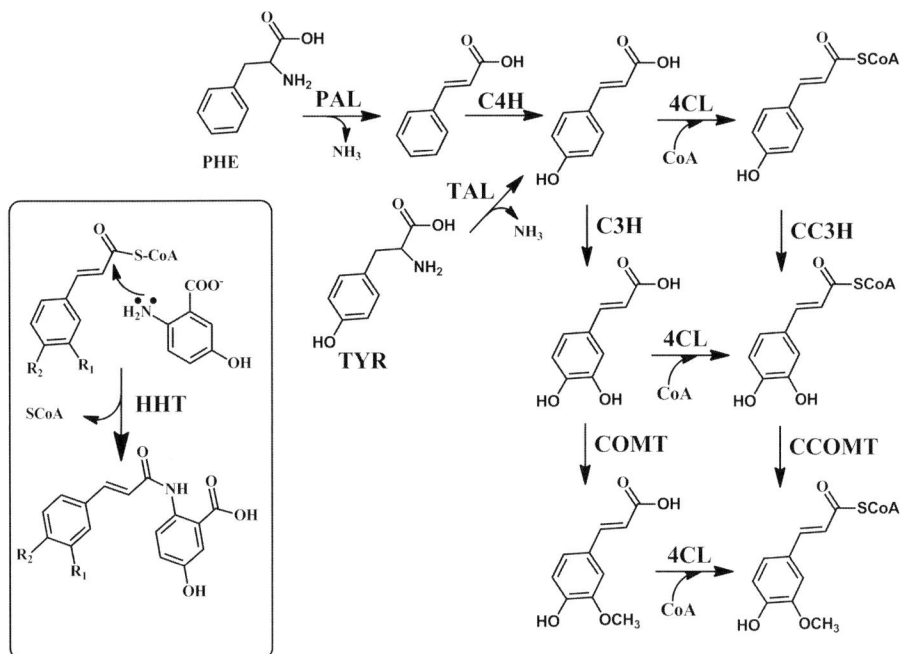

PAL = phenylalanine ammonia-lyase, TYR = tyrosine, TAL = tyrosine ammonia-lyase, C3H = p-coumarate 3-hydroxylase, C4H = cinnamic acid 4-hydroxylase, CC3H = p-coumarate-CoA 3-hydroxylase, COMT = caffeate O-methyl transferase, CCOMT = caffeoyl-CoA O-methyl transferase.

Figure 8.4   Biosynthesis of phenylpropanoids as precursors to avenanthramide biosynthesis. Inset shows the reaction of a cinnamoyl-CoA intermediate with an anthranilic acid derivative catalyzed by hydroxycinnamoyl-CoA:hydroxyanthranilate N-hydroxycinnamoyl transferase (HHT) in the final step of avenanthramide biosynthesis.

The next step in phenylpropanoid metabolism is hydroxylation of *trans*-cinnamic acid to p-coumaric acid catalyzed by a P-450 monooxygenase, cinnamic acid 4-hydroxylase (C4H, EC 1.14.13.11). C4H is induced by many of the environmental factors that influence PAL expression. Conversion of tyrosine directly to p-coumaric acid is also catalyzed by PAL in some monocots (Rösler *et al.*, 1997). p-Coumaric acid can be further modified by hydroxylation at the 3 position by p-coumarate 3-hydroxylase (C3H, EC 1.14.13) to yield caffeic acid. In contrast to PAL and 4-coumaryl CoA ligase (see below), C4H and C3H are each encoded by single genes in *Arabidopsis*. They also appear to have rate-limiting roles in monolignol biosynthesis, important precursors to both lignin and lignan production (Costa *et al.*, 2003). Ferulic acid is produced from either caffeic acid or caffeoyl CoA by caffeate O-methyl transferase (EC 2.1.1.68) or caffeoyl-CoA O-methyl transferase (EC 2.1.1.104), respectively.

Activation of the cinnamic acid carboxylate is usually required for further biochemical reactions, such as reduction to the corresponding aldehydes and subsequently to cinnamyl alcohols used in lignin biosynthesis or production of a multitude of natural products. This activation is mediated by the formation of a thioester bond to Coenzyme A catalyzed by ATP-requiring enzymes termed 4-coumarate CoA ligases (4CL, EC 6.2.1.12). Most vascular plants contain this enzyme family, whose isoforms likely play key roles in dictating the ultimate dispensation of the phenylpropanoid precursors. Most 4CLs accept only *p*-coumaric, caffeic, and ferulic acids as substrates (Schneider *et al.*, 2003), although there are reports of 4CL isoforms accepting sinapic acid (3,5-dimethoxy, 4-hydroxy cinnamic acid) as substrate (Lindermayr *et al.*, 2002; Hamberger and Hahlbrock, 2004). The diversity of the 4CL genes allows these enzymes to be induced by numerous developmental and environmental factors. An interest in altering the lignin content in various feedstocks for biofuel generation has stimulated considerable research into the specificities of this enzyme class (Vanholme *et al.*, 2008). Although much remains to be determined, it is clear that the 4CLs are critical in determining the fate of phenylpropanoids. The final step in avenanthramide biosynthesis is catalyzed by HHT, which is described in detail below.

One of the enigmas of biochemical processes in general and secondary metabolism in particular is: how are multienzyme-catalyzed biosynthetic pathways able to funnel certain metabolic intermediates through to their final product? The phenylpropanoid pathway to secondary metabolites has offered intriguing clues. Besides compartmentalization of metabolic pathways into subcellular organelles, metabolic channeling may be another mechanism by which cooperating enzymes are collocated to produce high concentrations of their reaction products available for the next biosynthetic step. Metabolic channeling may explain the ability of plants to direct metabolic flux through the myriad biosynthetic pathways involved in primary and secondary metabolism with minimal loss to the cellular milieu (Hrazdina and Jensen, 1992). A comprehensive review on this subject can be found elsewhere (Winkel, 2004). In brief, this concept suggests that many secondary metabolites result from transient associations of multienzyme complexes "channeling" intermediates from one enzymatic transformation to the next, which explains the efficient biosynthesis of compounds whose precursors exist at exceedingly low concentrations in the cell. There is increasing experimental evidence to support this concept, particularly as it relates to phenylpropanoid biosynthesis (Achnine *et al.*, 2004; Winkel, 2004). The specific protein associations likely reflect alterations in the primary structure of the enzyme isoforms that catalyze the same reactions in different biosynthetic pathways. For example, an *in silico* analysis of expressed sequence tag data for the PAL enzyme family by the Lewis group at Washington State University showed clear correlations between tissue types and expression of specific isozymes (Costa *et al.*, 2003). However, the involvement of metabolic channeling in avenanthramide biosynthesis has not been examined in any detail.

As briefly mentioned in the introduction of this chapter, early investigations into avenanthramide biosynthesis were precipitated by an interest in their phytoalexin properties. Thus, Mayama and his colleagues evaluated a series of oat cultivars with known Crown Rust-resistance genes (Mayama *et al.*, 1982). They challenged these cultivars with two races of *P. coronata* to which they had different susceptibilities and measured the length and rate at which the fungal hyphae grew. They also quantitated the levels of avenanthramide 2p and 2f (avenalumins I and II in the original text) in the leaves. During the first 24 h, hyphal growth rate was fairly uniform among the cultivars but subsequently slowed in highly to moderately resistant cultivars. Although there was some variation in susceptibility to the two different races among cultivars, avenanthramide 2p and 2f production was highly correlated with retardation of hyphae growth. This was observed in both the rate at which the avenanthramides were produced and their final concentrations (144 h after infection).

Mayama further investigated the genetics of *P. coronata* resistance by analyzing the crosses between a resistant cultivar (Shokan-1) and two susceptible cultivars (Kanota and CW-491-4) (Mayama *et al.*, 1982). F_1 progeny displayed an intermediate resistance, as determined by hyphae growth and appearance of uredia on infected leaves, whereas F_2 progeny of both crosses yielded the 3:1 ratio expected from a single semidominant gene. Moreover, 2p levels were again highly correlated with Crown Rust resistance.

8.9 Victorin sensitivity

An intriguing aspect of the single-gene resistance of oats was revealed by studying cultivars carrying the *Pc-2* allele, which confers resistance to certain races of Crown Rust. Oats carrying the *Pc-2* gene are susceptible to Victoria blight, caused by the fungus *Cochliobolus victoriae*, and exhibit sensitivity to victorin, a host-specific toxin produced by the fungus (*Vb* phenotype). Repeated attempts to separate these two characteristics have failed, suggesting that the *Pc-2* and *Vb* genes are either the same or very closely linked, likely the former (Rines and Luke, 1985; Mayama *et al.*, 1995; Navarre and Wolpert, 1999). Treating *Pc-2* oat leaves with very low concentrations of purified victorin induces avenanthramide 2p biosynthesis (Mayama *et al.*, 1986).

To determine the inheritability of Crown Rust resistance, victorin sensitivity, and avenanthramide biosynthesis, the cultivar Victoria (carrying *Pc-2*) was crossed with non-*Pc-2* cultivars (Kanota, CW-491-4, and Shokan-1). Victoria is incompatible with *P. coronata* races 202 and 226, Kanota and CW-491-4 are compatible with both races, and Shokan-1 is compatible with 202 and incompatible with 226. The F_1 Victoria × Kanota hybrids showed a gene-dosage effect in their ability to produce avenanthramide 2p following challenge with *P. coronata* race 226 (Mayama *et al.*, 1995). All three crosses showed the expected 1:2:1 ratio in terms of victorin sensitivity (sensitive, moderately sensitive, and insensitive) and a 3:1 ratio in terms of resistance to *P. coronata* race 202, with those showing susceptibility almost exclusively being insensitive to victorin toxicity. A small fraction of the victorin-insensitive plants showed partial resistance to *P. coronata* race 202, whereas none of the victorin-sensitive plants were susceptible. Analysis of 2p accumulation in F_2 hybrids of the Victoria × Shokan-1 cross, 72 h after inoculation with race 202, demonstrated that the victorin-sensitive (*Pc-2* homozygous) plants produced substantial amounts of 2p, the moderately sensitive (*Pc-2* heterozygous) plants produced about one-third as much 2p, and the insensitive plants produced only trace amounts of 2p. To further evaluate the segregation of victorin sensitivity and rust resistance, F_3 generations of two lines of the Victoria × CW-491-4 heterozygous plants were tested for victorin sensitivity and resistance to *P. coronata* race 202. Again a 1:2:1 ratio for victorin sensitivity and a 3:1 ratio for rust resistance were observed. Thus Crown Rust resistance, avenanthramide production, and victorin sensitivity all appeared to co-segregate, substantiating the association of the *Pc-2* gene with these traits.

The fact that victorin functions as both a toxin and as a phytoalexin elicitor is perplexing. In a study using fluorescein-labeled victorin and a bovine serum albumin (BSA)-victorin-fluorescein complex (BSA to prevent victorin transport across the cell membrane), Tada and colleagues (Tada *et al.*, 2005) demonstrated that victorin-mediated cell death preceded entry of the victorin into the mesophyll cells of oat leaves. They also showed that sublethal concentrations of BSA-victorin elicited high levels of avenanthramide 2p in the *Pc-2* carrying leaves. Pharmacological inhibitors of Ca^{2+} influx strongly inhibited victorin-induced cell death. Pretreatment with antimycin, which depletes ATP stores, likewise reduced cell death. From these results (for additional evidence, see Akimitsu *et al.*, 1993), they proposed that victorin exerts its effects through binding to a cell surface receptor, likely the gene product of *Pc-2/Vb*, which in turn stimulates ATP-dependent influx of Ca^{2+}, thereby providing a signal for both phytoalexin biosynthesis and apoptosis (Tada *et al.*, 2005). The response to victorin by oat leaves shows many of the hallmarks of apoptosis (Tada *et al.*, 2001) and is similar to that induced by infection with incompatible strains of *P. coronata*.

Much of the early work on the biosynthesis of avenanthramides focused on their production in vegetative tissue in response to fungal elicitation. Using detached leaves with epidermis layers peeled off and floated on solutions of chemical elicitors, several studies have shown that avenanthramide biosynthesis can be dramatically upregulated by chemical mimics of fungal infection and certain abiotic stressors (Bordin *et al.*, 1991; Ishihara *et al.*, 1996, 1997, 1998, 1999; Miyagawa *et al.*, 1996a, 1996b). Suspension cultures of oat callus also produce avenanthramides, predominately 2p and 4p (but others as well), in response to treatment with crab shell chitin (Wise *et al.*, 2009). Chitin, polymeric β-1,4 linked (2-acetylamino)-2-deoxy-D-glucose, is a component of fungal cell walls and crustacean exoskeletons. Chitin and its partially or fully deacetylated form, chitosan, serves as a signaling molecule to trigger plant defense mechanisms (Hahn, 1996). These elicitors of plant basal immunity are now termed microbial- (or pathogen-) associated molecular patterns (Boller and Felix, 2009). Ca^{2+} ionophores also induce avenanthramide 2p biosynthesis in oat leaf segments (Ishihara *et al.*, 1996).

8.10 Environment effects on avenanthramide production

In addition to oat leaves, avenanthramides are also found in oat grain (Collins, 1986, 1989), which appears to contain all the requisite enzymes necessary for their biosynthesis (Matsukawa *et al.*, 2000; Peterson and Dimberg, 2008; Dimberg and Peterson, 2009). However, unlike vegetative tissue, oat grain produces avenanthramides constitutively, although their levels tend to be highly variable (Dimberg *et al.*, 1996; Emmons and Peterson, 2001). Emmons and Peterson (2001) conducted a study on the relationship between environment and grain avenanthramide production. They grew three cultivars (Belle, Gem, and Dane), adapted for the United States upper Midwest, in seven locations throughout Wisconsin over 3 years and found a strong genotype × environment effect on avenanthramide levels in the grain. A comparison of total avenanthramides for each of the cultivars across all locations showed that Belle had significantly higher levels than Gem, and Gem had significantly higher levels than Dane (P = 0.05) during each of the three years of the study. Interestingly, when the total avenanthramides of the three cultivars were combined and their averaged values compared at two different locations (Sturgeon Bay and Arlington, WI), the Sturgeon Bay values far exceeded those of Arlington. The environmental factors responsible for the extraordinary levels of avenanthramides of grain grown in Sturgeon Bay have not been determined.

In an analysis of the agronomic traits of 33 oat genotypes grown in the western region of the United States, Peterson and colleagues (Peterson *et al.*, 2005) also observed a strong environmental effect on grain avenanthramide levels. However, in this study avenanthramide levels were far lower than those observed in the Midwest. This difference may be explained by the drier environment in the western United States, which does not favor Crown Rust. These results, and those from a previous study of Swedish oat cultivars grown in different environments (Mannerstedt-Fogelfors, 2001), suggest a relationship between Crown Rust and grain levels of avenanthramide (Peterson *et al.*, 2005).

Wise and colleagues (Wise *et al.*, 2008) evaluated 18 oat lines grown in three locations over 2 years for Crown Rust incidence and grain avenanthramide levels. These experiments were conducted in North Dakota, where rust pressure varies from high to nonexistent depending on seasonal variations in humidity. In this study, two environments, Fargo and Carrington in 2005, were subject to heavy Crown Rust infestations; the other four environments were essentially free from Crown Rust. As shown in Table 8.2, total avenanthramide content in all cultivars was dramatically higher in environments with rust infection than in the rust-free environment. For example, total avenanthramide levels in the cultivar Maida grown in Fargo were 17-fold higher under rust conditions than under nonrust conditions. Overall, observed Crown Rust resistance correlated with grain avenanthramide content (although some exceptions were noted). These reports suggest a relationship between environmental conditions, most notably (but likely not exclusively) Crown Rust incidence, and grain avenanthramide levels, which is intriguing because Crown Rust does not directly infect the oat grain. Two potential mechanisms may explain this. Either a mobile signaling mechanism similar to systemic acquired resistance upregulates avenanthramide biosynthesis in the grain, or avenanthramides are transported from the leaf tissue to the filling grain.

Table 8.2 Avenanthramide concentration and genetic Crown Rust resistance in the grain of 18 genotypes[a]

Genotype	Fargo 2005	Fargo 2006	Carrington 2005	Carrington 2006	Williston 2005	Williston 2006	Crown Rust resistance[b]
	Total avenanthramide concentration, mg/kg						
ND030291	79.7 a	5.5 b	70.0 a	7.0 b	6.4 d–f	6.3 c–h	5
HiFi	62.0 b	4.4 bc	32.5 c–e	4.9 cd	6.0 e–g	5.3 d–i	5
AC Assiniboia	44.5 c	3.3 c–e	36.4 c	9.5 a	11.5 a	33.2 a	4
Gem	37.8 cd	2.7 d–g	28.6 c–f	2.2 h	4.1 i–k	3.1 h–j	3
AC Pinnacle	37.2 cd	8.1 a	52.3 b	10.2 a	7.9 c	12.6 b	4
Maida	35.2 c–e	2.0 e–i	26.8 c–f	4.2 d–f	7.2 cd	6.8 c–f	4
Beach	32.2 d–f	2.8 d–g	17.4 f–i	5.0 cd	4.8 g–i	9.0 c	2
Brawn	30.6 d–g	2.3 d–i	19.8 f–i	3.2 e–h	3.8 i–k	4.9 e–i	1
CDC Weaver	27.1 d–g	2.5 d–h	34.1 cd	4.4 c–e	5.7 e–g	7.3 c–f	3
Killdeer	25.4 e–g	3.6 cd	24.2 d–g	5.8 bc	8.6 b	8.1 c–e	1
CDC Dancer	22.0 f–h	3.2 c–f	22.2 e–h	4.0 d–f	4.4 h–k	4.3 f–j	2
Ronald	21.3 f–i	1.9 f–i	13.0 g–i	3.8 d–g	4.2 l–k	5.1 d–i	3
Morton	20.5 g–i	1.4 hi	15.0 g–i	1.9 h	3.0 j–l	3.3 g–j	5
ND021612	12.6 h–j	1.5 g–i	10.2 i	2.2 h	2.2 l	1.8 i–j	5
AC Morgan	10.8 ij	2.7 d–g	17.3 f–i	3.2 e–h	6.8 c–e	8.4 cd	0
Leonard	10.4 ij	1.2 i	8.7 i	2.1 h	2.2 l	1.2 j	1
Otana	7.7 j	1.7 g–i	12.1 h–i	2.8 f–h	5.6 f–h	5.3 d–h	0
Triple Crown	6.6 j	2.5 d–h	9.9 i	2.4 gh	3.2 j–l	6.7 c–g	1
Mean	29.1	3.0	25.0	4.4	5.4	7.4	
Crown Rust environment	Yes	No	Yes	No	No	No	
Correlation with Crown Rust resistance	0.615**	0.295[ns]	0.475*	0.236[ns]	−0.011[ns]	−0.008[ns]	

[a]Values with different letters in a column differ significantly P < 0.05. [b]Genetic Crown Rust resistance as determined in the Fargo 2005 environment (0 indicates severe infection; 5 indicates complete resistance). **P < 0.001, *P < 0.05, ns = not significant.
Reprinted from Wise *et al.*, 2008. Courtesy of the USDA.

8.11 Hydroxycinnamoyl-CoA: Hydroxyanthranilate N-hydroxycinnamoyl transferase (HHT)

Ishihara and colleagues (Ishihara *et al.*, 1997) initially described the activity of HHT, the enzyme responsible for avenanthramide biosynthesis. Crude enzyme preparations were prepared from victorin C-induced oat leaves by ammonium sulfate precipitation (30–45% fraction) of phosphate buffer extractions, followed by desalting on Sephadex-25. These enzyme preparations were used to characterize the substrate specificity of HHT. The hydroxycinnamoyl-CoA esters used by HHT are not commercially available; hence, they were prepared by transesterification of the appropriate hydroxycinnamoyl-N-hydroxysuccinimide esters, as described by Stöckigt and Zenk (1975). The following substrates were used as acyl acceptors: anthranilate, 5-hydroxyanthranilate, tyramine, 3- and 4-hydroxyanthranilate. The first three of these are commercially available; the source for the last two was not provided in the report.

These substrates were analyzed using p-coumaryl-CoA as the common acyl donor. For the acyl acceptors, the maximum relative velocity (relV_{max}) was achieved with anthranilate as substrate, followed by 5-hydroxyanthranilate (59% relV_{max}) and 4-hydroxyanthranilate (41% relV_{max}). 3-Hydroxyanthranilate and tyramine did not appear to function as substrate. 5-Hydroxyanthranilate had the lowest K_m (12 μM), followed by 3-hydroxyanthranilate (120 μM) and anthranilate (340 μM). Thus, 5-hydroyxanthranilate proved to be the most efficiently used substrate with relV_{max}/K_m = 4.9 compared with 4-hydroxyanthranilate (relV_{max}/K_m = 0.34) and anthranilate (relV_{max}/K_m = 0.29). In terms of acyl donors, using 5-hydroxy anthranilic acid as substrate, feruloyl-CoA was the best substrate in terms of both relV_{max} and K_m (4 μM) followed by cinnamoyl-CoA (65% relV_{max}, K_m = 27 μM) and p-coumaroyl-CoA (24% relV_{max}, K_m = 16 μM). Interestingly, caffeoyl-CoA, the likely substrate for avenanthramide 2c, showed a relV_{max} only 9% of feruloyl-CoA and K_m = 18 μM. Avenanthramide 2c is usually not abundant in leaf extracts of fungus-infected or chemically-induced plants (Mayama *et al.*, 1981b; Ishihara *et al.*, 1997, 1999; Wise, 2011) but is typically the most abundant avenanthramide found in oat grain (Peterson and Dimberg, 2008; Ren and Wise, 2012). Avenalumoyl-CoA [5-(4'-hydroxyphenyl)-penta-2E,4E-dienoic acid-CoA] was also a reasonably good substrate (16% relV_{max}, K_m = 4.4 μM). The optimal pH for HHT was 7.0.

A similar study was conducted by the same group using various chitin oligomers, ranging from monomeric to hexamers, to elicit HHT in oat leaf segments and again using ammonium sulfate-precipitated protein as the enzyme source (Ishihara *et al.*, 1998). The chitin pentamer (penta-N-acetylchitopentose) proved to be the best elicitor and the enzyme characteristics were similar to those described in the previous report, although anthranilate had a far lower K_m (63 vs. 340 μM). The authors suggested that different isozymes, possessing different substrate affinities may have been elicited (Ishihara *et al.*, 1997). Although feruloyl-CoA proved to be the best substrate in both of these studies, avenanthramide 2f (the product of feruloyl-CoA) was not detected in victorin-treated tissue and comprised only 11% of the total avenanthramides in the chitin-treated leaves. Avenanthramide 2p was, by far, the most highly produced in the leaf segments, comprising 78% of the total avenanthramides (1p, 4p, and 4pd represented the rest). The following two explanations were offered for these results: (1) substrate availability may account for the prevalence of 2p, or (2) the metabolic fate of 2f may differ from that of 2p (Ishihara *et al.*, 1998). As discussed in section 8.13, the second rationale seems highly likely.

Substantiation that oats produce more than one isozyme of HHT came when Ishihara and his colleagues investigated HHT activity in oat grain (Matsukawa *et al.*, 2000). Both dry and germinated oat seeds were ground separately in liquid nitrogen, extracted into phosphate buffer, precipitated with ammonium sulfate, and, after desalting, fractionated by anion exchange chromatography (diethylaminoethanol [DEAE]-Sepharose followed by Mono Q). Two separate peaks of HHT activity were obtained from the DEAE column, with kinetic constants (K_m and relV_{max}) that were quite similar with respect to the acyl-CoA donors but were significantly different with respect to the anthranilate acceptors. One fraction had a K_m almost five times higher than the other for anthranilate. Analysis by gel filtration chromatography (Superdex 75) provided a molecular weight of approximately 40 kDa for both anion exchange HHT fractions, indicating that they did not represent multimeric forms of the enzyme. HHT isolated from seedling leaves elicited with penta-*N*-acetylchitopentose were found to occur as isozymes with similar elution profiles from the DEAE column. Seed germination was also shown to increase HHT activity. Analysis of the dissected germinating seedling indicated the HHT activity was localized primarily in the scutellum and endosperm. Little activity was found in the emerging shoots and none in the root tissue. Analysis of the grain avenanthramides showed that 2c was the predominate congener, and 2f and 2p were present in nearly equal concentrations, about half that of 2c. Imbibed seeds increased total avenanthramides by approximately 2.5-fold with a doubling in weight, but the relative profile of congeners remained essentially the same (Matsukawa *et al.*, 2000).

8.12 Cloning HHT

Confirmation for the existence of several isomers of HHT came with the cloning of three full-length HHT cDNAs and one partial fragment from oat. This was accomplished based largely on previous work in cloning a hydroxycinnamoyl/benzoyl-CoA:anthranilate *N*-hydroxycinnamoyl/benzoyltransferase (HCBT) from carnation (Yang *et al.*, 1997). Both HHT and HCBT belong to a large family of acyltransferases collectively termed BAHD transferases. BAHD is an acronym for the acronyms of the first four enzymes characterized in this family (BEAT, AHCT, HCBT, and DAT) (D'Auria, 2006). The carnation enzyme was cloned using peptide sequences from trypsin digested fragments of the enzyme responsible for dianthramide biosynthesis (Reinhard and Matern, 1989). Dianthramides are conjugates of anthranilic acid with benzoyl-CoA, which is catalyzed by HCBT.

Reverse transcription-polymerase chain reaction (RT-PCR) with degenerate primers and poly-A RNA isolated from elicited carnation suspension cultures produced a 0.8-kb cDNA. This fragment was used as a probe to screen a cDNA library from elicited carnation cultures. Three separate cDNAs encoding what appeared to be full-length acyltransferases were cloned and one was functionally expressed in *Escherichia coli* (Yang *et al.*, 1997). The three cDNAs showed high sequence identity (95–97%), suggesting that they were isoforms. Both native and cloned HCBT showed greater affinity for cinnamoyl- and 4-hydroxycinnamoyl-CoA than for benzoyl-CoA, suggesting that some forms of avenanthramides might be found in the plant. However, none of the avenanthramides were found in suspension cultures of carnation and they are only minor constituents of the native plant. Thus, these metabolites are not unique to oats (Ponchet *et al.*, 1988) and the enzymes necessary for their biosynthesis have been isolated and cloned from carnation. Interestingly, HCBT has a much greater affinity for anthranilate as substrate than for 3- or 4-hydroxyanthranilate, the derivatives found in the dianthramides. During their investigation of carnation phytoalexin biosynthesis, Reinhard and Matern (1989) showed that methoxydianthramide formation could be catalyzed using a microsomal preparation with 4-hydroxydianthramide as substrate, but not 4-hydroxyanthranilate. They were also unable to detect any hydroxylase activity toward anthranilic acid. This led them to conclude that modification of the anthranilate moiety in carnation phytoalexins occurs after formation of the *N*-benzoylanthranilic acid. As discussed in the following section, this does not appear to be the case with avenanthramide biosynthesis in oats.

Using conserved sequences of HCBT and related acyltransferases, Yang and colleagues designed degenerate primers for use on a cDNA library produced from oat leaves treated with an extract of a plant growth-promoting rhizobacterium, *Pseudomonas fluorescens* FPT 9601 (Yang *et al.*, 2004). They amplified a 0.4-kb DNA fragment whose translated sequence showed substantial sequence similarity to the predicted segment of HCBT. This PCR amplicon was used as a probe to screen the cDNA library, which resulted in three full-length HHT clones (AsHHT1-3) and a partial cDNA clone. The four clones showed 95–97% amino acid sequence identity with each other and 42–43% identity to the HCBT1 gene product (60% similarity) (Figure 8.5), although the DNA sequence identity was not nearly as great. The full-length clones represented proteins of 440 or 441 amino acids with molecular weights of 47.8 to 47.9 kDa; however, efforts to express AsHHT1 in *E. coli* have met with limited success. The recombinant plasmid resulted in massive overproduction of a soluble 48-kDa protein, but enzyme activity using 5-hydroxyanthranilate and feruloyl-CoA as substrate was quite low. Immunoblot analysis of crude protein extracts from elicited oat leaves using antibodies raised against AsHHT1 showed bands corresponding to proteins with molecular weights of approximately 28, 39, and 47 kDa. Thus, HHT appears to undergo posttranslational cleavage or other modifications in order to be fully functional (Yang *et al.*, 2004). A caffeoyl-CoA 3-*O*-methyl transferase gene from oats was also cloned in this study.

Investigation of the responses of Shokan-1 and PC-38 cultivars infected with *P. coronata* has produced intriguing results. Infection of Shokan-1 with the compatible race 203 and incompatible race 226 both resulted in the upregulation of HHT mRNA, as determined by RNA hybridization using an AsHHT1-based probe. The increase in mRNA occurred somewhat faster in the incompatible interaction, but by 24-h postinfection oat leaves infected with race 203 also showed increased HHT mRNA levels. Although virtually no avenanthramide was detected in leaves with the compatible infection, very high avenanthramide 2p levels (> 1600 mg/g fresh weight) were detected in leaves with the incompatible interaction. These results suggest that posttranscriptional factors regulate avenanthramide biosynthesis. In addition, hybridization analysis with an AsHHT4 probe (the partial sequence isolated with AsHHT1-3, see previously) indicated that production of that mRNA was constitutive; both compatible

```
                                                                          60
AsHHT1   (1)  MKITVRSSTVVVPAAETPRVRLWNANPDLVV--PRFHTPSVYFYRRGG-E----DGG AC
AsHHT2   (1)  MKITVRSTTVVVPAAETPRLRLWNANPDLVV--PRFHTPSVYFYRRGD------G GAAC
AsHHT3   (1)  MKITVRSSTVVVPAAETPRVRLWNANPDLVV--PRFHTPSVYFYRRGG------DGDA C
DcHCBT   (1)  M IQIK STMVRPAEETP KSLWLSNIDMILRTPYSHTGAVL YK PDNNEDNIHPSSSM
HvAGCT   (1)  MKITV SS  VKPEYG CGVAPGC ADVVPL-----T L KANF TYISV---IYAFH P
                                                                         120
AsHHT1  (54)  YFDAGRMRRALAEALVPFYPMAGRLAHDEDGRVEIDCNAEGVLFVEADAPDGAVDDFGDF
AsHHT2  (53)  YFDAARMRRALAEALV FYPMAGRLAHDEDGRVEIDCNAEGVLFLEADAPDG VDDFGDF
AsHHT3  (53)  YFDAARMRRALAEALVPFYPMAGRLAHDEDGRVEIDCNAEGVLFVEADAPDGAVDDFGDF
DcHCBT  (61)  YFDANILI ALS ALVPFYPMAGRLKI GD-RYEIDCNAEG LFVEAES- H LEDFGDF
HvAGCT  (53)  APPNAVL AGLG ALVDYREWAGRLGVDA G RAILLNDAG RFVEATA-DVALD VMP
                                                                         180
AsHHT1 (114)  VPTMGLKR-LIPTVDFTGGISSYPLLVVQVTHFKCGGVALGIAMQHHVADGFSGLHFINS
AsHHT2 (113)  APTMGLKR-LIPTVDFTGGISSYPLLV QVTHFKCGGVALGIGMQHHVADGFSGLHFINS
AsHHT3 (113)  APTMGLKR-LIPTVDFTGGISSYPLLVVQVTHFKCGGVALGIGMQHHVADGFSGLHFINS
DcHCBT (119)  RP DEL RVMVPTCDYSKGISSFPLLMVQLT FRCGGVSIG AQ HHV DG AH  FNNS
HvAGCT (112)  KPTSEVLS-----L PSG D  EELMLIQVT FACG L VG  AQHLVSDGRATS F LA
                                                                         240
AsHHT1 (173)  WSDLCRGVPIAVMPFIDRTI-LRARDPPVPTHPHIEYQPAPAMLGSEEPQALAGKPESPP
AsHHT2 (172)  WSDICRGVPIAVMPFIDRTL-LRARDPPVPTHPHIEYQPAPAMLGSEEPQALAGKPESPP
AsHHT3 (172)  WSDLCRGVPIAVMPFIDRTL-LRARDPPVPTHPHIEYQPAPAMLGSEEPQALAGKPESPP
DcHCBT (179)  WA I KGLLPALEPVHDRYLHLRPR PEQI Y HS F PFVPSLP ELL GK  K  T--
HvAGCT (167)  WS ATRGVAVDPVPVHDR SF  PREPLHVEYEHRGV FKPYEKAHDVVCGADG ED --
                                                                         300
AsHHT1 (232)  TAVDIFKLSRSDLGRLRAQLPTGEGAPRFSTYAVLGAHVWRCASLARGLAPEQPTKLYCA
AsHHT2 (231)  TAVDIFKLSRSDLGRLRAQLPTGEGAPRFSTYAVLGAHVWRCASMARGLAPEQPTKLYCA
AsHHT3 (231)  TAVDIFKLSRSDLGRLRAQLPTGEGAPRFSTYAVLGAHVWRCASLARGLAPEQPTKLYCA
DcHCBT (237)  ----LFILSR  I TLKQ LDL    R STYEVVAAHVWR  SKARGLSD   EIKLIMP
HvAGCT (225)  V KV  SR FI KLKAQA  GAPRP-CSTLQCVVAHLWR MTMARGLDGG  T VAIA
                                                                         360
AsHHT1 (292)  TDGRQRLT-PTHPDGYFGNVIFTATPI-AEAGKVTGSLADGATTIQDALEKMDDEYCHSA
AsHHT2 (291)  TDGRQRLT-PTLPDGYFGN IFTATPI-AEAGKVTGSLADGATTIQEALEKMDDEYCHSA
AsHHT3 (291)  TDGRQRLT-PTLPDGYFGNVIFTATPI-AEAGKVTGSLADGATTIQDALEKMDDEYCHSA
DcHCBT (293)  DGR RI NPSLP GYCGNVVFLA CTA VG LSC  L DTAG VQEAL GLDDDYL SA
HvAGCT (284)  DGRARMS-PQVPDGYTGNVI WARPTT AGELVDR VK A ELI  EVARI DGYF SF
                                                                         420
AsHHT1 (350)  LDYLELQPDLSALVRGA----HTFRCPNLGLTSWVRLPIHDADFGWGRPVFMGPGGIAYE
AsHHT2 (349)  LDYLELQPDLSALVRGA----HTFRCPNLGLTSWVRLPIHDADFGWGRPVFMGPGGIAYE
AsHHT3 (349)  LDYLELQPDLSALVRGA----HTFRCPNLGLTSWVRLPIHDADFGWGRPVFMGPGGIAYE
DcHCBT (353)  IDHTES P L  PYMGSP---  TLYPNVLV SWGRIPY AMDFGWGSP F GI  IFYD
HvAGCT (343)  IDFA SGAVE ERLVA ADAA MVL PNIEVDSWLRIP YDMDFGGGRPFF MP YLPVE
                                                                         480
AsHHT1 (406)  GLAFVLPSANRDGSLSVAISLQAEHMEKFRKMIFDF
AsHHT2 (405)  GLAFVLPSANRDGSLSVAISLQAEHMEKFRKMIFDF
AsHHT3 (405)  GLAFVLPSANRDGSLSVAISLQAEHMEKFRKMIFDF
DcHCBT (410)  GQ FLIPSR GDGSMTLAI LFS HL RFKKY YDF
HvAGCT (403)  GLL LLPSFLGDGSV  YV LFS  M  FK CCY  D
```

Figure 8.5 Deduced sequence alignment of AsHHT 1-3, the carnation hydroxycinnamoyl/benzoyl-CoA:anthranilate N-hydroxycinnamoyl/benzoyltransferase (DcHBTC) and a barley agmatine *p*-coumaryl-CoA transferase (HvAGCT). Underscored sequences were used to design degenerate primers for cloning oat hydroxycinnamoyl-CoA:hydroxyanthranilate N-hydroxycinnamoyl transferase.

PC-38/race 226 and incompatible Shokan-1/race 226 interactions produced protein bands before fungal infection, which increased in intensity after infection. The authors concluded that AsHHT4 may be involved in the biosynthesis of compounds other than phytoalexins. BADH acyltransferases are clearly an important gene family, with the list of recognized members constantly expanding. A 2006 review reported that the *Arabidopsis* genome is predicted to have at least 64 representatives, and the rice genome 119 representatives (D'Auria, 2006). Thus, the oat genome may have additional BAHD members with unknown functions. It is also possible that the HHTs reported may catalyze other acyltransferase activities *in planta*.

8.13   Metabolic flux of avenanthramides

Most studies of avenanthramide biosynthesis have shown that HHT most efficiently uses feruloyl-CoA as substrate, as determined by the relV_{max}/K_m ratio (Ishihara *et al.*, 1997, 1998; Matsukawa *et al.*, 2000). Even 4CL activity in crude extracts from elicited oat leaves reacts most efficiently with ferulic acid to produce feruloyl-CoA (Ishihara *et al.*, 1999). However, in induced leaves avenanthramide 2p is typically the most highly represented form, with 2f produced in lesser quantities. Moreover, avenanthramides appear to readily traverse the cell membrane. For example, excised oat leaves floated on elicitor solutions secrete avenanthramides into the solution (Miyagawa *et al.*, 1996a). Similarly, oat callus suspension cultures secrete avenanthramides 2p and 4p into the medium following chitin elicitation (Wise *et al.*, 2009). Interestingly, in both experimental systems avenanthramide levels in solution and in the tissue decrease over time.

In a detailed and incisive examination of avenanthramide metabolism, Okazaki (2004b) monitored the fate of labeled avenanthramides (2p and 2f). After adding ^{13}C-labeled avenanthramides to elicitor solutions upon which oat leaf (Shokan-1) segments were floated, ratios of unlabeled to labeled avenanthramide in the tissue very quickly matched those in the elicitor solution, indicating rapid equilibrium between the tissue and the external environment. Assuming that the addition of exogenous avenanthramides did not perturb normal metabolism and that the avenanthramide concentrations in solution mirrored those in the leaf, they determined the ratios of the labeled to unlabeled avenanthramides in solution over time to dissect avenanthramide biosynthesis (production of new avenanthramides) from avenanthramide metabolism (disappearance of free avenanthramides). They found that 2f was biosynthesized at a higher rate than 2p; it was also metabolized at a higher rate. These studies help explain the disparity between the *in vitro* kinetic data of enzymes involved in avenanthramide biosynthesis and the *in vivo* levels of the avenanthramides found in elicited leaves.

By floating chitin elicited leaf segments on solutions of ^{14}C-labeled 2f, these investigators also found that the some of the avenanthramide was converted into a dehydrodimer (Figure 8.6) previously described by Okazaki (2004a), some of which was putatively incorporated into the cell walls of the leaves. Although the identity of these metabolites was not determined, a portion of the radioactivity was released from the cell wall isolate by alkaline hydrolysis (saponification). Additional dimeric forms of avenanthramide 2f have subsequently been reported (Okazaki *et al.*, 2007). The ethyl acetate-extractable radiolabeled metabolites were eluted from both reverse-phase chromatography and Sephacryl gel filtration in multiple fractions and in a manner consistent with relatively high molecular weights. Adding peroxidase inhibitors to the elicitor solutions significantly reduced the rate of avenanthramide 2f metabolism. These findings indicate

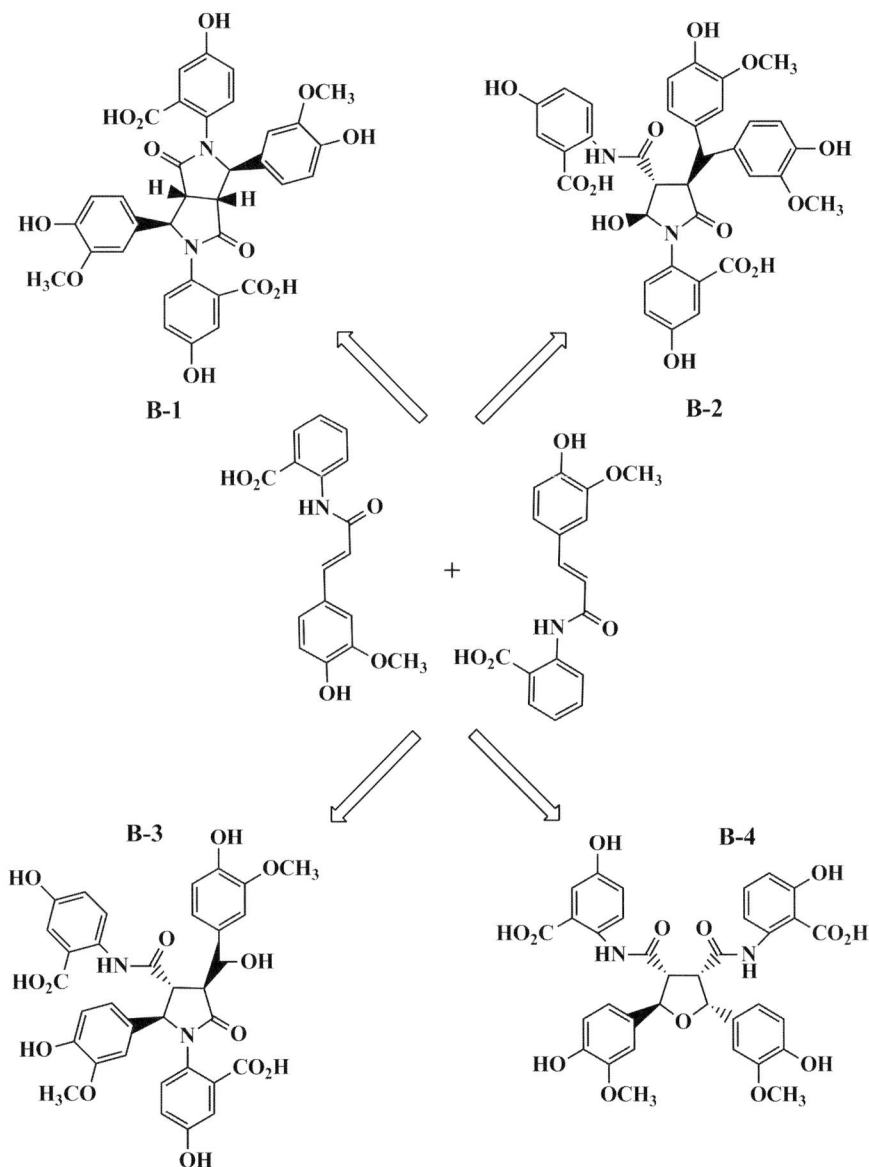

Figure 8.6   Structures of several dimers of avenanthramide 2f, termed bisavenanthramides by Ishihara (Okazaki *et al.*, 2004a, 2007). Shown are bisavenanthramides B-1 through B-4.

that avenanthramides are further metabolized, probably through a peroxidase-mediated radical mechanism, before incorporation into the cell wall. Phenoloxidase enzyme activity has also been reported in buffer extracts of oat grain and germinating oat seed (Bryngelsson *et al.*, 2003b; Skoglund *et al.*, 2008). The conversion of avenanthramide 2f into dehydrodimers and their incorporation into the cell wall may serve to mechanically reinforce the cell wall against pathogen invasion. A similar phenomenon was observed with diferulic acid in *P. coronata*-infected oat leaves (Ikegawa *et al.*, 1996).

8.14   Localization of avenanthramide biosynthesis

The temporal and spatial production of avenanthramides has recently been investigated using different methodologies. One study used line-scanning fluorescence microscopy in concert with laser micro sampling and nano-high performance liquid chromatography (Kajiyama *et al.*, 2006) to analyze avenanthramide concentrations in individual cells and determine their subcellular localization (Izumi *et al.*, 2009). Examining mesophyll cells in detached oat leaves (cv. Shokan-1) elicited with penta-*N*-acetylchitin, these investigators concluded that avenanthramide biosynthesis occurred in cells undergoing a hypersensitive response (HR). Taking advantage of the differences in the fluorescence spectrum of avenanthramides and chlorophyll, they observed that the spatial distribution of fluorescence from avenanthramides coincided with fluorescence attributed to chlorophyll, suggesting that avenanthramide biosynthesis takes place in the chloroplast.

In a different study conducted by Uchihashi *et al.* antibodies specific for avenanthramide 2p and AsHHT were used to immunostain tissues from oat leaves responding to *P. coronata* infection (Uchihashi *et al.*, 2011). Both avenanthramide 2p and AsHHT accumulated in close proximity to cells undergoing HR at the early stages of infection (36–48 h postinfection). At 120-h postinfection, 2p immunostaining showed a wider distribution around the HR, extending into some areas where AsHHT was not apparent. The avenanthramide appeared to accumulate within the cell walls of tissue undergoing HR and adjacent cells (Figure 8.7), as determined by transmission electron microscopy, suggesting incorporation into the cell walls or apoplastic transport of the avenanthramides. The latter seems more likely, because the antibodies used were highly specific for avenanthramide 2p. As discussed earlier, 2f appears to be the avenanthramide largely associated with cell wall fortification, likely in the form of bisavenanthramides. In contrast to the Izumi study, Uchihashi and colleagues did not observe localization of HHT or avenanthramides within organelles. Furthermore, the published sequences of the cloned avenanthramides do not contain a recognized transit peptide. However, the native enzyme appears to be truncated from the translated cDNA; thus, the compartmentalization of avenanthramide biosynthesis remains unclear.

Uchihashi and colleagues used quantitative RT-PCR to assess HHT mRNA levels in oat (cv Shokan-1) leaves infected with a compatible (203) and an incompatible race (226) of *P. coronata* (Uchihashi *et al.*, 2011). Using primers matching conserved sequences of AsHHT1-3, they found that by 12-h postinfection, HHT mRNA increased approximately fivefold in the incompatible interaction but did not increase in the compatible interaction. These results contradict those of previous studies using RNA hybridization to determine mRNA production (Yang *et al.*, 2004). By 24 h, HHT mRNA was reduced to baseline

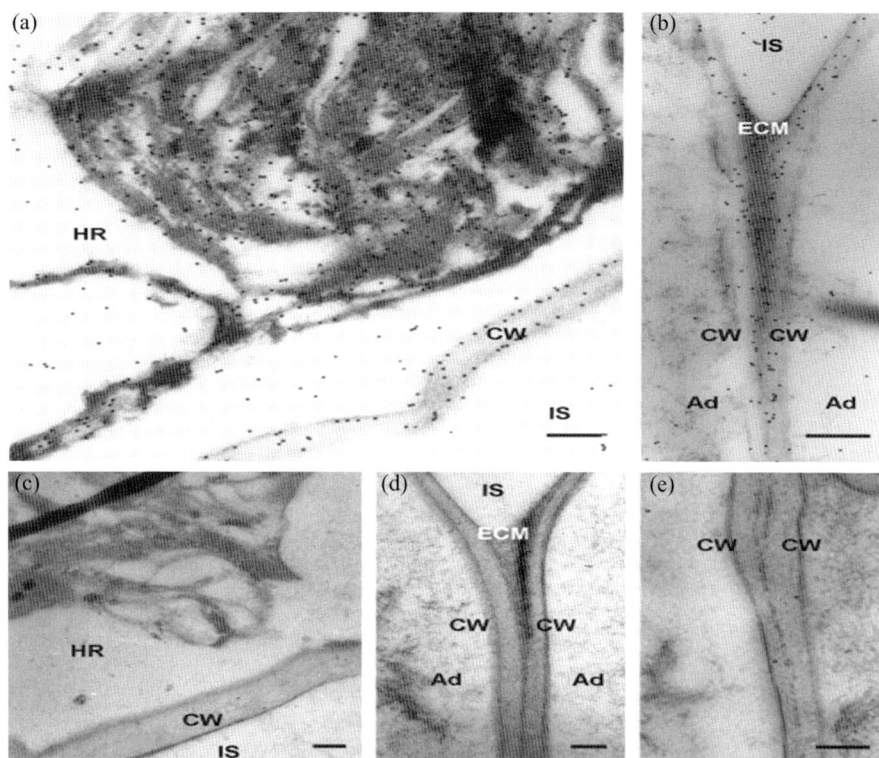

Ad = cell adjacent to HR cell, CW = cell wall, IS = intercellular space, ECM = extracellular matrix.
Bar = 200 nm

Figure 8.7 Immunolocalization of avenanthramides in ultra-thin sections of oat leaves infected with the incompatible race (226) of Crown Rust fungus. Infected oat leaves were sampled at 48 h postinoculation and analyzed by transmission electron microscopy with anti-avenanthramide antibody and gold-labeled secondary antibody. The experiment was repeated at least twice, and representative figures are presented. (a) Cell undergoing the hypersensitive response (HR). (b) HR-adjacent cells. (c–e) Negative controls: (c) HR cells treated without anti-avenanthramide antibody; (d) HR-adjacent cells treated without anti-avenanthramide antibody; (e) Mock inoculation control. *Source:* Uchihashi *et al.* (2011). Reproduced with permissions from Elsevier.

levels but increased again between 36- and 120-h postinfection. A similar analysis of pathogenesis-related protein-10 (PR-10) showed a 10-fold increase in both incompatible and compatible interactions; however, PR-10 mRNA fell to baseline levels and below by 120-h postinfection in the compatible interaction. In the incompatible system, PR-10 mRNA decreased slightly at 24 h but rebounded and remained high for the duration of the experiment. This likely reflects the zig-zag immune response model described by Jones and Dangl (2006), whereby plants upregulate certain biosynthetic pathways in response to pathogen-associated molecular pattern-triggered immunity (PTI). Pathogens respond by producing effectors to inhibit PTI. In some cases, the plants initiate a secondary immune response through the action of cytosolic receptors resulting in effector-triggered immunity (ETI). In the case of *P. coronata* infection, the PR-10 response is consistent with this model. In both incompatible and compatible interactions, the PTI appeared to upregulate production of this protein. The ETI failed in the compatible relationship presumably because the Shokan-1 cultivar did not respond adequately to the effector challenge. However, the incompatible race appears to have elicited an ETI. A similar response was observed with HHT, except that there was no PTI in the compatible interaction. These conclusions are somewhat speculative and much work is needed to firmly establish the nature of compatible versus incompatible relationships in oat Crown Rust infections.

The spatial location of avenanthramide biosynthesis in oat grain has not been as thoroughly investigated. Matsukawa described HHT activity in the endosperm and embryo in mature oat grain (Matsukawa *et al.,* 2000). The roles of HHT isoforms in the spatial and temporal regulation of avenanthramide biosynthesis, production of avenanthramide congeners, and metabolic channeling are unknown.

8.15 Plant defense activators

Plant defense activators are agrichemicals that initiate plant immunity by stimulating the systemic acquired response (Ryals *et al.,* 1996). They are, in essence, a means to "vaccinate" plants against potential pathogen attack. In the early 1990s investigators at Ciba-Geigy found that an analog of salicylic acid, 2,6-dichloro-isonicotinic acid, enhanced disease resistance in some crops if the plants were treated before pathogen attack (Kessmann *et al.,* 1994). This compound was never marketed because it was not well tolerated by many important crops. However, it did encourage research into this strategy for crop protection, resulting in the discovery of several benzothiadiazole derivatives such as benzo (1,2,3)-thiadiazole-7-carbothioic-*S*-methyl ester (BTH), marketed as Actigard^{TM}. Recent work in the author's laboratory has shown that, in greenhouse experiments, treatment with BTH as a root drench dramatically increased avenanthramide levels in the leaf tissue of oat seedlings (Wise, 2011). This increase was accompanied by upregulation of HHT and PR-10 mRNA, as determined by RNA hybridization analysis, consistent with a systemic acquired response. Anecdotally, mock-treated oat leaves also appeared to increase their avenanthramide levels relative to what is normally observed in greenhouse-grown (rust-free) seedlings. This may be due to an airborne signal from the BTH-treated plants. Elements of a systemic acquired response were clearly demonstrated in lima bean (*Phaseolus lunatus*) in response to airborne signaling from neighboring plants treated with BTH (Yi *et al.,* 2009). The role of airborne signaling in eliciting plant defense responses is increasingly recognized (Shulaev *et al.,* 1997; Heil and Ton, 2008), although little research on this phenomenon has been conducted with cereal crops.

Evaluation of several cultivars showed significant differences in the kinetics and magnitude of avenanthramide production in response to BTH treatment (Ren and Wise, 2012). Although all four cultivars tested in this study showed a marked increase in leaf avenanthramide levels, some responded much faster and the overall magnitude differed significantly between cultivars. As mentioned earlier in this chapter, oats normally produce avenanthramides constitutively in their grain, but concentrations range from 2 to 300 mg/kg or higher. There is also a strong environmental influence on grain avenanthramide levels, with Crown Rust infection likely being an important factor. Therefore, to test the effect of BTH on grain avenanthramide levels, plants were treated just before boot stage, and avenanthramide concentrations in the filling grain were analyzed two weeks after treatment. In all cases total avenanthramide levels in BTH-treated plants were substantially higher than in the mock-treated controls, but only by a statistically significant margin in one cultivar (Figure 8.8). For example, in Kame total avenanthramides were almost threefold greater in the treated grain than in the mock-treated grain, but very high variances obscured these differences in statistical terms.

It is interesting to note that the profile of avenanthramide congeners in the filling grain of BTH-treated plants differed from that of mature grain from the same cultivar (but not grown as part of the experiment). Specifically, avenanthramide 5p was far more abundant in the filling grain, whereas 2c was more abundant in the mature grain. Whether this reflects the normal maturation process or is a by-product of BTH treatment could not be assessed. A previous study by Peterson and Dimberg (2008) on the dynamics of avenanthramide production in filling grain showed that 2p was most abundant and some 2f was detected in the earliest stages of grain filling in all nine cultivars examined. Only at about 21 or 22 days after heading was 2c observed in any significant quantity. These authors did not report on 5p; it is not clear whether they did not observe 5p or simply did not recognize its presence. Interestingly, no HHT activity was detected in grain samples of any of the cultivars until at least 21 days after heading. These investigators also evaluated avenanthramides and HHT activity in 7-day-old greenhouse-grown seedlings. Unexpectedly, they found significant amounts of avenanthramides, particularly 2p and 2f, and smaller amounts of 2c, but no HHT activity. They suggested that avenanthramides were produced in the roots and transported to the leaves. Conversely, Wise detected avenanthramides in root tissue of BTH-treated seedlings, but only at seven days after treatment were they found in appreciable amounts and they continued to increase to the end of the 21-day experiment. There was no discernible HHT activity in the root tissue at any time (Wise, 2011). The lack of HHT activity in the roots, which were the first tissue to interact with the elicitor (thus, presumably the first tissue to respond) along with the delayed accumulation of avenanthramides, suggests that they were transported from leaf tissue to roots after BTH treatment.

8.16   False malting

Avenanthramide biosynthesis is known to be upregulated in germinating oat grain (Matsukawa *et al.*, 2000). Although avenanthramide concentrations increase in whole or milled/autoclaved grain as well as in germinating whole grain steeped under a range of temperatures and times, the greatest increases occurs by steeping and germinating whole grain under moderate (20°C) thermal conditions (Bryngelsson *et al.*, 2003a). There also appears to be a genetic component in the magnitude of avenanthramide production during germination (Skoglund *et al.*, 2008). A recent innovation in increasing avenanthramide levels in oat grain has

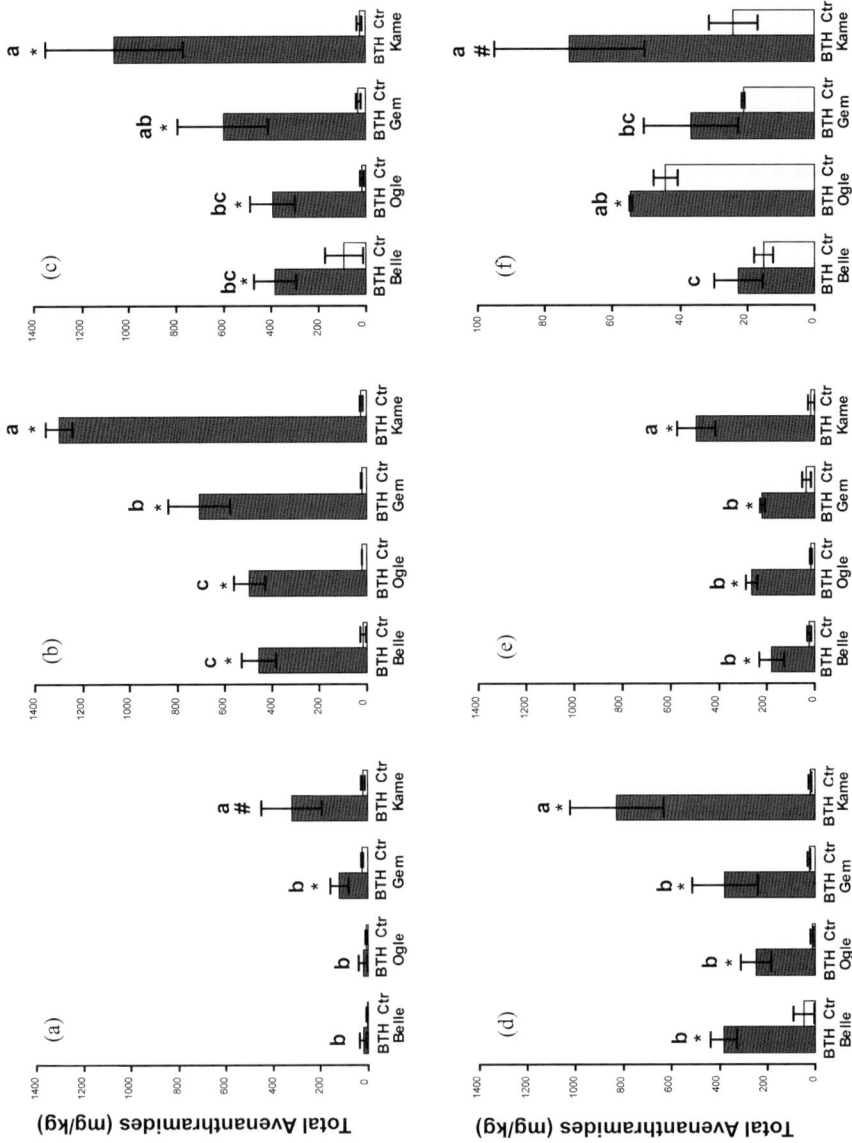

Figure 8.8 Total avenanthramide levels in mature oats (cultivars Belle, Ogle, Gem, Kame) after root soaking with 1.0 mM BTH (active ingredient). Total avenanthramide content in leaves harvested at (a) 24 h, (b) 48 h, (c) 96 h, (d) 168 h, and (e) 336 h after treatment. (f) Total avenanthramide content in the filling grain at 336 h after treatment. Source: Ren and Wise (2012). Reproduced with permission from Akademiai Press.

Bar heights represent mean total avenanthramide level. Error bars represent standard error of the mean (n = 3). Significant differences (α = 0.05) between BTH-treated cultivars at each time point are denoted by different letters above the bar. *P <0.05, treatment versus control (Ctr) for each cultivar; # P <0.10, treatment versus control for each cultivar.

resulted from work by Dr. F. William Collins. Collins and his colleagues developed a method which he termed "false malting". This term implies the onset of many of the biochemical processes associated with malting, but a "second dormancy" prevents the seed from germinating. Thus, nondormant grain is subjected to a two-phase heat treatment, 37°C for approximately 70 h followed by 70°C for 140 h. After cooling to room temperature, the grain is anaerobically steeped for 18 h at 32°C. This treatment reduces the germination rate to less than 1%. Following surface sterilization with 1% sodium hypochlorite the grain is malted at room temperature for 4 days. Adding 1–2% Ca^{2+} (as $CaCl_2$) during steeping enhances avenanthramide production in some cases. Depending on the variety of oat treated and modifications of the procedure outlined above, total avenanthramide content can be increased to nearly 2000 mg/kg, or 7.5-fold more than the starting levels (Collins and Burrows, 2010). After drying, the treated oat grain is suitable for most commercial food processing procedures.

8.17 Conclusion

Avenanthramides are important phytochemicals produced in oats. These novel metabolites provide important nutritional benefits and certain congeners possess strong antioxidant properties. Avenanthramides are expressed in vegetative tissue in response to pathogen invasion and appear to be important in protecting the plant from Crown Rust infection. The involvement of avenanthramides in response to other microbial pathogens or insect pests has not been reported, but they are strongly elicited by low levels of the host-specific toxin victorin from the fungal pathogen *Cochliobolus victoriae*. The specific function of avenanthramides in grain tissue is unknown.

The occurrence of these natural products in oats makes this cereal grain unique, but to exploit this characteristic a better understanding of the regulation of avenanthramide biosynthesis is needed. Efforts to breed oat germplasm with high avenanthramide levels are complicated by the fact that their production is strongly influenced by environmental factors. The use of plant defense activators may circumvent this obstacle. A number of plant defense activators are commercially available but their acceptance in the agricultural community has been limited, especially for use on cereal crops. Aside from the financial investment, this lack of acceptance is due to inconsistent results during field trials (Walters and Fountaine, 2009) and the fact that the induced responses can, under some circumstances, result in unacceptable allocation costs (Heil *et al.*, 2000), thus reducing yield. Continuing efforts to realize practical application of these agrichemicals may eventually prove worthwhile. For example, if vegetative responses can be correlated to grain levels, there could be an application in phenotyping oat seedlings for avenanthramide production by treating them with BTH (or some other plant defense activator).

The development of false malting by Collins is an exciting breakthrough. Provided that the nutritional benefits of avenanthramides continue to be demonstrated and publicized, the ability to convert oat grain to high avenanthramide-supplemented food products will undoubtedly improve the marketability of oats.

What genetic factors might influence the efficacy of false malting are currently unknown.

Naturally high levels of avenanthramides can be expected to extend the storage life of oats and may protect against mycotoxin-producing fungi. Thus, a more detailed understanding of the molecular processes regulating avenanthramide production will benefit the oat community. How the biosynthesis of avenanthramides is regulated and how they can be manipulated or improved through breeding are questions begging to be answered. Through the work of Mayama and Ishihara and their colleagues, the role of avenanthramides as phytoalexins is well established. Do avenanthramides have additional physiological functions in oats, and are they transported from one part of the plant to another? Furthermore, the function of avenanthramides in the grain is not easily explained. Do they function as radical scavengers? Their upregulation during germination suggests that possibility. Do they function as phytohormones or signaling molecules? Their ability to inhibit inflammatory processes in mammalian cells indicates that such a function is possible. Why are they found in root tissue after chemical elicitation? These are some of the questions on the role of avenanthramides in oat physiology that remain to be determined.

References

Achnine, L., *et al.* (2004) Colocalization of L-phenylalanine ammonia-lyase and cinnamate 4-hydroxylase for metabolic channeling in phenylpropanoid biosynthesis. *Plant Cell* **16**, 3098–3109.

Akimitsu, K., *et al.* (1993) Immunological evidence for a cell surface receptor of victorin using anti-victorin anti-idiotypic polyclonal antibodies. *Molecular Plant-Microbe Interactions* **6**, 429–433.

Bain, D.I. and Smalley, R.K. (1968) Synthesis of 2-substituted-4H-3,1-benzoxazin-4-ones. *Journal of the Chemical Society C* 1593–1597.

Blaakmeer, A., *et al.* (1994) Isolation, identification, and synthesis of miriamides, new host-markers from eggs of *Pieris brassicae*. *Journal of Natural Products* **57**, 90–99.

Boller, T. and Felix, G. (2009) A renaissance of elicitors: perception of microbe-associated molecular patterns and danger signals by pattern-recognition receptors. *Annual Review of Plant Biology* **60**, 379–406.

Bordin, A.P.A., *et al.* (1991) Potential elicitor for avenalumin accumulation in oat leaves. *Annals of the Phytopathological Society of Japan* **57**, 688–695.

Bratt, K., *et al.* (2003) Avenanthramides in oats (*Avena sativa* L.) and structure-antioxidant activity relationships. *Journal of Agriculture and Food Chemistry* **51**, 594–600.

Bryngelsson, S., *et al.* (2002) Effects of commercial processing on levels of antioxidants in oats (*Avena sativa* L.). *Journal of Agriculture and Food Chemistry* **50**, 1890–1896.

Bryngelsson, S., *et al.* (2003a) Levels of avenanthramides and activity of hydroxycinnamoyl-CoA:hydroxyanthranilate *N*-hydroxycinnamoyl transferase (HHT) in steeped or germinated oat samples. *Cereal Chemistry* **80**, 356–360.

Bryngelsson, S., *et al.* (2003b) Tentative avenanthramide-modifying enzyme in oats. *Cereal Chemistry* **80**, 361–364.

Cochrane, F.C., *et al.* (2004) The Arabidopsis phenylalanine ammonia lyase gene family: kinetic characterization of the four PAL isoforms. *Phytochemistry* **65**, 1557–1564.

Collins, F.W. (1986) Oat phenolics: structure, occurrence, and function. In: *Oats Chemistry and Technology* (ed. F.W. Webster). American Association of Cereal Chemists, St Paul, MN.

Collins, F.W. (1989) Oat phenolics: avenanthramides, novel substituted *N*-cinnamoylanthranilate alkaloids from oat groats and hulls. *Journal of Agriculture and Food Chemistry* **37**, 60–66.

Collins, F.W. and Burrows, V.D. (2010) Method for increasing concentration of avenanthramides in oats. *World Intellectual Property Organization.* Patent Cooperation Treaty PCT/CA2010/000458 Canada.

Collins, F.W. and Mullin, W.J. (1988) High-performance liquid chromatographic determination of avenanthramides, n-aroylanthranilic acid alkaloids from oats. *Journal of Chromatography A* **445**, 363–370.

Costa, M.A., *et al.* (2003) An *in silico* assessment of gene function and organization of the phenylpropanoid pathway metabolic networks in Arabidopsis thaliana and limitations thereof. *Phytochemistry* **64**, 1097–1112.

Crombie, L. and Mistry, J. (1990) The phytoalexins of oat leaves: 4-3,1-benzoxazin-4-ones or amides? *Tetrahedron Letters* **31**, 2647–2648.

D'Auria, J.C. (2006) Acyltransferases in plants: a good time to be BAHD. *Current Opinion in Plant Biology* **9**, 331–340.

Dimberg, L.H. and Peterson, D.M. (2009) Phenols in spikelets and leaves of field-grown oats (*Avena sativa*) with different inherent resistance to Crown Rust (*Puccinia coronata f. sp. avenae*). *Journal of the Science of Food and Agriculture* **89**, 1815–1824.

Dimberg, L.H., *et al.* (1996) Variation in oat groats due to variety, storage and heat treatment. I: Phenolic compounds. *Journal of Cereal Science* **24**, 263–272.

Dimberg, L.H., *et al.* (2001) Stability of oat avenanthramides. *Cereal Chemistry* **78**, 278–281.

Dixon, R.A. and Paiva, N.L. (1995) Stress-induced phenylpropanoid metabolism. *Plant Cell* **7**, 1085–1097.

Doehlert, D.C., *et al.* (2009) The green oat story: possible mechanisms of green color formation in oat products during cooking. *Journal of Food Science* **74**, S226–S231.

Emmons, C.L. and Peterson, D.M. (1999) Antioxidant activity and phenolic contents of oat groats and hulls. *Cereal Chemistry* **76**, 902–906.

Emmons, C.L. and Peterson, D.M. (2001) Antioxidant activity and phenolic content of oat as affected by cultivar and location. *Crop Science* **41**, 1676–1681.

Fagerlund, A., *et al.* (2009) Radical-scavenging and antioxidant activity of avenanthramides. *Food Chemistry* **113**, 550–556.

Girardet, N. and Webster, F.H. (2011) Oat milling: Specifications, storage, and processing. In: *Oats Chemistry and Technology* (eds F.H. Webster and P.J. Wood), 2nd edn. American Association of Cereal Chemists, St. Paul, MN.

Hahn, M.G. (1996) Microbial elicitors and their receptors in plants. *Annual Review of Phytopathology* **34**, 387–412.

Hamberger, B. and Hahlbrock, K. (2004) The 4-coumarate:CoA ligase gene family in Arabidopsis thaliana comprises one rare, sinapate-activating and three commonly occurring isoenzymes. *Proceedings of the National Academy of Sciences of the United States of America* **101**, 2209–2214.

Heil, M. and Ton, J. (2008) Long-distance signalling in plant defence. *Trends in Plant Science* **13**, 264–272.

Heil, M., *et al.* (2000) Reduced growth and seed set following chemical induction of pathogen defence: does systemic acquired resistance (SAR) incur allocation costs? *Journal of Ecology* **88**, 645–654.

Hrazdina, G. and Jensen, R. A. (1992) Spatial organization of enzymes in plant metabolic pathways. *Annual Review of Plant Physiology and Plant Molecular Biology* **43**, 241–267.

Huang, D., *et al.* (2005) The chemistry behind antioxidant capacity assays. *Journal of Agricultural and Food Chemistry* **53**, 1841–1856.

Ikegawa, T., *et al.* (1996) Accumulation of diferulic acid during the hypersensitive response of oat leaves to *Puccinia coronata f. sp. avenae* and its role in the resistance of oat tissues to cell wall degrading enzymes. *Physiological and Molecular Plant Pathology* **48**, 245.

Ishihara, A., *et al.* (1997) Induction of hydroxycinnamoyl-CoA:hydroxyanthranilate N-hydroxycinnamoyl transferase (HHT) activity in oat leaves by victorin C. *Zeitschrift fur Naturforschung* **52c**, 756–760.

Ishihara, A., *et al.* (1996) Involvement of Ca^{2+} ion in phytoalexin induction in oats. *Plant Science* **115**, 9–16.

Ishihara, A., *et al.* (1998) Induction of hydroxyanthranilate hydroxycinnamoyl transferase by oligo-*N*-acetylchitooligosaccharides in oats. *Phytochemistry* **47**, 969–974.

Ishihara, A., *et al.* (1999) Induction of biosynthetic enzymes for avenanthramides in elicitor-treated oat leaves. *Planta* **208**, 512–518.

Izumi, Y., *et al.* (2009) High-resolution spatial and temporal analysis of phytoalexin production in oats. *Planta* **229**, 931–943.

Jastrebova, J., *et al.* (2006) Selective and sensitive LC-MS determination of avenanthramides in oats. *Chromatographia* **63**, 419–423.

Jones, D.H. (1984) Phenylalanine ammonia-lyase: Regulation of its induction, and its role in plant development. *Phytochemistry*, 23, 1349–1359.

Jones, J.D.G. and Dangl, J.L. (2006) The plant immune system. *Nature* **444**, 323–329.

Kajiyama, S., *et al.* (2006) Single cell-based analysis of torenia petal pigments by a combination of ArF excimer laser micro sampling and nano-high performance liquid chromatography (HPLC)-mass spectrometry. *Journal of Bioscience and Bioengineering* **102**, 575–576.

Kessmann, H., *et al.* (1994) Induction of systemic acquired disease resistance in plants by chemicals. *Annual Review of Phytopathology* **32**, 439–459.

Kort, R., *et al.* (1996) Evidence for *trans-cis* isomerization of the *p*-coumaric acid chromophore as the photochemical basis of the photocycle of photoactive yellow protein. *FEBS Letters* **382**, 73–78.

Lindermayr, C., *et al.* (2002) Divergent members of a soybean (Glycine max L.) 4-coumarate:coenzyme A ligase gene family. *European Journal of Biochemistry* **269**, 1304–1315.

Maeda, H. and Dudareva, N. (2012) The shikimate pathway and aromatic amino acid biosynthesis in plants. *Annual Review of Plant Biology* **63**, 73–105.

Mann, J. (1987) *Secondary Metabolism.* Oxford University Press, USA.

Mannerstedt-Fogelfors, B. (2001) Antioxidants and lipids in oat cultivars as affected by environmental factors. Ph.D. Thesis, Swedish University of Agricultural Sciences, Uppsala.

Matsukawa, T., *et al.* (2000) Occurrence of avenanthramides and hydroxycinnamoly-CoA:hydroxyanthranilate N-hydroxycinnamoyltransferase activity in oat seeds. *Zeitschrift für Naturforschung C* **55**, 30–36.

Mayama, S., *et al.* (1981a) The production of phytoalexins by oat in response to Crown Rust, *Puccinia coronata* f. sp. *avenae*. *Physiological Plant Pathology* **19**, 217–226.

Mayama, S., *et al.* (1981b) Isolation and structure elucidation of genuine oat phytoalexin, avenalumin I. *Tetrahedron Letters* **22**, 2103–2106.

Mayama, S., *et al.* (1982) The role of avenalumin in the resistance of oat to Crown Rust, *Puccinia coronata* f. sp *avenae*. *Physiological Plant Pathology* **20**, 189–199.

Mayama, S., *et al.* (1986) The purification of victorin and its phytoalexin elicitor activity in oat leaves. *Physiological and Molecular Plant Pathology* **29**, 1–18.

Mayama, S., *et al.* (1995) Association of avenalumin accumulation with co-segregation of victorin sensitivity and Crown Rust resistance in oat lines carrying the Pc-2 gene. *Physiological and Molecular Plant Pathology* **46**, 263–274.

Meydani, M. (2009) Potential health benefits of avenanthramides of oats. *Nutrition Reviews* **67**, 731–735.

Miyagawa, H., *et al.* (1996a) Comparative studies of elicitors that induce phytoalexin in oats. *Journal of Pesticide Science* **21**, 203–207.

Miyagawa, H., *et al.* (1996b) A stress compound in oats induced by Victorin, a host-specific toxin from Helminthosporium victoriae. *Phytochemistry* **41**, 1473–1475.

Miyagawa, H., *et al.* (1995) Induction of avenanthramides in oat leaves inoculated with Crown Rust fungus, *Puccinia coronata* f. sp *avenae*. *Bioscience, Biotechnology and Biochemistry* **12**, 2305–2306.

Navarre, D.A. and Wolpert, T.J. (1999) Victorin induction of an apoptotic/senescence-like response in oats. *Plant Cell* **11**, 237–250.

Okazaki, Y., *et al.* (2004a) Identification of a dehydrodimer of avenanthramide phytoalexin in oats. *Tetrahedron* **60**, 4765–4771.

Okazaki, Y., *et al.* (2004b) Metabolism of avenanthramide phytoalexins in oats. *Plant Journal* **39**, 560–572.

Okazaki, Y., *et al.* (2007) New dimeric compounds of avenanthramide phytoalexin in oats. *Journal of Organic Chemistry* **72**, 3830–3839.

Peters, F.N. (1937) Oat flour as an antioxidant. *Industrial and Engineering Chemistry* **29**, 146–151.

Peterson, D.M. (2001) Oat antioxidants. *Journal of Cereal Science* **33**, 115–129.

Peterson, D.M. and Dimberg, L.H. (2008) Avenanthramide concentrations and hydroxycinnamoyl-CoA:hydroxyanthranilate N-hydroxycinnamoyltransferase activities in developing oats. *Journal of Cereal Science* **47**, 101–108.

Peterson, D. M., *et al.* (2002) Oat avenanthramides exhibit antioxidant activities in vitro. *Food Chemistry* **79**, 473–478.

Peterson, D.M., *et al.* (2005) Relationships among agronomic traits and grain composition in oat genotypes grown in different environments. *Crop Science* **45**, 1249–1255.

Ponchet, M., *et al.* (1988) Dianthramides (N-benzoyl and N-paracoumarylanthranilic acid derivatives) from elicited tissues of *Dianthus caryophyllus*. *Phytochemistry* **27**, 725–730.

Reinhard, K. and Matern, U. (1989) The biosynthesis of phytoalexins in *Dianthus caryophyllus* L. cell cultures: induction of benzoyl-CoA:anthranilate N-benzoyltransferase activity. *Archives of Biochemistry and Biophysics* **275**, 295–301.

Ren, Y. and Wise, M.L. (2012) Avenanthramide biosynthesis in oat cultivars treated with systemic acquired resistance elicitors. *Cereal Research Communications*. doi: 10.1556/CRC.2012.0035. http://www.akademiai.com/content/b7407t5672j8k6x1/?p=97ebb42381554769a8928d47b47a297d&pi=29 (last accessed 15 May 2013).

Rines, H.W. and Luke, H.H. (1985) Selection and regeneration of toxin-insensitive plants from tissue cultures of oats (*Avena sativa*) susceptible to Helminthosporium victoriae. *Theoretical and Applied Genetics* **71**, 16–21.

Rohde, A., *et al.* (2004) Molecular phenotyping of the pal1 and pal2 mutants of Arabidopsis thaliana reveals far-reaching consequences on phenylpropanoid, amino acid, and carbohydrate metabolism. *The Plant Cell* **16**, 2749–2771.

Rösler, J., *et al.* (1997) Maize phenylalanine ammonia-lyase has tyrosine ammonia-lyase activity. *Plant Physiology* **113**, 175–179.

Ryals, J.A., *et al.* (1996) Systemic acquired resistance. *Plant Cell* **8**, 1809–1819.

Schneider, K., *et al.* (2003) The substrate specificity-determining amino acid code of 4-coumarate:CoA ligase. *Proceedings of the National Academy of Sciences* **100**, 8601–8606.

Shulaev, V., *et al.* (1997) Airborne signalling by methyl salicylate in plant pathogen resistance. *Nature* **385**, 718–721.

Skoglund, M., *et al.* (2008) Avenanthramide content and related enzyme activities in oats as affected by steeping and germination. *Journal of Cereal Science* **48**, 294–303.

Stöckigt, J. and Zenk, M.H. (1975) Chemical synthesis and properties of hydroxycinnamoyl-coenzyme A derivatives. *Zeitschrift für Naturforschung C* **30**, 352–358.

Tada, Y., *et al.* (2001) Induction and signaling of an apoptotic response typified by DNA laddering in the defense response of oats to infection and elicitors. *Molecular Plant-Microbe Interactions* **14**, 477–486.

Tada, Y., *et al.* (2005) Victorin triggers programmed cell death and the defense response via interaction with a cell surface mediator. *Plant Cell Physiology* **46**, 1787–1798.

Tzin, V. and Galili, G. (2010) New insights into the shikimate and aromatic amino acids biosynthesis pathways in plants. *Molecular Plant* **3**, 956–972.

Uchihashi, K., *et al.* (2011) *In situ* localization of avenanthramide A and its biosynthetic enzyme in oat leaves infected with the Crown Rust fungus, *Puccinia coronata f. sp. avenae*. *Physiological and Molecular Plant Pathology* **76**, 173–181.

Vanholme, R., *et al.* (2008) Lignin engineering. *Current Opinion in Plant Biology* **11**, 278–285.

Walters, D.R. and Fountaine, J.M. (2009) Practical application of induced resistance to plant diseases: an appraisal of effectiveness under field conditions. *The Journal of Agricultural Science* **147**, 523–535.

Winkel, B.S. (2004) Metabolic channeling in plants. *Annual Review of Plant Biology* **55**, 85–107.

Wise, M.L. (2011) Effect of chemical systemic acquired resistance elicitors on avenanthramide biosynthesis in oat (*Avena sativa*). *Journal of Agricultural and Food Chemistry* **59**, 7028–7038.

Wise, M.L., *et al.* (2008) Association of avenanthramide concentration in oat (*Avena sativa* L.) grain with Crown Rust incidence and genetic resistance. *Cereal Chemistry* **85**, 639–641.

Wise, M.L., *et al.* (2009) Biosynthesis of avenanthramides in suspension cultures of oat (*Aven sativa*). *Plant Cell, Tissue and Organ Culture* **97**, 81–90.

Yang, Q., *et al.* (1997) Characterization and heterologous expression of hydroxycinnamoly/benzoyl-CoA:anthranilate N-hydroxycinnamoyl/benzoyltransferase from elicited cell cultures of carnation, *Dianthus caryophyllus L. Plant Molecular Biology* **35**, 777–789.

Yang, Q., *et al.* (2004) Analysis of the involvement of hydroxyanthranilate hydroxycinnamoyltransferase and caffeoyl-CoA 3-O-methyltransferase in phytoalexin biosynthesis in oat. *Molecular Plant-Microbe Interactions* **17**, 81–89.

Yi, H.-S., *et al.* (2009) Airborne induction and priming of plant defenses against a bacterial pathogen. *Plant Physiology* **151**, 2152–2161.

Part IV
Emerging Nutrition and Health Research

9
The Effects of Oats and Oat-β-Glucan on Blood Lipoproteins and Risk for Cardiovascular Disease

Tia M. Rains and Kevin C. Maki

Biofortis Clinical Research, Addison, IL, USA

9.1 Introduction

Coronary heart disease (CHD) is the leading cause of death and permanent disability in the United States (CDC, 2011). It is well established that elevated total cholesterol (TC) and low-density lipoprotein cholesterol (LDL-C) are major risk factors for CHD that can be modified by nutritional intervention (Roger *et al.*, 2012). According to the American Heart Association, an estimated 99 million American adults who are 20 years or older (44% of all US adults) have elevated TC (Roger *et al.*, 2012). LDL-C levels follow a similar pattern (Roger *et al.*, 2012). Data from the 2005–2008 National Health and Nutrition Examination Survey indicate that an estimated 71 million (33.5%) US adults have elevated LDL-C based on guidelines from the National Cholesterol Education Program Adult Treatment Panel III (CDC, 2011). In addition, 20.3% of American children aged 12 to 19 years have at least one abnormal lipid level (CDC, 2011).

Lowering LDL-C reduces coronary morbidity and mortality (LRC-CPPT Writing Group, 1984; Canner *et al.*, 1986; Scandinavian Simvastatin Survival Study Group, 1994; Shepherd *et al.*, 1995; LaRosa *et al.*, 2005). Evidence suggests that each 1% increment in LDL-C increases the lifetime risk of CHD as much as 3% (Brown and Goldstein, 2006; Cohen *et al.*, 2006). A 10% reduction in LDL-C maintained over an extended period has been estimated to reduce CHD risk by as much as 30% (Brown and Goldstein, 2006; Cohen *et al.*, 2006). These findings demonstrate that dietary interventions that help individuals lower LDL-C can have substantial public health impact.

Oats Nutrition and Technology, First Edition. Edited by YiFang Chu.
© 2014 John Wiley & Sons, Ltd. Published 2014 by John Wiley & Sons, Ltd.

9.2   Hypocholesterolemic effects of fiber

The lipid-lowering effects of viscous soluble dietary fibers have been well documented since 1963 (De Groot *et al.*, 1963). In 1999, a meta-analysis of 67 clinical studies provided a pooled estimate of a 2 mg/dL (0.05 mmol/L) reduction in LDL-C for every 1 g increase in viscous fiber consumption from oats, psyllium, guar, or pectin, without adverse effects on high-density lipoprotein cholesterol (HDL-C) or triglyceride (TG) levels (Brown *et al.*, 1999). This translates to a reduction of approximately 1.7% per gram of fiber consumed at the median LDL-C level in the US population (Carroll *et al.*, 2012). In 2006, the American Heart Association acknowledged that soluble fiber intake may decrease the risk of cardiovascular disease when consumed regularly as part of a diet low in saturated fat, trans fat, and cholesterol (Lichtenstein *et al.*, 2006). This was in agreement with the conclusions of regulatory agencies in the United States (Food and Drug Administration, 2003) and Europe (European Food Safety Authority, 2011), which approved the use of health claims related to cholesterol-lowering properties of products containing oat fiber.

Several mechanisms have been proposed to explain the cholesterol-lowering effects of viscous soluble fibers. Cholesterol homeostasis involves a balance between intestinal absorption and elimination, as well as endogenous cholesterol synthesis. An estimated 2000–3000 mg cholesterol enters the small intestine daily. This consists of exogenous cholesterol from food (10–15%) and endogenous cholesterol released with bile (85–90%) (Chen and Huang, 2009). Soluble fiber, such as that found in oats, forms a viscous solution in the small intestine, which disrupts micelle formation and reduces cholesterol transport to the intestinal brush border (Marlett, 2001). Cholesterol molecules are thereby trapped in the gut, preventing their entry into the enterocyte through the action of the Niemann-Pick C1-like 1 transporter (Kritchevsky and Story, 1974; Marlett, 1994; Chen and Huang, 2009). In addition, soluble fiber prevents the reabsorption of bile acids, which are made from cholesterol, interfering with their enterohepatic circulation (Lia *et al.*, 1995). This reduction in cholesterol and bile acid levels in the liver triggers an increase in LDL receptors, leading to a greater removal of LDL and other apolipoprotein B (Apo B)-containing lipoproteins from the circulation by the liver. The result is lower blood levels of LDL and other potentially atherogenic lipoprotein particles.

Another possible beneficial effect of viscous fiber is reduced endogenous cholesterol production, perhaps resulting from increased hepatic exposure to short chain fatty acids (SCFAs) produced by fermentation in the gut, particularly propionate (Schneeman, 2002). Acetate and propionate, two SCFAs produced by fermentation of indigestible carbohydrates by colonic microbiota, are thought to affect lipid levels. Acetate appears to enhance hepatic cholesterol synthesis but this effect is blocked or reversed by propionate (Wong *et al.*, 2006). The ratio of acetate to propionate in the blood appears to increase with age, which may partly explain the higher LDL-C levels associated with advancing age (Wolever *et al.*, 1996). Consumption of fermentable fibers, including oat β-glucan, may increase the circulating propionate to acetate ratio, favorably influencing LDL-C levels. However, little evidence exists for this mechanism; therefore, the role (if any) of SCFAs liberated by fermentation on the cholesterol-lowering effects of oats remains uncertain.

An additional mechanism whereby viscous fibers may lower cholesterol concentrations involves postprandial insulin concentration. Insulin stimulates hydroxymethylglutaryl coenzyme A reductase, the rate-limiting enzyme in cholesterol synthesis, and has a stimulatory effect on very low-density lipoprotein synthesis (Ness and Chambers, 2000; Reaven, 2003). Jenkins and colleagues (Jenkins *et al.*, 1989) showed that spreading the nutrient load over time by consuming the same diet as 17 snacks per day rather than three larger meals was associated with a 28% lower insulin concentration and 13.5% lower LDL-C concentration. The consumption of viscous fibers blunts the postprandial insulin response, which may contribute to their cholesterol-lowering effects. Furthermore, SCFAs liberated by colonic fermentation of some dietary fibers may inhibit the release of free fatty acids from adipose tissues and improve insulin sensitivity (Robertson *et al.*, 2003, 2005; Weickert *et al.*, 2008; Al-Lahham *et al.*, 2010). We and others have shown that a meal containing oat β-glucan blunts the postprandial insulin response, and chronic consumption may potentiate this effect (Maki *et al.*, 2007; Panahi *et al.*, 2007; Alminger and Eklund-Jonsson, 2008). These results are consistent with the hypothesis that two separate mechanisms account for the reduced glycemic and insulinemic responses: (i) increased gut viscosity creates a mechanical barrier that slows access of glucose molecules in the lumen to the brush border and access of digestive enzymes to their targets, thereby delaying glucose absorption; and (ii) SCFAs produced by colonic fermentation of β-glucan suppress circulating free fatty acid levels, thereby increasing peripheral insulin sensitivity. Additional research is needed to better understand the degree to which these mechanisms account for the cholesterol-lowering effects of oat products containing β-glucan.

9.3 Hypocholesterolemic effects of oats and oat β-glucan

Oat β-glucan is one of the best-studied viscous fibers with regard to cholesterol metabolism. Numerous studies have reported that consuming 2.8–6.0 g/day of oat β-glucan significantly lowers LDL-C concentrations, particularly in individuals with elevated baseline LDL-C levels (Pomeroy *et al.*, 2001; Saltzman *et al.*, 2001; Davy *et al.*, 2002; Keenan *et al.*, 2002; Pins *et al.*, 2002; Berg *et al.*, 2003; Kerckhoffs *et al.*, 2003; Maki *et al.*, 2003a, 2003b, 2007, 2010; Karmally *et al.*, 2005; Naumann *et al.*, 2006; Queenan *et al.*, 2007; Reyna-Villasmil *et al.*, 2007; Theuwissen and Mensink, 2007; Liatis *et al.*, 2009; Wolever *et al.*, 2010; Charlton *et al.*, 2012). Several meta-analyses have evaluated the relationship between oat β-glucan consumption and cholesterol reduction (Ripsin *et al.*, 1992; Brown *et al.*, 1999; Tiwari and Cummins, 2011). In a meta-analysis of 10 trials, 3 g/day of soluble fiber from oat products reduced TC by 5 mg/dL (–0.13 mmol/L; 95% CI: –0.19, –0.07), with hypercholesterolemic subjects showing larger reductions (Ripsin *et al.*, 1992). Similarly, a meta-analysis of 25 studies showed reductions of 1.5 mg/dL TC (–0.040 mmol/L; 95% CI: –0.054, –0.026) and 1.4 mg/dL LDL-C

(–0.037 mmol/L; 95% CI: –0.040, –0.034) per gram of soluble fiber consumed (Brown *et al.*, 1999). Most recently, Tiwari and Cummins (2011) conducted a meta-analysis of 14 intervention studies that evaluated the effects of consuming 2–10 g/day of oat β-glucan and reported reductions of 21 mg/dL TC (–0.55 mmol/L; 95% CI: –0.80, –0.30) and 22 mg/dL LDL-C (–0.58 mmol/L; 95% CI: –0.84, –0.33).

It is important to note that several investigations failed to show a hypocholesterolemic effect of oat β-glucan (Reyna *et al.*, 2003; Frank *et al.*, 2004; Biörklund *et al.*, 2005, 2008; Mårtensson *et al.*, 2005; Robitaille *et al.*, 2005; Chen *et al.*, 2006; Beck *et al.*, 2010; Cugnet-Anceau *et al.*, 2010; Tighe *et al.*, 2010). In general, greater reductions in LDL-C following β-glucan consumption have been observed in subjects with higher baseline cholesterol levels; therefore, participants with normal cholesterol levels may not have experienced significant changes in LDL-C (Ripsin *et al.*, 1992). Furthermore, the hypocholesterolemic properties of oat β-glucan may be modified by the food matrix or method of food processing, or both (Kerckhoffs *et al.*, 2003; El Khoury *et al.*, 2012). For example, hydrolyzed β-glucan, which has a lower molecular weight and low viscosity in solution, appears to have little effect on cholesterol levels (Wolever *et al.*, 2010).

Atherogenic lipoprotein particles, including LDL and remnants of TG-rich lipoproteins, each contain one molecule of Apo B, which is a strong predictor of CHD (Pischon *et al.*, 2005; Kastelein *et al.*, 2008; Mora *et al.*, 2009; Ramjee *et al.*, 2011). Therefore, Apo B concentration is a direct indicator of the total number of circulating atherogenic lipoprotein particles (Maki and Dicklin, 2008). Non-HDL-C correlates more strongly with Apo B than LDL-C because it reflects the cholesterol carried by all Apo B-containing lipoprotein particles (Ramjee *et al.*, 2011). Both Apo B and non-HDL-C are stronger predictors of CHD risk than LDL-C (Pischon *et al.*, 2005; Mora *et al.*, 2009; Ramjee *et al.*, 2011).

A small number of recent studies have reported favorable effects of oat bran and oat β-glucan on Apo B and non-HDL-C (Berg *et al.*, 2003; Liatis *et al.*, 2009; Kristensen and Bügel, 2011). In one study of overweight males, Berg and colleagues (Berg *et al.*, 2003) reported a 28% decrease in Apo B in subjects who consumed oat bran (35–50 g/day) as part of a low-fat diet. Likewise, the mean non-HDL-C level in subjects who consumed oat bran (102 g/day) was 16% lower than that of controls after a 2-week intervention (Kristensen and Bügel, 2011). In a study of subjects with type 2 diabetes mellitus, consuming β-glucan (3 g/day) resulted in a 15% decrease in non-HDL-C (Liatis *et al.*, 2009).

The circulating level of HDL-C is inversely associated with CHD risk; however, the effect of increasing HDL-C on CHD risk is less clear (Mahdy *et al.*, 2012). Reyna *et al.* (2003), Robitaille *et al.* (2005), and Reyna-Villasmil *et al.* (2007) all reported significant increases in HDL-C with daily consumption of oat-derived β-glucan (6 g), oat bran (28 g), or Oatrim[®] β-glucan-containing fat replacer. Similarly, Cugnet-Anceau (2010) reported significant increases in HDL-C when mildly obese subjects with type 2 diabetes consumed soup enriched with β-glucan (3.5 g) daily for eight weeks. HDL-C was significantly reduced in overweight adults after daily consumption of cereal and cereal bars with oat β-glucan (0, 1.5, or 3.0 g) for 6 weeks (Charlton *et al.*, 2012). However, when nonobese mildly hypercholesterolemic adults consumed an oat bran β-glucan preparation

(5.9 g/day added to bread and cookies or orange juice) for 4 weeks, the addition of oat bran did not significantly change HDL-C levels (Reyna *et al.*, 2003). Likewise, Beck and colleagues (Beck *et al.*, 2010) reported no changes in HDL-C when overweight subjects consumed β-glucan (5–6 g or 8–9 g daily) as part of a hypocaloric diet for 2 months. Thus, recently published studies have not consistently shown an effect of oat β-glucan consumption on HDL-C concentration and are generally consistent with meta-analyses suggesting that oat β-glucan intake does not increase HDL-C concentration (Brown *et al.*, 1999; Tiwari and Cummins, 2011)

The relationship between oat β-glucan intake and serum TG concentration is also unclear. A number of studies have reported that oat β-glucan ($\leq$ 6 g/d) has no effect on serum TG (Kerckhoffs *et al.*, 2003; Reyna *et al.*, 2003; Chen *et al.*, 2006; Reyna-Villasmil *et al.*, 2007; Charlton *et al.*, 2012). In contrast, Cugnet-Anceau and coworkers (Cugnet-Anceau *et al.*, 2010) observed a 3.8% reduction in TG levels following daily consumption of soup enriched with 3.5 g β-glucan for 6 weeks, which differed significantly from the 11.2% increase in TG levels in the control group. Recent data from a randomized, crossover study showed a 21% decrease in TG in subjects consuming a diet containing oat bran (102 g/day) for 2 weeks, but only a 10% decrease in TG in the control low-fiber group (Kristensen and Bügel, 2011). However, results of meta-analyses suggest no effect of oat fiber on TG concentrations (Brown *et al.*, 1999; Tiwari and Cummins, 2011).

9.4 Summary/Conclusions

The cholesterol-lowering effects of viscous fibers, particularly oat fiber and oat β-glucan, are well documented. Consuming $\geq$ 3 g oat β-glucan daily reduces TC and LDL-C levels by approximately 1.3–1.8% per gram consumed. Greater reductions are observed in hypercholesterolemic individuals than in those with normal cholesterol levels. However, oat fiber consumption does not appear to affect HDL-C or TG concentrations. Public health organizations and regulatory agencies in the United States and Europe have acknowledged the cholesterol-lowering properties of oat fiber.

References

Al-Lahham, S.H., *et al.* (2010) Biological effects of propionic acid in humans; metabolism, potential applications and underlying mechanisms. *Biochimica et Biophysica Acta* **1801**, 1175–1183.

Alminger, M. and Eklund-Jonsson, C. (2008) Whole-grain cereal products based on a high-fibre barley or oat genotype lower post-prandial glucose and insulin responses in healthy humans. *European Journal of Nutrition* **47**, 294–300.

Beck, E.J., *et al.* (2010) Oat beta-glucan supplementation does not enhance the effectiveness of an energy-restricted diet in overweight women. *British Journal of Nutrition* **103**, 1212–1222.

Berg, A., *et al.* (2003) Effect of an oat bran enriched diet on the atherogenic lipid profile in patients with an increased coronary heart disease risk. A controlled randomized lifestyle intervention study. *Annals of Nutrition and Metabolism* **47**, 306–311.

Biörklund, M., *et al.* (2005) Changes in serum lipids and postprandial glucose and insulin concentrations after consumption of beverages with beta-glucans from oats or barley: a randomized dose-controlled trial. *European Journal of Clinical Nutrition* **59**, 1272–1281.

Biörklund, M., *et al.* (2008) Serum lipids and postprandial glucose and insulin levels in hyperlipidemic subjects after consumption of an oat beta-glucan-containing ready meal. *Annals of Nutrition and Metabolism* **52**, 83–90.

Brown, M.S. and Goldstein, J.L. (2006) Biomedicine. Lowering LDL – not only how low, but how long? *Science* **311**, 1721–1723.

Brown, L., *et al.* (1999) Cholesterol-lowering effects of dietary fiber: a meta-analysis. *American Journal of Clinical Nutrition* **69**, 30–42.

Canner, P.L., *et al.* (1986) Fifteen-year mortality in Coronary Drug Project (CDP) patients: long-term benefit with niacin. *Journal of American College of Cardiology* **8**, 1245–1255.

Carroll, M.D., *et al.* (2012) Trends in lipids and lipoproteins in US adults, 1988–2010. *Journal of the American Medical Association* **308**, 1545–1554.

CDC (2011) Vital signs: prevalence, treatment, and control of high levels of low-density lipoprotein cholesterol. United States, 1999–2002 and 2005–2008. *Morbidity and Mortality Weekly Report* **60**, 109–114.

Charlton, K.E. *et al.* (2012) Effect of 6 weeks' consumption of β-glucan-rich oat products on cholesterol levels in mildly hypercholesterolaemic overweight adults. *British Journal of Nutrition* **107**, 1037–1047.

Chen, J. and Huang, X. (2009) The effects of diets enriched in beta-glucans on blood lipoprotein concentrations. *Journal of Clinical Lipidology* **3**, 154–158.

Chen, J., *et al.* (2006) A randomized controlled trial of dietary fiber intake on serum lipids. *European Journal of Clinical Nutrition* **60**, 62–68.

Cohen, J.C., *et al.* (2006) Sequence variations in PCSK9, low LDL, and protection against coronary heart disease. *New England Journal of Medicine* **354**, 1264–1272.

Cugnet-Anceau, C., *et al.* (2010) A controlled study of consumption of beta-glucan-enriched soups for 2 months by type 2 diabetic free-living subjects. *British Journal of Nutrition* **103**, 422–428.

Davy, B.M., *et al.* (2002) High-fiber oat cereal compared with wheat cereal consumption favorably alters LDL-cholesterol subclass and particle numbers in middle-aged and older men. *American Journal of Clinical Nutrition* **76**, 351–358.

De Groot, A.P., *et al.* (1963) Cholesterol-lowering effect of rolled oats. *Lancet* **1**, 303–304.

European Food Safety Authority (2011) Scientific Opinion on the substantiation of health claims related to beta-glucans from oats and barley and maintenance of normal blood LDL-cholesterol concentrations (ID 1236, 1299), increase in satiety leading to a reduction in energy intake (ID 851, 852), reduction of post-prandial glycaemic responses (ID 821, 824), and "digestive function" (ID 850) pursuant to Article 13(1) of Regulation (EC) No 1924/2006. *EFSA Journal* **9**, 2207.

El Khoury, D., *et al.* (2012) Beta glucan: health benefits in obesity and metabolic syndrome. *Journal of Nutrition and Metabolism.* **2012**, 851362.

Food and Drug Administration, HHS. (2003) Food labeling: health claims; soluble dietary fiber from certain foods and coronary heart disease. Final rule. *Federal Register* **68**, 44207–44209.

Frank, J. *et al.* (2004) Yeast-leavened oat breads with high or low molecular weight beta-glucan do not differ in their effects on blood concentrations of lipids, insulin, or glucose in humans. *Journal of Nutrition* **134**, 1384–1388.

Jenkins, D.J., *et al.* (1989) Nibbling versus gorging: metabolic advantages of increased meal frequency. *New England Journal of Medicine* **321**, 929–934.

Karmally, W. *et al.* (2005) Cholesterol-lowering benefits of oat-containing cereal in Hispanic Americans. *Journal of American Dietetic Association* **105**, 967–970.

Kastelein, J.J., *et al.* (2008) TNT Study Group. Lipids, apolipoproteins, and their ratios in relation to cardiovascular events with statin treatment. *Circulation* **117**, 3002–3009.

Keenan, M.D., *et al.* (2002) Oat ingestion reduces systolic and diastolic pressure in patients with mild or borderline hypertension: a pilot study. *Journal of Family Practice* **51**, 369.

Kerckhoffs, D.A., *et al.* (2003) Cholesterol-lowering effect of beta-glucan from oat bran in mildly hypercholesterolemic subjects may decrease when beta-glucan is incorporated into bread and cookies. *American Journal of Clinical Nutrition* **78**, 221–227.

Kristensen, M. and Bügel, S. (2011) A diet rich in oat bran improves blood lipids and hemostatic factors, and reduces apparent energy digestibility in young healthy volunteers. *European Journal of Clinical Nutrition* **65**, 1053–1058.

Kritchevsky, D. and Story, J.A. (1974) Binding of bile salts *in vitro* by nonnutritive fiber. *Journal of Nutrition* **104**, 458–462.

LaRosa, J.C., *et al.* (2005) Treating to New Targets (TNT) investigators. Intensive lipid lowering with atorvastatin in patients with stable coronary disease. *New England Journal of Medicine* **352**, 1425–1435.

Lia, A., *et al.* (1995) Oat beta-glucan increases bile acid excretion and a fiber-rich barley fraction increases cholesterol excretion in ileostomy subjects. *American Journal of Clinical Nutrition* **62**, 1245–1251.

Liatis, S., *et al.* (2009) The consumption of bread enriched with betaglucan reduces LDL-cholesterol and improves insulin resistance in patients with type 2 diabetes. *Diabetes and Metabolism* **35**, 115–120.

Lichtenstein, A.H., *et al.* (2006) Diet and lifestyle recommendations revision 2006: a scientific statement from the American Heart Association Nutrition Committee. *Circulation* **114**, 82–96. [Erratum in: *Circulation* 2006; **114**, e27.]

LRC-CPPT Writing Group. (1984) The Lipid Research Clinics Coronary Primary Prevention Trial results. I. Reduction in incidence of coronary heart disease. *Journal of the American Medical Association* **251**, 351–364.

Mahdy Ali, K., *et al.* (2012) Cardiovascular disease risk reduction by raising HDL cholesterol-current therapies and future opportunities. *British Journal of Pharmacology* **167**, 1177–1194

Maki, K.C. and Dicklin, M.R. (2008) How well do various lipids and lipoprotein measures predict cardiovascular disease morbidity and mortality. In: *Clinical Challenges in Lipid Disorders* (eds P.P. Toth and D. Sica), pp. 1–16. Clinical Publishing, Oxford.

Maki, K.C., *et al.* (2003a) Food products containing free tall oil-based phytosterols and oat beta-glucan lower serum total and LDL cholesterol in hypercholesterolemic adults. *Journal of Nutrition* **133**, 808–813.

Maki, K.C., *et al.* (2003b) Lipid responses to consumption of a beta-glucan containing ready-to-eat cereal in children and adolescents with mild-to-moderate primary hypercholesterolemia. *Nutrition Research* **23**,1527–1535.

Maki, K.C., *et al.* (2007) Effects of high-fiber oat and wheat cereals on postprandial glucose and lipid responses in healthy men. *International Journal for Vitamin and Nutrition Research* **77**, 347–356.

Maki, K.C., *et al* (2010) Whole-grain ready-to-eat oat cereal, as part of a dietary program for weight loss, reduces low-density lipoprotein cholesterol in overweight and obese adults more than a dietary program including low-fiber control foods. *Journal of American Dietetic Association* **110**, 205–214.

Marlett, J.A, *et al.* (1994) Mechanism of serum cholesterol reduction by oat bran. *Hepatology* **20**, 1450–1457.

Marlett, J.A. (2001) Dietary fiber and cardiovascular disease. In: *Handbook of dietary fiber* (eds S.S. Cho and M.L. Dreher), pp. 17–25. Marcel Dekker, New York.

Mårtensson, O., *et al.* (2005) Fermented, ropy, oat-based products reduce cholesterol levels. *Nutrition Research* **25**, 429–442.

Mora, S., *et al.* (2009) Lipoprotein particle profiles by nuclear magnetic resonance compared with standard lipids and apolipoproteins in predicting incident cardiovascular disease in women. *Circulation* **119**, 931–939.

Naumann, E., *et al.* (2006) Beta-glucan incorporated into a fruit drink effectively lowers serum LDL-cholesterol concentrations. *American Journal of Clinical Nutrition* **83**, 601–605.

Ness, G.C. and Chambers, C.M. (2000) Feedback and hormonal regulation of hepatic 3-hydroxy-3-methylglutaryl coenzyme A reductase: the concept of cholesterol buffering capacity. *Proceedings of the Society for Experimental Biology and Medicine* **224**, 8–19.

Panahi, S., *et al.* (2007) Beta-glucan from two sources of oat concentrates affect postprandial glycemia in relation to the level of viscosity. *Journal of the American College of Nutrition* **26**, 639–644.

Pischon, T., *et al.* (2005) Non-high-density lipoprotein cholesterol and apolipoprotein B in the prediction of coronary heart disease in men. *Circulation* **112**, 3375–3383.

Pins, J.J., *et al.* (2002) Do whole-grain cereals reduce the need for antihypertensive medications and improve blood pressure control? *Journal of Family Practice* **51**, 353–359

Pomeroy, S., *et al.* (2001) Oat beta-glucan lowers total and LDL-cholesterol. *Australian Journal of Nutrition and Dietetics* **58**, 51–55.

Queenan, K.M., *et al.* (2007) Concentrated oat beta-glucan, a fermentable fiber, lowers serum cholesterol in hypercholesterolemic adults in a randomized controlled trial. *Nutrition Journal* **6**, 6.

Ramjee, V., *et al.* (2011) Non-high-density lipoprotein cholesterol versus apolipoprotein B in cardiovascular risk stratification: do the math. *Journal of the American College of Cardiology* **58**, 457–463.

Reaven, G.M. (2003) Insulin resistance/compensatory hyperinsulinemia, essential hypertension and cardiovascular disease. *Journal of Clinical Endocrinology and Metabolism* **88**, 2399–2403.

Reyna, N.Y., *et al.* (2003) Sweeteners and beta-glucans improve metabolic and anthropometrics variables in well controlled type 2 diabetic patients. *American Journal of Therapeutics* **10**, 438–443.

Reyna-Villasmil, N., *et al.* (2007) Oat-derived beta-glucan significantly improves HDLC and diminishes LDLC and non-HDL cholesterol in overweight individuals with mild hypercholesterolemia. *American Journal of Therapeutics* **14**, 203–212.

Ripsin, C.M., *et al.* (1992) Oat products and lipid lowering. A meta-analysis. *Journal of the American Medical Association* **267**, 3317–3325.

Robertson, M.D., *et al.* (2003) Prior short-term consumption of resistant starch enhances postprandial insulin sensitivity in healthy subjects. *Diabetologia* **46**, 659–665.

Robertson, M.D., *et al.* (2005) Insulin-sensitizing effects of dietary resistant starch and effects on skeletal muscle and adipose tissue metabolism. *American Journal of Clinical Nutrition* **82**, 559–567.

Robitaille, J., *et al.* (2005) Effect of an oat bran-rich supplement on the metabolic profile of overweight premenopausal women. *Annals of Nutrition and Metabolism* **49**, 141–148.

Roger, V.L., *et al.* (2012) Executive summary: heart disease and stroke statistics – 2012 update: a report from the American Heart Association. *Circulation* **125**, 188–197. [Erratum in: *Circulation* 2012;**125**, e1001.]

Saltzman, E., *et al.* (2001) An oat-containing hypocaloric diet reduces systolic blood pressure and improves lipid profile beyond effects of weight loss in men and women. *Journal of Nutrition* **131**, 1465–1470.

Scandinavian Simvastatin Survival Study Group (1994) Randomised trial of cholesterol lowering in 4444 patients with coronary heart disease: the Scandinavian Simvastatin Survival Study (4S). *Lancet* **344**, 1383–1389.

Schneeman, B.O. (2002) Gastrointestinal physiology and functions. *British Journal of Nutrition* **88**, S159–S163.

Shepherd, J., *et al.* (1995) Prevention of coronary heart disease with pravastatin in men with hypercholesterolemia. West of Scotland Coronary Prevention Study (WOSCOPS) group. *New England Journal of Medicine* **333**, 1301–1307.

Theuwissen, E. and Mensink, R.P. (2007) Simultaneous intake of β-glucan and plant stanol esters affects lipid metabolism in slightly hypercholesterolemic subjects. *Journal of Nutrition* **137**, 583–588.

Tighe, P., *et al.* (2010) Effect of increased consumption of whole-grain foods on blood pressure and other cardiovascular risk markers in healthy middle-aged persons: a randomized controlled trial. *American Journal of Clinical Nutrition* **92**, 733–740.

Tiwari, U. and Cummins, E. (2011) Meta-analysis of the effect of β-glucan intake on blood cholesterol and glucose levels. *Nutrition* **27**, 1008–1016.

Weickert, M.O., *et al.* (2008) Impact of cereal fibre on glucose-regulating factors. *Diabetologia* **48**, 2343–2353.

Wolever, T.M.S., *et al.* (1996) Serum acetate:propionate ratio is related to serum cholesterol in men but not women. *Journal of Nutrition* **126**, 2790–2797.

Wolever, T.M., *et al.* (2010) Physicochemical properties of oat β-glucan influence its ability to reduce serum LDL cholesterol in humans: a randomized clinical trial. *American Journal of Clinical Nutrition* **92**, 723–732.

Wong, J.M., *et al.* (2006) Colonic health: fermentation and short chain fatty acids. *Journal of Clinical Gastroenterology* **40**, 235–243.

10
The Effects of Oats and β-Glucan on Blood Pressure and Hypertension

Tia M. Rains and Kevin C. Maki

Biofortis Clinical Research, Addison, IL, USA

10.1 Introduction

Hypertension, defined as an average systolic blood pressure (SBP) ≥ 140 mm Hg and/or diastolic blood pressure (DBP) ≥ 90 mm Hg, is a major risk factor for stroke, atherosclerotic cardiovascular disease, and early mortality (National High Blood Pressure Education Program, 2003; Appel *et al.*, 2006; Wright *et al.*, 2011). Approximately one in three adults (an estimated 76.4 million) in the United States have hypertension, with prevalence increasing over the lifespan such that more than two-thirds of adults over the age of 60 years are hypertensive (Wright *et al.*, 2011; Roger *et al.*, 2012). Approximately 21% of people have undiagnosed hypertension and another 25% have prehypertension, defined as SBP of 120–139 mm Hg and/or DBP of 80–89 mm Hg, with neither value exceeding the cut point for hypertension. It is projected that by 2030, an additional 27 million people will have high blood pressure, a 9.9% increase in prevalence from 2010 (Roger *et al.*, 2012).

Lifestyle modification is an effective approach to preventing and treating hypertension. Dietary modification, physical activity, and psychosocial factors have all been shown to modestly reduce blood pressure and lower the risk of related complications (Appel *et al.*, 2006). Even small reductions in blood pressure can reduce the risk for disease. For example, a 3-mm Hg reduction in SBP has been associated with an 8% reduction in stroke mortality and 5% reduction in mortality from coronary heart disease (Stamler, 1991). In general, lifestyle modification is more effective at reducing blood pressure in hypertensive individuals than in nonhypertensive individuals (Appel *et al.*, 2006).

Oats Nutrition and Technology, First Edition. Edited by YiFang Chu.
© 2014 John Wiley & Sons, Ltd. Published 2014 by John Wiley & Sons, Ltd.

10.2   Dietary patterns and blood pressure

Several dietary behaviors are associated with blood pressure. One of the best-studied dietary interventions for blood pressure control is the Dietary Approaches to Stop Hypertension (DASH) diet. The DASH eating plan, which is rich in whole-grain cereal products, fruits and vegetables, and low-fat dairy products, substantially lowers blood pressure in subjects with and without hypertension (Sacks *et al.*, 1995). Similarly, blood pressure is lower in vegetarian populations than in omnivores, and in individuals consuming greater amounts of fruits and vegetables than in those with low intakes (Sacks *et al.*, 1974; Haines *et al.*, 1980; Rouse *et al.*, 1984; Beilin *et al.*, 1987; Rodenas *et al.*, 2011). In addition, prospective epidemiological studies have shown that whole grain consumption is inversely associated with incident hypertension and blood pressure (Table 10.1) (Esmaillzadeh *et al.*, 2005; Wang *et al.*, 2007; Flint *et al.*, 2009). For example, in the Women's Health Study (n = 28 926 women), the relative risk of developing hypertension was 23% lower in women who reported eating at least four daily servings of whole grains compared with those eating less than half a daily serving during the 10-year follow-up (Wang *et al.*, 2007). Results from the Health Professionals Follow-up Study (n = 31 684 men) were similar. During the 18-year follow-up, whole grain intake $\geq$ 34 g/day was associated with a 21% lower risk for developing hypertension compared with whole grain intake $\leq$ 6.5 g/day (Flint *et al.*, 2009).

Since the late 1970s, observational studies have shown that increased intake of dietary fiber is favorably associated with blood pressure (Table 10.1). For example, in a sample of 30 861 male health professionals without diagnosed hypertension, dietary fiber intake was inversely associated with the development of hypertension during the 4-year follow-up after adjusting for age, weight, and alcohol and energy intakes (Ascherio *et al.*, 1992). Relative to men with a high fiber intake (> 24 g/day), those with a fiber intake <12 g/day had a relative risk of hypertension of 1.57 (95% CI, 1.20–2.05). In a 28-month follow-up study of 5880 men and women, Alonso and colleagues (Alonso *et al.*, 2006) found that subjects in the highest quintile of cereal fiber intake had a 40% lower risk of hypertension compared to those in the lowest quintile.

Controlled feeding studies have reported similar results. A meta-analysis of 24 randomized, controlled trials (N = 1404) published between 1966 and 2003 showed that a mean supplemental fiber intake of 11.5 g/day is associated with modest reductions in SBP (–1.1 mm Hg) and DBP (–1.3 mm Hg) (Streppel *et al.*, 2005). Furthermore, soluble fiber produced a greater effect on SBP (–1.3 mm Hg) and DBP (–0.8 mm Hg) than insoluble fiber (–0.2 and –0.6 mm Hg, respectively). Blood pressure reduction was greater in older subjects (> 40 y) and in those with hypertension. A similar meta-analysis of 25 randomized controlled trials (total of 1477 subjects) produced similar findings (Whelton *et al.*, 2005). Dietary fiber intakes ranging from 7.2 to 18.9 g/day were associated with non-significant reductions in SBP (–1.2 mm Hg) and significant reductions in DBP (–1.7 mm Hg). However, in subjects with hypertension, higher fiber intake was associated with significant reductions in both SBP (–6.0 mm Hg) and DBP (–4.2 mm Hg) (Whelton *et al.*, 2005).

Table 10.1 Observational studies examining the effects of oats, dietary fiber, and whole grains on blood pressure

Author, year	Design, follow up	Population	Intake of oats/whole grain/fiber	Results
Oats				
He et al., 1995	Lifelong dietary pattern	N = 850 Yi People (China)	• Q1: 0 g/d oats • Q4: >90 g/d oats	• SBP (Q1 vs. Q4): 109.7 vs. 100.4 mm Hg (p <0.05) • DBP (Q1 vs. Q4): 68.3 vs. 52.6 mm Hg (p <0.05) • 100 g/d oats associated with lower SBP (−3.1 mm Hg) and lower DBP (−1.3 mm Hg)
Fiber				
Lichtenstein et al., 1986	Cross-sectional	N = 2512 men	• Q1: ≤4.0 g/d cereal fiber • Q5: ≥10.5 g/d cereal fiber	• SBP (Q1 vs. Q5): 142.1 vs. 138.8 mm Hg • DBP (Q1 vs. Q5): 72.9 vs. 71.5 mm Hg Inverse relationship between cereal fiber intake and SBP (r = −0.053, p <0.01) and DBP (r = −0.057, p <0.01)
Jenner et al., 1988	Cross-sectional	N = 884 9-year-old children	• Q1 (boys): 12.7 g/d fiber • Q5 (boys): 27.0 g/d fiber	• No relationship between fiber intake and SBP in boys; inverse relationship between fiber intake and DBP in boys (p <0.01); Q5 DBP was 2.5 mm Hg lower than that of Q1 • No relationship between fiber intake and SBP or DBP in girls
Witteman et al., 1989	Prospective cohort 4 y	N = 58 218 women	• Q1: <10 g/d dietary fiber • Q5: ≥25 g/d dietary	• RR for hypertension Q5 vs. Q1: 0.76 (p = 0.002)[1] • RR for hypertension Q5 vs. Q1: 0.87 (p = 0.14)[2] [1]Adjusted for age, BMI, alcohol intake; [2]Further adjusted for intake of calcium, magnesium, and potassium.
Ascherio et al., 1992	Prospective cohort 4 y	N = 30 681 men	• Lowest quintile: <12 g fiber • Highest quintile: ≥24 g fiber	• RR for hypertension Q1 vs. Q5: 1.57 (p <0.0001)[1] • RR for hypertension Q1 vs. Q5: 1.46 (p = 0.015)[2] [1]Adjusted for age, BMI, and alcohol consumption; [2]Further adjusted for magnesium and potassium.

(continued)

Table 10.1   (*Continued*)

Author, year	Design, follow up	Population	Intake of oats/whole grain/fiber	Results
Ascherio et al., 1996	Prospective cohort 4 y	N = 121 700 women	• Lowest quintile: <10 g/d fiber • Highest quintile: ≥25 g/d fiber	• RR for hypertension Q5 vs. Q1: 1.01 (p = 0.75)[1] [1]Adjusted for age, BMI, and alcohol consumption.
Stamler et al., 1997	Prospective cohort 6 y	N = 11 342 men	• Not described	• SBP: R = −0.061 (NS)[1] • DBP: R = −0.070 (p <0.001)[1] [1]Adjusted for age, race, education, serum cholesterol, smoking, special diet status, BMI, and alcohol intake.
Ludwig et al., 1999	Prospective cohort 10 y	N = 1516 white men and women	• Q1: <5.9 g fiber/4184 kJ/d • Q5: >10.5 g fiber/4184 kJ/d	• SBP (Q1 vs. Q5): 109.1 vs. 106.9 mm Hg (p <0.01)[1,2] • DBP (Q1 vs. Q5): 72.4 vs. 69.7 mm Hg (p <0.001)[1,2] [1]Adjusted for sex, age, field center, education, energy intake, physical activity, cigarette smoking, alcohol intake, and vitamin supplement use; [2]Findings NS after adjusting for any of the following: total fat, saturated fat, unsaturated fat, carbohydrate, or protein intake
Ludwig et al., 1999	Prospective cohort 10 y	N = 1215 black men and women	• Q1: <5.9 g fiber/4184 kJ/d • Q5: >10.5 g fiber/4184 kJ/d	• SBP (Q1 vs. Q5): 111.6 vs. 111.5 mm Hg (p = 0.77)[1] • DBP (Q1 vs. Q5): 74.0 vs. 73.3 mm Hg (p = 0.70)[1] [1]Adjusted for sex, age, field center, education, energy intake, physical activity, cigarette smoking, alcohol intake, and vitamin supplement use

Bazzano et al., 2003	Prospective cohort 19 y	N = 9776 men and women	• Q1: <7.7 g fiber per 1735 kcal • Q4: >15.9 g fiber per 1735 kcal	• SBP (Q1 vs. Q5): 135.4 vs. 134.4 mm Hg (p <0.05)[1] • DBP (Q1 vs. Q5): 83.9 vs. 82.9 mm Hg (p <0.002)[1] [1]Adjusted for age
Alonso et al., 2006	Prospective cohort 2 y	N = 5880 men and women	• Q1: lowest cereal fiber intake (amount NR) • Q5: highest cereal fiber intake (amount NR)	• HR for hypertension Q5 vs. Q1: 0.7 (p = 0.18)[1] • HR for hypertension Q5 vs. Q1: 0.6 (p = 0.05)[2] [1]Adjusted for age, sex, BMI, physical activity, alcohol consumption, sodium intake, total energy intake, smoking, and hypercholesterolemia; [2]Further adjusted for fruit, vegetable, caffeine, magnesium, potassium, low-fat dairy, MUFA, and SFA intake.
Newby et al., 2007	Cross-sectional	N = 1516 men and women	• Q1: 2.2 g/d cereal fiber • Q5: 9.5 g/d cereal fiber	• DBP (Q1 vs. Q5): 80.8 vs. 77.8 (p = 0.009)[1] • DBP (Q1 vs. Q5): 79.5 vs. 79.2 (p = 0.90)[2] • SBP (Q1 vs. Q5): 130.8 vs. 127.6 (p = 0.06)[1] • SBP (Q1 vs. Q5): 128.7 vs. 129.7 (p = 0.27)[2] [1]Adjusted for age, age, sex, total energy, and decade of visit; [2]Further adjusted for race, education, vitamin supplement use, smoking, percentage of energy from saturated fat, alcohol, refined grain intake, BMI, use of lipid-lowering medications, and hypercholesterolemia

(continued)

Table 10.1 (*Continued*)

Whole grain

Author, year	Design, follow up	Population	Intake of oats/whole grain/fiber	Results
Esmaillzadeh et al., 2005	Cross-sectional	N = 827 men and women	• Q1: <10 g/d whole grain • Q2: 10 to <71 g/d whole grain • Q3: 71 to <143 g/d whole grain • Q4: ≥143 g/d whole grain	• SBP (Q1 vs. Q4): 115 vs. 115 (p >0.05) • DBP (Q1 vs. Q4): 81 vs. 77 (p <0.05) OR for hypertension: Q1: 1.00, Q5: 0.84 (95% CI, 0.73–0.99; p = 0.03)[1] [1]Adjusted for age, total energy intake, energy from fat, blood pressure medication use, estrogen use, smoking, physical activity, meat and fish consumption, and fruit/vegetable intake
Steffen et al., 2005	Prospective cohort 15 y	N = 4304 men and women	• Q1: <0.4 servings/d whole grain • Q2: 0.4–0.7 servings/d whole grain • Q3: 0.7–1.2 servings/d whole grain • Q4: 1.2–1.9 servings/d whole grain • Q5: >1.9 servings/d whole grain	HR for incident elevated blood pressure: • Q1: 1.00 • Q5: 0.83 (95% CI, 0.67–1.03; p = 0.03 for trend)
Newby et al., 2007	Cross-sectional	N = 1464 men and women	• Q1: 0.68 g/d whole grain • Q5: 45.8 g/d whole grain	• DBP (Q1 vs. Q5): 79.8 vs. 79.1 (p = 0.04)[1] • DBP (Q1 vs. Q5): 79.8 vs. 79.2 (p = 0.42)[2] • SBP (Q1 vs. Q5): 130.9 vs. 127.8 (p = 0.12)[1] • SBP (Q1 vs. Q5): 129.2 vs. 128.3 (p = 0.79)[2] [1]Adjusted for age, sex, total energy, and decade of visit; [2]Further adjusted for race, education, vitamin supplement use, smoking, percentage of energy from saturated fat, alcohol intake, refined grain intake, BMI, use of lipid-lowering medications, hypercholesterolemia

| Wang et al., 2007 | Prospective cohort 10 y | N = 28 926 women | • Q1: <0.5 servings/d whole grain
• Q5: ≥4 servings/d whole grain | • RR for hypertension Q5 vs. Q1: 0.86[1]
• RR for hypertension Q5 vs. Q1: 0.88 (p = 0.001)[2]
• RR for hypertension Q5 vs. Q1: 0.87 (p = 0.0009)[3]
• RR for hypertension Q5 vs. Q1: 0.89 (p = 0.007)[4]
[1]Adjusted for age, race, energy intake and treatment (aspirin, vitamin E, β-carotene, or placebo); [2]Adjusted as for model 1 plus smoking, alcohol use, exercise, family history of myocardial infarction, postmenopausal, postmenopausal hormone use, and multivitamin use; [3]Adjusted as for model 2 plus BMI, history of diabetes, and history of hypercholesterolemia; [4]Adjusted as for model 3 plus intake of fruit and vegetables, meats, and dairy products |
| Flint et al., 2009 | Prospective cohort 18 y | N = 31 684 men | • Q1: 0–6.5 g/d whole grain
• Q5: 34.3–326.4 g/d whole grain | • RR for hypertension Q5 vs. Q1: 0.72 (p <0.0001)[1]
• RR for hypertension Q5 vs. Q1: 0.81 (p <0.0001)[2]
[1]Adjusted for age and energy intake; [2]Adjusted for model 1 plus family history of coronary heart disease, family history of hypertension, smoking, alcohol, marital status, profession, height, fruit and vegetable intake, sodium intake, physical activity, multivitamin use, and cholesterol screening |

Abbreviations: BMI, body mass index; BP, blood pressure; DBP, diastolic blood pressure; M, men; MUFA, monounsaturated fatty acid; NR, not reported; OR, odds ratio; Q, quartile; RR, relative risk; SBP, systolic blood pressure; SFA, saturated fatty acid; W, women.

The mechanisms by which dietary fiber lowers blood pressure are not well understood but may involve the modulation of carbohydrate metabolism. Compounds that increase insulin sensitivity, such as thiazolidinediones and biguanides, can also lower blood pressure in animal models and humans (Landin *et al.*, 1991; Fonseca, 2003). Although the underlying mechanisms of this effect are uncertain, compensatory hyperinsulinemia may impair intracellular signaling pathways related to endothelium-dependent vascular relaxation, resulting in vasoconstriction and higher blood pressure (Kotchen *et al.*, 1996). Other processes that influence blood pressure may also be affected by hyperinsulinemia, including enhanced sodium and water reabsorption at the distal tubular nephron and stimulation of the sympathetic nervous system (Landsberg, 1999; Fonseca, 2003; Ferrinnini and Cushman, 2012).

Viscous fibers such as oat β-glucan delay glucose absorption from the gastrointestinal tract; thus, postprandial glucose and insulin levels are lower, which may improve insulin sensitivity over time as insulin receptors in peripheral tissues are upregulated (Wolever *et al.*, 1999; Keenan *et al.*, 2002; Maki *et al.*, 2007). In addition, colonic microbiota ferment dietary fiber to produce short chain fatty acids (SCFAs) (Wolever *et al.*, 1991). High blood levels of SCFAs suppress free fatty acid (FFA) release from adipose tissue (Wolever *et al.*, 1991). A large body of data from clinical investigations supports the view that chronic elevation of FFAs induces insulin resistance (Ferrannini *et al.*, 1983; Kashyap *et al.*, 2003; Homko *et al.*, 2003). For example, insulin resistance can be induced in healthy children with an infusion of Intralipid^R, which elevates FFA concentrations (Arslanian and Suprasongsin, 1997). Conversely, suppression of FFA release with the niacin analog acipimox improves insulin sensitivity (Ferrannini *et al.*, 1983). Thus, increased intake of fermentable fiber (e.g., resistant starch, guar gum, oat β-glucan) is associated with improved insulin sensitivity, as assessed by various methods including the euglycemic clamp technique and mathematical modeling of data from meal tolerance tests (Landin *et al.*, 1992; Robertson *et al.*, 2003, 2005; Weickert *et al.*, 2006).

Another possible mechanism may involve the proposed relationship between blood cholesterol and endothelial function. An elevated level of low-density lipoprotein cholesterol results in vasoconstriction, which may interfere with blood pressure regulation (Vakkilainen *et al.*, 2000). Accordingly, lower blood cholesterol concentration is associated with improved endothelial function and blood pressure (Anderson *et al.*, 1995). The beneficial effects of viscous fibers such as β-glucan on blood cholesterol concentrations are well documented (Chapter 9).

10.3   Oats and oat β-glucan: Effect on blood pressure and hypertension

A number of human intervention trials have assessed the effects of oat and β-glucan intake on blood pressure in both normotensive and hypertensive persons (Table 10.2). Saltzman and colleagues (Saltzman *et al.*, 2001), reported a 6-mm

Table 10.2 Randomized controlled trials examining the effects of oats/oat β-glucan on blood pressure

Authors, year	Design	Duration (weeks)	Population	Treatments	SBP (Δ from baseline)[1]	DBP (Δ from baseline)[1]
Studies of subjects not recruited based on blood pressure						
Saltzman et al., 2001	RCT, P	6	N = 43 weight-stable M/W	• 45 g oats/1000 kcal • Oat-free placebo foods *Hypocaloric dietary regimen	• ↓ (−6 vs. −1 mm Hg in oat vs. control group)	• ↔
Van Horn et al., 1991	RCT, P	8	N = 80 M/W	• 56.7 g/d instant oats • Control	• ↔	• ↔
Tighe et al., 2010	RCT, P	12	N = 206 M/W	• 1 serv/d wheat + 2 serv/d oats • 3 serv/d wheat • 3 serv/d refined foods	• ↓ (−6 vs. −1 mm Hg in oat vs. control group)	• ↔
Kestin et al., 1990	RCT, C	4	N = 24 hypercholesterolemic M	• 95 g/d oat bran (12 g NSP) • 35 g/d wheat bran (12 g NSP) • 60 g/d rice bran (12 g NSP)	• ↔	• ↔
Swain et al., 1990	RCT, C	6	N = 20 M/W	• 100 g/d oat bran (17 g fiber) • 100 g/d wheat (2.3 g fiber)	• ↔	• ↔
Jenkins et al., 2008	RCT, C	4	N = 28 hyperlipidemic M/W after 2.5 y on "portfolio diet"	• Oat bran bread (2 g/d β-glucan) • 454 g/d strawberries (0 g β-glucan)	• ↔	• ↔

(continued)

Table 10.2 (*Continued*)

Authors, year	Design	Duration (weeks)	Population	Treatments	SBP (Δ from baseline)[1]	DBP (Δ from baseline)[1]
Liatis *et al.*, 2009	RCT, P	3	N = 41 M/W with type 2 diabetes and hypercholesterolemia	• Oat bread (3 g/d β-glucan) • White bread	• ↔ • ↓ (−12 vs. − 2 mm Hg in oat vs. control group) in subset with hypertension at baseline (N = 12)	• ↔
Charlton *et al.*, 2012	RCT, P	6	N = 87 hypercholesterolemic and overweight M/W	Cereals/cereal bars with: • 3.0 g/d β-glucan • 1.5 g/d β-glucan • 0 g/d β-glucan	• ↔	• ↔
Wolever *et al.*, 2010	RCT, P	4	N = 367 M/W	2 serv/d RTE cereal with: • 3 g/d high MW β-glucan • 3 g/d medium MW β-glucan • 3 g/d low MW β-glucan • Wheat bran	• ↔	• ↔
Maki *et al.*, 2010	RCT, P	12	N = 204 hypercholesterolemic and overweight M/W	2 serv/d RTE cereal with: • 3 g/d β-glucan • 0 g/d β-glucan	• ↔	• ↔
Studies of subjects recruited based on elevated BP						
Davy *et al.*, 2002	RCT, P	6	N = 36 hypertensive M	• 60 g/d oatmeal + 76 g/d RTE oat bran cereal (14 g total fiber; 5.5 g β-glucan) • 60 g/d hot wheat cereal + 81 g/d RTE wheat cereal (14 g total fiber)	• ↔	• ↔

Study	Design	Duration (wk)	Subjects	Intervention	Outcomes
Pins et al., 2002	RCT, P	12	N = 88 hypertensive M/W undergoing treatment	• 60 g/d oatmeal + 77 g/d oat cereal bar (5.4 g β-glucan) • 65 g/d hot wheat cereal + 81 g/d RTE wheat cereal (0 g β-glucan).	• ↔ • 73% vs. 42% discontinued BP medications in the oat and wheat groups, respectively
Maki et al., 2007	RCT, P	12	N = 97 overweight M/W with untreated or suboptimally controlled hypertension	• 90 g/d oat bran RTE cereal + 60 g/d oatmeal + 20 g/d β-glucan powder (17 g total fiber; 7.7 g β-glucan) • 90 g/d low-fiber RTE cereal + 65 g/d hot cereal + 12 g/d maltodextrin (1.9 g total fiber)	• ↔ • ↓ (−8 mm Hg) in subset with BMI >31.5 kg/m² • ↔ • ↓ (−4 mm Hg) in subset with BMI >31.5 kg/m²
He et al., 2004	RCT, P	12	N = 110 M/W with untreated higher than optimal BP or stage 1 hypertension	• 60 g/d oat bran muffin + 84 g/d cereal bar (7.3 g β-glucan) • 93 g/d refined wheat muffin + 42 g/d corn flakes (0 g β-glucan)	• ↔

*Study outcomes are summarized as significantly lower (↓) or no significant change (↔) in systolic and diastolic blood pressure outcomes from baseline relative to the control condition.

Abbreviations: BMI, body mass index; BP, blood pressure; C, crossover; DBP, diastolic blood pressure; M, men; MW, molecular weight; NSP, nonstarch polysaccharide; P, parallel; RCT, randomized controlled trial; RTE, ready-to-eat; SBP, systolic blood pressure; serv, serving; W, women.

Hg decrease in SBP in healthy normotensive adults after a 6-week hypocaloric diet regimen (restricted by approximately 1000 kcal/day) that included 45 g/day of oats. In contrast, subjects who underwent calorie restriction but did not consume oats exhibited a 1-mm Hg decrease in SBP (Saltzman et al., 2001); DBP and weight loss did not differ between groups. Tighe and colleagues (Tighe et al., 2010) also reported greater reductions in SBP (−5 mm Hg) in healthy adults who consumed three servings/day of whole grain wheat and oats for 12 weeks, compared with adults who consumed refined cereal products (−1.3 mm Hg); DBP did not differ between groups. However, most other studies did not find that oat consumption improves blood pressure in normotensive subjects. For example, Jenkins and colleagues (Jenkins et al., 2008) provided hyperlipidemic individuals (N = 28) with oat bran bread (2 g/day β-glucan) for 4 weeks but observed no changes in SBP or DBP. In a similar study of 87 hypercholesterolemic and overweight men and women, daily intake of oat flake cereal (1.5 g/day β-glucan) or oat porridge (3.0 g/day β-glucan) for 6 weeks did not affect blood pressure (Charlton et al., 2012). Other human feeding trials with whole oats and oat bran concentrates also failed to demonstrate a blood pressure-lowering effect in subjects who had normal blood pressure at baseline (Kestin et al., 1990; Swain et al., 1990; Van Horn et al., 1991; Maki et al., 2007, 2010; Liatis et al., 2009; Wolever et al., 2010).

Although results from prospective cohort studies suggest that the blood pressure-lowering effect of oats may be more pronounced in individuals with hypertension or elevated blood pressure, human intervention studies in which blood pressure change was the primary outcome do not consistently support this conclusion. For example, Davy and colleagues (Davy et al., 2002) compared the effects of daily intake of hot oat cereal and ready-to-eat oat bran cereal (providing 5.5 g/day β-glucan) with daily intake of wheat products in 36 middle-aged hypertensive overweight or obese men. No differences in SBP or DBP were observed at the end of the 6-week intervention. Similarly, in a study of overweight adults with untreated or suboptimally controlled hypertension, Maki and colleagues (Maki et al., 2007) found no changes in blood pressure response in subjects who consumed three servings/day of oat products (oat bran cold cereal, oatmeal, and powdered oat β-glucan) versus controls who consumed a combination of wheat foods (low-fiber cold cereal, hot cereal, and maltodextrin powder). However, a subgroup analysis revealed significant decreases in both SBP (−8.3 mm Hg) and DBP (−3.9 mm Hg) in subjects with a body mass index above the median value for the study sample (>31.5 kg/m^2).

Pins and coworkers (Pins et al., 2002) compared the effects of oats (oatmeal and oat-containing snack product containing 5.4 g/day β-glucan) versus wheat (hot wheat cereal and ready-to-eat cereal containing 0.0 g/day β-glucan) in 88 hypertensive men and women. At the end of the 12-week intervention, SBP or DBP did not differ between groups. However, 73% of the subjects consuming oats were able to discontinue their blood pressure medication or reduce the dose by half, whereas only 42% of subjects consuming wheat were able to do so (Pins et al., 2002). Furthermore, subjects in the oat group who continued on the same dose of blood pressure medications showed a 7-mm Hg decrease in SBP and 4-mm Hg decrease in DBP.

Compared with whole oats, oat bran concentrates and β-glucan isolates contain a higher proportion of β-glucan and total fiber. However, feeding trials with these ingredients also yield conflicting results in hypertensive adults. He and coworkers (He *et al.,* 2004) compared the effects of 8 g/day of oat bran concentrate (7.3 g/day β-glucan) with a control (low-fiber wheat) in 110 adults with untreated higher than normal blood pressure. No differences in SBP or DBP were observed after the 12-week intervention period. Liatis and coworkers (Liatis *et al.,* 2009) compared the effects of β-glucan–enriched bread (3.0 g/day β-glucan) versus white bread on blood pressure in 41 patients with type 2 diabetes. Although changes in blood pressure did not differ between the two groups overall, β-glucan consumption was associated with lower SBP in a subset of patients with hypertension (n = 31) after 3 weeks (β-glucan –12.2 mm Hg vs. control –2.0 mm Hg).

10.4 Conclusion

Considerable evidence suggests that dietary factors affect blood pressure and/or hypertension risk. Several prospective cohort studies show an inverse association between blood pressure and consumption of whole grains and fiber. Human intervention trials with whole oats or concentrated β-glucan (3.0–7.7 g/day β-glucan) have yielded inconsistent results. However, blood pressure reduction appears to be more pronounced in individuals with hypertension and/or increased adiposity. Additional studies are needed to better define the effect of whole oat and β-glucan consumption on blood pressure and hypertension risk.

References

Alonso, A., *et al.* (2006) Vegetable protein and fiber from cereal are inversely associated with the risk of hypertension in a Spanish cohort. *Archives of Medical Research* **37**, 778–786.

Anderson, T.J., *et al.* (1995) Systemic nature of endothelial dysfunction in atherosclerosis. *American Journal of Cardiology* **75**, 71B–74B.

Appel, L.J., *et al.* (2006) Dietary approaches to prevent and treat hypertension: A scientific statement from the American Heart Association. *Hypertension* **47**, 296–308.

Arslanian, S. and Suprasongsin, C. (1997) Glucose-fatty acid interactions in prepubertal and pubertal children: effects of lipid infusion. *American Journal of Physiology* **272**, E523–E529.

Ascherio, A., *et al.* (1992) A prospective study of nutritional factors and hypertension among US men. *Circulation* **86**, 1475–1484.

Ascherio, A., *et al.* (1996) Prospective study of nutritional factors, blood pressure, and hypertension among US women. *Hypertension* **27**, 1065–1072.

Bazzano, L.A., *et al.* (2003) Dietary fiber intake and reduced risk of coronary heart disease in US men and women: the National Health and Nutrition Examination Survey I Epidemiologic Follow-up Study. *Archives of Internal Medicine* **163**, 1897–1904.

Beilin, L.J., *et al.* (1987) Vegetarian diet and blood pressure. *Nephron* **47**, 37–41.

Charlton, K.E., *et al.* (2012) Effect of 6 weeks' consumption of β-glucan-rich oat products on cholesterol levels in mildly hypercholesterolaemic overweight adults. *British Journal of Nutrition* **107**, 1037–1047.

Davy, B.M., *et al.* (2002) Oat consumption does not affect resting casual and ambulatory 24-h arterial blood pressure in men with high-normal blood pressure to stage I hypertension. *Journal of Nutrition* **132**, 394–398.

Esmaillzadeh, A., *et al.* (2005) Whole-grain consumption and the metabolic syndrome: a favorable association in Tehranian adults. *European Journal of Clinical Nutrition* **59**, 353–362.

Ferrannini, E., *et al.* (1983) Effect of fatty acids on glucose production and utilization in man. *The Journal of Clinical Investigation* **72**, 1737–1747.

Ferrannini, E. and Cushman, W.C. (2012) Diabetes and hypertension: the bad companions. *Lancet* **380**, 601–610.

Flint, A.J., *et al.* (2009) Whole grains and incident hypertension in men. *The American Journal of Clinical Nutrition* **90**, 493–498.

Fonseca, V.A. (2003) Management of diabetes mellitus and insulin resistance in patients with cardiovascular disease. *American Journal of Cardiology* **92**, 50J–60J.

Haines, A.P., *et al.* (1980) Haemostatic variables in vegetarians and non-vegetarians. *Thrombosis Research* **19**, 139–148.

He, J., *et al.* (1995) Dietary macronutrients and blood pressure in southwestern China. *Journal of Hypertension* **13**, 1267–1274.

He, J., *et al.* (2004) Effect of dietary fiber intake on blood pressure: a randomized, double-blind, placebo-controlled trial. *Journal of Hypertension* **22**, 73–80.

Homko, C.J., *et al.* (2003) Effects of free fatty acids on glucose uptake and utilization in healthy women. *Diabetes* **52**, 487–491.

Jenkins, D.J., *et al.* (2008) The effect of strawberries in a cholesterol-lowering dietary portfolio. *Metabolism* **57**, 1636–1644.

Jenner, D.A., *et al.* (1988) Diet and blood pressure in 9-year-old Australian children. *The American Journal of Clinical Nutrition* **47**, 1052–1059.

Kashyap, S., *et al.* (2003) A sustained increase in plasma free fatty acids impairs insulin secretion in nondiabetic subjects genetically predisposed to develop type 2 diabetes. *Diabetes* **52**, 2461–2474.

Keenan, J.M., *et al.* (2002) Oat ingestion reduces systolic and diastolic blood pressure in patients with mild or borderline hypertension: a pilot trial. *The Journal of Family Practice* **51**, 369.

Kestin, M., *et al.* (1990) Comparative effects of three cereal brans on plasma lipids, blood pressure, and glucose metabolism in mildly hypercholesterolemic men. *The American Journal of Clinical Nutrition* **52**, 661–666.

Kotchen, T.A., *et al.* (1996) Insulin and hypertensive cardiovascular disease. *Current Opinion in Cardiology* **11**, 483–489.

Landin, K., *et al.* (1991) Treating insulin resistance in hypertension with metformin reduces both blood pressure and metabolic risk factors. *Journal of Internal Medicine* **229**, 181–187.

Landin, K., *et al.* (1992) Guar gum improves insulin sensitivity, blood lipids, blood pressure, and fibrinolysis in healthy men. *The American Journal of Clinical Nutrition* **56**, 1061–1065.

Landsberg, L. (1999) Role of the sympathetic adrenal system in the pathogenesis of the insulin resistance syndrome. *Annals of the New York Academy of Sciences* **892**, 84–90.

Liatis, S., *et al.* (2009) The consumption of bread enriched with betaglucan reduces LDL-cholesterol and improves insulin resistance in patients with type 2 diabetes. *Diabetes and Metabolism* **35**, 115–120.

Ludwig, D.S., *et al.* (1999) Dietary fiber, weight gain, and cardiovascular disease risk factors in young adults. *JAMA: The Journal of the American Medical Association* **282**, 1539–1546.

Lichtenstein, M.J., *et al.* (1986) Heart rate, employment status, and prevalent ischaemic heart disease confound relation between cereal fibre intake and blood pressure. *Journal of Epidemiology and Community Health* **40**, 330–333.

Maki, K.C., *et al.* (2007) Effects of consuming foods containing oat B-glucan on blood pressure, carbohydrate metabolism and biomarkers of oxidative stress in men and women with elevated blood pressure. *European Journal of Clinical Nutrition* **61**, 786–795.

Maki, K.C., *et al.* (2010) Whole-grain ready-to-eat oat cereal, as part of a dietary program for weight loss, reduces low-density lipoprotein cholesterol in adults with overweight and obesity more than a dietary program including low-fiber control foods. *Journal of the American Dietetic Association* **110**, 205–214.

National High Blood Pressure Education Program (2003) *The seventh report of the Joint National Committee on Prevention, Detection, Evaluation, and Treatment of High Blood Pressure. National Institutes of Health, Publication Number* 03-5233.

Newby, P.K., *et al.* (2007) Intake of whole grains, refined grains, and cereal fiber measured with 7-d diet records and associations with risk factors for chronic disease. *The American Journal of Clinical Nutrition* **86**, 1745–1753.

Pins, J.J., *et al.* (2002) Do whole-grain cereals reduce the need for antihypertensive medications and improve blood pressure control? *The Journal of Family Practice* **51**, 353–359.

Robertson, M.D., *et al.* (2003) Prior short-term consumption of resistant starch enhances postprandial insulin sensitivity in healthy subjects. *Diabetologia* **46**, 659–665.

Robertson, M.D., *et al.* (2005) Insulin-sensitizing effects of dietary resistant starch and effects on skeletal muscle and adipose tissue metabolism. *The American Journal of Clinical Nutrition* **82**, 559–567.

Rodenas, S., *et al.* (2011) Blood pressure of omnivorous and semi-vegetarian postmenopausal women and their relationship with dietary and hair concentrations of essential and toxic metals. *Nutrición Hospitalaria* **26**, 874–883

Roger, V.L., *et al.* (2012) Heart disease and stroke statistics – 2012 update: a report from the American Heart Association. *Circulation* **125**, e2–e220.

Rouse, I.L., *et al.* (1984) Vegetarian diet, blood pressure and cardiovascular risk. *Australian and New Zealand Journal of Medicine* **14**, 439–443.

Sacks, F.M., *et al.* (1974) Blood pressure in vegetarians. *American Journal of Epidemiology* **100**, 390–398.

Sacks, F.M., *et al.* Rationale and design of the Dietary Approaches to Stop Hypertension trial (DASH). (1995) A multicenter controlled-feeding study of dietary patterns to lower blood pressure. *Annals of Epidemiology* **5**, 108–118.

Saltzman, E., *et al.* (2001) An oat-containing hypocaloric diet reduces systolic blood pressure and improves lipid profile effects of weight loss in men and women. *Journal of Nutrition* **131**, 1465–1470.

Stamler, J. (1991) Blood pressure and high blood pressure. Aspects of risk. *Hypertension* **18**, I95–I107.

Stamler, J., *et al.* (1997) Relation of body mass and alcohol, nutrient, fiber, and caffeine intakes to blood pressure in the special intervention and usual care groups in the Multiple Risk Factor Intervention Trial. *The American Journal of Clinical Nutrition* **65**, 338S–365S.

Steffen, L.M., *et al.* (2005) Associations of plant food, dairy product, and meat intakes with 15-y incidence of elevated blood pressure in young black and white adults: the Coronary Artery Risk Development in Young Adults (CARDIA) Study. *The American Journal of Clinical Nutrition* **82**, 1169–1177.

Streppel, M.T., *et al.* (2005) Dietary fiber and blood pressure: a meta-analysis of randomized placebo-controlled trials. *Archives of Internal Medicine* **165**, 150–156.

Swain, J.F., *et al.* (1990) Comparison of the effects of oat bran and low-fiber wheat on serum lipoprotein levels and blood pressure. *New England Journal of Medicine* **322**, 147–152.

Tighe, P., *et al.* (2010) Effect of increased consumption of whole-grain foods on blood pressure and other cardiovascular risk markers in healthy middle-aged persons: a randomized controlled trial. *The American Journal of Clinical Nutrition* **92**, 733–740.

Vakkilainen, J., *et al.* (2000) Endothelial dysfunction in men with small LDL particles. *Circulation* **102**, 716–721.

Van Horn, L., *et al.* (1991) Effects on serum lipids of adding instant oats to usual American diets. *American Journal of Public Health* **81**, 183–188.

Wang, L., *et al.* (2007) Whole- and refined-grain intakes and the risk of hypertension in women. *The American Journal of Clinical Nutrition* **86**, 472–479.

Weickert, M.O., *et al.* (2006) Cereal fiber improves whole-body insulin sensitivity in overweight and obese women. *Diabetes Care* **29**, 775–780.

Whelton, S.P., *et al.* (2005) Effect of dietary fiber intake on blood pressure: a meta-analysis of randomized, controlled clinical trials. *Journal of Hypertension* **23**, 475–481.

Witteman, J.C., *et al.* (1989) A prospective study of nutritional factors and hypertension among US women. *Circulation* **80**, 1320–1327.

Wolever, T.M., *et al.* (1991) Effect of method administration of psyllium on glycemic response and carbohydrate digestibility. *Journal of the American College of Nutrition* **10**, 364–371.

Wolever, T.M., *et al.* (1999) Day-to-day consistency in amount and source of carbohydrate intake associated with improved blood glucose control in type 1 diabetes. *Journal of the American College of Nutrition* **18**, 242–247.

Wolever, T.M., *et al.* (2010) Physicochemical properties of oat β-glucan influence its ability to reduce serum LDL cholesterol in humans: a randomized clinical trial. *The American Journal of Clinical Nutrition* **92**, 723–732.

Wright, J.D., *et al.* (2011) Mean systolic and diastolic blood pressure in adults aged 18 and over in the United States, 2001–2008. *National Health Statistics Reports* **25**, 1–22, 24.

11
Avenanthramides, Unique Polyphenols of Oats with Potential Health Effects

Mohsen Meydani

Vascular Biology Laboratory, Jean Mayer USDA Human Nutrition Research Center on Aging at Tufts University, Boston, MA, USA

11.1 Introduction

Epidemiological and clinical studies have suggested that a high intake of whole grain foods is associated with a lower risk for coronary heart disease (CHD), diabetes, colon cancer, inflammation, and all-cause mortality (Burkitt *et al.*, 1972; Reddy *et al.*, 1992; Burkitt, 1993; Gillman *et al.*, 1995; Ness and Powles, 1997; Slattery *et al.*, 1997; Anderson, 2000; Joshipura *et al.*, 2001). The high fiber content of whole grain foods is believed to be the major contributing factor to their beneficial health effects. An early meta-analysis of multiple controlled studies has suggested that consuming whole grains, including wheat, rice, maize, and oats, reduces the risk of CHD slightly better than consuming fruits or vegetables (Anderson, 2003).

Oats have been recognized as the cereal grain that most effectively promotes healthy heart function by lowering cholesterol. Although wheat bran and rice bran both contain high levels of fiber, they contain mainly insoluble fiber, which does not lower low-density lipoprotein (LDL) cholesterol. A recent comprehensive literature review by Kelly and colleagues (Kelly *et al.*, 2007) suggests that the beneficial effect of consuming whole grains on CHD in clinical intervention trials is primarily due to whole grain oats. The health-promoting effects of oats can be attributed primarily to their high soluble fiber content, namely β-glucan, which reduces blood LDL cholesterol levels by increasing the conversion of cholesterol to bile (Jenkins *et al.*, 2002). Accordingly, in 1997 the US Food and Drug Administration endorsed the use of the health claim that consuming oat and oat products containing soluble fiber can reduce the risk of heart disease. The required dose is 0.75 g of β-glucan per serving of food and consuming 3 g of oat β-glucan daily is thought to be effective in reducing blood LDL cholesterol levels. Although a meta-analysis of epidemiologic studies recently concluded that 3 g/day of oat or barley β-glucan is adequate to decrease blood cholesterol levels, several studies have reported that 3 g of β-glucan may not reduce blood cholesterol levels. Wolff and coworkers (Wolff *et al.*, 1998) suggested that this discrepancy may be associated with the degree of viscosity. In addition, results of a double-blind, parallel-design, multicenter clinical trial demonstrated that β-glucans with high and medium molecular weights are more effective in reducing blood LDL cholesterol levels than low molecular weight β-glucan.

In addition to promoting a healthy heart by reducing LDL cholesterol, consuming oat bran high in β-glucan has been shown to reduce glycemic, insulin, and glucagon responses after glucose or complex carbohydrate load in patients with type II diabetes (Hallfrisch *et al.*, 1995; Jenkins *et al.*, 2002). Further, soluble viscous fiber such as β-glucan improves gastrointestinal motility and gastric emptying and stimulates the secretion of gastric hormones involved in satiety signaling, such as cholecystokinin, glucagon-like peptide-1, peptide YY, and ghrelin (Juvonen *et al.*, 2009). Whole grains are also a rich source of nutrients, including complex carbohydrates, starch, oligosaccharides, minerals, vitamins, and phytochemicals (Slavin *et al.*, 1997; Anderson, 2000). Apart from fiber, the health benefits of these whole grain components and their mechanisms are not well understood.

In addition to its cholesterol-lowering effect, consuming oat bran together with vitamins E and C improves endothelial function (Katz *et al.*, 2001) and reduces blood pressure (Saltzman *et al.*, 2001). Although the mechanisms underlying these effects are not clear, these effects may be mediated by oats' ability to lower blood lipid levels through their β-glucan content or through their antioxidant or phytochemical components, which relax the smooth muscle cells lining the vascular wall. These phytochemicals can be divided into two types: free phenols, which are low molecular weight, soluble and absorbable, and possess antioxidant activities (e.g., tocopherols, tocotrienols, flavonoids, and hydroxycinnamates); and bound phenols, which are covalently linked to complex, high molecular weight insoluble cell components (e.g., lignin, cell wall polysaccharides, structural, and/or storage proteins) and require further metabolism in the gut to be effective.

11.2   Avenanthramides, the bioactive phenolics in oats

Oat antioxidants include tocopherols, tocotrienols, phytic acid, and several types of phenolic compounds including avenanthramides (Avns), which are low molecular weight, soluble phenolic compounds that are unique to oats (Collins and Mullin, 1988; Collins 1989). These compounds develop in the oat plant as phytoalexins (antipathogenic agents) and are produced in response to pathogens such as fungi (Matsukawa *et al.*, 2000; Okazaki *et al.*, 2004). Avns are conjugates of hydroxycinnamic acids (*p*-coumaric acid, caffeic acid, ferulic acid, and sinapic acid) with anthranilic acid or 5-hydroxy anthranilic acid. More than 20 different

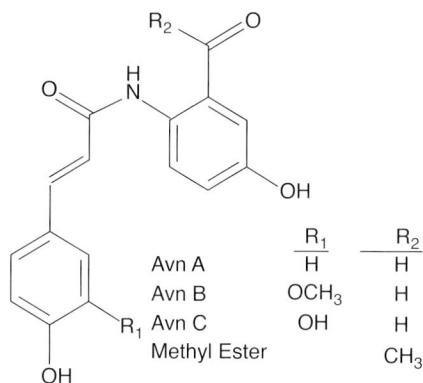

Figure 11.1 Chemical structure of avenanthramides (Avns).

forms of Avns are present when they are extracted from oats, three of which are major forms: Avn-A, Avn-B, and Avn-C (Figure 11.1) (Peterson *et al.*, 2002). However, Avn concentrations may change during commercial processing (Reddy *et al.*, 1992). For example, steaming and flaking moderately reduce the Avn-A content of dehulled oat groats, but steaming does not alter Avn-C and Avn-B levels. In addition, the autoclaving of oat grains and drum drying of steamed rolled oats decrease Avn content. Oat crop varieties with high Avn concentrations are being developed, as are methods to increase the concentration of these polyphenols in oat products and foods for human consumption.

Oats have higher antioxidant activity than other cereal grains. In fact, before the discovery of synthetic antioxidants, oat flour was used to prevent rancidity in other foods (Lee-Manion *et al.*, 2009). Avns are primarily responsible for the antioxidant activity of oats and display potent antioxidant activities *in vitro* and *in vivo* (Dimberg *et al.*, 1992; Peterson *et al.*, 2002; Bratt *et al.*, 2003; Chen *et al.*, 2004, 2005). The antioxidant activity of Avns in oats is 10–30 times higher than that of other phenolic antioxidants such as ferulic and caffeic acids (Dimberg *et al.*, 1992). When antioxidant activity was measured by reactivity toward DPPH (1,1-diphenyl-2-picrylhydrazyl), the antioxidant activity of Avns derived from 5-hydroxyanthranilic acid was found to be stronger than that of Avns derived from anthranilic acid (Bratt *et al.*, 2003). Among the Avns tested for antioxidant activity (using the DPPH and ferric-reducing antioxidant potential assays) and ability to prevent DNA damage, Avn-C (Avn-2c) showed the highest activity. Avn-C is one of the three major Avns found in oats and often comprises about one-third of the total Avn concentration in oat grain (Peterson *et al.*, 2002).

Ji and colleagues (Ji *et al.*, 2003) reported that an Avn-enriched oat extract (AvExO) fed to rats (100 mg/kg diet, providing approximately 20 mg Avn/kg body weight) increased superoxide dismutase activity in skeletal muscle, liver, and kidney and enhanced glutathione peroxidase activity in heart and skeletal muscles. Doubling the dose of AvExO in rats reduced exercise-induced production of reactive oxygen species (O'Moore *et al.*, 2005). Ren and coworkers (Ren *et al.*, 2011) reported that AvExO prevented D-galactose–induced oxidative damage and increased activity of antioxidant enzymes and upregulated their gene expression in mice. Although LDL isolated from Golden Syrian hamsters fed 0.25 g AvExO did not show significant resistance to *in vitro* copper-induced oxidation, the addition of 5 μmol ascorbic acid, which exerts antioxidant activity on hamster LDL, synergistically increased the resistance of LDL to oxidation (Chen *et al.*, 2004). In a randomized, placebo-controlled, crossover study, acute administration of an oral dose of AvExO also increased the reduced form of glutathione in blood, reflecting an increase in the antioxidant capacity of blood (Chen *et al.*, 2007). The copper-induced *in vitro* oxidation of LDL isolated from human subjects was dose-dependently suppressed by *in vitro* addition of oat phenolics, and adding ascorbic acid to the oat phenolics synergistically suppressed copper-induced LDL oxidation (Chen *et al.*, 2004). In a randomized, placebo controlled study, Sen and colleagues (Sen *et al.*, 2011) found that supplemental intake of 3.12 mg AvExO /d by 40 healthy subjects for 1 month significantly increased superoxide dismutase and glutathione levels in serum while reduced serum malondialdehyde levels as a marker of oxidative stress. They also found that intake of AvExO for one month significantly reduced total cholesterol, triglyceride and LDL cholesterol levels in serum.

The above-mentioned studies demonstrate the bioavailability of Avns directly in Golden Syrian hamsters and healthy human volunteers (Chen *et al.*, 2007) and indirectly in laboratory rats (Ji *et al.*, 2003). The peak plasma concentration of Avns in hamster appeared within 40 minutes after bolus administration of AvExO (Chen *et al.*, 2004), whereas the peak plasma concentration in humans appeared two hours after of bolus administration of AvExO (Chen *et al.*, 2007).

11.3   Anti-inflammatory and antiproliferative activity of avenanthramides

Many dietary components of foods that possess antioxidant activity (e.g., vitamin E, vitamin C, carotenoids, polyphenols) may modulate the functions of cells involved in inflammation. *In vitro* studies have suggested that Avns modulate cellular functions, not only through their antioxidant activity but also through interactions with molecular and signaling pathways that govern inflammatory responses. Using human aortic endothelial cells (HAECs) and smooth muscle cells (SMCs) in culture systems, we first examined the potential effects of oat Avns on the cell and molecular processes involved in artery inflammation and atherosclerosis development. We found that AvExO ($\leq$ 40 ng/mL) had no toxicity and did not affect HAEC viability. Adding AvExO to the culture medium dose-dependently inhibited interleukin (IL)-1β-induced expression of several vascular endothelial cell adhesion molecules, including intracellular adhesion molecule (ICAM)-1, vascular cell adhesion molecule (VCAM)-1, and E-selectin (Liu *et al.*, 2004). Suppression of these adhesion molecules by AvExO inhibited HAEC adhesiveness to U937 monocytic cells. In addition, AvExO dose-dependently reduced the IL-1β-induced production of several cytokines and chemokines (e.g., IL-6, IL-8, and monocyte chemotactic protein 1) that participate in inflammatory processes associated with the development of fatty streaks in arteries (Liu *et al.*, 2004).

Endothelial cell expression of adhesion molecules and proinflammatory cytokines and chemokines is regulated by redox-sensitive signal transduction, which involves the activation of the transcription factor nuclear factor-kappa B (NF-κB) (Collins, 1993; Collins and Cybulsky, 2001). We further investigated the molecular mechanism by which AvExO suppresses inflammatory cytokines and adhesion molecules and found that NF-κB suppression is mediated by attenu-ating inhibitor of κB (IκB) and IκB kinase phosphorylation, proteasome activ-ity, and IκB degradation in endothelial cells (Guo *et al.*, 2008). Consistent with our findings, Lv and colleagues (Lv *et al.*, 2009) recently reported that the syn-thetic Avn analog dihydroavenanthramide (DHAvn) inhibits cytokine-induced activation of the NF-κB pathway and protects against pancreatic β-cell damage. Most recently, Lee and colleagues (Lee *et al.*, 2011) reported that DHAvns inhibit breast MCF-7 cancer cell invasion by suppressing matrix metalloproteinase 9 and mitogen-activated protein kinase-mediated NF-κB activation.

Even without knowing the mechanism of these effects, the anti-inflammatory and anti-itch properties of oats have been known for many years. Oatmeal is a well-established remedy for poison ivy, poison oak, sunburn, eczema, and psori-asis. Colloidal oat extract, which also contains Avns, possesses antihistamine and anti-irritation activities (Kurtz and Wallo, 2007). More recently, several *in vitro* molecular and animal studies demonstrated the soothing effect of oat Avns on skin irritation. Sur and coworkers (Sur *et al.*, 2008) reported that applying Avns at concentrations as low as 1 ppb inhibits NF-κB activation in keratinocytes and reduces the release of IL-8, a proinflammatory chemokine. They also demon-strated that topical application of Avns (1–3 ppm) alleviated neurogenic inflam-mation and pruritogen-induced scratching in a murine model of itching, pre-sumably through its antihistamine and anti-inflammatory properties. Similarly, DHAvn reduced histamine-related skin disorders such as redness and itching (Heuschkel *et al.*, 2008).

Spiecker and colleagues (Spiecker *et al.*, 2002) reported that another synthetic drug, tranilast [N-(3′,4′-dimethoxycinnamoyl)-anthranilic acid], which is struc-turally similar to Avns (Figure 11.2), reduces inflammation by suppressing NF-κB activity and inhibiting the endothelial expression of ICAM-1, VCAM-1, E-selectin, and IL-6. This drug also inhibits the cytokine-induced expression of VCAM-1 and chemokines in corneal fibroblasts (Adachi *et al.*, 2010). In addi-tion, tranilast inhibits proliferation of vascular smooth muscle cells (VSMCs) and

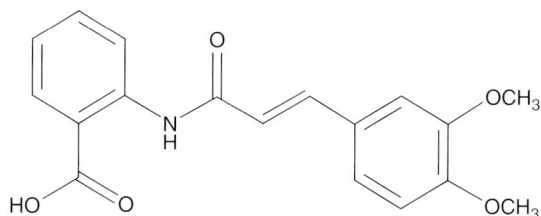

Figure 11.2 Chemical structure of tranilast.

dose-dependently prevents restenosis after angioplasty in animal models (Azuma *et al.*, 1976; Suzawa *et al.*, 1992; Tanaka *et al.*, 1994; Isaji *et al.*, 1998; Tamai *et al.*, 1999).

AvExO, synthetic Avn-C, and its methylated derivative also inhibit proliferation of rat A-10 SMCs and human VSMCs, important pathophysiological processes in the development of atherosclerosis and restenosis in arteries after angioplasty (Ross, 1999; Libby, 2002; Nie *et al.*, 2006a). We also discovered that the methylated form of Avn-C (Figure 11.1) reduces VSMC proliferation several fold. We then investigated the molecular mechanisms by which Avn-C inhibits SMC proliferation using synthetically prepared Avn-C. We found that Avn-C modulates several cell cycle regulatory proteins (e.g., p53, $p21^{Cip1}$, $p27^{Kip1}$, and cyclin D1), thereby inhibiting cell cycle signaling at the G1 to S phase progression (Nie *et al.*, 2006b). We also discovered that Avn-C inhibits hyperphosphorylation of retinoblastoma protein, an important biochemical process in G1 to S phase transition. Our findings suggest that the inhibitory effect of Avns on VSMC proliferation is mediated by upregulating $p21^{Cip1}$ cyclin-dependent kinase inhibitor and downregulating cyclin D1. In addition, Avn-C increases the expression and stability of p53 protein in VSMCs, which contributes to the increase in $p21^{Cip1}$ expression. This antiproliferative effect of Avns on VSMCs, together with their anti-inflammatory effects in HAECs, points to their potential antiatherosclerotic properties.

Our discovery that Avns suppress VSMC proliferation by modulating cell cycle proteins such as p53 (a major player in cancer cell proliferation) was consistent with previous epidemiological data and animal studies revealing that diets containing whole grains are associated with a reduced risk of colorectal cancer (Burkitt *et al.*, 1972; Burkitt, 1993; Reddy *et al.*, 1992; Slattery *et al.*, 1997); therefore, we decided to further evaluate the ability of Avns to suppress cancer cell proliferation. Soluble and insoluble fiber in whole grains is believed to produce its beneficial effects either directly (by diluting carcinogens in the colon or decreasing transient time) or indirectly (by regulating microbiome populations); however, it is possible that nonfiber and phenolic antioxidant components of whole grains exert beneficial effects on gut health (Burkitt *et al.*, 1972; Reddy *et al.*, 1987, 2000; Sakata, 1987; Howe *et al.*, 1992). In support of this hypothesis, a prospective study has reported no protective effect from fiber derived from fruits, vegetable, and cereals against colon cancer (Fuchs *et al.*, 1999). In addition, a large population-based cohort study has shown that, after adjusting for the effects of fiber, consuming whole grains still significantly decreased the risk of colon cancer (Larsson *et al.*, 2005). Therefore, other components of whole grains, such as those with anti-inflammatory properties, might be responsible for their protective effect against colorectal cancer. Approximately 30% of all cancers are associated with chronic inflammation; therefore, components of whole grains with anti-inflammatory properties may be important contributors to their potential anticancer effects.

After our discovery of the anti-inflammatory and antiproliferative properties of oat Avns (Liu *et al.*, 2004; Nie *et al.*, 2006a), we examined whether these compounds also exert anticancer actions by evaluating their effects on the expression of cyclooxygenase 2 (COX-2), which is involved in epithelial carcinogenesis, proliferation, and tumor growth in the colon epithelium (Guo *et al.*, 2010). We found that AvExO inhibited COX enzyme activity and prostaglandin E2 production in activated mouse peritoneal macrophages. AvExO, Avn-C, and the methyl-ester derivative of Avn-C did not suppress the proliferation of prostate or breast cancer cell lines but were effective in suppressing COX-2-positive Caco-2, HT29, and LS174T human colon cancer cell lines and the COX-2-negative HCT116 human colon cancer cell line (Guo *et al.*, 2009). However, Avns had no effect on COX-2 expression or prostaglandin E2 production in these cells. These findings suggest that Avns prevent colon cancer by inhibiting proliferation and inflammation but not COX-2 signaling in colon cancer cells, and through their effect on macrophages. It is also worth mentioning that Avns had no cytotoxic effects on differentiated colon cancer cells, which display the characteristics of normal colonic epithelial cells.

11.4 Summary and conclusion

Oats are a healthful cereal not only because of their ability to reduce blood cholesterol through β-glucan content but also because of the strong antioxidant activity of unique polyphenols, avenanthramides (Avns). These oat polyphenols exert anti-inflammatory and antiproliferative actions *in vitro* and *in vivo*, which add to the health benefits of consuming oats. In cultures of vascular cells, Avns attenuate the production of inflammatory cytokines and adhesion molecules, reduce the adhesion of monocytes to endothelial cells, and may prevent the proliferation of vascular SMCs, which participate in the development of atherosclerosis. Oat Avns also inhibit the proliferation of several colon cancer cell lines with no cytotoxic effect on normal colonic epithelial cells. *In vivo*, Avns appear to increase the endogenous antioxidant capacity of cells and display an anti-itch effect, presumably through antihistamine and anti-inflammatory actions. These biological activities of oat Avns may explain the well-known soothing effect of oats on irritated skin. The potential biological functions of oat Avns warrant further exploration for their potential to provide health benefits with regular oat consumption.

Acknowledgements

Supported by the US Department of Agriculture under agreement No. 58-1950-7-707. Any opinions, findings, conclusions, or recommendations expressed in this publication are those of the author and do not necessarily reflect the view of the US Department of Agriculture. I would also like to thank Stephanie Marco for her assistance in the preparation of this manuscript.

References

Adachi, T., *et al.* (2010) Inhibition by tranilast of the cytokine-induced expression of chemokines and the adhesion molecule VCAM-1 in human corneal fibroblasts. *Investigative Ophthalmology and Visual Science* **51**, 3954–3960.

Anderson, J.W. (2000) Dietary fiber prevents carbohydrate-induced hypertriglyceridemia. *Current Atherosclerosis Reports* **2**, 536–541.

Anderson, J.W. (2003) Whole grains protect against atherosclerotic cardiovascular disease. *Proceedings of the Nutrition Society* **62**, 135–142.

Azuma, H., *et al.* (1976) Pharmacological properties of N-(3′,4′-dimethoxycinnamoyl) anthranilic acid (N-5′), a new anti-atopic agent. *British Journal of Pharmacology* **58**, 483–488.

Bratt, K., *et al.* (2003) Avenanthramides in oats (*Avena sativa* L.) and structure–antioxidant activity relationships. *Journal of Agricultural Food Chemistry* **51**, 594–600.

Burkitt, D.P. (1993) Epidemiology of cancer of the colon and rectum. 1971. *Diseases of the Colon and Rectum* **36**, 1071–1082.

Burkitt, D.P., *et al.* (1972) Effect of dietary fibre on stools and the transit-times, and its role in the causation of disease. *Lancet* **2**, 1408–1412.

Chen, C.Y., *et al.* (2004) Avenanthramides and phenolic acids from oats are bioavailable and act synergistically with vitamin C to enhance hamster and human LDL resistance to oxidation. *The Journal of Nutrition* **134**, 1459–1466.

Chen, C.-Y., *et al.* (2005) Antioxidant capacity and bioavailability of oat avenanthramides. *The FASEB Journal* **19**, A1477.

Chen, C.Y., *et al.* (2007) Avenanthramides are bioavailable and have antioxidant activity in humans after acute consumption of an enriched mixture from oats. *The Journal of Nutrition* **137**, 1375–1382.

Collins, F.W. (1989) Oat phenolics: avenanthramides, novel substituted N-cinnamoylanthranilate alkaloids from oat groats and hulls. *Journal of Agricultural Food Chemistry* **37**, 60–66.

Collins, F.W. and Mullin, W.J. (1988) High-performance liquid chromatographic determination of avenanthramides, N-aroylanthranilic acid alkaloids from oats. *Journal of Chromatography* **45**, 363–370.

Collins, T. (1993) Biology of disease: Endothelial nuclear factor-κB and the initiation of the atherosclerotic lesion. *Laboratory Investigations* **68**, 499–508.

Collins, T. and Cybulsky, M.I. (2001) NF-kB: pivotal mediator or innocent bystander in atherosclerosis? *The Journal of Clinical Investigation* **107**, 255–264.

Dimberg, L.H., *et al.* (1992) Avenanthramides-a group of phenolic antioxidants in oats. *Cereal Chemistry* **70**, 637–641.

Fuchs, C.S., *et al.* (1999) Dietary fiber and the risk of colorectal cancer and adenoma in women. *The New England Journal of Medicine* **340**, 169–176.

Gillman, M.W., *et al.* (1995) Protective effect of fruits and vegetables on development of stroke in men. *JAMA* **273**, 1113–1117.

Guo, W., *et al.* (2008) Avenanthramides, polyphenols from oats, inhibit IL-1beta-induced NF-kappaB activation in endothelial cells. *Free Radical Biology and Medicine* **44**, 415–429.

Guo, W., *et al.* (2009) Dietary polyphenols, inflammation, and cancer. *Nutrition and Cancer* **61**, 807–810.

Guo, W., *et al.* (2010) Avenanthramides inhibit proliferation of human colon cancer cell lines *in vitro*. *Nutrition and Cancer* **62**, 1007–1016.

Hallfrisch, J., *et al.* (1995) Diets containing soluble oat extracts improve glucose and insulin responses of moderately hypercholesterolemic men and women. *The American Journal of Clinical Nutrition* **61**, 379–384.

Heuschkel, S., *et al.* (2008) Modulation of dihydroavenanthramide D release and skin penetration by 1,2-alkanediols. *European Journal of Pharmaceutics and Biopharmaceutics* **70**, 239–247.

Howe, G.R., *et al.* (1992) Dietary intake of fiber and decreased risk of cancers of the colon and rectum: evidence from the combined analysis of 13 case-control studies. *Journal of the National Cancer Institute* **84**, 1887–1896.

Isaji, M., *et al.* (1998) Tranilast: A new application in the cardiovascular field as an antiproliferative drug. *Cardiovascular Drug Reviews* **16**, 288–299.

Jenkins, D.J., *et al.* (2002) Soluble fiber intake at a dose approved by the US Food and Drug Administration for a claim of health benefits: serum lipid risk factors for cardiovascular disease assessed in a randomized controlled crossover trial. *The American Journal of Clinical Nutrition* **75**, 834–839.

Ji, L.L., *et al.* (2003) Effect of avenanthramides on oxidant generation and antioxidant enzyme activity in exercised rats. *Nutrition Research* **23**, 1579–1590.

Joshipura, K.J., *et al.* (2001) The effect of fruit and vegetable intake on risk for coronary heart disease. *Annals of Internal Medicine* **134**, 1106–1114.

Juvonen, K.R., *et al.* (2009) Viscosity of oat bran-enriched beverages influences gastrointestinal hormonal responses in healthy humans. *The Journal of Nutrition* **139**, 461–466.

Katz, D.L., *et al.* (2001) Effects of oat and wheat cereals on endothelial responses. *Preventive Medicine* **33**, 476–484.

Kelly, S.A., *et al.* (2007) Wholegrain cereals for coronary heart disease. *Cochrane Database of Systematic Reviews* **2** (Art. No.: CD005051). doi: 10.1002/14651858.CD005051.pub2.

Kurtz, E.S. and Wallo, W. (2007) Colloidal oatmeal: history, chemistry and clinical properties. *The Journal of Drugs in Dermatology* **6**, 167–170.

Larsson, S.C., *et al.* (2005) Whole grain consumption and risk of colorectal cancer: a population-based cohort of 60,000 women. *British Journal of Cancer* **92**, 1803–1807.

Lee, Y.R., *et al.* (2011) Dihydroavenanthramide D inhibits human breast cancer cell invasion through suppression of MMP-9 expression. *Biochemical and Biophysical Research Communications* **405**, 552–557.

Lee-Manion, A.M., *et al.* (2009) *In vitro* antioxidant activity and antigenotoxic effects of avenanthramides and related compounds. *Journal of Agricultural Food Chemistry* **57**, 10619–10624.

Libby, P. (2002) Inflammation in atherosclerosis. *Nature* **420**, 868–874.

Liu, L., *et al.* (2004) The antiatherogenic potential of oat phenolic compounds. *Atherosclerosis* **175**, 39–49.

Lv, N., *et al.* (2009) Dihydroavenanthramide D protects pancreatic beta-cells from cytokine and streptozotocin toxicity. *Biochemistry and Biophysical Research Communications* **387**, 97–102.

Matsukawa, T., *et al.* (2000) Occurrence of avenanthramides and hydroxycinnamoyl-CoA:hydroxyanthranilate N-hydroxycinnamoyltransferase activity in oat seeds. *Zeitschrift für Naturforschung C* **55**, 30–36.

Ness, A.R. and Powles, J.W. (1997) Fruit and vegetables, and cardiovascular disease: a review. *International Journal of Epidemiology* **26**, 1–13.

Nie, L., *et al.* (2006a) Avenanthramide, a polyphenol from oats, inhibits vascular smooth muscle cell proliferation and enhances nitric oxide production. *Atherosclerosis* **186**, 260–266.

Nie, L., *et al.* (2006b) Mechanism by which avenanthramide-c, a polyphenol of oats, blocks cell cycle progression in vascular smooth muscle cells. *Free Radical Biology and Medicine* **41**, 702–708.

O'Moore, K.M., *et al.* (2005) Effect of Avenanthramides on rat skeletal muscle injury induced by lengthening contraction. *Medicine and Science in Sports and Exercise* **37**, S466.

Okazaki, Y., *et al.* (2004) Metabolism of avenanthramide phytoalexins in oats. *The Plant Journal* **39**, 560–572.

Peterson, D.M., *et al.* (2002) Oat avenanthramides exhibit antioxidant activities in vitro. *Food Chemistry* **79**, 473–478.

Reddy, B.S., *et al.* (1987) Metabolic epidemiology of colon cancer: effect of dietary fiber on fecal mutagens and bile acids in healthy subjects. *Cancer Research* **47**, 644–648.

Reddy, B.S., *et al.* (1992) Effect of dietary fiber on colonic bacterial enzymes and bile acids in relation to colon cancer. *Gastroenterology* **102**, 1475–1482.

Reddy, B.S., *et al.* (2000) Preventive potential of wheat bran fractions against experimental colon carcinogenesis: implications for human colon cancer prevention. *Cancer Research* **60**, 4792–4797.

Ren, Y., *et al.* (2011) Chemical characterization of the avenanthramide-rich extract from oat and its effect on D-galactose-induced oxidative stress in mice. *Journal of Agricultural and Food Chemistry* **59**, 206–211.

Ross, R. (1999) Atherosclerosis: an inflammatory disease. *New England Journal of Medicine* **340**, 115–126.

Sakata, T. (1987) Stimulatory effect of short-chain fatty acids on epithelial cell proliferation in the rat intestine: a possible explanation for trophic effects of fermentable fibre, gut microbes and luminal trophic factors. *The British Journal of Nutrition* **58**, 95–103.

Saltzman, E., *et al.* (2001) An oat-containing hypocaloric diet reduces systolic blood pressure and improves lipid profile beyond effects of weight loss in men and women. *The Journal of Nutrition* **131**, 1465–1470.

Sen, L., *et al.* (2011) Antioxidant effects of oat avenanthramides on human serum. *Agricultural Sciences in China* **10**, 1301–1305.

Slattery, M.L., *et al.* (1997) Diet diversity, diet composition, and risk of colon cancer (United States). *Cancer Causes and Control: CCC* **8**, 872–882.

Slavin, J., *et al.* (1997) Whole-grain consumption and chronic disease: protective mechanisms. *Nutrition and Cancer* **27**, 14–21.

Spiecker, M., *et al.* (2002) Tranilast inhibits cytokine-induced nuclear factor kappaB activation in vascular endothelial cells. *Molecular Pharmacology* **62**, 856–863.

Sur, R., *et al.* (2008) Avenanthramides, polyphenols from oats, exhibit anti-inflammatory and anti-itch activity. *Archives Dermatological Research* **300**, 569–574.

Suzawa, H., *et al.* (1992) Inhibitory action of tranilast, an anti-allergic drug, on the release of cytokines and PGE2 from human monocytes-macrophages. *The Japanese Journal of Pharmacology* **60**, 85–90.

Tamai, H., *et al.* (1999) Impact of tranilast on restenosis after coronary angioplasty: tranilast restenosis following angioplasty trial (TREAT). *American Heart Journal* **138**, 968–975.

Tanaka, K., *et al.* (1994) Prominent inhibitory effects of tranilast on migration and proliferation of and collagen synthesis by vascular smooth muscle cells. *Atherosclerosis* **107**, 179–185.

Wolff, H., *et al.* (1998) Expression of cyclooxygenase-2 in human lung carcinoma. *Cancer Research* **58**, 4997–5001.

12

Effects of Oats on Obesity, Weight Management, and Satiety

Chad M. Cook, Tia M. Rains, and Kevin C. Maki
Biofortis Clinical Research, Addison, IL, USA

12.1 Introduction

In 2010, approximately two-thirds of the United States adult population was classified as overweight (body mass index [BMI] $\geq$25.0–29.99 kg/m^2) or obese (BMI $\geq$30.0 kg/m^2) (Flegal *et al.*, 2010). Furthermore, 86% of adults are predicted to be overweight or obese by 2030 if current trends continue (Wang *et al.*, 2008). Dietary strategies that reduce hunger or enhance satiation (sensations that determine meal size and duration) and satiety (sensations that determine the intermeal period) may help individuals lose weight or prevent weight gain (Blundell and MacDiarmid, 1997; Mattes *et al.*, 2005). One such dietary strategy is to increase consumption of whole grain, high-fiber foods.

Results of numerous cross-sectional and prospective cohort studies suggest that high habitual consumption of whole grain foods, such as oats, is inversely associated with BMI (Giacco *et al.*, 2011). For example, in the Nurses' Health Study (n = 74 091), women in the highest quintile of whole grain intake at baseline (1.62 servings/1000 kcal/d) weighed approximately 0.9 kg less on average than women in the lowest quintile (0.07 servings/1000 kcal/d) (Liu *et al.*, 2003). Increased consumption of whole grains has also been associated with lower body weight gain over time. In the Health Professionals Follow-up Study, 27 082 men gained an average of approximately 2 kg body weight over eight years (Koh-Banerjee *et al.*, 2004); however, multivariate analyses showed that for every 40-g/day increment in whole grain intake, weight gain was reduced by 0.49 kg. Similar results were reported in the all-male Physicians' Health Study (n = 17 881). Results of multivariate analyses showed that men who consumed $\geq$ 1 serving of whole grain breakfast cereal per day gained less body weight over the

8-year follow-up than men who rarely or never consumed these foods (1.13 kg vs 1.55 kg, respectively) (Bazzano *et al.*, 2005). The attenuation in weight gain with increased whole grain intake may be mediated by the high dietary fiber content of whole grains. For instance, although women in the Nurses' Health Study gained body weight during the 12-year follow-up period (1984–1996), increasing dietary fiber intake by 12 g/day was associated with approximately 3.5 kg (7.7 lb) less weight gain during this time. Results from the Coronary Artery Risk Development in Young Adults (CARDIA) study were similar. Adults aged 18–30 years who consumed more than 10.5 g dietary fiber/1000 kcal/day weighed 3.65 kg (8.1 lb) less than individuals who ate less than 5.9 g dietary fiber/1000 kcal/day after a 10-year follow-up (Ludwig *et al.*, 1999).

Taken together, these results suggest that increasing consumption of oats, a whole grain food that is a good source of dietary fiber (oat β-glucan), can reduce weight gain and may be a useful dietary strategy to help overweight or obese individuals regulate their body weight. This chapter summarizes the results of clinical intervention studies evaluating the effects of increased consumption of oats or oat dietary fiber (β-glucan) on body weight and measures of appetite.

12.2   Effects of oats and oat β-glucan on body weight

Numerous human intervention trials have assessed the effects of oat consumption on body weight using oat-based foods or by enriching foods, such as bread, with oat β-glucan (Table 12.1). An early study reporting successful weight loss with increased oat consumption showed that obese subjects who consumed an oat-based soup containing 4 g of dietary fiber as a main meal once or twice daily as part of a calorie-restricted diet lost approximately 6 kg over 23 weeks (Rytter *et al.*, 1996). The participants attributed the weight loss to increased satiety associated with the oat-based soup; however, it should be noted that a control group was not used in this study. More recently, Maki and colleagues evaluated the effect of oat consumption in overweight/obese individuals consuming a mildly energy-restricted diet (target 500 kcal/day deficit) over 12 weeks (Maki *et al.*, 2010). The results demonstrated that two daily servings of a ready-to-eat oat cereal (3 g/day β-glucan) was associated with a larger reduction in waist circumference than an energy-matched low-fiber control cereal (−3.3 cm vs. −1.9 cm, respectively, p = 0.01). Although both groups lost a similar amount of body weight (oat cereal, −2.2 kg; control, −1.7 kg), the effect of oat consumption on waist circumference is consistent with data from population studies in which high habitual consumption of whole grains, such as oats, is inversely associated with waist circumference (Newby *et al.*, 2007; Williams *et al.*, 2008).

Additional data from intervention trials are in line with results from major prospective cohort studies suggesting that whole grain high-fiber foods attenuate weight gain (Liu *et al.*, 2003; Koh-Banerjee *et al.*, 2004; Bazzano *et al.*, 2005). Results of a study by He and colleagues suggested that consuming oat-based foods may reduce weight gain (He *et al.*, 2004). In this randomized, double-blind, controlled trial, hypertensive individuals incorporated into their usual diets either

Table 12.1   Summary of studies on the effect of oats or oat β-glucan on body weight

Reference	Design	Oats treatment(s)	Control treatment(s)	Duration	Sample size n = M/F	Baseline body weight (kg)	Body weight △ from baseline (kg)
Oat foods							
Rytter et al., 1996	Single group, non-randomized	Oat-based soup 1× or 2× daily	No control	23 wk	O: 9/23	O: 83.5 ± 2.2	O: -6.0 ± NR
Maki et al., 2010	R, P	2 servings/d WG-RTE oat cereal (3 g/d oat β-glucan)	Energy-matched low-fiber foods	12 wk CRD	O: 19/58 C: 12/55	O: 88.7 ± 1.9 C: 87.6 ± 1.8	BW O: -2.2 ± 0.3* C: -1.7 ± 0.3* WC (cm) O: -3.3 ± 0.4*† C: -1.9 ± 0.4*
He et al., 2004	R, DB, PC	Oat-based muffin and cereal (8 g/d β-glucan)	Wheat muffin and corn flake cereal (no β-glucan)	12 wk UD + muffin and cereal	O: 22/34 C: 22/32	O: 82.1 ± 2.2 C: 83.6 ± 2.1	O: 0.1 (-0.4, 0.7)[a] C: 0.6 (0.1, 1.1)
Saltzman et al., 2001a	R, P	Diet with rolled oats; 45 g/1000 kcal/d	Oat-free, low-soluble fiber diet	2 wk WMD 6 wk CRD	O: 11/11 C: 9/12	O: 74.9 ± 3.1 C: 78.0 ± 3.2	O: -3.90 ± 0.34* *C: -4.00 ± 0.05*
Saltzman et al., 2001b	R, P	Diet with rolled oats; 45 g/1000 kcal/d	Oat-free, low-soluble fiber diet	2 wk WMD 6 wk CRD	O: 11/10 C: 9/11	O: 75.7 ± 3.2 C: 77.8 ± 3.3	O: -4.35 ± 0.37* C: -4.43 ± 0.27*
De Oliveira et al., 2003[b]	R, SB, P	3 oat cookies/d (60 g/d)	3 apples/d or 3 pears/d (300 g/d)	12 wk CRD	O: 9 F C: 26 F	O: 78.9 ± 3.2 C: 77.7 ± 2.1	O: -0.88 ± 0.65 C: -1.21 ± 0.38*†
Robitaille et al., 2005	R, P	2 oat bran muffins/d (28 g oats/d)	No muffins	2 wk NCEP step I diet (run-in) 4 wk NCEP step 1 diet + muffins	O: 18 F C: 16 F	O: 76.5 ± 3.1 C: 79.9 ± 3.0	O: 0.36 ± 0.33 C: -0.66 ± 0.27*

(continued)

Table 12.1   (*Continued*)

Reference	Design	Oats treatment(s)	Control treatment(s)	Duration	Sample size n = M/F	Baseline body weight (kg)	Body weight Δ from baseline (kg)
Charlton et al., 2012	R, SB, P	*Group 1 (G1)* RTE oat cereal • 1.5 g/d β-glucan *Group 2 (G2)* Oatmeal and oat-based cereal bar • 3.2 g/d β-glucan	Cornflakes, puffed rice, and wheat cereal bars	6 wk WMD + education promoting healthy dietary habits	G1: 11/15 G2: 15/15 C: 15/16	G1: 77.9 ± 3.1 G2: 77.1 ± 2.0 C: 81.1 ± 2.5	G1: 3.9 ± NR* G2: 1.5 ± NR* C: 0.8 ± NR*
Maki et al., 2007	R, DB, P	*1 serving/d:* • RTE oat bran cereal • Oatmeal • Powdered oat β-glucan	*1 serving/d:* • RTE wheat-based cereal • Low-fiber hot cereal • Maltodextrin powder	12 wk NCEP TLC diet; study foods consumed 3×/wk	O: 14/12 C: 19/15	O: 92.9 ± 3.5 C: 93.1 ± 2.9	Data NR; authors state no Δ within or between groups[c]
Oat fiber							
Reyna-Villasmil et al., 2007	R, P	Bread (6 g β-glucan) + 60 min/d walking	Whole wheat bread + 60 min/d walking	1 wk run-in 8 wk AHA step II diet	O: 19 M C: 19 M	O: 76.8 ± 2.6 C: 76.0 ± 2.2	O: −5.8 ± NR*† C: −3.8 ± NR*
Liatis et al., 2009	R, DB, P	30 g/d bread w/ oat β-glucan (3 g/d)	30 g/d white wheat bread	3 wk UD	O: 12/11 C: 11/7	O: 81.6 ± 2.5 C: 74.4 ± 2.9	O: −1.03 ± 1.64* C: −0.39 ± 1.18
Biörklund et al., 2008	R, PC, SB, P	Soup with oat β-glucan (4 g)	Soup without oat β-glucan	3 wk lead-in UD + control soup 5 wk UD + soup	O: 22 sex NR C: 21 sex NR	*BMI (mg/k²)* O: 25.3 ± 3.5 C: 24.7 ± 2.7	*BMI (% Δ)* O: 0.41 C: 0.40

Study	Design	Intervention	Control	Duration	N	Baseline	Change
Wolever et al., 2010	R, DB, P	Oat bran cereal with 3–4 g β-glucan (varying MW)	Wheat bran cereal	4 wk UD + cereal	3M: 36/51[d] 3H: 27/37 4L: 43/43 4M: 33/34 C: 87	Baseline BW was NR	NR; authors state no Δ within or between groups
Beck et al., 2010	R, P	Group 1 (G1) • High fiber diet + 5–6 g/d oat β-glucan Group 2 (G2) • High fiber diet + 8–9 g/d oat β-glucan	High-fiber diet, no oat β-glucan	3 months	G1: 21 F G2: 19 F C: 16 F	G1: 80.9 ± 1.8 G2: 77.6 ± 1.5 C: 77.0 ± 2.0	G1: −4.3 ± 0.8 G2: −3.9 ± 0.7 C: −4.0 ± 0.6
Cugnet-Anceau et al., 2010	R, DB, PC, P	Soup containing 3.5 g oat β-glucan	Soup with no oat β-glucan	3 wk run-in C soup 8 wk UD + soup	O: 29 MandF C: 24 MandF	BMI (mg/k²) O: 30.5 ± 0.8 C: 29.0 ± 0.8	BMI (mg/k²) O: 0.2 ± 0.2 C: 0.4 ± 0.3

Data are presented as mean ± standard error of the mean.

*Significant change in body weight from baseline within group (p < 0.05).

† Body weight change from baseline differed significantly between the oat and control group (p < 0.05).

[a]The primary outcome was change in blood pressure; weight loss was not encouraged or discouraged. Data presented as mean and 95% confidence intervals.

[b]Data from the apple and pear treatment groups were combined (no difference in subject baseline characteristics).

[c]Primary outcomes included changes in blood pressure, postprandial glucose and insulin levels, and biomarkers of oxidative stress. Subjects were advised to maintain their body weight and usual level of physical activity throughout the trial.

[d]3H, 3 g high molecular weight (MW) oat β-glucan/d; 4M, 4 g medium MW oat β-glucan/d; 3M, 3 g medium MW oat β-glucan/d; 4L, 4 g low MW oat β-glucan/d.

Abbreviations: AHA, American Heart Association; BMI, body mass index; BW, body weight; C, control treatment; CRD, caloric-restricted diet; DB, double blind; F, females; H, high; LF, low fiber; M, males; MW, molecular weight; N, normal; NCEP, National Cholesterol Education Program; NR, not reported; O, oats treatment; P, parallel; PC, placebo-controlled; R, randomized; RTE, ready-to-eat; SB, single blind; TLC, Therapeutic Lifestyle Changes diet; UD, usual diet; WC, waist circumference; WG, whole grain; WMD, weight maintenance diet.

high-fiber oat bran muffins plus an oat-based cereal daily (7.3 g β-glucan) or low-fiber refined wheat muffins plus a corn flake cereal daily (0 g β-glucan). Over a 12-week period, the control group gained 0.8 kg (95% confidence interval [CI]: 0.2, 1.4 kg), whereas weight remained stable in the group consuming high-fiber oat-based foods (weight gain 0.1 kg, 95% CI: –0.5, 0.7 kg; $\triangle$ body weight, oats – control = –0.7 kg; 95% CI: –1.6, 0.1 kg, p = 0.07). However, not all intervention studies have demonstrated beneficial effects of increased consumption of oats on body weight.

In two similar studies, Saltzman and coworkers evaluated the effects of oat consumption on body weight in men and women with BMIs ranging from 20.5 to 33.9 kg/m^2, but observed no effect of a 6-week calorie-restricted diet high in whole rolled oats (45 g/1000 kcal/d) compared with an oat-free low-fiber control diet (Saltzman et al., 2001a, 2001b). De Oliveira and coworkers found that incorporating three oat cookies per day in place of foods with a similar caloric value did not enhance the effectiveness of a calorie-restricted diet in overweight, hypercholesterolemic women during a 12-week period (De Oliveira et al., 2003). A four-week intervention evaluated the effects of oat bran-rich muffins (28 g/day oat bran, 6.2 g/day total dietary fiber, 376 kcal/day) on fasting lipids in premenopausal overweight women following the National Cholesterol Education Program (NCEP) Step I diet. The oat bran-enriched food improved lipid profiles but did not alter body weight (0.36 kg, p > 0.05); however, the control group consuming the NCEP diet with no additional intervention did lose body weight (–0.66 kg; p < 0.001) (Robitaille et al., 2005).

In contrast, two studies that examined the effects of consuming bread enriched with oat β-glucan reported positive effects. In a study by Liatis and colleagues (Liatis et al., 2009), men and women with type 2 diabetes and elevated low-density lipoprotein cholesterol (> 130 mg/dL) incorporated enriched bread made with wheat and oat flour (3 g/day oat β-glucan) into their usual diets for three weeks. Participants who consumed the β-glucan-enriched bread showed significant reductions in body weight (–1.0 kg; p = 0.006) and waist circumference (–1.6 cm; p = 0.02), whereas participants who consumed white bread showed no changes in weight or waist circumference. In a longer study, overweight and obese men incorporating bread enriched with oat β-glucan (6 g/day) into the American Heart Association Step II diet for 8 weeks lost more body weight (–7.9% change from baseline) than similar individuals on the Step II diet alone (–4.9% change from baseline) (Reyna-Villasmil et al., 2007).

Beck and coworkers studied the effects of an energy-restricted diet that incorporated 0, 5–6, or 8–9 g/day of a commercially available oat bran (containing 22% oat β-glucan) in cold and hot cereals in 66 overweight women (Beck et al., 2010). In all groups, body weight was approximately 5% lower than baseline after three months and waist circumference was reduced (p < 0.001), with no significant difference among the groups. Studies using oat β-glucan-enriched soup also failed to demonstrate an effect on body weight or waist circumference after 5 weeks of consuming β-glucan (4 g/day) in hypercholesterolemic men and women (Biörklund et al., 2008) or 8 weeks of consuming β-glucan (3.5 g/day) in obese men and women with type 2 diabetes (Cugnet-Anceau et al., 2010). Charlton and coworkers evaluated the effects of a diet that included cereal bars containing β-glucan (0, 1.5, or 3.0 g/day) in mildly hypercholesterolemic and overweight men and women (Charlton *et al.*, 2012). After 6 weeks, all subjects lost weight, but no significant difference in body weight change was detected among the groups. Maki and coworkers also reported no effect of consuming a high β-glucan diet (7.7 g/day) for 12 weeks on body weight or caloric intake in overweight men and women with elevated blood pressure (Maki *et al.*, 2007). However, the primary objective of this study was to assess the effects of oat β-glucan intake on blood pressure; subjects were asked to maintain a constant body weight while following the NCEP Therapeutic Lifestyle Changes (TLC) diet. It should be noted that outcomes such as change in body weight, BMI, or waist:hip ratio were secondary outcomes in many of the aforementioned trials, which were designed to determine the effects of increased oats or oat β-glucan consumption on fasting lipid profiles or blood pressure.

12.3   Effects of oats on appetite

Subjective appetite (e.g., hunger or fullness) is often assessed using a visual analog scale (VAS), which consists of a straight 100-mm vertical line with two extreme states anchored at either end. For example, a scale to assess hunger asks the question "how hungry are you?" and is anchored with "not at all hungry" at one end and "as hungry as I have ever felt" at the other end of the line. A participant places a mark on the 100-mm line to indicate their level of hunger. To quantify the sensation of hunger, the distance from the end of the line anchored with "not at all hungry" to the participant's mark on the line is measured with a ruler. Other measures of acute appetite responses include measuring self-selected food intake after a predefined test meal, often referred to as the preload-test meal paradigm, along with VAS assessments.

One of the early studies providing evidence that oats (in the form of oatmeal) enhance short-term satiety came from researchers at the University of Sydney in Australia. Holt and coworkers fed 38 isocaloric foods (240 kcal each) from six food categories to groups of 11–13 university students classified as nonrestrained eaters, each with a healthy BMI (Holt *et al.*, 1995). Participants rated subjective satiety using a VAS every 15 minutes after eating for 120 minutes. The satiety index score calculated for each food was expressed as a percentage of the satiety response produced by white bread (reference food). The lowest and highest scores for each category are shown in Figure 12.1. Oatmeal produced the highest satiety index score in the breakfast cereal category and the third highest satiety index score overall behind white fish and boiled potatoes. No additional acute trials examining the effects of oatmeal on subjective appetite could be located in the published literature, but studies using oat-based cereals, breads and pastas, and meal replacement bars have been published.

In a randomized, double-blind, crossover study, Hlebowicz and colleagues (Hlebowicz *et al.*, 2007) found that the effect of whole-meal oat flakes (50 g; 4 g total fiber) on short-term (120 min) satiety did not differ from that of cornflakes (50 g; 1.5 g total fiber) or all-bran flakes (50 g; 7.5 g total fiber). A similar study by this group found no difference in short term (90 min) satiety between a meal of yogurt plus oat bran flakes and a meal of yogurt plus corn flakes in healthy

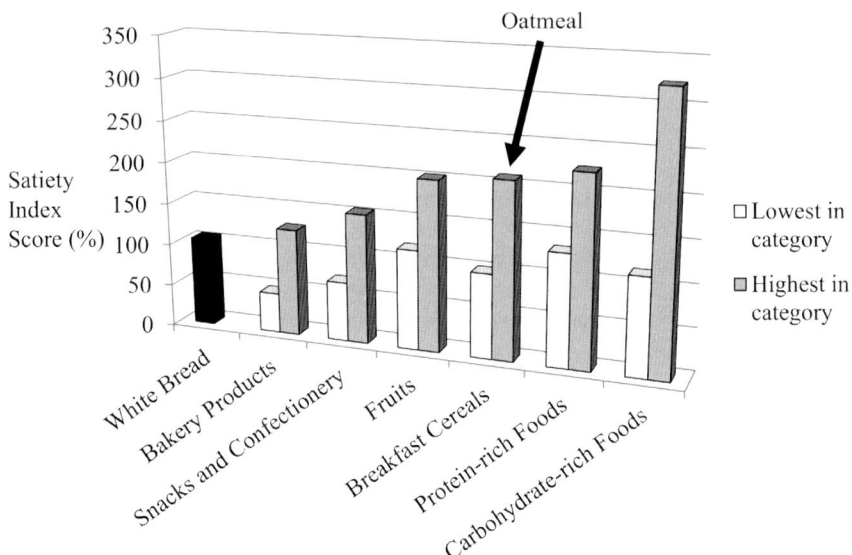

Figure 12.1 Satiety index score of oatmeal relative to other foods in six different food categories. The satiety index score of each test food was expressed as a percentage of the reference food; white bread = 100%. (Figure adapted from data published by Holt *et al.*, 1995.)

individuals (Hlebowicz *et al.*, 2008). However, these two studies were limited by small sample sizes (12 subjects each). Two separate experiments using the preload/test meal paradigm found that oat bread, but not oat pasta, was more satiating than the respective wheat-based counterparts (wheat bread or wheat pasta). However, the observed increase in satiety with oat bread did not reduce food intake two hours after the preload at an *ad libitum* test meal (Berti *et al.*, 2005).

In a randomized, single-blind, crossover study (Weickert *et al.*, 2006), 14 women consumed three isoenergetic and macronutrient matched portions of low-fiber white bread (control), oat hull fiber-enriched bread (10.6 g oat fiber/portion), and wheat fiber-enriched bread (10.5 g wheat fiber/portion). Subjective appetite ratings were assessed, and plasma concentrations of the appetite-suppressing hormone peptide tyrosine tyrosine (PYY) and the appetite-stimulating hormone ghrelin were measured for 5 hours after eating. The results showed that wheat fiber, but not oat fiber, decreased postprandial plasma concentrations of both PYY and ghrelin compared with the control. However, subjective satiety sensations were not affected by either treatment. Changes in PYY and ghrelin may have offset each other, because reduced PYY would be expected to increase satiety and lower plasma ghrelin levels would be expected to reduce appetite. These findings are difficult to interpret because of the limited sample size and lack of a subsequent *ad libitum* test meal to determine the effects of oat or wheat fiber on self-selected food intake. Peters and coworkers evaluated the effect of isocaloric breakfast meal replacement bars containing 0.3 g β-glucan from 6.8 g oats; 0.9 g β-glucan from 8 g barley, 8 g fructo-oligosaccharides, or 0.9 g β-glucan from barley plus 8 g fructo-oligosaccharides in healthy individuals

(Peters *et al.*, 2009). No significant differences were observed in appetite ratings or food intake at an *ad libitum* lunch consumed 240 minutes after the test meal.

Additional acute test meal studies have been conducted to determine the influence of oat β-glucan on short term subjective appetite. In two separate papers reporting results from the same randomized, crossover, acute test meal study, 14 overweight men and women consumed five different breakfast meals on five different occasions (Beck *et al.*, 2009a, 2009b). Each meal contained 200 mL reduced-fat milk along with a corn-based control cereal or one of four oat-based cereals containing varying concentrations and molecular weights of oat β-glucan. Plasma levels of appetite-regulating hormones were assessed and subjective appetite was evaluated for 240 minutes postprandially, followed by an *ad libitum* buffet lunch. This study showed that doses of β-glucan > 3 g significantly decreased insulin and increased subjective satiety ratings. Furthermore, plasma levels of PYY and cholecystokinin (CCK), another gastrointestinal hormone with appetite-suppressing properties, increased in a linear fashion with increasing doses of β-glucan (R^2 = 0.994 and 0.970, respectively) in the first 4 hours after a meal, suggesting that oat β-glucan modulates plasma concentrations of these hormones. In addition, participants who consumed the cereal highest in β-glucan (approximately 5.5 g) reduced their energy intake at the lunch buffet by approximately 110 kcal compared with controls ($p = 0.033$).

Lyly and coworkers provided 19 healthy individuals with isocaloric beverages (approximately 239 kcal) containing fiber (oat β-glucan, wheat bran, guar gum) or a control beverage with no fiber (Lyly *et al.*, 2009). Compared with the control beverage, the beverage containing guar gum (7.8 g fiber) was the only treatment that increased satiety and decreased the desire to eat, based on 120-minute area under the curve values. The oat β-glucan beverage appeared to increase satiety and decrease desire to eat but these responses did not differ significantly from those associated with the control beverage. The authors concluded that differences in beverage viscosity were partly responsible for differences in satiety responses. In a follow-up study, the same group evaluated the satiating effects of beverages with varying combinations of oat fiber (0, 5, or 10 g total fiber, 50% β-glucan), energy content (700 kJ [167 kcal] or 1400 kJ [335 kcal]), and viscosity (high or low) in 29 healthy men and women (Lyly *et al.*, 2010). Energy content did not influence subjective appetite ratings but higher viscosity was associated with less hunger and increased satiety during the 180-minute postprandial period. These findings are consistent with previous studies showing that beverage viscosity is inversely related to postprandial hunger (Mattes and Rothacker, 2001) and with a recent systematic review that found that more viscous fibers (e.g., β-glucan) reduced appetite and acute energy intake more often than less viscous fibers (Wanders *et al.*, 2011).

Juvonen and colleagues (Juvonen *et al.*, 2009) directly compared two isocaloric oat bran-based liquid meals that had identical chemical composition but different viscosities (low or high). In this study, the low-viscosity liquid meal induced a greater postprandial increase in satiety and blood levels of appetite-suppressing hormones (CCK, glucagon-like peptide-1 [GLP-1], and PYY) and decreased blood levels of the appetite-stimulating hormone ghrelin. No significant differences were observed in the other subjective appetite ratings (hunger, desire to eat, or fullness) or in self-selected food intake during an *ad libitum* meal served three hours after beverage intake. Biörklund and colleagues found that a beverage containing 5 g β-glucan (from oat or barley) for 5 weeks had the same effect on postprandial satiety as a fiber-free control beverage, although it should be noted that beverage viscosity was not measured (Biörklund *et al.*, 2008). Besides viscosity, the type of food matrix may influence the effect of β-glucan on body weight. Wolever and colleagues reported that oat β-glucan of varying doses and molecular weights (low vs high viscosities) incorporated into a ready-to-eat cereal had no effect on changes in body weight over a 4-week period (Wolever *et al.*, 2010).

Acute test meal trials investigating relationships between oat or oat β-glucan consumption and satiety-related outcomes have provided mixed results and do not allow definitive conclusions to be made. The available data suggest that changes in perceived satiety or circulating gastrointestinal appetite-regulating hormones associated with oats or oat β-glucan may not translate into corresponding reductions in self-selected food intake. That said, differences in study design, such as variations in the way subjective appetite is assessed or the way oats or oat fibers are administered (e.g., as whole oats or a supplement mixed into liquid, baked into bread, or part of a whole food like cooked oats) likely contributes to variability in results across studies. Furthermore, the investigations discussed previously have differed markedly in terms of the amount of whole oats or oat fibers used, use of a control group (or lack thereof), duration of the intervention, basal diet, and study population. This may explain why acute studies do not clearly show that oat consumption affects satiety-related outcomes. In addition, one of the proposed mechanisms for increased satiety is related to increased colonic fermentation of oat dietary fibers, which would be missed by acute studies. Thus, any potential benefit of increasing oat consumption on measures of appetite may not be reflected in single test meal studies.

12.4 Possible mechanisms of action

Although the available data show mixed results, physiological mechanisms have been proposed to explain how increased oat consumption may modulate appetite and ultimately body weight. Oats are a good source of total dietary fiber (approximately 10–15 g/100 g) and contain viscous soluble fiber, providing approximately 5 g (oatmeal) to 7 g (oat bran) of oat β-glucan per 100 g (Glore *et al.*, 1994; Anderson and Bridges, 1988). Two recent literature reviews examining the satiating properties of dietary fibers concluded that viscosity is responsible for fiber's ability to suppress appetite (Kristensen and Jensen, 2011; Wanders *et al.*, 2011). Consuming soluble fibers, such as oat β-glucan, creates a viscous solution that slows gastric emptying, creates a barrier to digestive enzymes, and slows translocation of glucose molecules from the intestinal lumen to the brush border (Marciani *et al.*, 2000, 2001). In concert, these effects reduce the rate at which glucose is absorbed, blunting postprandial glucose and insulin excursions. Because insulin is a lipogenic hormone, lowering day-long insulin levels by incorporating oat β-glucan into multiple meals may reduce lipid formation/storage and enhance lipid oxidation, thus reducing weight gain over time (Figure 12.2). However, insulin has acute appetite-suppressing effects during the postprandial

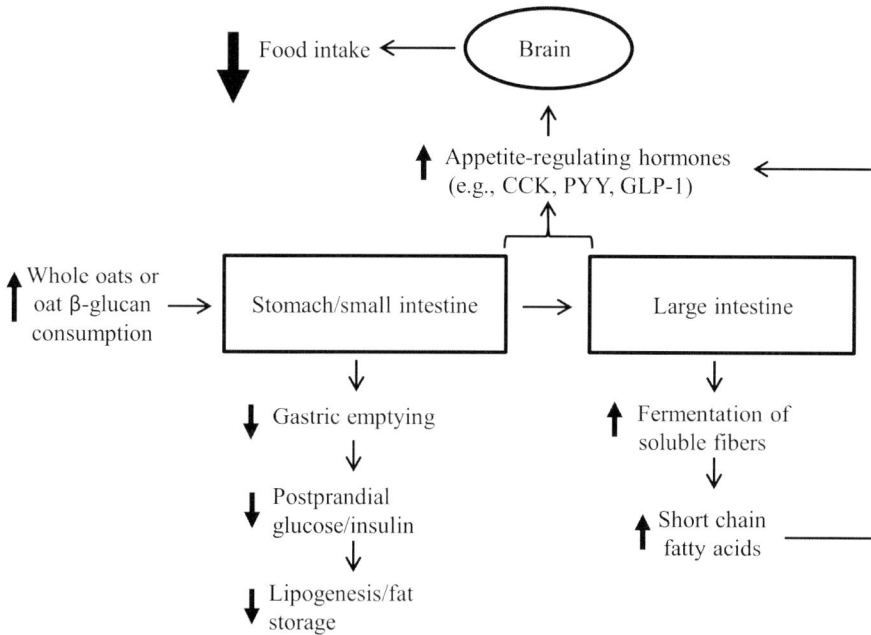

Figure 12.2 Possible mechanisms of action whereby oats affect appetite and body weight.

period (Menéndez and Atrens, 1991; Air *et al.*, 2002); therefore, the acute and longer-term effects of plasma insulin levels on body weight regulation remain to be elucidated.

In addition, oat fibers that escape digestion in the small intestine provide material for colonic fermentation. This process results in the production of short-chain fatty acids (SCFAs), such as acetate, butyrate, and propionate, and can be detected as increased breath hydrogen. An increase in SCFA production with increased dietary fiber intake is associated with increased postprandial concentrations of appetite-suppressing hormones GLP-1 and PYY, which are derived from endocrine L-cells of the ileum (distal portion of the small intestine) and colon (Figure 12.2) (Wen *et al.*, 1995; Zhou *et al.*, 2008; Cani *et al.*, 2009). The free fatty acid receptors FFAR2 and FFAR3, which are colocalized within L-cells in the human gastrointestinal tract and activated by SCFAs, may facilitate PYY and GLP-1 release with increased fiber consumption (Chambers *et al.*, 2011). Although the increase in GLP-1 after consuming fermentable fibers has been observed consistently in animal studies, this response has been difficult to confirm in humans, possibly because humans have more highly developed neural pathways that respond to GLP-1, which has a short half-life (< 2 min) in circulation. Furthermore, interindividual differences in gut microbiota may modulate the effect of oat consumption on appetite and body weight/adiposity, because intake of fermentable fibers alters the species composition of the gut microbiota in the short and long term (Flint *et al.*, 2012).

FFAR2 and FFAR3 have also been identified in human fat cells (adipocytes). Evidence suggests that circulating SCFAs cause the following: (i) decreased plasma levels of nonesterified (free) fatty acids by inhibiting lipolysis in adipose tissue; (ii) decreased proinflammatory adipokine secretion; and (iii) increased adipocyte differentiation (Chambers *et al.*, 2011). These events may improve whole body insulin sensitivity and facilitate weight control by reducing day-long insulin levels, thereby decreasing lipogenesis and fat storage.

Several other hormones secreted by the gastrointestinal tract (e.g., ghrelin and CCK) influence appetite but few long-term studies have evaluated the influence of oat or oat fiber intake on these hormones. Data from acute test meal studies suggest that viscous fibers promote a greater release of CCK, possibly by slowing the rate of digestion after a high-fiber meal, which prolongs lipid contact with intestinal cells. For example, in healthy men, a meal that contained barley pasta high in β-glucan fiber (Bourdon *et al.*, 1999) or one that contained white beans (Bourdon *et al.*, 2001) produced greater peak CCK concentrations and/or a longer-lasting elevation in postprandial CCK levels compared with low-fiber control meals. It is reasonable to infer similar acute responses would occur with oat consumption. In many of the previously discussed trials, appetite-regulating hormones were not measured; therefore, causal mechanisms underlying changes in subjective appetite are difficult to determine. Future studies using oats should measure blood levels of gastrointestinal-derived hormones to determine whether changes in circulating hormone concentrations after oat or oat β-glucan consumption correlate with changes in subjective appetite ratings or actual measured food intake at an *ad libitum* meal.

12.5   Summary

In summary, clinical intervention studies investigating the effects of oats or oat fibers (principally β-glucan) have provided mixed results for the effect of increased consumption of whole oats or oat β-glucan on subjective appetite, food intake, or body weight loss in adults. However, increasing intake of fermentable dietary fibers appears to increase plasma levels of appetite-suppressing gastrointestinal hormones, which may reduce energy intake and promote long-term body weight regulation. Confounding factors not controlled for in many studies may mask any effect of oats on weight management. For example, individuals who habitually consume more oats may be more likely to make other healthy lifestyle choices, such as eating a nutritionally balanced diet and participating in regular physical activity. Additional studies in this area should carefully consider the energy and fiber content of the control diets, potential ethnic/racial differences in the study populations, and the possibility that age (especially old age) or other dietary factors (e.g., dietary fat) might be important effect modifiers. In general, more well-controlled clinical trials are needed to clarify the relationship between increased oat consumption and modulation of appetite and body weight.

References

Air, E.L., *et al.* (2002) Acute third ventricular administration of insulin decreases food intake in two paradigms. *Pharmacology Biochemistry and Behavior* **72**, 423–429.

Anderson, J.W. and Bridges, S.R. (1988) Dietary fiber content of selected foods. *American Journal of Clinical Nutrition* **47**, 440–447.

Bazzano, L.A., *et al.* (2005) Dietary intake of whole and refined grain breakfast cereals and weight gain in men. *Obesity Research* **13**, 1952–1960.

Beck, E.J., *et al.* (2009a) Increases in peptide Y-Y levels following oat beta-glucan ingestion are dose-dependent in overweight adults. *Nutrition Research* **29**, 705–709.

Beck, E.J., *et al.* (2009b) Oat beta-glucan increases postprandial cholecystokinin levels, decreases insulin response and extends subjective satiety in overweight subjects. *Molecular Nutrition and Food Research* **53**, 1343–1351.

Beck, E.J., *et al.* (2010) Oat beta-glucan supplementation does not enhance the effectiveness of an energy-restricted diet in overweight women. *British Journal of Nutrition* **103**, 1212–1222.

Berti, C., *et al.* (2005) Effect on appetite control of minor cereal and pseudocereal products. *British Journal of Nutrition* **94**, 850–858.

Biörklund, M., *et al.* (2008) Serum lipids and postprandial glucose and insulin levels in hyperlipidemic subjects after consumption of an oat beta-glucan-containing ready meal. *Annals of Nutrition and Metabolism* **52**, 83–90.

Blundell, J.E. and MacDiarmid, J.I. (1997) Fat as a risk factor for overconsumption: satiation, satiety, and patterns of eating. *Journal of the American Dietetic Association* **97**, S63–S69.

Bourdon, I., *et al.* (1999) Postprandial lipid, glucose, insulin, and cholescyctokinin responses in men fed barley pasta enriched with β-glucan. *American Journal of Clinical Nutrition* **69**, 55–63.

Bourdon, I., *et al.* (2001) Beans, as a source of dietary fiber, increase cholescystokinin and apolipoprotein B48 response to test meals in men. *Journal of Nutrition* **131**, 1485–1490.

Cani, P.D., *et al.* (2009) Gut microbiota fermentation of prebiotics increases satietogenic and incretin gut peptide production with consequences for appetite sensation and glucose response after a meal. *American Journal of Clinical Nutrition* **90**, 1236–1243.

Chambers, E., *et al.* (2011). Dietary starch and fiber: potential benefits to body weight and glucose metabolism. *Diabetes Management* **1**, 521–528.

Charlton, K.E., *et al.* (2012) Effect of 6 weeks' consumption of beta-glucan-rich oat products on cholesterol levels in mildly hypercholesterolaemic overweight adults. *British Journal of Nutrition* **107**, 1037–1047.

Cugnet-Anceau, C. *et al.* (2010) A controlled study of consumption of beta-glucan-enriched soups for 2 months by type 2 diabetic free-living subjects. *British Journal of Nutrition* **103**, 422–428.

De Oliveira, M., *et al.* (2003) Weight loss associated with a daily intake of three apples or three pears among overweight women. *Nutrition* **19**, 253–256.

Flegal, K.M., *et al.* (2010) Prevalence and trends in obesity among US adults, 1999–2008. *Journal of the American Medical Aassociation* **303**, 235–241.

Flint, H.J., *et al.* (2012) The role of the gut microbiota in nutrition and health. *Nature Reviews. Gastroenterology and Hepatology* **9**, 577–589.

Giacco, R., *et al.* (2011) Whole grain intake in relation to body weight: from epidemiological evidence to clinical trials. *Nutrition, Metabolism, and Cardiovascular Diseases* **21**, 901–908.

Glore, S.R., *et al.* (1994) Soluble fiber and serum lipids: a literature review. *Journal of the American Dietetic Association* **94**, 425–436.

He, J., *et al.* (2004) Effect of dietary fiber intake on blood pressure: a randomized, double-blind, placebo-controlled trial. *Journal of Hypertension* **22**, 73–80.

Hlebowicz, J., *et al.* (2007) Effect of commercial breakfast fibre cereals compared with corn flakes on postprandial blood glucose, gastric emptying and satiety in healthy subjects: a randomized blinded crossover trial. *Nutrition Journal* **6**, 22.

Hlebowicz, J., *et al.* (2008) Effect of muesli with 4 g oat beta-glucan on postprandial blood glucose, gastric emptying and satiety in healthy subjects: a randomized crossover trial. *Journal of the American College of Nutrition* **27**, 470–475.

Holt, S.H., *et al.* (1995) A satiety index of common foods. *European Journal of Clinical Nutrition* **49**, 675–690.

Juvonen, K.R., *et al.* (2009) Viscosity of oat bran-enriched beverages influences gastrointestinal hormonal responses in healthy humans. *Journal of Nutrition* **139**, 461–466.

Koh-Banerjee, P., *et al.* (2004) Changes in whole-grain, bran, and cereal fiber consumption in relation to 8-y weight gain among men. *American Journal of Clinical Nutrition* **80**, 1237–1245.

Kristensen, M. and Jensen, M.G. (2011) Dietary fibres in the regulation of appetite and food intake. Importance of viscosity. *Appetite* **56**, 65–70.

Liatis, S., *et al.* (2009). The consumption of bread enriched with betaglucan reduces LDL-cholesterol and improves insulin resistance in patients with type 2 diabetes. *Diabetes and Metabolism* **35**, 115–120.

Liu, S., *et al.* (2003) Relation between changes in intakes of dietary fiber and grain products and changes in weight and development of obesity among middle-aged women. *American Journal of Clinical Nutrition* **78**, 920–927.

Ludwig, D.S., *et al.* (1999) Dietary fiber, weight gain, and cardiovascular disease risk factors in young adults. *Journal of the American Medical Aassociation* **282**, 1539–1546.

Lyly, M., *et al.* (2009) Fibre in beverages can enhance perceived satiety. *European Journal of Nutrition* **48**, 251–258.

Lyly, M., *et al.* (2010) The effect of fibre amount, energy level and viscosity of beverages containing oat fibre supplement on perceived satiety. *Food and Nutrition Research* **54**, 2149. doi: 10.3402/fnr.v54i0.2149.

Maki, K.C., *et al.* (2007) Effects of consuming foods containing oat beta-glucan on blood pressure, carbohydrate metabolism and biomarkers of oxidative stress in men and women with elevated blood pressure. *European Journal of Clinical Nutrition* **61**, 786–795.

Maki, K.C., *et al.* (2010) Whole-grain ready-to-eat oat cereal, as part of a dietary program for weight loss, reduces low-density lipoprotein cholesterol in adults with overweight and obesity more than a dietary program including low-fiber control foods. *Journal of the American Medical Aassociation* **110**, 205–214.

Marciani, L., *et al.* (2000) Gastric response to increased meal viscosity assessed by echo-planar magnetic resonance imaging in humans. *Journal of Nutrtion* **130**, 122–127.

Marciani, L., *et al.* (2001) Effect of meal viscosity and nutrients on satiety, intragastric dilution, and emptying assessed by MRI. *American Journal of Physiology Gastrointestinal and Liver Physiology* **280**, G1227–G1233.

Mattes, R.D. and Rothacker, D. (2001) Beverage viscosity is inversely related to postprandial hunger in humans. *Physiology and Behavior* **74**, 551–557.

Mattes, R.D., *et al.* (2005) Appetite: measurement and manipulation misgivings. *Journal of the American Dietetic Association* **105**, S87–S97.

Menéndez, J.A. and Atrens, D.M. (1991) Insulin and the paraventricular hypothalamus: modulation of energy balance. *Brain Research* **555**, 193–201.

Newby, P.K., *et al.* (2007) Intake of whole grains, refined grains, and cereal fiber measured with 7-d diet records and associations with risk factors for chronic disease. *American Journal of Clinical Nutrition* **86**, 1745–1753.

Peters, H.P., *et al.* (2009) No effect of added beta-glucan or of fructooligosaccharide on appetite or energy intake. *American Journal of Clinical Nutrition* **89**, 58–63.

Reyna-Villasmil, N., *et al.* (2007) Oat-derived beta-glucan significantly improves HDLC and diminishes LDLC and non-HDL cholesterol in overweight individuals with mild hypercholesterolemia. *American Journal of Ttherapeutics* **14**, 203–212.

Robitaille, J., *et al.* (2005) Effect of an oat bran-rich supplement on the metabolic profile of overweight premenopausal women. *Annals of Nutrition and Metabolism* **49**, 141–148.

Rytter, E., *et al.* (1996) Changes in plasma insulin, enterostatin, and lipoprotein levels during an energy-restricted dietary regimen including a new oat-based liquid food. *Annals of Nutrition and Metabolism* **40**, 212–220.

Saltzman, E., *et al.* (2001a) An oat-containing hypocaloric diet reduces systolic blood pressure and improves lipid profile beyond effects of weight loss in men and women. *Journal of Nutrtion* **131**, 1465–1470.

Saltzman, E., *et al.* (2001b) Effects of a cereal rich in soluble fiber on body composition and dietary compliance during consumption of a hypocaloric diet. *Journal of the American College of Nutrition* **20**, 50–57.

Wanders, A.J., *et al.* (2011) Effects of dietary fibre on subjective appetite, energy intake and body weight: a systematic review of randomized controlled trials. *Obesity Reviews* **12**, 724–739.

Wang, Y., *et al.* (2008) Will all Americans become overweight or obese? Estimating the progression and cost of the US obesity epidemic. *Obesity* **16**, 2323–2330.

Weickert, M.O., *et al.* (2006) Wheat-fibre-induced changes of postprandial peptide YY and ghrelin responses are not associated with acute alterations of satiety. *British Journal of Nutrition* **96**, 795–798.

Wen, J., *et al.* (1995) PYY and GLP-1 contribute to feedback inhibition from the canine ileum and colon. *American Journal of Physiology* **269**, G945–G952.

Williams, P.G., *et al.* (2008) Cereal grains, legumes, and weight management: a comprehensive review of the scientific evidence. *Nutrition Reviews* **66**, 171–182.

Wolever, T.M., *et al.* (2010) Physicochemical properties of oat beta-glucan influence its ability to reduce serum LDL cholesterol in humans: a randomized clinical trial. *American Journal of Clinical Nutrition* **92**, 723–732.

Zhou, J., *et al.* (2008) Dietary resistant starch upregulates total GLP-1 and PYY in a sustained day-long manner through fermentation in rodents. *American Journal of Physiology Endocrinology and Metabolism* **295**, E1160–E11666.

13
Effects of Oats on Carbohydrate Metabolism

Susan M. Tosh

Agriculture and Agri-Food Canada, Guelph, ON, Canada

13.1   Introduction

Oats have long been a part of the European and North American diet. Oatmeal and oat flour are used widely, alone or in combination with other grains, in breakfasts cereals, crackers, bread, snack foods, cookies, and other baked goods. Although oats have long been considered healthful, there is now scientific evidence to support this idea. In addition to lowering serum cholesterol concentrations, oat foods have the ability to significantly lower blood glucose concentrations following a meal.

Based on the strength of the scientific evidence, the European Food Safety Authority (EFSA) published an opinion allowing health claims for postprandial blood glucose response reduction on oat and barley foods (EFSA, 2011). The document states that, "a cause and effect relationship has been established between the consumption of beta-glucans from oats and barley and a reduction of the postprandial glycemic response. In order to obtain the claimed effect, 4 g of beta-glucans from oats or barley for each 30 g of available carbohydrate should be consumed per meal. The target population is individuals who wish to reduce their postprandial glycemic responses."

Many clinical trials have been conducted to measure the postprandial response and the results are summarized below. In addition to comparing the effects of specific foods and food processes on the glycemic response, many of the studies investigated the mechanism of action of oat foods.

13.2   Epidemiology

There have been no epidemiological studies dealing specifically with oat consumption and risk for development of type 2 diabetes. An investigation of vegetarians, who consumed more oat products than the general population, showed

Oats Nutrition and Technology, First Edition. Edited by YiFang Chu.
© 2014 John Wiley & Sons, Ltd. Published 2014 by John Wiley & Sons, Ltd.

lower glucose, insulin levels, and insulin resistance than non-vegetarians (Vala-chovičová *et al.*, 2006). Whole grain consumption, including whole grain oats, has been associated with reduced risk of type 2 diabetes (Meyer *et al.*, 2000; Mon-tenen *et al.*, 2003). Increased consumption of dietary fiber from cereals has also been shown to reduce risk of type 2 diabetes (Meyer *et al.*, 2000; Montenen *et al.*, 2003; Hodge *et al.*, 2004; Schulze *et al.*, 2004, 2007; Krishnan *et al.*, 2007; Hop-ping *et al.*, 2010). Intake of low glycemic index (GI) foods has been correlated with reduced risk of developing type 2 diabetes (Hodge *et al.*, 2004; Schulze *et al.*, 2004; Krishnan *et al.*, 2007). These studies were carried out in diverse populations across the world, including the United States (Meyer *et al.*, 2000; Schulze *et al.*, 2004; Krishnan *et al.*, 2007; Hopping *et al.*, 2010), Europe (Montenen *et al.*, 2003; Schulze *et al.*, 2007), and Australia (Hodge *et al.*, 2004).

13.3   Mechanisms of postprandial blood glucose reduction

An elegant experiment demonstrated that the active component of oats was the mixed linkage β-glucan, $(1\rightarrow3)(1\rightarrow4)$-β-D-glucan (Braaten *et al.*, 1991). After consumption of oat bran or cream of wheat cereals, blood glucose and insulin concentrations were followed for two hours. Both glucose and insulin levels were significantly lower after consumption of the oat bran cereal than after wheat cereal. However, when purified oat β-glucan was added to the cream of wheat cereal at the same level as in the oat bran porridge, the blood glucose and insulin levels were similar to those observed after the oat bran porridge meal. This result clearly demonstrated that oat β-glucan was responsible for postprandial blood glucose attenuation. Since then, many clinical trials have been conducted which show that oat foods and β-glucan ameliorate blood glucose levels by a number of different mechanisms.

13.3.1   Influence on food microstructure and physicochemical characteristics

In intact grains and rolled oats, the cell walls remain intact, acting as a phys-ical barrier to enzyme activity to break starch down into sugar. In one study, boiled oat kernels were found to have lower glycemic and insulin responses than the same oats after they were steamed, rolled, and served either raw (muesli) or cooked as porridge (Granfeldt *et al.*, 1995). Additionally, thick oat flakes were found to have a lower glycemic response than thin oat flakes (Granfeldt *et al.*, 2000). During processing of foods using oat bran, β-glucan competes with starch for water, which may result in a lower degree of starch gelatinization and slower digestibility of starch (Regand *et al.*, 2011).

13.3.2   In the upper gastrointestinal tract

After consumption of oat foods, β-glucan influences glycemic and insulinemic responses in a number of ways. In the stomach, oat β-glucan can develop high viscosity. High-viscosity polysaccharides slow the mixing of the meal with gastric juices, including digestive enzymes (Marciani *et al.*, 2001). It also slows the rate of gastric emptying (Hlebowicz *et al.*, 2007). High-viscosity oat beverages have been shown to slow the rate of gastric emptying more than low-viscosity beverages (Juvonen *et al.*, 2009), which corresponded with lower glycemic and insulinemic responses. The fat content of oat porridge does not significantly affect the glycemic or insulin response (Tuomasjukka *et al.*, 2007).

However, this development of viscosity is dependent on the dose and characteristics of the β-glucan. While intact cell walls provide a physical barrier to starch digestion (Granfeldt *et al.*, 2000), dissolved β-glucan increases viscosity (Wood *et al.*, 1990; Panahi *et al.*, 2007). As with other polymer solutions, the viscosity of β-glucan increases exponentially with concentration and molecular weight (Ren *et al.*, 2003). Decreasing the solubility of β-glucan reduces its effective dose in processed foods. The glycemic response after consumption of oat bran muffins treated to vary the β-glucan solubility decreased as the solubility increased (Lan-Pidhainy *et al.*, 2007). Similarly, increasing the length of the β-glucan polymers increases the incidence of entanglements in solution, reducing the solution's tendency to flow and mix. In liquid systems, the glycemic response has been correlated with beverage viscosity (Wood 2000; Panahi *et al.*, 2007). In solid foods, a strong relationship between the glycemic response and β-glucan molecular weight has also been demonstrated. Oat bran muffins (Tosh *et al.*, 2008), extruded cereal (Brummer *et al.*, 2012), granola-type products (Regand *et al.*, 2011), and other products (Regand *et al.*, 2009) showed that as molecular weight decreased, meals with equivalent doses of β-glucan show increased glycemic response. Increasing β-glucan dose can compensate for reduction in solubility (Lan-Pidhainy *et al.*, 2007) or molecular weight (Tosh *et al.*, 2008).

To estimate the development of viscosity in the upper gastrointestinal tract after ingestion of solid oat β-glucan-containing foods, *in vitro* digestion protocols have been developed. The most widely used method roughly simulates human digestion from the mouth to the small intestine (Beer *et al.*, 1997). Food is coarsely ground and dispersed in a phosphate buffer (pH 6.8) with salivary α-amylase. The pH is adjusted to 2, to mimic the pH of the stomach, and pepsin is added. After 30 minutes of mixing at low speed (30 rpm), the pH is adjusted to 6.8 and pancreatin is added. At the end of an additional 90 minutes of mixing, the dissolved β-glucan is removed by centrifugation of the slurry. The concentration and molecular weight of the β-glucan in the physiological extract has been measured in porridge (Regand *et al.*, 2009), extruded breakfast cereals (Brummer *et al.*, 2012), muffins (Lan-Pidhainy *et al.*, 2007; Tosh *et al.*, 2008), and other products (Mäkeläinen *et al.*, 2007; Regand *et al.*, 2009, 2011). The viscosity of the β-glucan in solution is significantly related to the human glycemic response to the foods tested, as shown in Figure 13.1. Recently, a simplified version of the protocol was developed. The ground food is mixed with phosphate buffer, α-amylase, and protease, and mixed in a rotational viscometer for 2 hours (Gamel *et al.*, 2012). The final viscosity correlates well with the longer protocol. Similar *in vitro* protocols have also been used (Granfeldt *et al.*, 1992; Östman *et al.*, 2006), where the contribution of starch and protein to viscosity is minimized by enzymatic hydrolysis of these components and viscosity is measured.

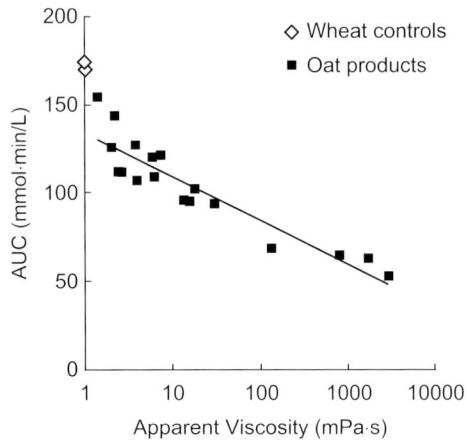

Figure 13.1 Relationship between glycemic responses of human subjects (area under the curve of the postprandial blood glucose curve) and the apparent viscosity (at 30/s) of the β-glucan extracted by simulated digestion. AUC = −25 log(η) + 134 (r^2 = 0.85). (Pooled data from Lan-Pidhainy *et al.*, 2007; Tosh *et al.*, 2008, Brummer *et al.*, 2012.)

13.3.3　In the lower intestinal tract

Although the acute effects of oat foods on postprandial blood glucose are realized in the upper gastrointestinal tract (Batalina *et al.*, 2001), they continue to have effects in the colon. Oat β-glucan is a fermentable fiber. Because β-glucan is at least partially soluble in food matrices, it is readily broken down by bacterial enzymes. *In vitro* fermentation studies have shown that oat bran fiber and β-glucan are depolymerized more quickly by colonic bacteria than wheat bran fiber and are digested similarly to inulin (Wood *et al.*, 2002). Fermentation of oat bran or β-glucan isolate by colonic bacteria results in increased acetate, propionate, and butyrate production, and decreased pH compared to wheat bran. Another *in vitro* study (Queenan *et al.*, 2007) confirmed that production of acetate, propionate, and butyrate occurred when oat β-glucan, guar gum, and inulin were fermented by colonic bacteria.

Consumption of oat foods affects postprandial glucose, not only during the meal when they are consumed but also during subsequent meals. When subjects consumed 5.0 g oat β-glucan as part of an evening meal, postprandial blood glucose area under the curve (AUC) was 18% lower after a standardized breakfast (Nilsson *et al.*, 2008a). Plasma concentrations of short-chain fatty acids, particularly acetate, were elevated following the breakfast meal, and breath hydrogen was lower, suggesting a relationship between fermentation and glycemic response.

13.4　Clinical studies using whole oat products

Several studies have investigated the glycemic response to rolled oat porridge, the most-recognizable whole oat product. Table 13.1 lists studies conducted on whole

Table 13.1 Clinical studies conducted with whole oat foods. Food format, dose of β-glucan (βG), and available carbohydrate (AC) are shown; the change in area under the postprandial blood glucose curve (AUC) and an indication of whether the AUC and postprandial insulin changed significantly are indicated

Reference	Food format	βG dose (g)	AC dose (g)	Change in AUC (mmol·min/L) sig.[a]		Insulin change sig.[a]
Alminger and Eklund-Jonsson, 2008	oat tempe	1.8	25	−79	yes	yes
Behall et al., 2005	pudding with oat flour	3.23[b]	73.7[c]	−62	yes	no
	pudding with oatmeal	3.23[b]	73.7[c]	−49	yes	no
Granfeldt et al., 1995	boiled rolled oats	3.3[b]	50	−12	no	no
	boiled oat kernels	3.5[b]	50	−80	yes	no
Hätönen et al., 2006	oatmeal porridge	4	50	−29	yes	
Hlebowicz et al., 2007	wholemeal oatflake cereal	0.5	31.5	−23	no	
Liljeberg et al., 1992	bread with coarse boiled oats	2.1[b]	50	−16	yes	yes
Liljeberg et al., 1996	oat porridge	2.1[b]	35.5	7	no	no
Nilsson et al., 2008b	boiled oat kernels	2.9	50	−40	no	

[a]yes means significant reduction detected ($p < 0.05$).
[b]Soluble fiber measured.
[c]1 g AC/kg body weight, average given.

oat products where β-glucan or soluble fiber content was measured. Oat porridge containing 4 g β-glucan (Hätönen *et al.*, 2006) significantly reduced the postprandial glucose response, whereas porridges containing 3.3 g β-glucan (Granfeldt *et al.*, 1994) or 2.1 g β-glucan (Liljeberg *et al.*, 1996) did not. The GI website (University of Sydney, 2013) lists the results of 18 GI tests for rolled and steel cut-oat porridges. The average GI was 57 ± 11, at the low end of the range for medium GI foods (GI range, 55–70). The four test results for instant oatmeal porridges had an average GI of 79. A commercial whole grain-extruded oat cereal had a GI of 74 (Wolever *et al.*, 1994). Another whole meal oat cereal containing only 0.5 g β-glucan per serving did not significantly reduce postprandial blood glucose (Hlebowicz *et al.*, 2007). Boiled, intact oat kernels lower the AUC more than oat porridge (Granfeldt *et al.*, 1994), probably because the starch is less accessible when the bran layer and cell walls are intact. An innovative oat tempe product made by fermenting intact grains (Alminger and Eklund-Jonsson, 2008) significantly lowered glucose and insulin responses.

Whole oat products are limited by the concentration of β-glucan in the grain. In the whole oat studies, the β-glucan dose ranged from 0.5–4 g per serving. Although 60% of these studies significantly reduced the glycemic response, none of these products would meet the EFSA requirement of 4 g β-glucan per 30 g available carbohydrate.

13.5   Clinical studies using oat bran products

Oat bran has higher β-glucan content and lower starch content than whole oats, making it possible to formulate high-fiber foods. A large number of clinical studies has been conducted on a wide range of processed foods made from oat bran. The results of 16 studies with normal subjects, where β-glucan or soluble fiber content was measured, are summarized in Table 13.2. The β-glucan content in these studies ranged from 3 to 12 g per serving. Oat bran was boiled to make porridge (Wood *et al.*, 1990; Regand *et al.*, 2009) and extruded to make breakfast cereal (Brummer *et al.*, 2012) or muesli (Granfeldt *et al.*, 1994, 2008; Hlebowicz *et al.*, 2008). Muffins were used to establish the dose-response (Lan-Pidhainy *et al.*, 2007; Tosh *et al.*, 2008) as well as effects of β-glucan solubility (Lan-Pidhainy *et al.*, 2007) and molecular weight (Tosh *et al.*, 2008). Oat bran was also added to beverages (Mäkeläinen *et al.*, 2007; Ulmius *et al.*, 2011) and soup (Biorkland *et al.*, 2008). A granola product was used to examine the effects of β-glucan molecular weight and starch dose (Regand *et al.*, 2011).

Five of the seven treatments known to have low molecular weight β-glucan ($< 250\,000$ g/mol) did not significantly lower the glycemic response. It appears that even at low molecular weight, 8 g β-glucan produces sufficient viscosity to reduce the glycemic response. Of the 35 remaining treatments, the average reduction in glucose AUC was 45 ± 22 mmol·min/L. Considering that the AUC for 50 g glucose control is in the range of 120–240 mmol·min/L (Tosh, 2013), this finding represents a considerable reduction.

Clinical trials using oat bran foods have also been conducted involving subjects with type 2 diabetes. A meal containing 3 g β-glucan in an extruded muesli product showed a significant reduction in postprandial blood glucose in diabetic subjects (Kabir *et al.*, 2002) whereas a similar product did not shown significant results in subjects with a normal glycemic response (Granfeldt *et al.*, 2008). Oat bran cereal products containing 3.7, 6.2, and 7.3 g β-glucan per serving resulted in a linear decrease in GI (14, 48, and 57 mmol·min/L, respectively) with increasing dose ($r^2 = 0.989$) (Jenkins *et al.*, 2002). Similar reductions in GI (29, 59, and 65 mmol·min/L) were observed for extruded oat bran cereals containing 4, 6, and 8.4 g β-glucan ($r^2 = 0.834$) (Tappy *et al.*, 1996). In another study, 3 and 9.4 g doses of β-glucan from oat bran flour or extruded cereal demonstrated significant reductions in postprandial blood glucose compared to control (Tapola *et al.*, 2005). Overall, nine of ten treatments containing 3–9.4 g oat β-glucan showed significant reduction in postprandial AUC. Thus, oat foods appear to be more effective for blood glucose control in persons with type 2 diabetics than in those with normal responses.

Table 13.2 Clinical studies conducted with oat bran foods; food format, dose of β-glucan (βG), and available carbohydrate (AC) are shown; the change in area under the postprandial blood glucose curve (AUC) and an indication of whether the AUC and postprandial insulin changed significantly are indicated

Reference	Food format	βG dose (g)	AC dose (g)	Change in AUC (mmol·min/L) sig.[a]	Insulin change sig.[a]
Biorkland et al., 2008	soup (LMW)[b]	4	25.7	0	no
Brummer et al., 2012	oat bran cereal (HMW)[b]	8.6	31	−56	yes
	oat bran cereal (LMW)[b]	8.3	31	−46	yes
	oat bran cereal (MMW)[b]	8.7	31	−64	yes
	oat bran cereal (MMW)[b]	8.4	31	−65	yes
Granfeldt et al., 1995	oat bran muesli	3.3[c]	50	−23.6	no	no
Granfeldt et al., 2008	oat bran muesli	3[c]	50	−16.7	no	no
	oat bran muesli	4[c]	50	−29.3	yes	yes
Hallfrisch et al., 2003	pudding with oat bran (raw)	3.7[c]	83.9[d]	−71	yes	yes
Hlebowicz et al., 2008	oat bran muesli	4	32.7	−15.5	yes	
Holm et al., 1992	oat bran fettucini	5.2	54.2	−4.5	yes	yes
Juntunen et al., 2002	rye bread with oat bran	5.4	50	−48	yes	yes
Lan-Pidhainy et al., 2007	oat bran muffin	8	50	−79	yes	
	previously frozen oat bran muffin	8	50	−66	yes	
	previously frozen oat bran muffin	8	50	−48	yes	
	oat bran muffin	12	50	−73	yes	
	previously frozen oat bran muffin	12	50	−68	yes	
	previously frozen oat bran muffin	12	50	−63	yes	
Mäkeläinen et al., 2007	oat bran drink	2	50	−26.9	n/s	no
	oat bran drink	4	50	−68.8	n/s	no
	previously frozen oat bran drink	4	50	−58.9	n/s	no
	oat bran drink	6	50	−60.6	n/s	no

(continued)

Table 13.2 (*Continued*)

Reference	Food format	βG dose (g)	AC dose (g)	Change in AUC (mmol·min/L) sig.[a]	Insulin change sig.[a]
Regand *et al.*, 2009	oat porridge (HMW)[b]	4	43	−37	no	
	oat crisp bread (LMW)[b]	4	64	−11	no	
	oat granola (HMW)[b]	4	44	−29	no	
	oat pasta (MMW)[b]	4	42	−7	no	
Regand *et al.*, 2011	oat granola product (HMW)[b]	6.2	38	−35	yes	
	oat granola product (MMW)[b]	6.2	38	−28	yes	
	oat granola product (LMW)[b]	6.2	38	2	no	
	oat granola product (HMW)[b]	6.3	60	−33	yes	
	oat granola product (MMW)[b]	6.3	60	0	no	
	oat granola product (LMW)[b]	6.3	60	−5	no	
Tosh *et al.*, 2008	oat bran muffin (HMW)[b]	4	50	−50	yes	
	oat bran muffin (MMW)[b]	4	50	−44	yes	
	oat bran muffin (MMW)[b]	4	50	−26	no	
	oat bran muffin (LMW)[b]	4	50	−15	no	
	oat bran muffin (HMW)[b]	8	50	−76	yes	
	oat bran muffin (MMW)[b]	8	50	−74	yes	
	oat bran muffin (MMW)[b]	8	50	−49	yes	
	oat bran muffin (LMW)[b]	8	50	−58	yes	
Ulmius *et al.*, 2011	beverage with oat bran	5	75	−40	no	yes
Wood *et al.*, 1990	oat bran porridge	8.8	60	−54	yes	

[a] n/s not specified.
[b] yes means significant reduction detected (p < 0.05).
[c] HMW = high molecular weight, MMW = medium molecular weight, LMW = low molecular weight.
[d] Soluble fiber measured.
[e] 1 g AC/kg body weight, average given.

13.6   Clinical studies using oat-derived β-glucan preparations

Oat β-glucan extracts have been used to demonstrate efficacy and the role of viscosity in blood glucose control as well as in prototype food products. Some of these are shown in Table 13.3. High concentrations of β-glucan served as viscous solutions or mixed with wheat porridge were initially used to show that β-glucan alone was responsible for the effect observed with oat foods (Wood *et al.*, 1990; Braaten *et al.*, 1991). An oat extract was shown to be as effective as oat porridge containing an equivalent amount of β-glucan in reducing GI (Hallfrisch *et al.*, 2003). A dose-response study showed a reduction in the glycemic response, as dose was increased from 1.8 to 7.2 g (Wood *et al.*, 1994). Two studies (Wood *et al.*, 1994; Panahi *et al.*, 2007) demonstrated that low-molecular weight β-glucan does not affect the glycemic response. In addition to forming viscous liquids, oat β-glucan will form gels when allowed to stand for a period of days. Low molecular weight β-glucan forms gels that are more elastic in nature than medium molecular weight β-glucan (Lazaridou *et al.*, 2003). Low molecular weight β-glucan solutions and gels resulted in a glycemic response similar to that of glucose solutions (Kwong *et al.*, 2013). When 50% or 75% of the low molecular weight β-glucan in the gel were replaced by high molecular weight β-glucan, holding the total concentration of β-glucan constant at 4 g per serving, the gels still did not significantly reduce the postprandial blood glucose response.

An oat fiber extract was used to establish dose response in muffins (Behall *et al.*, 2006). Adding resistant starch (high-amylose cornstarch) to the muffins further reduced the postprandial glycemic response. Sourdough bread containing 5% oat fiber was produced and found to have a GI of 54, making it a low GI food, whereas white wheat bread is a high GI food (De Angelis *et al.*, 2007).

13.7   Dose response

Combining the data from the different studies in Tables 13.1, 13.2, and 13.3, it is possible to derive a dose-response relationship. The data were analyzed as described previously (Tosh, 2013). Treatments with low molecular weight β-glucan were excluded. There is a large degree of scatter in the data, shown in Figure 13.2. There could be several reasons for this finding, including differences in methodology for glycemic response testing. However, differences in food processing accounts for much of the variability. During processing, heat, water activity, enzyme activity, and mechanical force all influence the solubility and molecular weight of β-glucan and the accessibility of starch to enzymatic digestion. These changes in the food matrix greatly affect the glycemic response (Granfeldt *et al.*, 1995; Tosh *et al.*, 2008). Despite the scatter in the data, a clear dose response was detected ($r^2 = 0.41$, p < 0.0001). The data predict a reduction in the AUC of 5.5 ± 0.8 mmol·min/L for each gram of oat β-glucan consumed. The lowest responses were achieved for glucose drinks containing purified β-glucan compared to a 50-g glucose beverage (Wood *et al.*, 1994; Kwong *et al.*, 2013).

Table 13.3   Clinical studies conducted with oat β-glucan isolates; food format, dose of β-glucan (βG), and available carbohydrate (AC) are shown; the change in area under the postprandial blood glucose curve (AUC) and an indication of whether the AUC and postprandial insulin changed significantly are indicated

Reference	Food format	βG dose (g)	AC dose (g)	Change in AUC (mmol·min/L)	sig.[a]	Insulin change sig.[a]
Behall et al., 2006	oat bran muffins	0.3[b]	72[c]	3	no	no
	oat bran muffins	0.9[b]	72[c]	5	no	no
	oat bran muffins	3.7[b]	72[c]	−26	no	yes
Braaten et al., 1991	oat β-glucan isolate (HMW)[d]	11.3	50	−64	yes	yes
De Angelis et al., 2007	sourdough bread with oat fibre	3.9	50	−35	yes	
Hallfrisch et al., 2003	oat extract	3.8	83.9[c]	−44	yes	yes
Kwong et al., 2013	β-glucan beverage (HMW)[d]	4	50	−94	yes	
	β-glucan beverage (LMW)[d]	4	50	−65	no	
	β-glucan gel (LMW)[d]	4	50	−14	no	
	β-glucan gel (mixed MW)[d]	4	50	−42	no	
	β-glucan gel (mixed MW)[d]	4	50	−36	no	
Panahi et al., 2007	beverage with extract (HMW)[d]	6	75	−39	yes	
	beverage with extract (LMW)[d]	6	75	8	no	
Wood et al., 1994	beverage with isolate (HMW)[d]	1.8	50	−19	no	no
	beverage with isolate (HMW)[d]	3.6	50	−38	no	no
	beverage with isolate (HMW)[d]	7.2	50	−42	no	no
	beverage with isolate (LMW)[d]	7.2	50	1	no	no
	beverage with isolate (MMW)[d]	7.2	50	−29	no	no
	beverage with isolate (HMW)[d]	7.2	50	−33	no	yes
Wood et al., 1990	beverage with isolate	8.6	60	−93	yes	
	cream of wheat with isolate	8.6	60	−57	yes	

[a]yes means significant reduction detected (p < 0.05).
[b]Soluble fibre measured.
[c]1 g AC/kg body weight, average given.
[d]HMW = high molecular weight, MMW = medium molecular weight, LMW = low molecular weight.

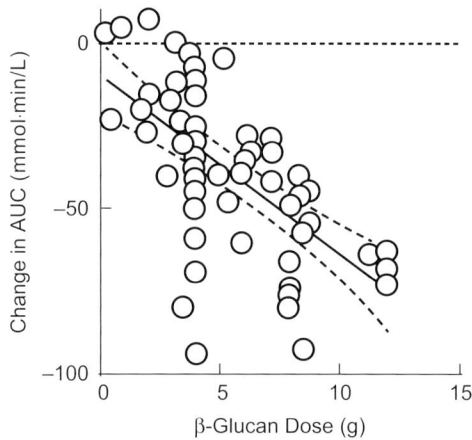

Figure 13.2   Dose-response relationship between changes in glycemic response from control (difference between area under the curve of the postprandial blood glucose curves) and β-glucan dose in oat treatment. $\Delta AUC = -5.5$ (βG dose) $- 9.5$ ($r^2 = 0.41$, $p < 0.0001$).

13.8   Longer-term glucose control

Although no studies have been done with the primary focus of following long-term blood glucose control in normal subjects, fasting glucose has been measured as a secondary outcome in a number of studies lasting 3–12 weeks. The results are summarized in Table 13.4. In a group of healthy men who consumed oat foods as part of a hypocaloric diet for six weeks, no significant differences in fasting glucose, fasting insulin, or insulin resistance (as measured by the homeostatic model assessment of insulin resistance method) were observed (Saltzman *et al.*, 2001). When healthy men and women consumed oat bread containing high molecular weight (but not low molecular weight) oat β-glucan for 3 weeks, an increase in fasting serum insulin and a decrease in fasting glucose compared to baseline were detected (Frank *et al.*, 2004). No control was included in this study. Overweight men who consumed oat cereals for 12 weeks did not have significant changes in fasting glucose or insulin concentrations, insulin sensitivity, or acute insulin response to glucose as measured using an intravenous glucose tolerance test (Davy *et al.*, 2002). In a group of men and women with elevated blood pressure, fasting glucose was significantly lower after consuming 2.8 g β-glucan every day for 4 weeks (Pins *et al.*, 2002). However, other groups of mildly hypertensive men and women did not have significant changes in fasting glucose after consuming oat foods (Keenan *et al.*, 2002; Maki *et al.*, 2007). Studies including subjects with mildly elevated serum cholesterol concentrations who consumed oat bread for 6 weeks (Queenan *et al.*, 2007) or oat bran cereal for four weeks (Wolever *et al.*, 2010) did not experience changes in fasting plasma glucose levels.

Studies including diabetic subjects have had somewhat different results. A study of subjects with uncontrolled type 2 diabetes suggested that a diabetes-adapted diet including oatmeal may lower insulin requirements (Lammert *et al.*,

Table 13.4 Longer-term clinical studies conducted with oat foods; food format, dose of β-glucan (βG), and time of interventions are shown; whether fasting glucose, fasting insulin, and insulin sensitivity changed significantly[a] are indicated

Reference	Cohort	Food format	βG dose (g/day)	Time (weeks)	Fasting glucose	Fasting insulin	Insulin sensitivity
Beck et al., 2010	OW[b] women	various	5 or 8	12	no	no	
Charlton et al., 2012	HC[c], OW[b]	various	1.5 or 3	6	no	no	no[d]
Cugnet-Anceau et al., 2010	obese T2D[e]	soup	3.5 (LMW)[f]	8	no		
Davy et al., 2002	OW[b] men	cereal	5.5	12	no	no	no[g]
Frank et al., 2004	healthy	bread	6 (HMW)[f]	3	yes[h]	yes[h]	
			6 (LMW)[f]		no	no	
Keenan et al., 2002	high BP[i]	cereal	5.5	6	no	no	
Liatis et al., 2009	T2D[e]	bread	3	3	yes[h]	yes	yes
Maki et al., 2007	high BP[i]	various	7.7	12	no		
Pins et al., 2002	high BP[i]	cereal	2.8	4	yes		
Queenan et al., 2007	HC	bread	6	6	no	no	
Reyna-Villasmil et al., 2007	overweight	bread	6	8	yes[h]		
Saltzman et al., 2001	healthy	various	3.7	6	no	no	no[d]
Tighe et al., 2010	healthy	various	12		no	no	no[d]
Wolever et al., 2010	HC[c]	cereal	3	4	no		

[a]yes means significant reduction detected ($p < 0.05$).
[b]OW = overweight.
[c]HC = mildy elevated cholesterol.
[d]Measured by homeostatic insulin response (HOMA-IR).
[e]T2D = type 2 diabetes.
[f]HMW = high molecular weight, LMW = low molecular weight.
[g]Measured by intravenous glucose tolerance test.
[h]Significantly different from baseline only.
[i]BP = blood pressure.

2008). Consuming oat bran bread for three weeks reduced fasting glucose, fasting insulin, and insulin sensitivity (Liatis *et al.*, 2009) in diabetic subjects. Glycosylated hemaglobin and fasting glucose were not reduced when type 2 diabetic subjects were given 3.5 g per day of low molecular weight β-glucan for 8 weeks (Cugnet-Anceau *et al.*, 2010).

13.9 Summary

Taking all of the currently available information into consideration, there is sufficient evidence to substantiate claims that mixed linkage β-glucan, the bioactive component of oats, significantly attenuates postprandial blood glucose and insulin concentrations. Food processing appears to influence the efficacy of oat foods somewhat. Disruption of the bran layer and cell walls, or depolymerization of the β-glucan can reduce efficacy. Alternatively, increased solubilization of the β-glucan by heat and moisture appears to increase efficacy. Nevertheless, a dose-response relationship has been established. The influence of oats on long term glucose control is less well established. The limited number of studies conducted in type 2 diabetics suggests a beneficial influence on fasting blood glucose and insulin levels. In normal subjects, fasting blood glucose and insulin concentrations tended not to change significantly. However, it is not clear that normal subjects would necessarily benefit from lower fasting blood glucose concentrations.

Despite the evidence of health benefits, oats are not widely used in bread and other staple foods. In the crop year 2011/2012, oats represented only 1% of the grains consumed worldwide (USDA, 2013). The 22.6 million metric tons of oats produced was 30 times less than the 696.4 million metric tons of wheat produced (USDA, 2013). Thus, there is an imbalance of cereal consumption in the world. Shifting consumption patterns toward increased consumption of oat and barley, which has been shown to ameliorate postprandial blood glucose levels, has the potential to lower the risk of developing type 2 diabetes in the general population.

References

Alminger, M. and Eklund-Jonsson, C. (2008) Whole-grain cereal products based on a high-fibre barley or oat genotype lower postprandial glucose and insulin responses in healthy humans. *European Journal of Nutrition* **47**, 294–300.

Battilana, P. *et al.* (2001) Mechanisms of action of β-glucan in postprandial glucose metabolism in healthy men. *European Journal of Clinical Nutrition* **55**, 327–333.

Beck, E.J. *et al.* (2010) Oat β-glucan supplementation does not enhance the effectiveness of an energy-restricted diet in overweight women. *British Journal of Nutrition* **103**, 1212–1222.

Beer, M.U. *et al.* (1997) Effect of cooking and storage on the amount and molecular weight of $(1\to3)(1\to4)$-β-D-glucan extracted from oat products by an in vitro digestion system. *Cereal Chemistry* **74**, 705–709.

Behall, K.M. *et al.* (2005) Comparison of hormone and glucose responses of overweight women to barley and oats. *Journal of the American College of Nutrition* **24**, 182–188.

Behall, K.M. *et al.* (2006) Comparison of both resistant starch and β-glucan improves postprandial plasma glucose and insulin in women. *Diabetes Care* **29**, 976–981.

Biörklund, M. *et al.* (2008) Serum lipids and postprandial glucose and insulin levels in hyperlipidemic subjects after consumption of an oat beta-glucan-containing ready meal. *Annals of Nutrition and Metabolism* **52**, 83–90.

Braaten, J.T. *et al.* (1991) Oat gum, a soluble fiber which lowers glucose and insulin in normal individuals after an oral glucose load: comparison with guar gum. *American Journal of Clinical Nutrition* **53**, 1425–1430.

Brummer, Y. *et al.* (2012) Glycemic response to extruded oat bran cereals processed to vary in molecular weight. *Cereal Chemistry* **89**, 255–261.

Charlton, K.E. *et al.* (2012) Effect of 6 weeks' consumption of β-glucan-rich oat products on cholesterol levels in mildly hypercholesterolaemic overweight adults. *British Journal of Nutrition* **3**, 1–11.

Cugnet-Anceau, C. *et al.* (2010) A controlled study of consumption of beta-glucan-enriched soups for 2 months by type 2 diabetic free-living subjects. *British Journal of Nutrition* **103**, 422–428.

Davy, B.M. *et al.* (2002) High-fiber oat cereal compared with wheat cereal consumption favorably alters LDL-cholesterol subclass and particle numbers in middle-aged and older men. *American Journal of Clinical Nutrition* **76**, 351–358.

De Angelis, M. *et al.* (2007) Use of sourdough lactobacilli and oat fibre to decrease the glycaemic index of white wheat bread. *British Journal of Nutrition* **98**, 1196–1205.

EFSA (European Food Safety Authority) (2011) Scientific opinion on the substantiation of health claims related to beta-glucans from oats and barley and maintenance of normal blood LDL-cholesterol concentrations (ID 1236, 1299), increase in satiety leading to a reduction in energy intake (ID 851, 852), reduction of postprandial glycaemic responses (ID 821, 824), and "digestive function" (ID 850) pursuant to Article 13(1) of Regulation (EC) No 1924/2006 *European Food Safety Authority Journal* **9**, 2207–2228.

Frank, J. *et al.* (2004) Yeast-leavened oat breads with high or low molecular weight β-glucan do not differ in their effects on blood concentrations of lipids, insulin, or glucose in humans. *Journal of Nutrition* **134**, 1384–1388.

Gamel, T.H. *et al.* (2012) Application of the rapid visco analyzer (RVA) as an effective rheological tool for measurement of β-glucan viscosity. *Cereal Chemistry* **89**, 52–58.

Granfeldt, Y.E. *et al.* (1992) An *in vitro* procedure based on chewing to predict metabolic response to starch in cereal and legume product. *European Journal of Clinical Nutrition* **46**, 649–660.

Granfeldt, Y. *et al.* (1994) Glucose and insulin responses to barley products: Influence of food structure and amylose-amylopectin ratio. *American Journal of Clinical Nutrition* **59**, 1075–1082.

Granfeldt, Y. *et al.* (1995) Metabolic responses to starch in oat and wheat products. On the importance of food structure, incomplete gelatinization or presence of viscous dietary fibre. *European Journal of Clinical Nutrition* **49**, 189–199.

Granfeldt, Y.E. *et al.* (2000) An examination of the possibility of lowering the glycemic index of oat and barley flakes by minimal processing. *Journal of Nutrition* **130**, 2207–2214.

Granfeldt, Y. *et al.* (2008) Muesli with 4 g oat β-glucans lowers glucose and insulin responses after a bread meal in healthy subjects. *European Journal of Clinical Nutrition* **62**, 600–607.

Hallfrisch, J. *et al.* (2003) Physiological responses of men and women to barley and oat extracts (Nu-trimX). II. Comparison of glucose and insulin responses. *Cereal Chemistry* **80**, 80–83.

Hätönen, K.A. *et al.* (2006) Methodologic considerations in the measurement of glycemic index: Glycemic response to rye bread, oatmeal porridge, and mashed potato. *American Journal of Clinical Nutrition* **84**, 1055–1061.

Hlebowicz, J. *et al.* (2007) Effect of commercial breakfast fibre cereals compared with corn flakes on postprandial blood glucose, gastric emptying and satiety in healthy subjects: A randomized blinded crossover trial. *Nutrition Journal* **6**, 22.

Hlebowicz, J. *et al.* (2008) Effect of muesli with 4 g oat β-glucan on postprandial blood glucose, gastric emptying and satiety in healthy subjects: A randomized crossover trial. *Journal of the American College of Nutrition* **27**, 470–475.

Hodge, A.M. *et al.* (2004) Glycemic index and dietary fiber and the risk of type 2 diabetes. *Diabetes Care* **27**, 2701–2706.

Holm, J. *et al.* (1992) Influence of sterilization, drying and oat bran enrichment of pasta on glucose and insulin responses in healthy subjects and on the rate and extent of *in vitro* starch digestion. *European Journal of Clinical Nutrition* **46**, 629–640.

Hopping, B.N. *et al.* (2010) Dietary fiber, magnesium, and glycemic load alter risk of type 2 diabetes in a multiethnic cohort in Hawaii. *Journal of Nutrition* **140**, 68–74.

Jenkins, A.L. *et al.* (2002) Depression of the glycemic index by high levels of β-glucan fiber in two functional foods tested in type 2 diabetes. *European Journal of Clinical Nutrition* **56**, 622–628

Juntunen, K.S. *et al.* (2002) Postprandial glucose, insulin, and incretin responses to grain products in healthy subjects. *American Journal of Clinical Nutrition* **75**, 254–262.

Juvonen, K.R. *et al.* (2009) Viscosity of oat bran-enriched beverages influences gastrointestinal hormonal responses in healthy humans. *Journal of Nutrition* **139**, 461–466,

Kabir, M. *et al.* (2002) Four-week low-glycemic index breakfast with a modest amount of soluble fibers in type 2 diabetic men *Metabolism* **51**, 819–826.

Keenan, J.M. *et al.* (2002) Oat ingestion reduces systolic and diastolic blood pressure in patients with milk or borderline hypertension: A pilot trial. *Journal of Family Practice* **51**, 369.

Krishnan, S. *et al.* (2007) Glycemic index, glycemic load, and cereal fiber intake and risk of type 2 diabetes in US black women. *Archives of Internal Medicine* **167**, 2304–2309.

Kwong, M.G.Y. *et al.* (2013) Attenuation of glycemic responses by oat β-glucan solutions and viscoelastic gels is dependent on molecular weight distribution. *Food and Function.* doi: 10.1039/C2FO30202K.

Lammert, A. *et al.* (2008) Clinical benefit of a short term dietary oatmeal intervention in patients with type 2 diabetes and severe insulin resistance: A pilot study. *Experimental and Clinical Endocrinology and Diabetes* **116**, 132–134.

Lan-Pidhainy, X. *et al.* (2007) Reducing beta-glucan solubility in oat bran muffins by freeze-thaw treatment attenuates its hypoglycemic effect. *Cereal Chemistry* **84**, 512–517.

Lazaridou, A. *et al.* (2003) Molecular size effects on rheological properties of oat β-glucans in solution and gels. *Food Hydrocolloids* **17**, 693–712.

Liatis, S. *et al.* (2009) The consumption of bread enriched with betaglucan reduces LDL-cholesterol and improves insulin resistance in patients with type 2 diabetes. *Diabetes and Metabolism* **35**, 115–120.

Liljeberg, H. *et al.* (1992) Metabolic responses to starch in bread containing intact kernels versus milled flour. *European Journal of Clinical Nutrition* **46**, 561–575.

Liljeberg, H.G.M. *et al.* (1996) Products based on a high fiber barley genotype, but not on common barley or oats, lower postprandial glucose and insulin responses in healthy humans. *Journal of Nutrition* **126**, 458–466.

Mäkeläinen, H. *et al.* (2007) The effect of β-glucan on the glycemic and insulin index. *European Journal of Clinical Nutrition* **61**, 779–785.

Maki, K.C. *et al.* (2007) Effects of consuming foods containing oat β-glucan on blood pressure, carbohydrate metabolism and biomarkers of oxidative stress in men and women with elevated blood pressure. *European Journal of Clinical Nutrition* **61**, 786–795.

Marciani, L. *et al.* (2001) Effect of meal viscosity and nutrients on satiety, intragastric dilution, and emptying assessed by MRI *American Journal of Physiology. Gastrointestinal and Liver Physiology* **280**, G1227–G1233.

Meyer, K.A. *et al.* (2000) Carbohydrates, dietary fiber, and incident type 2 diabetes in older women. *American Journal of Clinical Nutrition* **71**, 921–930.

Montonen, J. *et al.* (2003) Whole-grain and fiber intake and the incidence of type 2 diabetes. *American Journal of Clinical Nutrition* **77**, 622–629.

Nilsson, A. *et al.* (2008a) Effects of GI vs content of cereal fibre of the evening meal on glucose tolerance at a subsequent standardized breakfast. *European Journal of Clinical Nutrition* **62**, 712–720.

Nilsson, A.C. *et al.* (2008b) Effect of cereal test breakfasts differing in glycemic index and content of indigestible carbohydrates on daylong glucose tolerance in healthy subjects. *American Journal of Clinical Nutrition* **87**, 645–654.

Östman, E. *et al.* (2006) Glucose and insulin responses in healthy men to barley bread with different levels of $(1\rightarrow3;1\rightarrow4)$-β-glucans; predictions using fluidity measurements of *in vitro* enzyme digests. *Journal of Cereal Science* **43**, 230–235.

Panahi, S. *et al.* (2007) β-Glucan from two sources of oat concentrates affect postprandial glycemia in relation to the level of viscosity. *Journal of the American College of Nutrition* **26**, 639–644.

Pins, J.J. *et al.* (2002) Do whole-grain oat cereals reduce the need for antihypertensive medications and improve blood pressure control? *Journal of Family Practice* **51**, 353–359.

Queenan, K.M. *et al.* (2007) Concentrated oat β-glucan, a fermentable fiber, lowers serum cholesterol in hypercholesterolemic adults in a randomised controlled trial. *Nutrition Journal* **6**, 6.

Regand, A. *et al.* (2009) Physicochemical properties of glucan in differently processed oat foods influence glycemic response. *Journal of Agricultural and Food Chemistry* **57**, 8831–8838.

Regand, A. *et al.* (2011) The molecular weight, solubility and viscosity of oat beta-glucan affect human glycemic response by modifying starch digestibility. *Food Chemistry* **129**, 297–304.

Ren, Y. *et al.* (2003) Dilute and semi-dilute solution properties of (1-3)(1-4)-β-D-glucan, the endosperm cell wall polysaccharide of oats (*Avena sativa* L.). *Carbohydrate Polymers* **53**, 401–408.

Reyna-Villasmil, N. *et al.* (2007) Oat-derived beta-glucan significantly improves HDLC and diminishes LDLC and non-HDL cholesterol in overweight individuals with mild hypercholesterolemia. *American Journal of Therapeutics*. **14**, 203–212.

Saltzman, E. *et al.* (2001) An oat-containing hypocaloric diet reduces systolic blood pressure and improves lipid profile beyond effects of weight loss in men and women. *Journal of Nutrition* **131**, 1465–1470.

Schulze, M.B. *et al.* (2004) Glycemic index, glycemic load, and dietary fiber intake and incidence of type 2 diabetes in younger and middle-aged women. *American Journal of Clinical Nutrition* **80**, 348–356.

Schulze, M.B. *et al.* (2007) Fiber and magnesium intake and incidence of type 2 diabetes: A prospective study and meta-analysis. *Archives of Internal Medicine* **167**, 956–965.

Tapola, N. *et al.* (2005) Glycemic responses of oat bran products in type 2 diabetic patients. *Nutrition Metabolism and Cardiovascular Diseases* **15**, 255–261.

Tappy, L. *et al.* (1996) Effects of breakfast cereals containing various amounts of beta-glucan fibers on plasma glucose and insulin responses in NIDDM subjects. *Diabetes Care* **19**, 831–834.

Tighe, P. *et al.* (2010) Effect of increased consumption of whole-grain foods on blood pressure and other cardiovascular risk markers in healthy middle-aged persons: a randomized controlled trial. *American Journal of Clinical Nutrition* **92**, 733–740.

Tosh, S.M. (2013) Review of human studies investigating the postprandial blood glucose lowering ability of oat and barley food products. *European Journal of Clinical Nutrition* **67**, 310–317. doi:10.1038/ejcn.2013.25.

Tosh, S.M. *et al.* (2008) Glycemic response to oat bran muffins treated to vary molecular weight of β-glucan. *Cereal Chemistry* **85**, 211–217.

Tuomasjukka, S. *et al.* (2007) The glycaemic response to rolled oat is not influenced by the fat content. *British Journal of Nutrition* **97**, 744–748.

Ulmius, M. *et al.* (2011) An oat bran meal influences blood insulin levels and related gene sets in peripheral blood mononuclear cells of healthy subjects. *Genes and Nutrition* **6**, 429–439.

University of Sydney. (2013) GI database. http://www.glycemicindex.com/index.php (last accessed 7 May 2013).

USDA (United States Department of Agriculture) (2013) Grain: World Markets and Trade. Foreign Agricultural Service Circular Series FG 12-12. http://www.fas.usda.gov/psdonline/circulars/grain.pdf. Accessed December 27, 2012.

Valachovičová, M. *et al.* (2006) No evidence of insulin resistance in normal weight vegetarians. *European Journal of Nutrition* **45**, 52–54.

Wolever, T.M.S. *et al.* (1994) Glycaemic index of 102 complex carbohydrate foods in patients with diabetes. *Nutrition Research* **14**, 651–669.

Wolever, T.M., *et al.* (2010) Physicochemical properties of oat β-glucan influence its ability to reduce serum LDL cholesterol in humans: a randomized clinical trial. *American Journal of Clinical Nutrition* **92**, 723–732.

Wood, P.J. *et al.* (1990) Comparisons of viscous properties of oat and guar gum and the effects of these and oat bran on glycemic index. *Journal of Agricultural and Food Chemistry* **38**, 753–757.

Wood, P.J. *et al.* (1994) Effect of dose and modification of viscous properties of oat gum on plasma glucose and insulin following an oral glucose load. *British Journal of Nutrition* **72**, 731–743.

Wood, P.J. *et al.* (2000) Evaluation of role of concentration and molecular weight of oat beta-glucan in determining effect of viscosity on plasma glucose and insulin following an oral glucose load. *British Journal of Nutrition* **84**, 19–23.

Wood, P.J. *et al.* (2002) Fermentability of oat and wheat fractions enriched in β-glucan using human fecal inoculation. *Cereal Chemistry* **79**, 445–454.

14
Effects of Oats and β-Glucan on Gut Health

Renee Korczak and Joanne Slavin

Department of Food Science and Nutrition, University of Minnesota, St. Paul, MN, USA

14.1   Oats and β-glucan

Avena sativa L. (common oat) is the most important of the cultivated oats (Butt *et al.*, 2008). The outermost layer of the oat kernel, oat bran, can be isolated from the kernel and consumed as a supplement. Oat bran contains vitamins, minerals, carbohydrate (68%), protein (17%), and fat (9%) (Butt *et al.*, 2008). Oat bran consists of 15–22% dietary fiber and 10.4% β-glucan, which is a linear, unbranched polysaccharide. Oat β-glucan is highly viscous; its viscosity is dependent on the length of the polysaccharide chain. In the 1980s, specific health benefits of oats were recognized (i.e., ability to lower serum cholesterol and reduce the risk of cardiovascular disease) (Anderson and Gustafson, 1988).

14.2   Digestive health

Grains, including oats, have been shown to support gut health in a variety of ways. For example, dietary fiber increases stool weight and promotes normal laxation (Grabitske and Slavin, 2008). The term laxation refers to a wide range of gastrointestinal effects, including stool weight, transit time, bloating and distention, flatus, constipation, and diarrhea; however, no standardized, accepted definitions exist for constipation and diarrhea. Constipation is a chronic condition that is common in Western society but poorly studied. It has been defined as fewer than three bowel movements per week, but most people define constipation as less than one bowel movement per day. However, frequency of defecation is only one aspect of constipation; ease of stool passage (lack of straining) is another component of normal laxation.

Bowel habits are affected by medications, stress, physical activity, food volume, type of food, fluid intake, hormones, and other environmental factors. Although subjective measures of bowel function are important, objective measures such as wet and dry stool weight, and gastrointestinal transit time are useful biomarkers. Increased bulk, softness or pliability of colonic contents, and increased intestinal motility protect against constipation. Stool weight increases as fiber intake increases, which tends to normalize defecation frequency to once daily and gastrointestinal transit time to 2–4 days. The increase in stool weight is due not only to the presence of the fiber but also by the water held by the fiber and partial fermentation of the fiber by intestinal microbiota, which increase the amount of bacteria in the stool. Bacteria also bind water, so bacterial mass increases stool weight but generally not as much as the undigested fiber.

Diarrhea is an unpleasant digestive disorder that almost everyone experiences at some time. When food is consumed, it typically remains in liquid form during most of the digestive process. When food residue passes through the large intestine, most of the remaining fluid is absorbed, leaving a semisolid stool. In diarrhea, the ingested food and fluid pass too quickly or in too large an amount (or both) through the intestine. Fluid is not sufficiently absorbed, and the result is a watery bowel movement. Commonly accepted criteria for clinical diarrhea are: stool output > 200 g/day; watery, difficult to control bowel movements, and more than three bowel movements per day (Bliss et al., 1992). Colonic fermentation of dietary fiber may improve gastrointestinal tolerance and decrease diarrhea by protecting the intestine from bacterial overgrowth. Although a meta-analysis of randomized, controlled trials found no evidence that dietary fiber is effective in treating diarrhea (Homann et al., 1994), the addition of fiber to enteral diets is well accepted in clinical practice (Klosterbuer et al., 2011).

Irritable bowel syndrome (IBS) is defined as a group of functional bowel disorders characterized by chronic or recurrent abdominal pain or discomfort, usually in the lower abdomen. The discomfort is associated with disturbed bowel function (i.e., diarrhea or constipation alone or alternating) and abdominal distention and bloating (Drossman et al., 2002). The persistent symptoms of IBS result in a considerable negative effect on health-related quality of life. The prevalence of IBS has been estimated to be 10–20% of adults in the United States and Europe (Drossman et al., 1993); however, the actual prevalence is certainly higher because 70% of symptomatic adults do not seek medical evaluation. Women with IBS report symptoms of constipation and abdominal discomfort more often, whereas men with IBS are more likely to report diarrhea. Psychological disturbances (e.g., anxiety and depression) are more common in individuals with IBS who seek treatment for their symptoms than in those who do not seek treatment. This indicates that psychological disturbances may exacerbate IBS symptoms and influence health care-seeking behavior.

Results of clinical trials assessing the role of grains in IBS have reported inconsistent findings. Austin and colleagues reported that a very low carbohydrate diet can improve symptoms and quality of life in patients with diarrhea-predominant IBS (Austin et al., 2009). However, a systematic review of dietary interventions for children with IBS concluded that there is a lack of high-quality data on the effectiveness of dietary interventions and no evidence that fiber supplements, lactose-free diets, or Lactobacillus supplements are effective in children with recurrent abdominal pain (Huertas-Ceballos et al., 2009). Bijkerk and colleagues conducted a systematic review to determine the effect of dietary fiber on global IBS symptoms, IBS-related abdominal pain, and IBS-related constipation (Bijkerk *et al.*, 2004). Only two of the six trials evaluating insoluble wheat fiber reported improvement in global symptoms. Miller bran improved IBS-related constipation but was no better than placebo in the treatment of global symptoms. A meta-analysis conducted by Longstreth and colleagues showed that general fiber supplementation alleviates global IBS symptoms but does not alleviate abdominal pain, which is the most important feature that distinguishes IBS from functional constipation or functional diarrhea (Longstreth *et al.*, 2006). Neither probiotics nor prebiotics were found to be effective in the treatment of IBS (Spiller, 2008).

Although it is accepted that grain fiber plays a role in digestive health (Slavin, 2008), few studies have assessed individual grains and their effects on gut health. In addition, most studies have evaluated the effects of dietary fiber only in diseased populations.

14.3   Short chain fatty acids and fiber fermentability

Fiber is fermented by anaerobic intestinal bacteria that generate short chain fatty acids (SCFAs), which serve as an energy source for colonic mucosal cells (Slavin, 2008). Acetate, propionate, and butyrate are the SCFAs produced in the highest concentrations (Topping and Clifton, 2001). Acetate is a fuel for skeletal and cardiac muscle, kidney, and the brain. Butyrate is the preferred fuel of the colonic epithelium, in particular, the distal colon and rectum. Propionate is metabolized by the liver and may play a role in lowering cholesterol. Dietary fibers produce varying proportions of individual SCFAs and different amounts of total SCFA. In addition, SCFAs produced by fermentation of undigested carbohydrates are associated with a decreased risk of cancer (Topping and Clifton, 2001). Particle size, solubility, surface area, and other factors also affect the extent of fermentation and the nature of the SCFAs produced.

Because fermentable fiber is beneficial to health, it is important to understand the fermentability of each type of fiber. Fiber fermentability is difficult to study *in vivo* because of the invasiveness of colon studies and the dynamic nature of the colon, Fermentation can be estimated by measuring the amount of fiber consumed and subtracting the amount of fiber remaining in the feces. However, this approach is difficult because bacterial cell walls are isolated from the feces along with the fiber. No biomarkers exist to easily determine fiber fermentation *in vivo*; therefore, *in vitro* models are generally used.

SCFAs are absorbed from the colonic lumen shortly after they are produced. Because no method has been developed to accurately measure SCFA absorption *in vivo*, measuring the SCFAs excreted in the feces is the best estimate of SCFA production. However, 95–99% of SCFAs are absorbed; therefore, studying the amount of SCFAs in the feces of human volunteers provides only a partial picture. A closed laboratory system can estimate fiber fermentability without losing SCFAs to colonic absorption. *In vitro* fermentation with representative human colonic microflora is a well established, noninvasive, time efficient means to esti-mate fiber fermentability. Indeed, batch fermentation has been shown to degrade nonstarch polysaccharides to a similar extent as the human colon, based on resid-ual nonstarch polysaccharides in fecal samples and fermentation flasks (Wisker *et al.*, 1998).

14.4   Large bowel effects of whole grains

Whole grains are rich sources of fermentable carbohydrates, including resistant starch and oligosaccharides, as well as dietary fiber (Slavin, 2004). These nondi-gestible carbohydrates increase fecal weight (wet and dry) and speed intestinal transit. The dietary fiber content of individual grains varies greatly, ranging from 15% in rye to 4% in rice (Charalampopoulos *et al.*, 2002). Dietary fiber occurs in decreasing amounts from the outer pericarp to the endosperm, except for ara-binoxylan, which is also a major component of endosperm cell walls. Approxi-mately one-third of the fiber in oats, rye, and barley is soluble and the rest is insol-uble. Soluble fiber is associated with cholesterol lowering and improved glucose response, whereas insoluble fiber is associated with improved laxation. Wheat is lower in soluble fiber than most grains, and rice contains virtually no soluble fiber. The processing of grains changes the carbohydrate composition. Refined grains are low in total dietary fiber and the refining process removes more of the insoluble fiber than soluble fiber.

Disrupting the cell walls can increase the fermentability of dietary fiber. In addition, coarse wheat bran has a greater fecal bulking effect than finely ground wheat bran at the same dosage (Heaton *et al.*, 1988). Coarse bran delays gas-tric emptying and accelerates small bowel transit. The effect of coarse bran was similar to that of inert plastic particles, suggesting that the coarse nature of whole grains exerts a physiological effect beyond the differences in composition between whole and refined grains (McIntyre *et al.*, 1997).

McIntosh and coworkers measured markers of bowel health in overweight middle-aged men fed refined or whole grain foods that were high in either rye or wheat (McIntosh *et al.*, 2003). In addition to the baseline diet containing 14 g of dietary fiber, the refined grain foods provided 5 g of dietary fiber and the whole grain foods provided 18 g of dietary fiber. The high-fiber rye and wheat foods increased fecal output by 33–36% and reduced fecal β-glucuronidase activity by 29%. Postprandial plasma insulin was decreased by 46–49% and postprandial plasma glucose by 16–19%. In addition, high-fiber rye foods were associated with significantly increased plasma enterolactone and fecal butyrate. The authors con-cluded that rye appears to be more effective than wheat in improving biomarkers of bowel health.

Chen and coworkers compared the mechanisms by which wheat bran and oat bran increase stool weight in humans (Chen *et al.*, 1998). Bacteria and lipids were found to be the major contributors to increased stool weight with oat bran, whereas undigested plant fiber accounted for much of the increase in stool weight with wheat bran consumption. The results suggest that oat bran increases stool weight by providing rapidly fermented soluble fiber in the proximal colon for bacterial growth, which is sustained until excretion by fermentation of the insoluble fiber.

14.5   Fermentation of individual dietary fibers

To compare the fermentation of dietary fibers, an *in vitro* fermentation method
is employed (Pylkas *et al.*, 2005) using fecal samples obtained from three donors
who are disease-free and consume a typical, low-fiber American diet. Fecal samples from the three subjects are mixed together, and fibers are fermented in the
fecal slurry. SCFA production is measured as an endpoint of fiber fermentation
at time points ranging from 0 to 48 hours.

Wheat dextrin, inulin (degree of polymerization about 10), and partially
hydrolyzed guar gum (PHGG) were evaluated using this system (Stewart *et al.*,
2009), and concentrations of acetate, propionate, butyrate, and total SCFAs were
determined. Glucose was fermented as a positive control to ensure active fermentation of the system and no fiber was the negative control. Statistical analyses
were performed with a SAS statistical software package, version 8.0 (SAS Institute, Cary, NC). Fibers were compared by analysis of variance and Tukey pairwise comparison. The results showed that all three fibers were fermentable but
PHGG produced only low levels of all SCFAs at all time points. Wheat dextrin
and inulin produced similar amounts of total SCFA at 24 hours, significantly more
than that produced by PHGG ($p = 0.002$). However, SCFA production by inulin
fermentation peaked at four hours and decreased by eight hours. In contrast,
SCFA production by wheat dextrin fermentation increased steadily throughout
the 24-hour period. The rapid fermentation of inulin suggests that it may cause
excess gas production. Acetate was the main SCFA produced by these fibers,
accounting for roughly 50% of total SCFA production.

Few *in vitro* fermentation studies have been conducted with oats. Kim and
White (2010) compared the fermentation of high, medium, and low molecular weight β-glucan from oats. They found that low molecular weight β-glucan
produced greater amounts of SCFA than high molecular weight β-glucan after
24 hours. Connolly and colleagues compared the fermentation of different sized
oat flakes and found that larger oat flakes produced more propionate and
butyrate than smaller oat flakes (Connolly *et al.*, 2010).

Karppinen and colleagues compared the *in vitro* fermentation of polysaccharides from rye bran, wheat bran, oat bran, and inulin (Karppinen *et al.*, 2000).
After enzymatic digestion of the brans to remove starch and protein, the brans
and inulin were fermented with human fecal inoculum. Inulin, a short fructose
polymer, was consumed significantly faster than the more complex carbohydrates
of the cereal brans. Carbohydrates of oat bran (rich in β-glucan) were consumed
faster than those of rye and wheat brans (rich in arabinoxylan). In all brans, glucose was consumed faster than the other main sugars, arabinose and xylose. Formation of gases was fastest and greatest with inulin.

14.6   Prebiotics

A prebiotic is defined as "a nondigestible food ingredient that beneficially affects
the host by selectively stimulating the growth and/or activity of one or a limited number of bacteria in the colon and thus improves host health" (Gibson
and Roberfroid, 1995). This definition was recently updated as "a selectively fermented ingredient that allows specific changes, both in the composition and/or

activity in the gastrointestinal microbiota that confers benefits upon host well-being and health" (Gibson *et al.*, 2004). The following nondigestible oligosac-charides appear to be prebiotics: inulin, fructo-oligosaccharides (FOS), galacto-oligosaccharides, soya-oligosaccharides, xylo-oligosaccharides, pyrodextrins, and isomalto-oligosaccharides (Macfarlane *et al.*, 2006).

Although all nondigestible carbohydrates that reach the gut have the potential to alter microflora, prebiotics selectively stimulate the growth of bifidobacteria and lactobacilli at the expense of other bacteria such as *Bacteroides*, clostridia, eubacteria, enterobacteria, and enterococci. Although no standard exists to allow a substance to be called a prebiotic, there is evidence that the nondigestible oligosaccharides described above alter colonic microflora in a beneficial manner. Nevertheless, microbial growth response varies widely in healthy human subjects, and microflora composition is influenced by numerous factors, including drugs, antibiotics, age, and overall diet.

To assess the prebiotic effects of dietary oligosaccharides, a method using a prebiotic index (PI) was developed based on changes in key bacterial groups during fermentation (Palframan *et al.*, 2003). The bacterial groups incorporated into this PI equation include bifidobacteria, lactobacilli, clostridia, and bacteroides. This concept was validated using data from previously published studies; quantitative PI scores generally supported the qualitative conclusions drawn from these studies. It is hoped that the PI equation can be used to screen prebiotic carbohydrates for prebiotic effects *in vitro*.

Not all fermentable carbohydrates in the gut are nondigestible carbohydrates obtained from the diet. The mucus layer of the gut also provides fermentable oligosaccharides that influence bacterial growth. Although most studies of microflora composition measure the microbial content of feces, bacteria that grow adjacent to the colonic mucosa may be particularly important for the host immune function. Langlands and coworkers found that bifidobacteria numbers could be increased more than 10 fold in the mucosa of the proximal and distal colon in patients fed 15 g/day of a prebiotic mixture (7.5 g inulin and 7.5 g FOS) for two weeks before undergoing a colonoscopy (Langlands *et al.*, 2004).

14.6.1   Prebiotics and immunity

The immune system protects the body from foreign substances and the largest immune organ is the gut, where these subsatances are continuously exposed to diverse antigens. The gut immune system inhibits the growth of harmful pathogens while promoting the growth of beneficial organisms (Hooper, 2004). Nondigestible carbohydrates influence the immune system, not only by altering the number and composition of intestinal microflora but also by affecting the gut-associated lymphoid tissue, which contains about 60% of all lymphocytes in the body (Watzl *et al.*, 2005).

Prebiotics have been shown to improve acute disorders and chronic disorders (Bruck, 2006) but few human studies demonstrate an immune effect of prebiotics other than the alteration of microbial composition. Bruck (2006) concludes that the primary action of prebiotics is to improve resistance to pathogens by increasing the numbers of bifidobacteria and lactobacilli, which lowers gut pH to a level at which pathogens are no longer able to compete. Adverse effects of prebiotics have been found *in vitro* and in animal models and must be considered when developing prebiotics for human use.

14.6.2 Prebiotic potential of grains

Costabile and coworkers conducted a double-blind, randomized, crossover study of in 31 volunteers to compare the prebiotic effects of whole grain and wheat bran (Costabile *et al.*, 2008). Numbers of fecal bifidobacteria and lactobacilli were significantly higher after whole grain ingestion than after wheat bran ingestion. However, fecal SCFA levels were similar after ingestion of either cereal.

Although the amount of FOS, considered to be a prebiotic, has not been systematically determined in all grains, wheat is known to be particularly high in this carbohydrate (0.8–4.0% FOS in fresh material) (Vernazza *et al.*, 2006). Biesiekierski and colleagues measured fructans in a wide range of foodstuffs and found that oats contained 0.11 g/portion and rye contained 0.6 g/portion (Biesiekierski *et al.*, 2011). At least two types of oligosaccharides exist in cereal grains, galactosyl derivatives of sucrose (stachyose and raffinose) and fructosyl derivatives of sucrose (FOS) (Henry and Saini, 1989); however, the distributions of these compounds in cereal grains have not been fully established. Oligosaccharides have been detected in wheat bran (Yamada *et al.*, 1993) and wheat germ (Pomeranz, 1988), and wheat germ is particularly high in raffinose oligosaccharides (7.2% on a dry basis) (Charalampopoulos *et al.*, 2002).

Oligosaccharides can be isolated from cereal grains and purified. However, this process is difficult because of the complexity of these compounds and their connections with other molecules. Cereal grains are also concentrated sources of resistant starch. Although results of animal studies provide evidence for resistant starch as a prebiotic, few human studies have evaluated its effect on gut microbiota (Crittenden, 2006).

Arabinoxylans and β-glucans are the major cereal fibers fermented by bacteria in the human gastrointestinal tract (Crittenden, 2006). Bifidobacteria and lactobacilli cannot ferment cereal β-glucans well *in vitro* (Crittenden *et al.*, 2002) but metabolize the oligosaccharides resulting from the partial hydrolysis of β-glucan. However, results of *in vitro* studies show that *Bifidobacterium longum* and *B. adolescentis* are able to ferment arabinoxylan from cereal sources, as well as arabinoxylan oligosaccharides (Crittenden *et al.*, 2002). In addition, rye bran rich in arabinoxylan has been shown to promote the growth of bifidobacteria in mice (Oikarinen *et al.*, 2003). In contrast, potentially harmful bacteria such as *Escherichia coli*, *Clostridium perfringens*, or *C. difficile* do not directly ferment these substrates.

The effect of sourdough fermentation on the prebiotic properties of grains has also been evaluated. Sourdough fermentation of rye bran has been shown to enhance its bioactivity (Katina *et al.*, 2007). Indigenous lactobacillus and enzymes concentrated in the outer layers of grains contribute to changes seen in bran during sourdough fermentation; the extent of these changes can be modulated by modifying the milling process during separation of the bran. Wheat and rye contain kestose, nystose, and other FOS similar to inulin (Campbell *et al.*, 1997);

however, few data are available on the metabolism of wheat and rye polysaccharides by bifidobacteria and lactobacilli. Strains of *Lactobacillus sanfranciscensis* used in bread making produce an exopolysaccharide (high molecular weight fructan of the levan type) (Korakli *et al.*, 2002). During dough fermentation, the polyfructans and starch of wheat and rye are degraded by cereal enzymes, whereas the exopolysaccharides are retained in the bread dough.

The effect of specific grain fractions on gut microflora has been evaluated in both animal and human studies. Matteuzzi and colleagues conducted a double-blind, placebo-controlled study to evaluate the prebiotic activity of a wheat germ preparation in 32 healthy subjects (Matteuzzi *et al.*, 2004). After 20 days of supplementation, coliform populations and pH decreased significantly, and the number of lactobacilli and bifidobacteria increased significantly in subjects who had low basal levels; other bacterial groups and total bacteria were unchanged. An arabinoxylan-rich germinated barley product was also found to simulate bifidobacteria proliferation in the human intestine (Kanauchi *et al.*, 1999). The barley product was compared with probiotics and antibiotics as a treatment for experimental colitis in rats. The authors concluded that modification of intestinal microflora by prebiotics, including germinated barley fiber, may be a useful adjunct in the treatment of ulcerative colitis (Fukuda *et al.*, 2002). Additional animal studies supporting the use of germinated barley for ulcerative colitis have been described by Bamba and colleagues (Bamba *et al.*, 2002).

14.7   Other mechanisms underlying the effect of oats on gut function

The viscosity of dietary fibers is thought to alter the physiological function of the gut. Dikeman and coworkers evaluated the viscosities of various dietary fibers in an *in vitro* system (Dikeman *et al.*, 2006). Guar gum, psyllium, and oat bran exhibited viscous characteristics throughout small intestinal simulation, indicating possible glucose- and lipid-lowering effects. In addition, the gut uses hormones to communicate with the brain to regulate eating habits (Badman and Flier, 2005); however, the importance of gut microflora in preventing obesity has only recently been considered. Juvonen and coworkers evaluated the effect of viscosity on gastrointestinal hormonal responses of healthy human subjects consuming oat bran-enriched beverages (Juvonen *et al.*, 2009). The low-viscosity beverage induced a greater postprandial increase in satiety and increased levels of cholecystokinin, glucagon-like peptide, and peptide YY. In addition, gastric emptying was faster after consuming the low-viscosity beverage. These results suggest that polysaccharide structure is important in the modulation of postprandial satiety.

14.8   Conclusion

Oats add dietary fiber, resistant starch, and oligosaccharides to the diet. High in soluble fiber, oats increase stool weight, primarily by increasing bacterial mass. Carbohydrates that escape digestion and absorption in the small intestine and are fermented in the large intestine may function as prebiotics. In addition, the viscosity of oats appears to influence important physiological properties, such as regulation of satiety-related gut hormones. However, few clinical trials have evaluated the specific effects of oats or oat bran on gut health; thus further research is needed to better understand the role of oats in digestive health.

References

Anderson, W.J. and Gustafson, N.J. (1988) Hypocholesterolemic effects of oat and bean products. *American Journal of Clinical Nutrition* **48**, 749–753.

Austin, G.L., *et al.* (2009) A very-low-carbohydrate diet improves symptoms and quality of life in diarrhea-predominant irritable bowel syndrome. *Clinical Gastroenterology and Hepatology* **7**, 706–708.e1.

Badman, M.K. and Flier, J.S. (2005) The gut and energy balance: visceral allies in the obesity wars. *Science* **307**, 1909–1914.

Bamba, T., *et al.* (2002) A new prebiotic from germinated barley for nutraceutical treatment of ulcerative colitis. *Journal of Gastroenterology and Hepatology* **17**, 818–824.

Biesiekierski, J.R., *et al.* (2011) Quantification of fructans, galacto-oligosaccharides and other short-chain carbohydrates in processed grains and cereals. *Journal of Human Nutrition and Dietetics* **24**, 154–176.

Bijkerk, C.J., *et al.* (2004) Systematic review: the role of different types of fibre in the treatment of irritable bowel syndrome. *Alimentary Pharmacology and Therapeutics* **19**, 245–251.

Bliss, D.Z., *et al.* (1992) Defining and reporting diarrhea in tube fed patients: what a mess! *The American Journal of Clinical Nutrition* **55**, 753–759.

Bruck, W.M. (2006) Dietary intervention for improving human health: Acute disorders. In: *Prebiotics: Development and Application* (eds G.R. Gibson and R.A Rastall), pp. 157–79. John Wiley and Sons Ltd, Chichester, UK.

Butt, M.S., *et al.* (2008) Oat: unique among the cereals. *European Journal of Nutrition* **47**, 68–79.

Campbell, J.M., *et al.* (1997) Selected fructooligosaccharide (1-kestose, nystose, and 1-beta-fructofuranosylnystose) composition of foods and feeds. *Journal of Agricultural and Food Chemistry* **45**, 3076–3082.

Charalampopoulos, D., *et al.* (2002) Application of cereals and cereal components in functional foods: a review. *International Journal of Food Microbiology* **79**, 131–141.

Chen, H.L., *et al.* (1998) Mechanisms by which wheat bran and oat bran increase stool weight in humans. *The American Journal of Clinical Nutrition* **68**, 711–719.

Costabile, A., *et al.* (2008) Whole-grain wheat breakfast cereal has a prebiotic effect on the human gut microbiota: a double-blind, placebo-controlled, crossover study. *British Journal of Nutrition* **99**, 110–120.

Crittenden, R. (2006) Emerging prebiotic carbohydrates. In: *Prebiotics: Development and Application* (eds G.R. Gibson and R.A. Rastall), pp. 111–33. John Wiley and Sons Ltd, Chichester, UK.

Crittenden, R., *et al.* (2002) *In vitro* fermentation of cereal dietary fibre carbohydrates by probiotic and intestinal bacteria. *Journal of the Science of Food and Agriculture* **82**, 781–789.

Connolly, M.L., *et al.* (2010) *In vitro* evaluation of the microbiota modulation abilities of different sized whole oat grain flakes. *Anerobe* **16**, 483–488.

Dikeman, C.L., *et al.* (2006) Dietary fibers affect viscosity of solutions and simulated human gastric and small intestinal digesta. *Journal of Nutrition* **136**, 913–919.

Drossman, D.A., *et al.* (1993) US householder survey of functional gastrointestinal disorders: prevalence, sociodemography, and health impact. *Digestive Diseases and Sciences* **38**, 1569–1580.

Drossman, D.A., *et al.* (2002) AGA technical review on irritable bowel syndrome. *Gastroenterology* **123**, 2108–2131.

Fukuda, M., *et al.* (2002) Prebiotic treatment of experimental colitis with germinated barley foodstuff: a comparison with probiotic or antibiotic treatment. *International Journal of Molecular Medicine* **9**, 65–70.

Gibson, G.R. and Roberfroid, M. (1995) Dietary modulation of the human colonic microbiota: introducing the concept of prebiotics. *Journal of Nutrition* **125**, 1401–1412.

Gibson, G.R., *et al.* (2004) Dietary modulation of the human colonic microbiota: updating the concept of prebiotics. *Nutrition Research Reviews* **17**, 259–275.

Grabitske, H.A. and Slavin, J.L. (2008) Laxation and like: Assessing digestive health. *Nutrition Today* **43**, 193–198.

Heaton, K.W., *et al.* (1988) Particle size of wheat, maize, and oat test meals: effects on plasma glucose and insulin responses and on the rate of starch digestion *in vitro*. *The American Journal of Clinical Nutrition* **47**, 675–682.

Henry, R.J. and Saini, H.S. (1989) Characterization of cereal sugars and oligosaccharides. *Cereal Chemistry* **66**, 362–365.

Homann, H.H., *et al.* (1994) Reduction in diarrhea incidence by soluble fiber in patients receiving total or supplemental enteral nutrition. *Journal of Parenteral and Enteral Nutrition* **18**, 486.

Hooper, L.V. (2004) Bacterial contributions to mammalian gut development. *Trends in Microbiology* **12**, 129–134.

Huertas-Ceballos, A.A., *et al.* (2009) Dietary interventions for stomach ache with no identifiable cause in children. *Cochrane Database of Systematic Reviews* 1 (Art. No.: CD003019). doi: 10.1002/14651858.CD003019.pub3.

Juvonen, K.R., *et al.* (2009) Viscosity of oat bran-enriched beverages influences gastrointestinal hormonal responses in healthy humans. *Journal of Nutrition* **139**, 461–466.

Kanauchi, O., *et al.* (1999) Increased growth of Bifidobacterium and Eubacterium by germinated barley food stuff, accompanied by enhanced butyrate production in healthy volunteers. *International Journal of Molecular Medicine* **3**, 175–179.

Katina, K., *et al.* (2007) Bran fermentation as a means to enhance technological properties and bioactivity of rye. *Food Microbiology* **24**, 175–186.

Karppinen, S., *et al.* (2000) *In vitro* fermentation of polysaccharides of rye, wheat and oat brans and inulin by human faecal bacteria. *Journal of the Science of Food and Agriculture* **80**, 1469–1476.

Kim, H.J. and White, P.J. (2010) *In vitro* bile-acid binding and fermentation of high, medium, and low molecular weight beta-glucan. *Journal of Agricultural and Food Chemistry* **58**, 628–634.

Klosterbuer, A., *et al.* (2011) Benefits of dietary fiber in clinical nutrition. *Nutrition in Clinical Practice* **26**, 625–635.

Korakli, M., *et al.* (2002) Metabolism by bifidobacteria and lactic acid bacteria of polysaccharides from wheat and rye, and exopolysaccharides produced by *Lactobacillus sanfranciscensis*. *Journal of Applied Microbiology* **92**, 958–965.

Langlands, S.J., *et al.* (2004) Prebiotic carbohydrates modify the mucosa-associated microflora of the human large bowel. *Gut* **53**, 1610–1616.

Longstreth, G.F., *et al.* (2006) Functional bowel disorders. *Gastroenterology* **130**, 1480–1491.

Macfarlane, S., *et al.* (2006) Review article: prebiotics in the gastrointestinal tract. *Alimentary Pharmacology and Therapeutics* **24**, 701–714.

Matteuzzi, D., *et al.* (2004) Prebiotic effects of a wheat germ preparation in human healthy subjects. *Food Microbiology* **21**, 119–125.

McIntosh, G.H., *et al.* (2003) Whole-grain rye and wheat foods and markers of bowel health in overweight middle-aged men. *The American Journal of Clinical Nutrition* **77**, 967–974.

McIntyre, A., *et al.* (1997) Effect of bran, ispaghula, and inert plastic particles on gastric emptying and small bowel transit in humans: the role of physical factors. *Gut* **40**, 223–227.

Oikarinen, S., *et al.* (2003) Plasma enterolactone or intestinal Bifidobacterium levels do not explain adenoma formation in multiple intestinal neoplasia (Min) mice fed with two different types of rye-bran fractions. *British Journal of Nutrition* **90**, 119–125.

Palframan, R., *et al.* (2003) Development of a quantitative tool for the comparison of the prebiotic effect of dietary oligosaccharides. *Letters in Applied Microbiology* **37,** 218–284.

Pomeranz, Y. (1988) Chemical composition of kernel structures. In: *Wheat: Chemistry and Technology* (ed. Y. Pomeranz), Vol 1, 3rd edn, pp. 91–158. AACC, St. Paul, MN.

Pylkas, A.M., *et al.* (2005) Comparison of different fibers for in vitro production of short chain fatty acids by intestinal microflora. *Journal of Medicinal Food* **8**, 113–116.

Slavin, J.L. (2004) Whole grains and human health. *Nutrition Research Reviews* **17**, 100–112.

Slavin, J.L. (2008) Position of the American Dietetic Association: Health implications of dietary fiber. *Journal of the American Dietetic Association* **108**, 1716–1731.

Spiller, R. (2008) Review article: probiotics and prebiotics in irritable bowel syndrome. *Alimentary Pharmacology and Therapeutics* **28**, 385–397.

Stewart, M.L., *et al.* (2009) Assessment of dietary fiber fermentation: Effect of *Lactobacillus reuteri* and reproducibility of short-chain fatty acids concentrations. *Molecular Nutrition and Food Research* **53**, S114–S120.

Topping, D.L. and Clifton, P.M. (2001) Short-chain fatty acids and human colonic function: Roles of resistant starch and nonstarch polysaccharides. *Physiology Reviews* **81**, 1031–1064.

Vernazza, C.L., *et al.* (2006) Human colonic microbiology and the role of dietary intervention: introduction to prebiotics. In: *Prebiotics: Development and Application.* (eds G.R. Gibson and R.A. Rastall), pp. 1–8. John Wiley and Sons Ltd, Chichester, UK.

Watzl, B., *et al.* (2005) Inulin, oligofructose and immunomodulation. *British Journal of Nutrition* **93** (Suppl 1), S49–S55.

Wisker, E., *et al.* (1998) Fermentation of non-starch polysaccharides in mixed diets and single fibre sources: Comparative studies in human subjects and in vitro. *British Journal of Nutrition* **80**, 253–261.

Yamada, H., *et al.* (1993) Structure and properties of oligosaccharides from wheat bran. *Cereal Foods World* **38**, 490–492.

15
Oats and Skin Health

Joy Makdisi[1]*, Allison Kutner[1]*, and Adam Friedman[1,2]
[1]*Division of Dermatology, Department of Medicine, Montefiore Medical Center, Bronx, New York, USA*
[2]*Department of Physiology and Biophysics, Albert Einstein College of Medicine, Bronx, New York, USA*
*These authors contributed equally to this chapter.

15.1   History of colloidal oatmeal use

The use of whole oats (*Avena sativa*) to treat both internal and external ailments dates back to ancient Egypt and the Arabian peninsula (Dohil, 2011) and has been noted in the medical literature dating back to the Roman times. Oats not only served a nutritional role but their use as a topical cleanser and treatment for a variety of dermatologic conditions was well known (Kurtz and Wallo, 2007).

Colloidal oatmeal has a long history of use for dermatologic conditions including atopic dermatitis and other inflammatory dermatoses (Cerio *et al.*, 2010). For example, in the traditional medicine of Lebanon, a macerate of oat grains was used as an antirheumatic agent (El Beyrouthy *et al.*, 2008). Today, people of the southern Appalachia region in the United States continue to use *A. sativa* oats based on medicinal folklore, believing that topical application alleviates symptoms of chickenpox, poison ivy, and other rashes (Cavender, 2006). *A. sativa*, known as the common oat, was first brought to North America in the early seventeenth century. In the late nineteenth century oats began to be used in products other than food. Literature was published describing the cosmetic benefits of oatmeal-containing bath oils and face masks, and oat flour was marketed as an antioxidant. In addition, oatmeal baths were used to ameliorate a variety of pruritic skin conditions (Miller, 1979). However, oats in this form do not dissolve well and in the 1940s colloidal oatmeal was manufactured from ground dehulled oats. Colloidal oatmeal maintains its moisturizing effects but disperses well in baths and can be incorporated into lotions and creams for topical application. By 1945 colloidal oatmeal was commercially available for individual use and in the 1950s it was combined with emollient oils for use as a lubricant.

Oats Nutrition and Technology, First Edition. Edited by YiFang Chu.

In 1989, the United States Food and Drug Administration (FDA) approved colloidal oatmeal as a safe and effective over-the-counter drug, provided that its concentration and composition were standardized (Kurtz and Wallo, 2007). In 2003, the FDA noted that colloidal oatmeal could relieve irritation and itching due to a number of dermatoses, providing temporary skin protection (FDA, 2003). At this time, the FDA approved colloidal oatmeal as a skin protectant, defining it as "a drug product that temporarily protects injured or exposed skin or mucous membrane surfaces from harmful or annoying stimuli, and may help provide relief to such surfaces" (FDA, 2003). In fact, colloidal oatmeal is one of the few natural products that the FDA recognizes as a safe over-the-counter skin treatment (Cerio *et al.*, 2010). Today colloidal oatmeal can be found in creams, lotions, emollients, shampoo, cleansing bars, body wash, bath treatments, and shaving gels (FDA, 2003; Aries *et al.*, 2005; Kurtz and Wallo 2007).

15.2   Oat structure and composition

15.2.1   Processing

Colloidal oatmeal is produced without the use of chemical solvents by grinding oats and boiling them to extract the colloidal component. After harvesting oat grains with high protein content, they are cleaned and purified to remove imperfect grains and foreign components, such as weed seeds and other grains. Then, they are dehulled to remove the shell surrounding the grain, yielding groats (Schrickel, 1986). Lipase, the most abundant enzyme in oats, is responsible for rancidity in stored oats, which can occur within hours of dehulling (Ekstrand *et al.*, 1993). This enzyme system must therefore be inactivated by steam (Lehtinen *et al.*, 2003). The oats can then be rolled and pulverized to a fine powder (Cerio *et al.*, 2010). The particles in colloidal oatmeal are smaller than 75 microns, which results in better dispersion in solution and more effective topical application (Dohil, 2011). The small particles are able to deposit on the skin to form an occlusive barrier. Colloidal oatmeal must be stored in a temperature- and humidity-controlled environment.

15.2.2   Composition

Colloidal oatmeal consists of sugars and amino acids (65%), proteins (15–20%), lipids (11%), and fiber (5%). There are small amounts of vitamins, saponins, flavonoids, prostaglandin inhibitors, ash, and a very small amount of avenanthramides (0.06%) (Brown and Dattner, 1998; Kurtz and Wallo, 2007; Cerio *et al.*, 2010). The lipid component of oats is larger than that of other cereal grains, with the most abundant lipid being highly unsaturated triglycerides (Åman and Hesselman, 1984). To protect these lipids from oxidation, oats also contain a variety of antioxidant components (Collins, 1986). This antioxidant activity is due primarily to phenolic ester compounds (Collins, 1986; Graf, 1992), the amount of which depends on the stage of plant growth. These compounds include all major classes of phenolic esters: benzoic and cinnamic acids, quinines, flavones, flavonols, chalcones, flavanones, anthocyanidines, and aminophenolics

(Collins, 1986). The most important antioxidant phenols in oats are the glyceryl esters of hydroxycinnamic, ferulic, p-coumaric, and caffeic acids (Emmons and Peterson, 1999). Oats also contain flavonoids, phenolics that strongly absorb ultraviolet A radiation (UVA, 320–370 nm) (Collins, 1986). Other phenolic esters isolated from oats include avenacins, which structurally belong to the saponins. Avenacins contain a large lipophilic region and a short sugar chain and thus are able to interact with both lipid and nonlipid components, imparting soap-like properties. Finally, oats contain a wide range of minerals and vitamins (Lockhart and Hurt, 1986), of which vitamin E (α-tocopherol) is the most important and clinically relevant (Collins, 1986; Emmons and Peterson, 1999).

Avenanthramides are soluble phenolic metabolites that are unique to oats (Meydani, 2009; Wise 2011). Of the 25 structural varieties of avenanthramides, three forms are most abundant (Peterson et al., 2002; Sur et al., 2008). Avenanthramides are substituted N-cinnamoylanthranilic acids (i.e., alkaloids) specific to A. sativa (Sur et al., 2008). Evidence for the de novo production of avenanthramides in oats includes the discovery of hydroxycinnamoyl-CoA:hydroxyanthranilate N-hydroxycinnamoyltransferase, which catalyzes the final step in avenanthramide biosynthesis, in oat leaves and seeds (Ishihara et al., 1998; Matsukawa et al., 2000). Furthermore, introducing labeled L-phenylalanine and anthranilic acid into oat leaves in the presence of an elicitor resulted in de novo incorporation into avenanthramides (Ishihara et al., 1999).

Avenanthramides are present at 300 parts per million (0.03%). Therefore, they are considered minor constituents of oats based on chemical composition. Yet, this bioactive element is the primary source of the grain's antioxidant and anti-inflammatory properties (Kurtz and Wallo, 2007). They exert their effects by inactivating nuclear factor-kappa B (NF-κB) and inhibiting inflammatory cytokines (e.g., such as tumor necrosis factor α) in keratinocytes (Pazyar et al., 2012). They also inhibit histamine release, prostaglandin biosynthesis (Sur et al., 2008), and cleavage of arachidonic acid from phospholipids in keratinocytes (Aries et al., 2005).

15.3 Clinical properties

Given that its polymorphic structure includes carbohydrates, lipids, and other moieties, colloidal oatmeal displays many different functions and beneficial dermatologic properties. Oats produce well-defined cardioprotective effects, including inhibition of atherosclerosis and hypertension (Meydani, 2009; Wise, 2011), and can function as a cleanser, moisturizer, and skin protectant. Oats also demonstrate anti-inflammatory, antipruritic, antioxidant, antifungal, and buffering properties when applied topically (Dimberg et al., 1993; Peterson et al., 2002; Meydani, 2009) (Figure 15.1).

15.3.1 Anti-inflammatory properties

The presumed anti-inflammatory properties of oats have led to the centuries-long use of oats to prevent and treat skin irritations. Current research has now provided evidence for the anti-inflammatory effects of topically applied oatmeal

```
                          ┌─────────────────┐
                          │  Oat Properties │
                          └─────────────────┘
        ┌──────────────┬──────────┴──────────┬──────────────┐
┌────────────────┐ ┌──────────────┐ ┌──────────────┐ ┌──────────────┐
│ Anti-inflammatory│ │ Antipruritic │ │  Antioxidant │ │  Antifungal  │
└────────────────┘ └──────────────┘ └──────────────┘ └──────────────┘
```

Anti-inflammatory	Antipruritic	Antioxidant	Antifungal
• Inhibition of phospholipase A2 in keratinocytes decreases arachidonic acid release, which in turn decreases proinflammatory eicosanoid formation • Inhibition of NF-κB decreases IL-8 release	• Inhibition of neurogenic inflammation decreases neuropeptide release by C fibers, thereby decreasing pruritus	• Phenols act as H atom donors to inhibit radical chain reactions • Metal ion chelators • Lipoxygenase inhibition decreases fatty acid oxidation	• *Pc-2* gene confers resistance to *Puccinia coronate* (crown rust fungus)

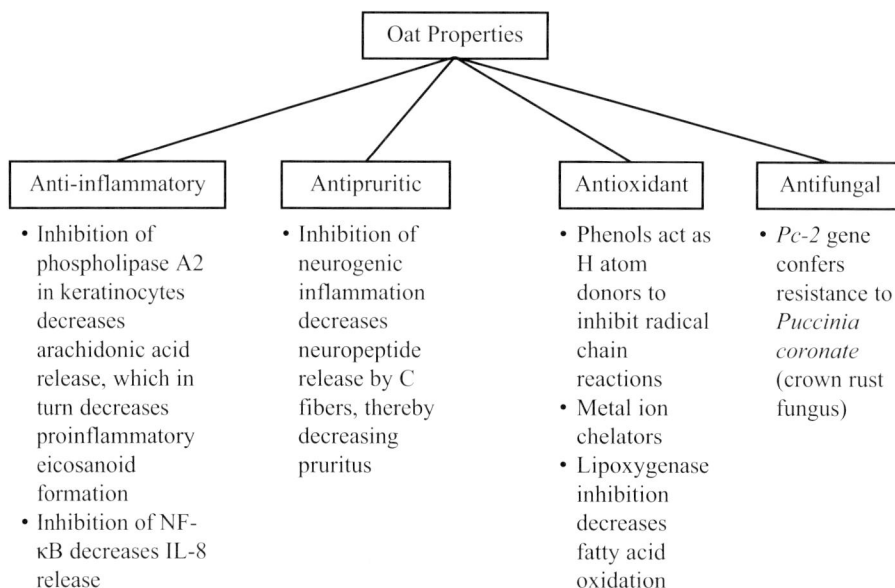

Figure 15.1 Clinical properties of oats.

extracts (Vie *et al.*, 2002). Arachidonic acid and its metabolites (i.e., eicosanoids) are involved in physiologic functions in the epidermis but also play a role in pathologic skin states (Aries *et al.*, 2005). Inflammatory skin disorders, including acute guttate psoriasis, chronic plaque psoriasis, and atopic dermatitis (Ruzicka *et al.*, 1986; Fogh *et al.*, 1989; Fogh and Kragballe, 2000), exhibit high cutaneous levels of arachidonic acid, eicosanoids, and phospholipase A2, the enzyme that mobilizes arachidonic acid (Aries *et al.*, 2005). In contrast, the skin of healthy individuals and areas of uninvolved skin in patients with dermatoses contain very low or even undetectable levels of arachidonic acid and the eicosanoid leukotriene B4 (LTB4). Relatively high levels of LTB4 are seen in areas of skin inflammation, such as the blister fluid from atopic dermatitis and psoriatic plaque lesions (Ruzicka *et al.*, 1986). LTB4 promotes inflammation (Fogh and Kragballe, 2000) as a potent chemotactic factor that stimulates neutrophil degranulation and binding to the endothelium and induces keratinocyte proliferation (Fogh and Kragballe, 2000; Aries *et al.*, 2005). It may also play a role in postinflammatory pigment alteration, because it has been shown to induce melanogenesis in a dose-dependent manner (Morelli *et al.*, 1992; Fogh and Kragballe, 2000). When applied to normal human skin under an occlusive dressing, LTB4 induces a dose-dependent wheal and flare reaction (Fogh and Kragballe, 2000).

Another proinflammatory eicosanoid, 12-hydroxy-eicosatetraenoic acid, which is predominantly synthesized in the skin (Woollard, 1986; Fogh and Kragballe 2000), is found in higher concentrations in psoriatic lesions than in controls (Woollard, 1986). It is similar to LTB4 in that it is a neutrophil chemoattractant and induces DNA synthesis in keratinocytes (Fogh and Kragballe, 2000). Prostaglandin E2 is another eicosanoid found in the skin; it exhibits both pro- and anti-inflammatory properties. Intradermal injection of prostaglandin E2 into healthy skin induces erythema and edema (Crunkhorn and Willis, 1971; Flower et al., 1976) and increases vascular permeability (Crunkhorn and Willis, 1971). Many other eicosanoids in the skin may contribute to inflammatory cutaneous conditions. It is likely that inhibition of these signaling molecules would ameliorate the inflammation seen in dermatoses.

A species of *A. sativa* dose-dependently inhibits phospholipase A2 in keratinocytes, thereby decreasing arachidonic acid release from phospholipids and subsequent eicosanoid formation (Saeed et al., 1981; Aries et al., 2005). This *A. sativa* extract exerts anti-inflammatory actions *in vitro* by inhibiting the arachidonic acid cascade (Vie et al., 2002). Avenanthramides also inhibit NF-κB signaling in keratinocytes, thereby blocking production of the proinflammatory cytokine interleukin (IL)-8 (Sur et al., 2008). This finding is consistent with research demonstrating that synthetic avenanthramides reduce the production of proinflammatory cytokines in endothelial cells. This effect is also mediated by the inhibition of NF-κB, demonstrating that synthetic avenanthramides exert effects similar to those of avenanthramides isolated from oats (Guo et al., 2008). Taken together, these findings demonstrate that oats can inhibit progression of the proinflammatory arachidonic acid cascade and may alleviate the inflammation in various dermatoses.

Topical application of avenanthramides has also been shown to reduce neurogenic inflammation. Neurogenic inflammation is a process by which unmyelinated C fibers release neuropeptides, which subsequently act on peripheral blood vessels and immune cells, resulting in erythema, warmth, edema, and hypersensitivity (Richardson and Vasko, 2002). Neurogenic dermatitis can be induced by topical application of resiniferatoxin (Liebel et al., 2006), a naturally occurring ultra-potent analog of capsaicin that activates sensory neurons. The degree of edema in neurogenic inflammation and response to treatment can be quantified. Topical application of avenanthramides was found to inhibit resiniferatoxin-induced ear edema (Liebel et al., 2006), providing additional evidence for the anti-inflammatory properties of avenanthramides.

The anti-inflammatory effects of oat components were further investigated in a murine model of contact hypersensitivity (Sur et al., 2008). In this study, mice were initially sensitized with oxazolone, a chemical allergen frequently used in hypersensitivity experiments (Nakano, 2004). Five days later, the mice were challenged with oxazolone applied to the left ear. One hour after the challenge, avenanthramides were applied to the left ear, and inhibition of contact hypersensitivity was evaluated by comparing ear weight (i.e., degree of edema) of avenanthramide-treated mice compared with controls (Sur et al., 2008). A 43–67% dose-dependent reduction in edema ($P < 0.05$) in the avenanthramide-treated mice demonstrated the anti-inflammatory effect of these oat components (Sur et al., 2008). Given that avenanthramides inhibit proinflammatory cytokine release, the authors concluded that this mechanism also mediates the anti-inflammatory effects seen in this murine model of contact hypersensitivity.

Research has also examined how the anti-inflammatory action of avenanthramides modulates cellular and molecular processes that underlie arterial inflammation and resultant atherosclerosis (Meydani, 2009). *In vitro* experiments have revealed that, in addition to reducing monocyte adhesion to human aortic endothelial cells and expression of cellular adhesion molecules, avenanthramides also decrease the secretion of proinflammatory cytokines IL-6 and IL-8 (Liu *et al.*, 2004). IL-6 is a potent promoter of the T-helper cell type 2 immune response, which is upregulated in eczematous dermatoses. IL-8 is a key chemotactic agent for neutrophils and plays an important role in the pathogenesis of inflammatory dermatoses such as psoriasis when overexpressed. This cardiovascular research provides valuable insight into the benefits of avenanthramide use in skin disease.

15.3.2   Antipruritic properties

Pruritus (itching) is described as an uncomfortable feeling that stimulates a desire to scratch; therefore, the amount of scratching can be used as a measure of the degree of itching (Bernhard, 1994; Sur *et al.*, 2008). Pruritus is a feature of many dermatoses and ranges in severity from mild and acute to intractable. A vicious itch–scratch cycle can occur, potentially disrupting the skin barrier and increasing susceptibility to superimposed infections (Schmelz, 2003).

As previously described, topically applied avenanthramides can inhibit neurogenic inflammation. Pruritus is mediated by the same neuropeptides released by C fibers in neurogenic inflammation, triggering nociceptive responses (e.g., pain, itching, stinging). This suggests that avenanthramides may also have anti-itch activity (Sur *et al.*, 2008). One study examined the antipruritic effects of avenanthramides using compound 48/80, a substance synthesized by a condensation reaction of formaldehyde and N-methyl-p-methoxyphenethylamine. Compound 48/80 induces mast cell degranulation and histamine release; intradermal injection in mice therefore generates an itch-associated response (Koibuchi *et al.*, 1985). In this murine itch model, avenanthramide-treated animals scratched 40.7% less than untreated controls ($P < 0.05$) (Sur *et al.*, 2008). Because specific inflammatory cytokines, particularly IL-8, are elevated in various pruritic skin diseases (Lippert *et al.*, 1998), Sur and colleagues evaluated the effect of avenanthramides on IL-8 and observed that avenanthramides suppress IL-8 release (Sur *et al.*, 2008).

Given that avenanthramides successfully decrease pruritus in murine models (Sur *et al.*, 2008) and inhibit the release of inflammatory cytokines that contribute to pruritic skin diseases (Lippert *et al.*, 1998; Sur *et al.*, 2008), it was hypothesized that colloidal oatmeal might be able to mitigate pruritus in various skin conditions. In a study of burn wound patients, a 5% colloidal oatmeal shower oil significantly decreased itch compared with paraffin oil alone (Matheson *et al.*, 2001; Kurtz and Wallo, 2007). In another clinical study, 139 patients with a variety of pruritic dermatoses were treated with a colloidal oatmeal bath and cleanser for 3 months. More than 71% of the patients achieved complete or near-complete relief of pruritus during the study period (Grais, 1953). Another clinical study evaluated the efficacy of colloidal oatmeal baths impregnated with emollient oils as adjuvant therapy for a variety of pruritic dermatoses, including atopic dermatitis, contact dermatitis, ichthyosis, infantile eczema, insect bites, lichen planus, pityriasis rosea, seborrheic dermatitis, psoriasis, and urticaria. In all infants and children studied, rapid symptomatic relief was achieved with the daily use of oatmeal baths, whereas patients who did not use daily oatmeal baths as part of their treatment regimen complained of recurrent pruritus and required longer treatment (O'Brasky, 1959).

15.3.3 Antioxidant properties

In addition to anti-inflammatory and antipruritic properties, colloidal oatmeal possesses antioxidant properties. Antioxidants play a vital role in preventing and ameliorating chronic diseases by decreasing intracellular oxidative damage induced by reactive oxygen species (Hollman, 2001; Peterson *et al.*, 2002). Because free radical damage is present in inflammatory reactions, it follows that oat antioxidants may protect against inflammation (Chawla *et al.*, 1987; Saija *et al.*, 1999).

Phenols exert antioxidant activity through several mechanisms. Some are hydrogen atom donors that inhibit the cascade of radical chain reactions (Bratt *et al.*, 2003), whereas others function as metal ion chelators (Bratt *et al.*, 2003). Some phenols have dual mechanisms of action with synergistic effects. Because phenolic avenanthramides exhibit antioxidant activity *in vitro*, their potential antioxidant activity in oats was investigated (Dimberg *et al.*, 1993; Bratt *et al.*, 2003). Dimberg and colleagues extracted and fractionated phenolic compounds from a variety of oat samples using ethanol at room temperature (Dimberg *et al.*, 1993). The antioxidant activity of two avenanthramides was determined by measuring oxygen consumption in a linoleic acid system (Dimberg *et al.*, 1993). The results revealed the considerable antioxidative capacity of oats that is derived from the avenanthramides. In another study by Bratt and colleagues, eight avenanthramides identified in oat extracts were synthesized and assessed for antioxidant activity by determining reactivity toward 1,1-diphenyl-2-picrylhydrazyl and linoleic acid (Bratt *et al.*, 2003). Both of these methods measure the efficiency of hydrogen atom transfer from a phenol to a radical (Bratt *et al.*, 2003). The avenanthramides in oats demonstrated antioxidant activity and showed higher antioxidant activity than other oat phenolic compounds, possibly due to the resonance structure of an amide bond (Bratt *et al.*, 2003).

Other phenols in oats confer additional functional properties. For example, the flavonoids in oats protect against UVA radiation (Collins, 1986), and tocopherol (vitamin E) protects against both inflammation and photodamage. Vitamin E is a naturally occurring lipid-soluble antioxidant that protects the skin from free radical damage. Various *in vitro* and *in vivo* studies have demonstrated its ability to inhibit UV-induced cutaneous damage and prostaglandin E biosynthesis (Nachbar and Korting, 1995). By modulating prostaglandin E biosynthesis, the lipo-oxygenase function of thrombocytes and neutrophils is altered, resulting in a visible anti-inflammatory effect (Panganamala and Cornwell, 1982; Hong and Chow 1988).

15.3.4 Antifungal properties

Avenanthramides are phytoalexins, which are substances synthesized by plants that protect against pathogens, including fungi. The *Pc-2* gene in *A. sativa* plants confers resistance to *Puccinia coronate*, the crown rust fungus (Mayama *et al.*, 1986). Inoculation with spores of this fungus induces avenanthramide production in oat plants, which inhibits further fungal growth (Miyagawa *et al.*, 1995). In contrast, healthy leaves do not produce these metabolites. Despite these findings, no *in vivo* studies have been carried out in mice or humans to determine the usefulness of oats as antifungal agents. These findings highlight the need for future research in this area.

15.3.5   Hydration and moisture retention

Hydration is important in maintaining the integrity of the skin barrier. An adequate amount of water in the stratum corneum is directly related to the normal structure of the skin, and alterations in the lipid bilayer or a decrease in "natural moisturizing factor" can reduce the water content (Fartasch, 1997; Rawlings and Harding, 2004). Insufficient moisture prevents the natural breakdown of desmosomes and leads to an accumulation of scales, which can become quite severe, resulting in itch and inflammation.

Colloidal oatmeal suspension deposits small particles on the skin to form a viscous occlusive barrier (Grais, 1953). The high concentration of hydrophilic polysaccharides present in colloidal oatmeal is responsible for its viscous nature (Boussault *et al.*, 2007). In particular, the hydrocolloid β-D-glucan displays greater viscosity in solution than other similar biological compounds and confers emollient properties (Wood *et al.*, 1978). The increased viscosity of colloidal oatmeal prolongs its contact with the skin. This barrier prevents the entry of external irritants and retains moisture in the stratum corneum (Grais, 1953). This makes it an effective adjunct treatment for dry skin conditions (Nebus *et al.*, 2004).

15.3.6   Oats and skin pH

The skin is a slightly acidic microenvironment with a pH of approximately 5.5 (Grais, 1953). Acidity enhances the skin's barrier function, protecting against pathogen entry and maintaining the integrity of the superficial keratin layer. When skin is inflamed, the pH increases from acidic to basic according to the degree of skin inflammation. Soap increases the alkalinity of the skin, but perspiration decreases the skin's pH, bringing it back to its acidic baseline. Topical application of oat extract is also able decrease skin pH to normal (Grais, 1953). These results indicate that colloidal oatmeal applied to inflamed skin may act as a buffer to restore the physiologic pH (Grais, 1953).

15.4   Clinical applications of oats

15.4.1   Atopic dermatitis

Atopic dermatitis, also known as atopic eczema, is a common chronic skin condition that occurs often in children. The pathogenesis of this disease is complex but appears to have an important genetic component. Mutations in filaggrin, which encodes a protein that is abundant in the outer layers of the epidermis, are strong genetic determinants in the etiology and phenotypic presentation of atopic dermatitis (Henderson *et al.*, 2008; Irvine *et al.*, 2011). For example, one study reported that 42% of individuals heterozygous for mutations in the filaggrin gene developed atopic dermatitis (Irvine *et al.*, 2011). Filaggrin is an important component of the stratum corneum. When it is deficient, the protective barrier is disrupted, and other epidermal abnormalities develop (Henderson *et al.*, 2008). Both atopic dermatitis lesions and areas of clinically uninvolved skin exhibit increased transepidermal water loss, elevated pH, and diminished skin hydration (Berardesca *et al.*, 1990; Seidenari and Giusti, 1995). Disruption of the skin barrier allows the infiltration of irritants and antigens, which induces the release of inflammatory cytokines, further exacerbating the inflammation characteristic of stratum corneum dehydration (Denda *et al.*, 1998; Elias *et al.*, 2008). Severe pruritus develops as a result of this initial xerosis, triggering a intense itch–scratch cycle. This cascade of events amplifies the immune response, ultimately leading to the development of eczematous lesions, a central feature of atopic dermatitis.

It follows from this pathophysiological sequence of events that restoring the skin barrier, ensuring adequate stratum corneum hydration, and decreasing inflammation are necessary to improve atopic dermatitis symptoms. Moisturizing and enhancing the skin barrier decrease the primary inflammation of xerosis and entry of harmful irritants two-fold, diminishing pruritus and scratching (Eichenfield *et al.*, 2003; Abramovits, 2005). Emollients have been suggested by several consensus organizations as first-line treatment for atopic dermatitis to reduce or avoid steroid use and the associated side effects (Lodén, 2003).

Colloidal oatmeal has been approved by the FDA as an over-the-counter skin protectant. One of its approved uses is to "protect and relieve minor skin irritation and itching due to eczema" (Kurtz and Wallo, 2007). As discussed earlier, colloidal oatmeal contains a variety of dermatologically active compounds with anti-inflammatory and antioxidant properties and also forms an occlusive barrier (Kurtz and Wallo, 2007, Cerio *et al.*, 2010). High levels of β-glucan and starches contribute to the protective seal, which helps the stratum corneum retain water.

Oatmeal extracts appear to confer special benefits to patients with atopic dermatitis (Vie *et al.*, 2002). For this reason, oat-containing emollients and moisturizers are commonly used as adjuvant therapy to topical corticosteroids. Several recent studies demonstrated the efficacy of colloidal oatmeal for this use (Aries *et al.*, 2005). One study enrolled 25 patients with mild to moderate atopic dermatitis (at least 5% body surface area involvement) for an 8-week period (Nebus *et al.*, 2009). Patients using systemic therapies with a potential effect on eczema were excluded; otherwise, patients were able to continue their prescribed regimens. Treatment included twice-daily application of colloidal oatmeal-based cream and once-daily use of oat-based body wash. Several clinical factors were assessed, including body surface area, pruritus, erythema, excoriation, and quality of life (Hanifin *et al.*, 2001). The results demonstrated that daily use of oatmeal-based adjunct therapies in patients with atopic dermatitis improved measured outcomes at all time points (Nebus *et al.*, 2009). Furthermore, the oat regimen was well tolerated and did not interfere with other prescribed topical therapies (Nebus *et al.*, 2009).

Another study evaluated adjuvant treatment of mild to moderate atopic dermatitis in a pediatric population (mean age of 2.4 years) (Fowler *et al.*, 2012). The adjunct therapy was a combination of colloidal oatmeal cream and a colloidal oatmeal cleanser. Patients continued their topical prescription medications but discontinued use of any other cleansers or moisturizing creams. Composite outcomes consisting of affected body surface area, itch and inflammatory symptoms, and quality of life were assessed (Fowler *et al.*, 2012). The oatmeal-based treatment was well tolerated, reduced itching, and progressively improved the subjects' skin condition (Fowler *et al.*, 2012).

Another clinical study assessed oatmeal-based body wash and body cream in patients with atopic dermatitis aged 15–60 years (Nebus *et al.*, 2007; Fowler *et al.*, 2012). The colloidal oatmeal cream also contained glycerin, ceramides, and petrolatum for moisture retention and occlusion. Patients used each product twice daily and were evaluated at the end of 2 weeks (Nebus *et al.*, 2007). 62% of patients showed an improvement in mean eczema severity score ($P < 0.05$) and decreased itching ($P < 0.05$) (Nebus *et al.*, 2007), and the adjunct therapy was well tolerated. Finally, Grimalt and colleagues evaluated infants with atopic dermatitis severe enough to require high-potency topical corticosteroid treatment and found that those who underwent adjuvant therapy with oat-containing emollients required 42% less corticosteroid during a 6-week period and showed a dramatic improvement in clinical symptoms (Grimalt *et al.*, 2007).

Taken together, these studies confirm that colloidal oatmeal is an effective and safe adjunct therapy for atopic dermatitis in both pediatric and adult populations. Daily use significantly improved clinical outcomes including pruritus, affected body surface area, and quality of life. Pruritus was attenuated as early as the first week of treatment, thus reducing scratching and barrier compromise. Further clinical improvement was seen with continued treatment, with more than 60% of children and 30% of adults becoming clear or almost clear by the fourth week (Nebus *et al.*, 2007; Fowler *et al.*, 2012).

15.4.2   Contact dermatitis

Allergic contact dermatitis is a common health concern that presents in two phases. Exposure to an irritant in a susceptible host initially causes sensitization. In the later phase, exposure to the same allergen induces a cutaneous inflammatory reaction. Sometimes the initial exposure does not cause a skin reaction but repeated exposure leads to sensitivity and eventual reaction to the substance. This is considered a form of delayed-type hypersensitivity. Although the pathophysiologic cascade is instigated by the allergen, it is dependent on the activation and clonal expansion of T lymphocytes. Common allergens that can induce contact dermatitis include nickel and other metals, rubber and latex, fragrances, fabrics, antibiotics, adhesives, and poison ivy, poison oak, and poison sumac plants.

Irritant contact dermatitis is the most common type of contact dermatitis. It was defined in the past as a nonspecific, nonimmunological skin reaction to an irritant (Slodownik *et al.*, 2008). A more complex pathophysiology that has an immune component has been proposed that involves multiple endogenous and exogenous factors, such as skin irritants, environmental factors, age, sex, and existing skin disease (Berardesca and Distante, 1994; Marks *et al.*, 2002). In fact, the pathophysiologic changes of irritant damage overlap with those of the immune response. The irritant and modifying factors trigger a cascade of events, resulting in disruption of the skin barrier, injury to epidermal cells, and release of inflammatory cytokines from damaged keratinocytes. Almost any chemical has the potential to trigger this reaction, including detergents, soaps, acids, and solvents.

The mainstay of treatment for irritant contact dermatitis is to avoid exposure to the irritant and to use moisturizing creams. Topical corticosteroids are sometimes employed but recent studies suggest that this may further compromise the skin barrier (Simon *et al.*, 2001). Although oatmeal creams and baths have been used to treat pediatric and adult contact dermatitis, few recent studies have confirmed their efficacy. Much of the current literature relating to the efficacy of oatmeal in the treatment of contact dermatitis cites several studies from the late 1950s. In a much-cited study by Dick (1958), 30 patients with contact dermatitis were treated with colloidal oatmeal baths. The patients benefited from the soothing and cleansing properties of the colloidal oatmeal oil and did not experience additional irritation due to the treatment. In two similar studies by O'Brasky (1959) and Smith (1958), patients with various pruritic dermatoses, including contact dermatitis, underwent treatment with colloidal oatmeal emollient baths, which greatly diminished pruritus and irritation. In contrast, untreated patients had prolonged courses and recurrent itching. A more recent clinical study used the sodium lauryl sulfate irritation model to evaluate the ability of oatmeal extracts to prevent skin irritation (Vie *et al.*, 2002). In 12 healthy individuals, pretreatment with oatmeal extracts reduced inflammation, protected barrier function, and prevented the increased blood flow that is commonly seen with sodium lauryl sulfate-induced skin irritation (Vie *et al.*, 2002). None of these studies were randomized controlled trials that specifically evaluated contact dermatitis; no such studies exist. However, these studies and a study by Franks (1958) demonstrate the utility of colloidal oatmeal as an adjunct therapy for treatment of contact dermatitis.

15.4.3 Wound healing

Wound healing is a complex process with many phases, including inflammation, granulation tissue formation, re-epithelialization, and matrix formation. Many studies have shown that oxidation reactions, often induced by free radicals, are involved in the pathophysiology of various inflammatory conditions. More recently, free radicals and oxidative stress have been implicated in impaired wound healing. *A. sativa* extracts contain a wide variety of flavonoid-type compounds thought to be involved in the antioxidant activities of oats (Chopin, 1977). As previously discussed, vitamin E, phytic acid, and avenanthramides also confer strong antioxidant properties to oats (Peterson, 2001). The relationship between oxidative damage and wound healing suggests that oats may also promote wound healing.

Murine incisional and excisional wound models were used to evaluate the wound healing activity of oat extracts (Akkol *et al.*, 2011). Topical application of the oat extracts on incisional wounds significantly increased wound tensile strength. In addition, excisional wound contraction was significantly higher in oat-treated mice than in controls, which showed no significant wound contraction. Three phases of wound healing (inflammation, proliferation, and remodeling) were evaluated. The oat extract-treated groups showed a significant increase in remodeling and epithelialization. Histopathological results confirmed improved wound contracture and healing time after topical treatment with oat extracts (Akkol *et al.*, 2011).

Several other studies have confirmed the beneficial effect of *A. sativa* on wound healing. One study showed that β-glucan from *A. sativa* increased resistance to infection by *Staphylococcus aureus* and *Eimeria vermiformis* in an *in vivo* mouse model. Antimicrobial agents such as β-glucan enhance the skin barrier against microbial invasion and thus promote rapid healing (Yun *et al.*, 2003). Furthermore, vasodilatation and invasion of inflammatory mediators during wound healing can result in erythema and edema. When human skin fragments stimulated with a neuromediator were treated with oatmeal extracts, the mean surface of dilated vessels and edema were significantly decreased; this mechanism may underlie the wound healing effect of oats (Boisnic *et al.*, 2003). While providing further evidence for the wound healing effects of *A. sativa*, these studies have elucidated an additional mechanism for this activity. In addition to the well-known antioxidant activity of oats, oat extracts containing β-glucans promote wound healing by stimulating collagen deposition, tissue granulation, and re-epithelialization; activating immune cells and increasing macrophage infiltration into the wound area; and increasing wound tensile strength (Cerci *et al.*, 2008).

15.4.4   Sunburn

Sunburn is the acute reaction of skin to excessive UV radiation exposure and can range from mild, painless erythema to vesicle and bullae formation. UV-mediated erythema develops within 3–5 hours after exposure, peaks at 12–24 hours, and fades over the subsequent 72 hours. The pathogenesis of sunburn is likely secondary to vasodilation of blood vessels in the upper dermis, which leads to edema. Inflammatory mediators such as cytokines, prostaglandins, free oxygen radicals, mast cell mediators, and histamine also appear to participate in UV-induced erythema. Although protective clothing, sun avoidance, and UVA/UVB sunblocks can protect the skin, sunburns are common, particularly in fair-skinned individuals. In fact, up to 47% of the United States Caucasian population is unlikely to use these methods to protect against sunburn (Hall *et al.*, 1997). Numerous studies have evaluated the use of nonsteroidal anti-inflammatory drugs, corticosteroids, antioxidants, antihistamines, and emollients to treat sunburn and reported no improvement or only mild amelioration of symptoms with these treatments. The most effective and practical approach to managing acute sunburn may be to focus on the UV light-induced symptoms of erythema, pain, and pruritus (Han and Maibach, 2004). Importantly, colloidal oatmeal has been found to relieve the itching and inflammation of sunburn (Saeed *et al.*, 1981; Barr, 2002; Aburjai and Natsheh 2003).

15.4.5 Drug eruptions

Colloidal oatmeal can also reduce the dermatologic side effects associated with medications such as epidermal growth factor receptor (EGFR) inhibitors and tyrosine kinase inhibitors, which are used to treat EGFR-positive tumors. Despite their therapeutic efficacy, 60–70% of patients taking these drugs experience papulopustular rashes. This rash typically occurs within the first 2 weeks of therapy and can be severe enough to cause patients to discontinue treatment (Alexandrescu *et al.*, 2007; Talsania *et al.*, 2008). Therefore, amelioration of this common side effect is crucial to prevent treatment interruption. In a study of patients with an EGFR inhibitor or tyrosine kinase inhibitor-induced rash, 11 patients were treated with a colloidal oatmeal lotion three times a day for 7 days. The overall response rate was 100%, with a complete response rate of 60% (Alexandrescu *et al.*, 2007). A recent case report describes papulopustular eruption and intractable pruritus due to erlotinib therapy that were not relieved by a variety of topical and oral therapies; colloidal oatmeal cream was the only successful treatment for this side effect (Talsania *et al.*, 2008). Unlike more potent topical treatments, such as steroids (Talsania *et al.*, 2008), metronidazole, erythromycin, salicylic acid, and benzoyl peroxide (Alexandrescu *et al.*, 2007), colloidal oatmeal is not associated with any toxicity (Alexandrescu *et al.*, 2007; Cerio *et al.*, 2010). However, it is not known whether the beneficial effects of oat-containing lotion are due to its anti-inflammatory properties or another mechanism.

15.5 Side effects of oats

15.5.1 Oat-induced contact dermatitis and contact urticaria

Protein contact with the skin can cause protein contact dermatitis (PCD) and contact urticaria (Amaro and Goossens, 2008). PCD frequently presents as chronic or recurrent eczema and contact urticaria is often immune-mediated (i.e., immune-mediated contact urticaria [ICU]). The clinical manifestations of ICU are associated with antigen dose and route of exposure; therefore, the reaction can be limited to the contact site(s) or can be more generalized (diffuse urticaria). ICU can also manifest with respiratory and/or gastrointestinal tract involvement or symptoms of allergic rhinoconjunctivitis, asthma, and oral allergy syndrome (Amaro and Goossens, 2008). The pathogenetic features of PCD and ICU are similar: both are type 1 hypersensitivity reactions mediated by IgE antibodies to specific proteins in sensitized individuals (Amaro and Goossens, 2008). PCD can also demonstrate type 4 delayed-hypersensitivity reactions, as assessed by patch tests (Amaro and Goossens, 2008). The gold standard for diagnosing PCD and ICU are skin prick tests, which reveal IgE-mediated allergy (immediate hypersensitivity). Open testing may also aid in diagnosis but results are often negative unless the offending substance is applied to eczematous or otherwise damaged skin. Serum IgE levels may also be useful in diagnosing PCD and ICU (Amaro and Goossens, 2008).

Even though oats exert anti-inflammatory effects and alleviate symptoms of damaged skin, particularly atopic dermatitis (Aries *et al.*, 2005), a subset of patients with atopic dermatitis experiences contact allergy with exposure to oat proteins (Boussault *et al.*, 2007; Amaro and Goossens, 2008). In fact, oat proteins are one of the many causes of PCD and ICU (Amaro and Goossens, 2008). The stratum corneum of intact epidermal skin serves as a physical barrier to the penetration of allergenic proteins (Smith Pease *et al.*, 2002; Vansina *et al.*, 2010), decreasing the likelihood of percutaneous sensitization to allergens. However, the epidermis of patients with atopic dermatitis shows reduced stratum corneum barrier function, making it more vulnerable to environmental irritants (Amaro and Goossens, 2008; Goujon *et al.*, 2009) and increasing the risk for cutaneous sensitization to allergens, including oats.

A case report of a 3-year-old girl with atopic dermatitis detailed a flare in her condition after application of an oat-containing moisturizer. Further investigation into this patient's reaction found that she had positive prick test and patch test results for *A. sativa* (Pazzaglia *et al.*, 2000). Similarly, a 7-year-old girl with atopic dermatitis developed allergic contact urticaria to an oat-containing cream. The urticarial lesions developed within minutes of applying the cream and dissipated within an hour. Skin test results for oats were later found to be positive and oat-specific IgE antibodies were detected (De Paz Arranz *et al.*, 2002). Another case report detailed the case of a 33-year-old woman with atopic dermatitis who presented with an erythematous and pruritic cutaneous reaction immediately after applying an oat-containing moisturizer. The reaction dissipated a few hours after application. Results of prick test and an enzyme-linked immunosorbent assay (ELISA) showed a positive reaction to *A. sativa* extract and the patient had elevated serum levels of oat-specific IgE antibodies. However, this patient did not demonstrate immunoreactivity against a new protein-free oat extract (Vansina *et al.*, 2010). Finally, a case series of three children with atopic dermatitis and long history of treatment with an oatmeal-based bath developed exacerbations of eczema after their baths. Their eczema rapidly improved after discontinuing this treatment. The results of radioallergosorbent and patch tests for oats were positive for all three children, and none had consumed oats (Riboldi *et al.*, 1988).

In an effort to determine the frequency of oat sensitization in patients with atopic dermatitis, Boussault and colleagues examined the prevalence of oat sensitization in 302 children with atopic dermatitis who were referred for allergy testing (Boussault *et al.*, 2007). Results of atopy patch test and skin prick test revealed oat sensitization in 32.5% of the children. Almost 75% with cutaneous allergy to oats had previously used oat-containing creams, and upon application, 16% experienced cutaneous symptoms including eczema flares, pruritus, and diffuse erythema. The study concluded that oat sensitization in children with atopic dermatitis is higher than expected and may be due to the impaired epidermal barrier. The authors of this study suggested that topical products containing oat proteins should not be used to treat infants with atopic dermatitis (Boussault *et al.*, 2007).

An ongoing debate exists about the safety of oat-containing topical treatments in patients with atopic dermatitis (Vansina *et al.*, 2010). Although some authors argue that infants with atopic dermatitis should not be treated with oat-containing skin products (Boussault *et al.*, 2007), others believe that the evidence of sensitization to oats after such use is insufficient (Pigatto *et al.*, 1997; Goujon-Henry *et al.*, 2008; Goujon *et al.*, 2009; Vansina *et al.*, 2010). For example, a 2009 study reported that daily use of an oat-based cleansing bar and moisturizing cream did not induce cutaneous hypersensitivity reactions in cereal-sensitized adults with atopic dermatitis. The authors concluded that although these oat-containing cosmetics are not allergenic to oat-sensitized patients, it is still possible that other oat-containing skin products can trigger an allergic response in atopic individuals (Goujon *et al.*, 2009). A randomized controlled trial evaluated the frequency of allergic skin reactions to topical application of oat and rice colloidal grain suspensions in healthy children and those with atopic dermatitis. Neither group developed urticarial or allergic skin reactions; however, 12% of the patients with atopic dermatitis developed irritant reactions to the testing materials and 23% of the patients had positive radioallergosorbent test results. The authors concluded that topical application of colloidal oat products is an appropriate adjunct therapy for children with atopic dermatitis, because there was no evidence of sensitization in this group (Pigatto *et al.*, 1997).

Given the discordance among study results, it remains unclear whether individuals with atopic dermatitis are likely to develop cutaneous sensitization to oat proteins or other allergens. Thus, further investigation is needed to evaluate the safety profile of oat-containing products in this patient population. Additionally, the risk of contact dermatitis to oats in patients without an atopic history must be considered. Although described less frequently in the literature, contact dermatitis has occurred in non-atopic individuals. A case report of a 33-year-old woman with no documented atopic history describes an eczematous rash that erupted when she began working at a garden shop; the rash subsided on weekends and holidays. Her responsibilities included weighing bran, oats, bird seed, and grass seed. Lesions were most prominent on the flexor aspect of both forearms but were also present on her face, neck, ankles, waist, and the dorsal aspect of both hands. Patch test results revealed contact allergy to both oats and bran, and her rash subsided once she avoided contact with these irritants (Dempster, 1981).

15.5.2 Percutaneous sensitization and allergy

Cutaneous sensitization to oat proteins in patients with atopic dermatitis is an established phenomenon. In their study of patients with atopic dermatitis and oat sensitization, Boussault and coworkers found that 15.6% of children with positive atopy patch test or prick test results to oats also had an oral oat allergy, which was previously unknown to parents (Boussault *et al.*, 2007). The sensitivity to oats manifested as a variety of symptoms, including abdominal pain, vomiting, diarrhea, cough, asthma attack, facial erythema, pruritus, and aggravation of atopic dermatitis lesions (Boussault *et al.*, 2007). Additionally, in a case report a 33-year-old woman with atopic dermatitis and cutaneous allergy to oats, as demonstrated by prick test, began to experience an urticarial reaction on her lips and trunk after consuming oat-containing food products. Upon further investigation, she was found to have elevated serum levels of oat-specific IgE antibodies and positive ELISA test results for *Avena sativa* oat extract (Vansina *et al.*, 2010).

The findings suggest that sensitization to food allergens can occur via the cutaneous route as well as the oral route. This phenomenon has been previously described for other antigens. Hsieh and colleagues demonstrated for the first time that food allergy can be induced by cutaneous allergen exposure with the cutaneous application of ovalbumin to BALB/c mice, which induced a high level of ovalbumin-specific IgE; a subsequent oral challenge caused anaphylaxis, increased plasma histamine levels, and histologic changes in the lungs and intestines (Hsieh *et al.*, 2003). This effect has also been observed with peanuts. One study found that 84% of preschool-age children with peanut allergy had been treated with creams containing peanut oil during the first 6 months of life to treat diaper rash, eczema, dry skin, and other cutaneous inflammatory conditions. These results suggest that exposure to peanut antigens through inflamed skin results in allergic sensitization (Lack *et al.*, 2003).

To better understand the effect of oat-containing skin products in children with atopic dermatitis, a randomized controlled trial should be undertaken, comparing the rates of oat sensitization in these children after treatment with oat-containing or oat-free emollients (Goujon-Henry *et al.*, 2008). Similar to ovalbumin and peanuts, oats may be able to induce allergy through the percutaneous route (Boussault *et al.*, 2007).

15.6　Conclusions

Oats have been used for centuries to treat a variety of dermatoses and current research continues to demonstrate this grain's utility in treating a broad range of skin diseases. Advances in research and medical technology provide state-of-the-art, targeted therapies that are effective but often associated with unwanted side effects. It is important to recognize the multifaceted benefits of this naturally abundant plant, including anti-inflammatory, antipruritic, and antioxidant effects, as well as its lack of toxicity. The adjunctive use of topical oat-containing products for a variety of cutaneous diseases may reduce the need for prolonged or repeated use of other therapies with undesirable side effects. Although oat-containing products are often employed on a daily basis as cosmeceuticals, more research is warranted to determine their efficacy in the treatment of skin disorders. Large-scale studies are needed to better understand the benefits and potential side effects of oat-containing products used to treat specific dermatologic conditions.

References

Abramovits, W. (2005). A clinician's paradigm in the treatment of atopic dermatitis. *Journal of the American Academy of Dermatology* **53**, S70–S77.

Aburjai, T. and Natsheh, F.M. (2003) Plants used in cosmetics. *Phytotherapy Research* **17**, 987–1000.

Akkol, E.K., *et al.* (2011) Assessment of dermal wound healing and in vitro antioxidant properties of *Avena sativa* L. *Journal of Cereal Science* **53**, 285–290.

Alexandrescu, D., *et al.* (2007) Effect of treatment with a colloidal oatmeal lotion on the acneform eruption induced by epidermal growth factor receptor and multiple tyrosine-kinase inhibitors. *Clinical and Experimental Dermatology* **32**, 71–74.

Åman, P. and Hesselman, K. (1984) Analysis of starch and other main constituents of cereal grains. *Swedish Journal of Agricultural Research* **14**, 135–139.

Amaro, C. and Goossens, A. (2008) Immunological occupational contact urticaria and contact dermatitis from proteins: a review. *Contact Dermatitis* **58**, 67–75.

Aries, M., *et al.* (2005) Avena Rhealba inhibits A23187-stimulated arachidonic acid mobilization, eicosanoid release, and cPLA2 expression in human keratinocytes: potential in cutaneous inflammatory disorders. *Biological and Pharmaceutical Bulletin* **28**, 601–606.

Barr, T.L. (2002) Oat protein complex and method of use. US Patent 6416788.

Berardesca, E. and Distante, F. (1994) The modulation of skin irritation. *Contact Dermatitis* **31**, 281–287.

Berardesca, E., *et al.* (1990) *In vivo* hydration and water-retention capacity of stratum corneum in clinically uninvolved skin in atopic and psoriatic patients. *Acta Dermato-Venereologica*, **70**, 400.

Bernhard, J.D. (1994) *Itch: Mechanisms and Management of Pruritus*. McGraw-Hill, New York.

Boisnic, S., *et al.* (2003) Inhibitory effect of oatmeal extract oligomer on vasoactive intestinal peptide-induced inflammation in surviving human skin. *International Journal of Tissue Reactions* **25**, 41–46.

Boussault, P., *et al.* (2007) Oat sensitization in children with atopic dermatitis: prevalence, risks and associated factors. *Allergy* **62**, 1251–1256.

Bratt, K., *et al.* (2003) Avenanthramides in oats (*Avena sativa* L.) and structure-antioxidant activity relationships. *Journal of Agricultural and Food Chemistry* **51**, 594–600.

Brown, D. J. and Dattner, A. M. (1998) Phytotherapeutic approaches to common dermatologic conditions. *Archives of Dermatology* **134**, 1401–1404.

Cavender, A. (2006) Folk medical uses of plant foods in southern Appalachia, United States. *Journal of Ethnopharmacology* **108**, 74–84.

Cerci, C., *et al.* (2008) The effects of topical and systemic beta glucan administration on wound healing impaired by corticosteroids. *Wounds* **20**, 341–346.

Cerio, R., *et al.* (2010) Mechanism of action and clinical benefits of colloidal oatmeal for dermatologic practice. *Journal of Drugs in Dermatology* **9**, 1116–1120

Chawla, A., *et al.* (1987) Anti-inflammatory action of ferulic acid and its esters in carrageenan induced rat paw oedema model. *Indian Journal of Experimental Biology* **25**, 187.

Chopin, J. (1977) C-Glycosylflavones from *Avena sativa*. *Phytochemistry (Oxford)* **16**, 2041–2043.

Collins, F.W. (1986) Oat phenolics: structure, occurrence, and function. In: *Oats: Chemistry and Technology* (ed. F.W. Webster), pp. 227–95. Americal Association of Cereal Chemists, St. Paul, MN.

Crunkhorn, P. and Willis, A.L. (1971) Cutaneous reactions to intradermal prostaglandins. *British Journal of Pharmacology* **41**, 49–56.

De Paz Arranz, S., *et al.* (2002) Allergic contact urticaria to oatmeal. *Allergy* **57**, 1215.

Dempster, J.G. (1981) Contact dermatitis from bran and oats. *Contact Dermatitis* **7**, 122.

Denda, M., *et al.* (1998) Low humidity stimulates epidermal DNA synthesis and amplifies the hyperproliferative response to barrier disruption: implication for seasonal exacerbations of inflammatory dermatoses. *Journal of Investigative Dermatology* **111**, 873–878.

Dick, L.A. (1958) Colliodal emollient baths in pediatric dermatoses. *Archives of Pediatrics* **75**, 506–508.

Dimberg, L., *et al.* (1993) Avenanthramides – a group of phenolic antioxidants in oats. *Cereal Chemistry* **70**, 637–641.

Dohil, M. (2011) Atopic dermatitis and other inflammatory skin disease – Natural ingredients for skin care and treatment. *Journal of Drugs in Dermatology* **10**, S6–S9.

Eichenfield, L.F., *et al.* (2003) Consensus conference on pediatric atopic dermatitis. *Journal of the American Academy of Dermatology* **49**, 1088–1095.

Ekstrand, B., *et al.* (1993) Lipase activity and development of rancidity in oats and oat products related to heat treatment during processing. *Journal of Cereal Science* **17**, 247–254.

El Beyrouthy, M., *et al.* (2008) Plants used as remedies antirheumatic and antineuralgic in the traditional medicine of Lebanon. *Journal of Ethnopharmacology* **120**, 315–334.

Elias, P.M., *et al.* (2008) Basis for the barrier abnormality in atopic dermatitis: outside-inside-outside pathogenic mechanisms. *Journal of Allergy and Clinical Immunology* **121**, 1337–1343.

Emmons, C.L. and Peterson, D.M. (1999) Antioxidant activity and phenolic contents of oat groats and hulls. *Cereal Chemistry* **76**, 902–906.

Fartasch, M. (1997) Epidermal barrier in disorders of the skin. *Microscopy Research and Technique* **38**, 361–372.

FDA (Food and Drug Administration) (2003) Skin protectant drug products for over-the-counter human use; final monograph. *Federal Register* **68**, 33362–33381.

Flower, R.J., *et al.* (1976) Inflammatory effects of prostaglandin D2 in rat and human skin. *British Journal of Pharmacology* **56**, 229–233.

Fogh, K. and Kragballe, K. (2000) Eicosanoids in inflammatory skin diseases. *Prostaglandins and Other Lipid Mediators* **63**, 43–54.

Fogh, K., *et al.* (1989) Eicosanoids in acute and chronic psoriatic lesions: leukotriene B4, but not 12-hydroxy-eicosatetraenoic acid, is present in biologically active amounts in acute guttate lesions. *Journal of Investigative Dermatology* **92**, 837–841.

Fowler, J., *et al.* (2012) Colloidal oatmeal formulations as adjunct treatments in atopic dermatitis. *Journal of Drugs in Dermatology: JDD* **11**, 804–807.

Franks, A. (1958) Dermatologic uses of baths. *American Practitioner and Digest of Treatment* **9**, 1998.

Goujon, C., *et al.* (2009) Tolerance of oat-based topical products in cereal-sensitized adults with atopic dermatitis. *Dermatology* **218**, 327–333.

Goujon-Henry, C., *et al.* (2008) Do we have to recommend not using oat-containing emollients in children with atopic dermatitis? *Allergy* **63**, 781–782.

Graf, E. (1992) Antioxidant potential of ferulic acid. *Free Radical Biology and Medicine* **13**, 435–448.

Grais, M. (1953) Role of colloidal oatmeal in dermatologic treatment of the aged. *AMA Archives of Dermatology and Syphilology* **68**, 402–407.

Grimalt, R., *et al.* (2007) The steroid-sparing effect of an emollient therapy in infants with atopic dermatitis: a randomized controlled study. *Dermatology* **214**, 61–217.

Guo, W., *et al.* (2008) Avenanthramides, polyphenols from oats, inhibit IL-1 [beta]-induced NF-[kappa] B activation in endothelial cells. *Free Radical Biology and Medicine* **44**, 415–429.

Hall, H.I., *et al.* (1997) Sun protection behaviors of the US white population. *Preventive Medicine* **26**, 401–407.

Han, A. and Maibach, H.I. (2004) Management of acute sunburn. *American Journal of Clinical Dermatology* **5**, 39-47.

Hanifin, J., *et al.* (2001) The eczema area and severity index (EASI): assessment of reliability in atopic dermatitis. *Experimental Dermatology* **10**, 11–18.

Henderson, J., *et al.* (2008) The burden of disease associated with filaggrin mutations: a population-based, longitudinal birth cohort study. *Journal of Allergy and Clinical Immunology* **121**, 872–877. e9.

Hollman, P. (2001) Evidence for health benefits of plant phenols: local or systemic effects? *Journal of the Science of Food and Agriculture* **81**, 842–852.

Hong, C. and Chow, C. (1988) Induction of eosinophilic enteritis and eosinophilia in rats by vitamin E and selenium deficiency. *Experimental and Molecular Pathology* **48**, 182–192.

Hsieh, K., *et al.* (2003) Epicutaneous exposure to protein antigen and food allergy. *Clinical and Experimental Allergy* **33**, 1067–1075.

Irvine, A.D., *et al.* (2011) Filaggrin mutations associated with skin and allergic diseases. *New England Journal of Medicine* **365**, 1315–1327.

Ishihara, A., *et al.* (1998) Induction of hydroxyanthranilate hydroxycinnamoyl transferase activity by oligo-N-acetylchitooligosaccharides in oats. *Phytochemistry* **47**, 969–974.

Ishihara, A., *et al.* (1999) Biosynthesis of oat avenanthramide phytoalexins. *Phytochemistry* **50**, 237–242.

Koibuchi, Y., *et al.* (1985) Histamine release induced from mast cells by active components of compound 48/80. *European Journal of Pharmacology* **115**, 163–170.

Kurtz, E.S. and Wallo, W. (2007) Colloidal oatmeal: history, chemistry and clinical properties. *Journal of Drugs in Dermatology* **6**, 167–170.

Lack, G., *et al.* (2003) Factors associated with the development of peanut allergy in childhood. *New England Journal of Medicine* **348**, 977–985.

Lehtinen, P., *et al.* (2003) Effect of heat treatment on lipid stability in processed oats. *Journal of Cereal Science* **37**, 215–221.

Liebel, F., *et al.* (2006) Anti-inflammatory and anti-itch activity of sertaconazole nitrate. *Archives of Dermatological Research* **298**, 191–199.

Lippert, U., *et al.* (1998) Role of antigen-induced cytokine release in atopic pruritus. *International Archives of Allergy and Immunology* **116**, 36–39.

Liu, L., *et al.* (2004) The antiatherogenic potential of oat phenolic compounds. *Atherosclerosis* **175**, 39–49.

Lockhart, H.B. and Hurt, H.D. (1986) Nutrition of oats. In: *Oats: Chemistry and Technology* (ed. F.W. Webster), pp. 297–308. American Association of Cereal Chemists, St. Paul, MN.

Lodén, M. (2003) Role of topical emollients and moisturizers in the treatment of dry skin barrier disorders. *American Journal of Clinical Dermatology* **4**, 771–788.

Marks, J., *et al.* (2002) Allergic and irritant contact dermatitis. In: *Contact and Occupational Dermatology*, 3rd edn (eds J. Marks, V. Deleo, and P. Elsner), pp. 3–12. Mosby, St. Louis, MO.

Matheson, J., *et al.* (2001) The reduction of itch during burn wound healing. *Journal of Burn Care and Research* **22**, 76.

Matsukawa, T., *et al.* (2000) Occurrence of avenanthramides and hydroxycinnamoyl-CoA: hydroxyanthranilate N-hydroxycinnamoyltransferase activity in oat seeds. *Zeitschrift fur Naturforschung C* **55**, 30–36.

Mayama, S., *et al.* (1986) The purification of victorin and its phytoalexin elicitor. *Physiological and Molecular Plant Pathology* **29**, 1–18.

Meydani, M. (2009) Potential health benefits of avenanthramides of oats. *Nutrition Reviews* **67**, 731–735.

Miller, A. (1979) Oat derivatives in bath products. *Cosmetics and Toiletries* **94**, 72–80.

Miyagawa, H., *et al.* (1995) Induction of avenanthramides in oat leaves inoculated with crown rust fungus, Puccinia coronata f. sp. avenae. *Bioscience, Biotechnology, and Biochemistry* **59**, 2305–2306.

Morelli, J.G., *et al.* (1992) Leukotriene B4-induced human melanocyte pigmentation and leukotriene C4-induced human melanocyte growth are inhibited by different isoquino-linesulfonamides. *Journal of Investigative Dermatology* **98**, 55–58.

Nachbar, F. and Korting, H. (1995) The role of vitamin E in normal and damaged skin. *Journal of Molecular Medicine* **73**, 7–17.

Nakano, Y. (2004) Stress-induced modulation of skin immune function: two types of antigen-presenting cells in the epidermis are differentially regulated by chronic stress. *British Journal of Dermatology* **151**, 50–64.

Nebus, J., *et al.* (2004) Alleviating dry, ashen skin in patients with skin of color. *Journal of the American Academy of Dermatology* **50**, 77.

Nebus, J., *et al.* (2007) Evaluating the safety and tolerance of a body wash and moisturizing regimen in patients with atopic dermatitis. *Journal of the American Academy of Dermatology* **56**, AB71.

Nebus, J., *et al.* (2009) A daily oat-based skin care regimen for atopic skin. *Journal of the American Academy of Dermatology* **60**, AB67.

O'Brasky, L. (1959) Management of extensive dry skin conditions. *Connecticut Medicine* **23**, 20.

Panganamala, R.V. and Cornwell, D.G. (1982) The effects of vitamin E on arachidonic acid metabolism. *Annals of the New York Academy of Sciences* **393**, 376–391.

Pazyar, N., *et al.* (2012) Oatmeal in dermatology: A brief review. *Indian Journal of Dermatology Venereology and Leprology* **78**, 142–145.

Pazzaglia, M., *et al.* (2000) Allergic contact dermatitis due to avena extract. *Contact Dermatitis* **42**, 364.

Peterson, D.M. (2001) Oat antioxidants. *Journal of Cereal Science* **33**, 115–129.

Peterson, D., *et al.* (2002) Oat avenanthramides exhibit antioxidant activities in vitro. *Food Chemistry* **79**, 473–478.

Pigatto, P., *et al.* (1997) An evaluation of the allergic contact dermatitis potential of colloidal grain suspensions. *American Journal of Contact Dermatitis* **8**, 207–209.

Rawlings, A. and Harding, C. (2004) Moisturization and skin barrier function. *Dermatologic Therapy* **17**, 43–48.

Riboldi, A., *et al.* (1988) Contact allergic dermatitis from oatmeal. *Contact Dermatitis* **18**, 316–317.

Richardson, J.D. and Vasko, M.R. (2002) Cellular mechanisms of neurogenic inflammation. *Journal of Pharmacology and Experimental Therapeutics* **302**, 839–845.

Ruzicka, T., *et al.* (1986) Skin levels of arachidonic acid-derived inflammatory mediators and histamine in atopic dermatitis and psoriasis. *Journal of Investigative Dermatology* **86**, 105–108.

Saeed, S., *et al.* (1981) Inhibitor(s) of prostaglandin biosynthesis in extracts of oat (*Avena sativa*) seeds. *Biochemical Society Transactions* **9**, 44.

Saija, A., *et al.* (1999) Ferulic and caffeic acids as potential protective agents against photooxidative skin damage. *Journal of the Science of Food and Agriculture* **79**, 476–480.

Schmelz, M. (2003) Neurophysiologic basis of itch. In: *Itch Basic Mechanisms and Therapy* (eds G. Yosipovitch, M.W. Greaves, A.B. Fleischer, and F. McGlone), pp. 5–12. Marcel Dekker, New York.

Schrickel, D.J. (1986) Oats production, value, and use. In: *Oats: Chemistry and Technology* (ed. F.W. Webster), pp. 1–11. American Association of Cereal Chemists, St. Paul, MN.

Seidenari, S. and Giusti, G. (1995) Objective assessment of the skin of children affected by atopic dermatitis: a study of pH, capacitance and TEWL in eczematous and clinically uninvolved skin. *Acta Dermato-Venereologica* **75**, 429.

Simon, M., *et al.* (2001) Persistence of both peripheral and non-peripheral corneodesmosomes in the upper stratum corneum of winter xerosis skin versus only peripheral in normal skin. *Journal of Investigative Dermatology* **116**, 23–30.

Slodownik, D., *et al.* (2008) Irritant contact dermatitis: a review. *Australasian Journal of Dermatology* **49**, 1–11.

Smith, G. (1958) The treatment of various dermatoses associated with dry skin. *Journal of the South Carolina Medical Association* **54**, 282–283.

Smith Pease, C.K., *et al.* (2002) Skin as a route of exposure to protein allergens. *Clinical and Experimental Dermatology* **27**, 296–300.

Sur, R., *et al.* (2008) Avenanthramides, polyphenols from oats, exhibit anti-inflammatory and anti-itch activity. *Archives of Dermatological Research* **300**, 569–574.

Talsania, N., *et al.* (2008) Colloidal oatmeal lotion is an effective treatment for pruritus caused by erlotinib. *Clinical and Experimental Dermatology* **33**, 108.

Vansina, S., *et al.* (2010) Sensitizing oat extracts in cosmetic creams: is there an alternative? *Contact Dermatitis* **63**, 169–171.

Vie, K., *et al.* (2002) Modulating effects of oatmeal extracts in the sodium lauryl sulfate skin irritancy model. *Skin Pharmacology and Applied Skin Physiology* **15**, 120–124.

Wise, M. (2011) Effect of chemical systemic acquired resistance elicitors on avenanthramide biosynthesis in oat (*Avena sativa*). *Journal of Agricultural Food Chemistry* **59**, 7028–7038.

Wood, P., *et al.* (1978) Extraction of high-viscosity gums from oats. *Cereal Chemistry* **55**, 1038–1049.

Woollard, P.M. (1986) Stereochemical difference between 12-hydroxy-5,8,10,14-eicosatetraenoic acid in platelets and psoriatic lesions. *Biochemical and Biophysical Research Communications* **136**, 169–176.

Yun, C.H., *et al.* (2003) β-Glucan, extracted from oat, enhances disease resistance against bacterial and parasitic infections. *FEMS Immunology and Medical Microbiology* **35**, 67–75.

Part V
Public Health Policies and Consumer Response

16
Health Claims for Oat Products: A Global Perspective

Joanne Storsley[1], Stephanie Jew[2], and Nancy Ames[1]
[1] *Agriculture and Agri-Food Canada, Winnipeg, MB, Canada*
[2] *Agriculture and Agri-Food Canada, Ottawa, ON, Canada*

16.1   Introduction

Cardiovascular disease (CVD), a group of disorders of the heart and blood vessels that includes coronary heart disease (CHD), is the number one cause of death in the world (WHO, 2012). Mortality rates due to CHD and its associated risk factors are highest in the United States but are among the top causes of death in many countries; Statistics Canada reported that the second leading cause of death in Canada in 2008 was due to cardiovascular disease (68 342 deaths), close behind the leading cause of death, cancer (71 125 deaths) (Statistics Canada, 2009). The risk factors associated with CVD include high levels of total and low-density lipoprotein (LDL) cholesterol, and high blood pressure; type 2 diabetes (another risk factor for CVD) has also been a major contributor to the increased prevalence of CVD. Approximately 80% of all CVDs and type 2 diabetes can be prevented by implementing a healthy diet, engaging in regular physical activity, and avoiding tobacco smoke (Webster and Wood, 2011). Thus, CVD and type 2 diabetes are largely preventable, with diet potentially playing a role in their prevention (Health Canada, 2010). The World Health Organization (WHO) has acknowledged a need for increased investment in national programs aimed at the prevention of cardiovascular diseases (WHO, 2012). Modifying risk factors through dietary intervention offers great potential for reducing the incidence of CVD.

Oats Nutrition and Technology, First Edition. Edited by YiFang Chu.
© 2014 John Wiley & Sons, Ltd. Published 2014 by John Wiley & Sons, Ltd.

Consumers all over the world have been taking greater control of their health, with an increased interest in the role that nutrition plays in their well-being. They are becoming more aware of the growing body of evidence from epidemiology, clinical trials, and modern nutritional biochemistry that underline the connection between diet and health. As consumer knowledge of this field has evolved, manufacturers have sought to fulfill the appetite for food products that could be used to promote good health.

Oatmeal was the first government-certified health food (Fitzsimmons, 2012) and continues to be recognized for its health benefits. A considerable amount of research has been conducted on the health benefits of oats in the last 30 years (Webster and Wood, 2011), the majority of which has focused on its cholesterol-lowering properties; thus, the association between oat consumption and hypocholesterolemia has been well established (Behall and Hallfrisch, 2006; Andon and Anderson, 2008; Anderson *et al.*, 2009; Cheickna and Hui, 2012). Some of the more recent oat nutritional studies suggest that the health benefits of oats may be further reaching, by positively impacting glycemia, satiety/body weight, and blood pressure (Swain *et al.*, 1990; Pick *et al.*, 1996; Howarth *et al.*, 2001; Keenan *et al.*, 2002; Pins *et al.*, 2002; Kim *et al.*, 2006; Tosh *et al.*, 2008; Regand *et al.*, 2011). While oats are a good source of fiber, including β-glucan, arabinoxylan, and cellulose, the positive impact of oats on cholesterol and the glycemic response has been mainly attributed to β-glucan (Tapola and Sarkkinen, 2009), the primary soluble dietary fiber component in oats. The consumption of soluble dietary fiber from various sources (including oats) has been shown to be inversely associated with CHD (Anderson, 1995; Pereira *et al.*, 2004). Moreover, studies examining whole grain consumption (including whole oats) have shown associations with reduced risks of diabetes, CHD, and certain types of cancers (Gordon, 2003; Fardet, 2010).

Extensive research on the health benefits of oats has led to approved health claims for oat products around the world. These claims vary from one country to another, and within any country from one target group in the population to another, depending on the needs of the target groups and the country's educational policy/framework. The data required to substantiate a health claim also varies by country or jurisdiction.

This chapter defines nutritional and health claims, reviews the scientific evidence required to substantiate oat health claims according to Codex guidelines, discusses the benefits of health claims, and examines health claims for oat products currently permitted in Canada, the United States, the European Union (EU), and other countries around the world. Considerations for conducting clinical trials in support of a health claim, with emphasis on the food component tested, is also discussed.

16.2   Definition of health claims

Claims on foods can be generally categorized into two groups: (i) nutrition claims and (ii) health claims (Table 16.1). Nutrition claims are statements that describe the presence, absence, or level of a nutrient, whereas health claims

Table 16.1 Definition and examples of nutrition and health claims[a]

Type of claim	Definition	Example
Nutrition claim	Any representation that states, suggests, or implies that a food has particular nutritional properties including but not limited to the energy value and the content of protein, fat, and carbohydrates, as well as the content of vitamins and minerals.	Nutrient content claims and nutrient comparative claims; see definitions and further examples below.
Nutrient content claim	A claim that describes the level of a nutrient within a food.	"high in fiber" "low in fat" "source of calcium"
Nutrient comparative claim	A claim that compares the nutrient levels and/or energy value of two or more foods.	"reduced" "less than" "fewer" "increased" "more than"
Health claim	Any representation that states, suggests, or implies that a relationship exists between a food or a constituent of that food and health.	Nutrient function claims, other function claims, and reduction of disease risk claims; see definitions and further examples below.
Nutrient function claim	A claim that describes the physiological role of the nutrient in growth, development, and normal functions of the body.	Nutrient A (naming a physiological role of nutrient A in the body in the maintenance of health and promotion of normal growth and development). Food X is a source high in nutrient A.
Other function claim	A claim that concerns specific beneficial effects of the consumption of foods or their constituents in the context of the total diet on normal functions or biological activities of the body; such claims relate to a positive contribution to health or to the improvement of a function or to modifying or preserving health.	Substance A (naming the effect of substance A on improving or modifying a physiological function or biological activity associated with health). Food Y contains X grams of substance A.
Reduction of disease risk claim	A claim relating the consumption of a food or food constituent in the context of the total diet to the reduced risk of developing a disease or health-related condition. Risk reduction means significantly altering a major risk factor(s) for a disease or health-related condition. Diseases have multiple risk factors and altering one of these risk factors may or may not have a beneficial effect. The presentation of risk reduction claims must ensure, for example by use of appropriate language and reference to other risk factors, that consumers do not interpret them as prevention claims.	"A healthful diet low in nutrient or substance A may reduce the risk of disease D. Food X is low in nutrient or substance A." "A healthful diet rich in nutrient or substance A may reduce the risk of disease D. Food X is high in nutrient or substance A."

[a]Definitions and examples from Codex (Codex Alimentarius, 2011).
Source: Food and Agriculture Organisation of the United Nations. Reproduced with permission.

are statements that connect a food, food component, or nutrient to a state of desired health.

16.3   Substantiation of health claims

According to Codex guidelines (Codex Alimentarius, 2011), "Health claims must be based on current relevant scientific substantiation and the level of proof must be sufficient to substantiate the type of claimed effect and the relationship to health." Codex guidelines further state that the health claim must be made up of two parts: "information on the physiological role of the nutrient or an accepted diet–health relationship, followed by information on the composition of the product relevant to the physiological role of the nutrient or the accepted diet–health relationship unless the relationship is based on a whole food or foods whereby the research does not link to specific constituents of the food."

Codex has also issued recommendations on the scientific substantiation of health claims intended to assist national authorities in the evaluation of health claims (Codex Alimentarius, 2011). Accordingly, various jurisdictions around the world have published guidance for the submission of food health claims, including the United States (US Department of Health and Human Services *et al.*, 2009), Canada (Health Canada, 2009), Europe (EFSA, 2011a), and Australia/New Zealand (Food Standards Australia New Zealand, 2013). Although the regulatory frameworks differ in these various jurisdictions, there are several common guiding principles for food health claims substantiation (Table 16.2). A key element that all the guidance documents emphasize is the need for a comprehensive review of all the scientific evidence in a systematic and transparent manner. This comprehensive review is necessary to ensure that the totality of evidence is examined; that is, that all relevant scientific data should be analyzed whether the evidence is in favor or not in favor of the health claim. Another commonality is the requirement for human studies in the substantiation of the health claim; generally, animal and/or *in vitro* studies will be accepted but only as supportive evidence, since there are scientific uncertainties in extrapolating results from these types of studies to humans. Some examples in the use of animal and/or *in vitro* studies would be to provide evidence related to mechanisms of action(s) or to demonstrate biological plausibility. In terms of the types of human studies used for the substantiation of health claims, the "gold standard" is generally randomized, controlled, double-blinded studies because they offer the best assessment of a causal relationship between a food and health effect.

Table 16.2   Common guiding principles for health claim substantiation[1]

- Systematic and transparent methodology for comprehensively evaluating data
- Substantiation of health claim based on totality of evidence
- Use of human studies to substantiate health claim (animal and/or *in vitro* studies generally acceptable as supportive evidence)

---

[1]Based on EFSA, 2011a; US Department of Health and Human Services *et al.*, 2009; Health Canada, 2009.

16.4   Health claims and dietary recommendations for oat products

16.4.1   Approved health claims for soluble oat fiber

Various global jurisdictions have approved health claims related to oat products including the EU (European Commission, 2011, 2012), Canada (Health Canada, 2010), the United States (Electronic Code of Federal Regulations, 2012c), Malaysia (Malaysia MoH, 2010), Australia and New Zealand (Food Standards Australia New Zealand, 2013) (Table 16.3). Of the currently approved health claims, the majority links oat β-glucan with cholesterol lowering and/or reduced risk of heart disease; the minimum quantity of oat β-glucan required per serving size for these claims ranges from 0.75 to 1 g, with some jurisdictions also requiring the food to state that the beneficial effect is based on 3 g β-glucan/day. The effect of oat β-glucan on postprandial glycemia has also been approved in some jurisdictions, with the minimum quantity of oat β-glucan varying from 4 g oat β-glucan per 30 g available carbohydrate to 6.5 g oat β-glucan per 100 g serving. A final health effect approved specifically for oat fiber has been in the European Union, where a claim for oat grain fiber and an increase in fecal bulk has been authorized for use in foods that are high in oat fiber.

16.4.2   Whole grain claims and dietary recommendations

In addition to oat β-glucan health claims, oat products and/or ingredients may be able to use whole grain-related health claims. Some countries, including the United States (US FDA 1999, 2003, Electronic Code of Federal Regulations, 2012a, 2012b) and Singapore (Agri-Food & Veterinary Authority of Singapore, 2011), have whole grain claims related to the reduction of risk of heart disease and some cancers (Table 16.4). Meanwhile, some other jurisdictions including Canada (Health Canada, 2012b), the European Union (EFSA, 2010), and Australia/New Zealand (Food Standards Australia New Zealand, 2007) have not authorized whole grain claims. Reasons given by these jurisdictions for not approving whole grain claims include the following: the definition of whole grain in available studies was inconsistent and vague (Food Standards Australia New Zealand, 2007); some studies measured intake of fiber rather than whole grains and thus may have included intake from other non-whole grain sources, since intake of whole grains was low in many studies—therefore, it was unclear whether the results could be attributed to the intake of whole grains (Food Standards Australia New Zealand, 2013). The effect of whole grains on total and LDL cholesterol from controlled clinical trials was largely attributed to trials examining grains high in β-glucan fiber that cannot be generalized to other grains such as wheat and, therefore, a whole grain claim would be misleading if applied to grains not high in β-glucan fiber (Health Canada, 2012b).

Table 16.3 Summary of internationally approved β-glucan health claims

Country	Health claim	Eligible food source	Conditions — Amount of β-glucan required
Europe (European Commission 2011, 2012)	"Oat β-glucan has been shown to lower/reduce blood cholesterol. High cholesterol is a risk factor in the development of coronary heart disease."	• Oat β-glucan	• At least 1 g oat beta glucan per quantified portion. • Information shall be given to the consumer that the beneficial effect is obtained with a daily intake of 3 g oat β-glucan.
	"β-glucan contributes to the maintenance of normal blood cholesterol levels."	• Oats, oat bran, barley, barley bran, or mixtures of these sources	• At least 1 g β-glucan from eligible food sources or from mixtures of these sources per quantified portion. • Information shall be given to the consumer that the beneficial effect is obtained with a daily intake of 3 g β-glucan from eligible food sources, or from a mixtures of these sources
	"Consumption of β-glucan from oats or barley as part of a meal contributes to the reduction of the blood glucose rise after that meal."	• Oats or barley	• At least 4 g β-glucan from the eligible foods sources for each 30 g of available carbohydrates in a quantified portion as part of the meal. • To bear the claim, information shall be given to the consumer that the beneficial effect is obtained by consuming β-glucan from the eligible food sources as part of the meal.
	"Oat grain fiber contributes to an increase in fecal bulk."	• Oats	• The claim may be used only for food that is high in oat fiber as referred to in the claim HIGH FIBER as listed in the Annex to Regulation (EC) No 1924/2006.

Canada (Health Canada, 2010)	*Primary statement:* "[serving size from Nutrition Facts table in metric and common household measures] of (Brand name) [name of food] [with name of eligible fiber source] supplies/provides [X% of the daily amount][1] of the fiber shown to help reduce/lower cholesterol." *Additional statement:* "Oat fiber helps reduce/lower cholesterol." "High cholesterol is a risk factor for heart disease." "Oat fiber helps reduce/lower cholesterol, (which is) a risk factor for heart disease." *Examples:* If the eligible fiber source is a food itself: "1 cup (X g) of Quaker Oatmeal supplies X% of the daily amount of the fibres shown to help reduce cholesterol." If the eligible fiber source is an ingredient: "1 muffin (X g) with oat bran provides X% of the daily amount of the fibres shown to help lower cholesterol."	• Oat bran, rolled oats (also known as oatmeal), and whole oat flour, either as food themselves (oat bran and rolled oats) or as ingredients (oat bran, rolled oats, and whole oat flour) in formulated foods	• The food must contain at least 0.75 g β-glucan oat fiber per reference amount and per serving of stated size from the eligible sources.

United States (Electronic Code of Federal Regulations, 2012c)	"Soluble fiber from certain foods and risk of coronary heart disease."
	Model claims:
	(1) "Soluble fiber from foods such as [name of soluble fiber source from paragraph (c)(2)(ii) of this section and, if desired, the name of food product], as part of a diet low in saturated fat and cholesterol, may reduce the risk of heart disease. A serving of [name of food] supplies ____ grams of the [grams of soluble fiber specified in paragraph (c)(2)(i)(G) of this section] soluble fiber from [name of the soluble fiber source from paragraph (c)(2)(ii) of this section] necessary per day to have this effect."
	(2) "Diets low in saturated fat and cholesterol that include [____ grams of soluble fiber specified in paragraph (c)(2)(i)(G) of this section] of soluble fiber per day from [name of soluble fiber source from paragraph (c)(2)(ii) of this section and, if desired, the name of the food product] may reduce the risk of heart disease. One serving of [name of food] provides ____ grams of this soluble fiber."
		• β-glucan-soluble fiber from
		  ○ The following whole oat and barley sources: oat bran, rolled oats, whole oat flour, Oatrim, whole grain and dry milled barley, barley β-fiber;
		  ○ Psyllium husk
			• The food product shall contain one or more of the whole oat or barley foods, and the whole oat or barley foods shall at least 0.75 g soluble fiber per reference amount customarily consumed of the food product.[2]
			• The food containing the Oatrim or barley β-fiber shall contain at least 0.75 g β-glucan-soluble fiber per reference amount customarily consumed of the food product.[2]
			• The psyllium food shall contain at least 1.7 g soluble fiber per reference amount customarily consumed of the food product.[2]

(continued)

Table 16.3 (*Continued*)

Country	Health claim	Eligible food source	Conditions — Amount of β-glucan required
Malaysia (Malaysia MoH, 2010)	"β-glucan from (state the source) helps lower or reduce cholesterol."	• Oats and barley	• Minimum amount of 0.75 g β-glucan per serving. • The following statement is must be provided on the label: "Amount recommended for cholesterol lowering effect is 3 g/day."
	"Oat-soluble fiber (β-glucan) helps to lower the rise of blood glucose provided it is not consumed together with other food."	• Oats	• Minimum amount of 6.5 g oat β-glucan per 100 g. • The following statement is must be provided on the label: "For advice regarding consuming this product, consult your medical professional."
Australia/New Zealand[1] (Food Standards Australia New Zealand, 2013)	"β-glucan reduces blood cholesterol."	• Oat bran, whole grain oats, whole grain barley	• The food must contain at least 1 g per serving of β-glucan from the oat or barley food.[3]

[1]The "daily amount" referred to in the primary statement is 3 g β-glucan oat fibre.
[2]A daily dietary intake of 3 g or more per day of β-glucan soluble fiber from either whole oats or barley, or a combination of whole oats and barley has been associated with reduced risk of coronary heart disease.
[3]The dietary context for this claim is for a diet containing 3 g of β-glucan per day.
Note: Food Standards Australia New Zealand (FSANZ) approved a high level claim as it is shown in the table, they also approved a general level claim that has the same conditions but wording is "β-glucan reduces dietary and biliary cholesterol absorption".

Table 16.4 Summary of internationally approved whole grain health claims

Country	Health claim	Criteria
United States (US FDA 1999, 2003; Electronic Code of Federal Regulations, 2012a, 2012b)	"Fruits, vegetables, and grain products that contain fiber, particularly soluble fiber, and the risk of coronary heart disease." *Model health claims:* (1) Diets low in saturated fat and cholesterol, and rich in fruits, vegetables, and grain products that contain some types of dietary fiber, particularly soluble fiber, may reduce the risk of heart disease, a disease associated with many factors. (2) Development of heart disease depends on many factors. Eating a diet low in saturated fat and cholesterol, and high in fruits, vegetables, and grain products that contain fiber may lower blood cholesterol levels and reduce your risk of heart disease.	• *Nature of the claim:* ○ The claim states that diets low in saturated fat and cholesterol, and high in fruits, vegetables, and grain products that contain fiber "may" or "might" reduce the risk of heart disease; ○ In specifying the disease, the claim uses the following terms: "heart disease" or "coronary heart disease;" ○ The claim is limited to those fruits, vegetables, and grains that contain fiber; ○ In specifying the dietary fiber, the claim uses the term "fiber," "dietary fiber," "some types of dietary fiber," "some dietary fibers," or "some fibers;" the term "soluble fiber" may be used in addition to these terms; ○ In specifying the fat component, the claim uses the terms "saturated fat" and "cholesterol;" ○ The claim indicates that development of heart disease depends on many factors; ○ The claim does not attribute any degree of risk reduction for coronary heart disease to diets low in saturated fat and cholesterol, high in fruits, vegetables, and grain products that contain fiber. • *Nature of the food:* ○ The food shall be or shall contain a fruit, vegetable, or grain product; ○ The food shall meet the nutrient content requirements of § 101.62 for a "low saturated fat," "low cholesterol," and "low fat" food; ○ The food contains, without fortification, at least 0.6 g soluble fiber per reference amount customarily consumed; ○ The content of soluble fiber shall be declared in the nutrition information panel, consistent with § 101.9(c)(6)(i)(A).

(continued)

Table 16.4 (Continued)

Country	Health claim	Criteria

	"Fiber-containing grain products, fruits, and vegetables, and cancer."
	
	Model health claims:
	(1) Low-fat diets rich in fiber-containing grain products, fruits, and vegetables may reduce the risk of some types of cancer, a disease associated with many factors.
	(2) Development of cancer depends on many factors. Eating a diet low in fat and high in grain products, fruits, and vegetables that contain dietary fiber may reduce your risk of some cancers.
		• *Nature of the claim:*
		  ○ The claim states that diets low in fat and high in fiber-containing grain products, fruits, and vegetables "may" or "might" reduce the risk of some cancers;
		  ○ In specifying the disease, the claim uses the following terms: "some types of cancer," or "some cancers;"
		  ○ The claim is limited to grain products, fruits, and vegetables that contain dietary fiber;
		  ○ The claim indicates that development of cancer depends on many factors;
		  ○ The claim does not attribute any degree of cancer risk reduction to diets low in fat and high in fiber-containing grain products, fruits, and vegetables;
		  ○ In specifying the dietary fiber component of the labeled food, the claim uses the term "fiber," "dietary fiber," or "total dietary fiber";
		  ○ The claim does not specify types of dietary fiber that may be related to risk of cancer.
		
		• *Nature of the food:*
		  ○ The food shall be or shall contain a grain product, fruit, or vegetable;
		  ○ The food shall meet the nutrient content requirements of § 101.62 for a "low fat" food;
		  ○ The food shall meet, without fortification, the nutrient content requirements of § 101.54 for a "good source" of dietary fiber.

	"Diets rich in whole grain foods and other plant foods, and low in total fat, saturated fat, and cholesterol may help reduce the risk of heart disease and certain cancers."	• "Whole grain foods:" foods that contain 51% or more whole grain ingredient(s) by weight per reference amount customarily consumed.
	"Diets rich in whole grain foods and other plant foods, and low in saturated fat and cholesterol may help reduce the risk of heart disease."	• "Whole grain foods:" foods that contain 51% or more whole grain ingredient(s) by weight per reference amount customarily consumed.

| Singapore (Agri-Food & Veterinary Authority of Singapore, 2011) | "A healthy diet rich in whole grains, fruits, and vegetables that contain dietary fiber may reduce the risk of heart disease. (The name of the food) is low in/free of fat and high in dietary fiber." | • A product from these food groups—whole grains, fruit, vegetables, or fiber-fortified foods; and
• Low in fat (not more than 3 g fat per 100 g or not more than 1.5 g fat per 100 ml), or fat free (not more than 0.15 g fat per 100 g or 100 ml); and
• High in dietary fiber (not less than 3 g per 100 kcal or not less than 6 g per 100 g or 100 ml); and
• With at least 25% of the dietary fiber comprising soluble fiber. |
| | "A healthy diet rich in fiber-containing foods such as whole grains, fruits, and vegetables may reduce the risk of some types of cancers. (The name of the food) is free/ low in fat and high in dietary fiber." | • A product from these food groups – whole grains, fruit, vegetables, or fiber-fortified foods; and
• Low in fat (not more than 3 g fat per 100 g or not more than 1.5 g fat per 100 ml), or fat free (not more than 0.15 g fat per 100 g or 100 ml); and
• High in dietary fiber (not less than 3 g per 100 kcal or not less than 6 g per 100 g); and
• Reference quantity of the food product should not contain sodium in an amount exceeding 25% of sodium RDA, which is taken as 2000 mg. |

Table 16.5   Examples of international whole grain dietary recommendations

Country	Dietary recommendation
Australia: Food for Health – Dietary Guidelines for Australian Adults (Australian Government, 2003)	"Eat plenty of cereals (including breads, rice, pasta, and noodles), preferably whole grain."
Canada: Eating Well with Canada's Food Guide (Health Canada, 2007)	"Make at least half of your grain products whole grain each day. Eat a variety of whole grains such as barley, brown rice, oats, quinoa, and wild rice. Enjoy whole grain breads, oatmeal, or whole wheat pasta."
Germany: 10 Guidelines of the German Nutrition Society (DGE) for a Wholesome Diet (Deutsche Gesellschaft fur Ernahrung e.V., 2005)	"Bread, pasta, rice, grain flakes, preferably from whole grain, as well as potatoes contain hardly any fat but plenty of vitamins, minerals, and dietary fiber and phytochemicals. Consume these foods preferably with low-fat ingredients. At least 30 grams of dietary fiber daily, especially from whole grain products, are recommended. A high intake lowers the risk of various nutrition-related diseases."
United Kingdom: The Eatwell Plate (UK NHS, 2011)	"Based on the eatwell plate, you should try to eat plenty of potatoes, bread, rice, pasta, and other starchy foods. Choose whole grain varieties whenever you can."
United States: My Plate (USDA, 2011)	"Make at least half of your grains whole grains. Substitute whole grain choices for refined-grain breads, bagels, rolls, breakfast cereals, crackers, rice, and pasta. Check the ingredients list on product labels for the words "whole" or "whole grain" before the grain ingredient name. Choose products that name a whole grain first on the ingredients list."

Regardless of whether a health claim has been approved for whole grains, many countries promote their intake through dietary recommendations because whole grains provide important nutrients, such as fiber, vitamins, and minerals (Table 16.5) (Australian Government, 2003; Deutsche Gesellschaft fur Ernahrung e.V., 2005; Health Canada, 2007; UK NHS, 2011; USDA, 2011).

16.5   Benefits of health claims

Food health claims can be beneficial to all stakeholders, including producers, industry, and consumers. Government-approved health claims can help educate and improve the health of consumers, as well as help develop regulatory guidelines for the food industry in using claims on product labels (Ames and Rhymer, 2008; Ames *et al.*, 2011). A health claim should provide the consumer with information regarding proven health benefits associated with a specific functional food or component like β-glucan-soluble fiber, including details on the target group and the effective dose-response relationship. Health claims can also serve the food industry as advertising or marketing tools to increase consumption of food products. In the case of oats, health claims stimulate marketing and processing aspects of oat products that in turn increase use and production.

Health claims have been shown to assist consumers in making better food choices for a healthier diet and gaining a better understanding of diet-disease relationships (Ippolito and Mathios, 1991; Kim et al., 2001; Webster and Wood, 2011). Kim and colleagues investigated whether food labels would improve diet quality as measured by the Healthy Eating Index (Kim et al., 2001). Of food label components, including nutritional panels, serving sizes, nutrient content claims, ingredient lists, and health claims, health claims on food labels provided the highest level of improvement in diet quality. To assess consumer understanding of the relationship between diet and disease, Ippolito and Mathios (1991) investigated a period of time when producers in the United States were banned from advertising cereal health benefits but were then subsequently allowed to make health claims. The authors examined the period 1978–1984 (that is, pre-health claims) and 1985–1987 (that is, when health claims were permitted). Ippolito and Mathios (1991) found that only 8.5% of adults reported having knowledge of the fiber–cancer relationship in 1984, whereas in 1986 there was a dramatic increase in adults who reported knowledge of such a relationship, up to 32.0%. This also led to an increase in most segments of the population increasing their fiber cereal consumption after health claims were added to the market.

Positive impacts of health claims on industry and producers have also been documented with an increase in sales of the targeted foods (Paul et al., 1999; Marchonie, 2009). Paul and colleagues investigated the impact of the approval of the Quaker Oats health claim that consisted of the relationship between consumption of oat products and the risk of CHD in the United States (Paul et al., 1999). Prior to approval of the health claim, oatmeal sales were decreasing at an annual rate of 3–4%. In January 1996, the announcement of the proposed claim was followed by extensive print and electronic media coverage. From January to June 1996, oatmeal sales increased by 5%. Furthermore, subsequent to the US Food and Drug Administration (FDA) announcing approval of the oat health claim in January 1997, health claims started to appear on product labels around March/April 1997, and oatmeal sales continued to increase at a rate of 4–5% from January to June 1997 (Paul et al., 1999).

Another potential advantage of health claims is increased product innovation. Development of new foods with known health benefits, such as cholesterol lowering, may stimulate new processing opportunities for functional foods and provide consumers with reliable, healthy food choices. This was seen with the introduction of fiber-cancer health claims from 1985–1987 in the United States (Ippolito and Mathios, 1991). During this time, cereals introduced into the market were significantly higher in fiber than new cereals introduced between 1979 and 1984, when health claims were banned. While new cereal products had an average of 1.70 g fiber per ounce during the earlier period, the 1985–1987 products averaged 2.59 g fiber per ounce. Furthermore, other healthy aspects of the cereal products were enhanced, with sodium and fat content of the cereals decreasing in the period of 1985–1987 (Ippolito and Mathios, 1991).

Since the Quaker Oat health claim approval in the United States, other juris-dictions, such as the European Union and Canada, have approved the use of health claims (EFSA, 2011b, c, d, e, f, g, h, i; Health Canada, 2010, 2012a).

16.6   Nutritional information and health claims: How can health claims ensure clarity versus confusion?

There is no doubt that health claims influence consumer choice and that con-sumers are increasingly interested in getting more fiber in their diets. However, some critics express concern that the increased interest in fiber as a marketing claim is not necessarily a good thing, since health claims can potentially be mis-leading. For example, according to the results of a recent study published in the *Journal of Consumer Affairs* (Zank and Kemp, 2012), when one treatment group was shown the front panel of a cereal product with a fiber nutritional claim and a second group of subjects was shown the same panel but without the claim, those individuals shown the claim perceived the product to have significantly more fiber. The authors suggest that even though nutritional information on pack-ages has led to increased awareness of the relationship between diet and disease risk, many consumers do not carry out a thoughtful processing of the nutritional information.

Not only do health claims and nutritional labeling have the potential to mis-lead consumers by influencing which products they choose to purchase based on perceived nutritional content, but they also could be misleading with respect to the dosage needed to achieve the desired/claimed health benefit. For example, most of the research on β-glucan from oats and cholesterol lowering shows that a minimum consumption of 3 g β-glucan per day is required to achieve the effect; for many oat products, this would require consuming more than one serving of the food. Expressing the β-glucan amount per serving as a percentage of the min-imum daily requirement to achieve cholesterol lowering may assist individuals in getting sufficient oats in their diet to realize such health benefits.

Consumers may only reap the benefits of oat consumption if they are informed of the healthful benefits of oat products through truthful labeling and advertis-ing by the food industry. However, it is also apparent that health claims must be strictly regulated by government. In addition, consumers must be well educated on the use and interpretation of food labeling information to make informed choices; this will likely require the help and/or implementation of national pro-grams directed at disease prevention. If these objectives can be achieved, food health claims can achieve public health objectives while being valuable for pro-ducers, industry, and consumers.

Labeling provides consumers with information about the nutritional prop-erties of a food, thereby facilitating the selection of a healthy diet (Hawkes, 2004). Listing nutrients is also a means to provide evidence for any nutritional claim made on the label and to encourage food manufacturers to improve the nutritional attributes of their products. Health claims provide information to consumers about the nutritional and health advantages of particular foods or nutrients. If appropriately applied, health claims may help consumers choose foods associated with good nutrition and health. Health claims are also a valuable marketing technique for food companies, since they are more visible on food packages than nutritional labels and a point of differentiation between one product and another.

Another aspect of labeling that plays a role in food choice is the quantitative ingredient declaration, whereby the percentages of specific ingredients are listed (Hawkes, 2004). Quantitative ingredient declaration can be perceived as a public health measure because it helps consumers assess the amount of healthy ingredients that are present in foods. A published WHO-Food and Agriculture Organization report on diet, nutrition, and prevention of chronic diseases suggested that nutritional labels are an important means of facilitating choice of and access to nutrient-dense foods. The WHO global strategy on diet, physical activity, and health endorsed in May 2004 by the World Health Assembly states that providing accurate, standardized, and comprehensible information on the content of food items is conducive to consumers making healthy choices (Hawkes, 2004).

The regulation of nutritional labels and health claims partly determines the extent to which the potential benefits of those labels or claims can be realized. Regulations can dictate or recommend when labeling should be mandatory and in what form nutritional information should appear. Regulations on health claims can be implemented to promote the use of responsible health claims, guiding which health claims should be used on which foods and how they should appear on the label. In the United States, to ensure that health claims do not appear on "junk foods," foods with a health claim must contain at least 10% of the daily value for one of six nutrients: protein, vitamin A, vitamin C, calcium, iron, or dietary fiber, without fortification. Furthermore, products containing high amounts of total fat, saturated fat, cholesterol, or sodium are not permitted to make health claims.

16.7   Considerations in conducting research for health claim substantiation

Preserving the properties of β-glucan in foods to maintain beneficial physiological effects is an important aspect of a proposed health claim. Clinical trials for health claim substantiation must, therefore, take great care in ensuring that both the properties and amounts of β-glucan in test foods are properly controlled. Numerous studies have shown that the processing techniques used (e.g., extrusion, heat/moisture treatments such as cooking or autoclaving, extraction, and concentration) may affect the physiochemical properties of the β-glucan (Ikegami *et al.*, 1996; Izydorczyk *et al.*, 2000; Tosh *et al.*, 2008, 2010; Regand *et al.*, 2009; Brummer *et al.*, 2012), and may ultimately physiologically change the product, either positively or negatively. Furthermore, exposure of β-glucan to endogenous or exogenous β-glucanases can markedly degrade β-glucan over time. This has been shown with barley β-glucan when endogenous β-glucanases are not inactivated or when oats are combined with wheat containing β-glucanase in oat pasta.

Studies involving eating oats in its various forms, such as oat flakes or oat bran, have shown relatively consistent cholesterol-lowering effects. The greatest inconsistencies and conflicting results have been reported in clinical trials using extracted β-glucans. Some studies have shown that β-glucan can undergo depolymerization in mixed beverages or in the presence of ascorbic acid (Kivelä *et al.*, 2009a, 2009b, 2011, 2012). A few studies have reported cholesterol lowering with consumption of low molecular weight β-glucan (Bae *et al.*, 2010), which calls into question the role of molecular weight and/or viscosity in cholesterol lowering.

Another proposed mechanism by which β-glucan may lower LDL cholesterol is through binding of bile acids. One recent study looked at low, medium, and high molecular weight β-glucan created by treating extracted high molecular weight oat β-glucan with lichenase and found that the low molecular weight material had the greatest bile acid binding *in vitro* (Kim and White, 2010); more research is required to validate this observation and determine whether low molecular weight β-glucan has the greatest effect on LDL cholesterol lowering in a clinical trial. Another study looked at the effect of a concentrated β-glucan extract on cholesterol levels and found no significant reduction (Keogh *et al.*, 2003). The authors concluded that structural changes might have occurred during extraction and storage. However, Keenan and colleagues reported effective lowering of cholesterol when subjects were fed foods containing concentrated barley β-glucan extract (Keenan *et al.*, 2002). In a recent summary of the current literature, Newman and Newman (2008) suggest that there are insufficient data on the effects of processing to draw meaningful conclusions regarding the influence of processing on the efficacy of extracted β-glucan on blood lipid composition currently.

Nevertheless, it is apparent that processing can have a large effect on dietary fiber, but by selecting a suitable cultivar and appropriate processing conditions, it may be possible to achieve specific health effects.

The European Union has listed a number of criteria for substantiation of health claims that can serve as guidelines for researchers in designing and carrying out experiments to support health claims (Asp and Bryngelsson, 2008), the first of which states that the food or food component to which the claimed effect is attributed should be characterized. Studies involving test foods, such as those made from oats, should include compositional as well as physicochemical properties, taking into account the effects of processing and storage on these attributes.

Nutritional composition (protein, fat, starch, total dietary fiber, soluble dietary fiber, and β-glucan content) is often vital in interpreting the results of a study to support a health claim. In the case of oats and cholesterol lowering, there are limited numbers of well-conducted studies investigating the dose-response effect of β-glucan. β-glucan analysis to confirm content of both ingredients and final products is desirable. The American Association for Cereal Chemistry method 32-23 (AACC, 1999), AOAC method 992.28 (American Association of Analytical Chemistry, 2000), and International Association for Cereal Science and Technology standard method 166 (International Association for Cereal Science and Technology, 1998) are the preferred methods for measuring barley β-glucan content. In addition, it is good to test the β-glucan content of the test foods both before and after processing, over the time of storage, and periodically throughout the course of the study, to ensure that the dosage remains consistent. For example, if β-glucan was undergoing depolymerization and was broken down sufficiently (i.e., into glucose) during processing or storage, a decrease in the content of β-glucan would be observed. With the β-glucan content of the test foods established, developing test foods to contain different amounts of β-glucan is achievable.

Since the physicochemical properties of β-glucan are believed to be important in determining their health benefits, characterization of the β-glucan within the test foods is imperative. Viscosity is one such means of characterization and is often measured *in vitro*. Some researchers have attempted to mimic human digestion by adding appropriate enzymes and buffers, and mixing and incubating for the appropriate time, to obtain a slurry containing soluble β-glucan that is then used for viscosity measurements (Beer *et al.*, 1997; Tosh *et al.*, 2010). However, the methods used to measure viscosity vary from one study to the next, indicating a need to have a standardized method to measure β-glucan viscosity for health claim substantiation. Furthermore, since the two main contributors to the viscosity of β-glucan are concentration (solubility) and molecular weight, it is also helpful to measure the solubility of β-glucan in the food product being tested as well as the molecular weight distribution of the solubilized β-glucan to determine the mechanism behind the measured health outcome.

Ames and colleagues, in their effort to design test foods for clinical trials and to meet the requirements of a health claim, noted that there is a need to characterize the range of effects that processing has on the physicochemical properties of β-glucan in various food matrices, to establish standardized food processing and preparation methodologies to create a series of oat products and formulations with a defined and reproducible food matrix and establish realistic molecular weight and viscosity ranges for treatment effects (Ames *et al.*, 2011). Considerable time, effort, and collaboration are needed to ensure that the test foods (e.g., those made from oats) used in studies to support health claims are thoroughly evaluated and are consistent. This has significant implications in our understanding of the relationship between consumption of a food and its health benefits.

References

AACC (1999) β-Glucan Content of Barley and Oats – Rapid Enzymatic Procedure, AACC International Method. AACC International, St. Paul, MN.

Agri-Food & Veterinary Authority of Singapore (2011) Food labelling and advertisements [online]; available: http://www.ava.gov.sg/FoodSector/FoodLabelingAdvertisement/ [last accessed 9 May 2013].

American Association of Analytical Chemistry (ed.) (2000) Method 992.28: (1-3)(1-4) Beta-D-Glucans in oat and varley fractions and ready-to-eat cereals. In: *Official Methods of Analysis of AOAC International*, 17th edn. AOAC International, Gaithersburg, MD.

Ames, N. P. and Rhymer, C. R. (2008) Issues surrounding health claims for barley. *The Journal of Nutrition*, **138**(6), 1237S–1243S.

Ames, N., Storsley, J., Gamel, T., and Tosh, S. (2011) Validating the health benefits of barley foods: effect of processing on physicochemical properties of beta-glucan in barley test foods. *Cereal Foods World*, **56**(4), A28.

Anderson, J. W. (1995) Dietary fibre, complex carbohydrate and coronary artery disease. *The Canadian Journal of Cardiology*, **11** (Suppl G), 55G–62G.

Anderson, J. W., Baird, P., Davis Jr, R. H., *et al.* (2009) Health benefits of dietary fiber, *Nutrition Reviews*, **67**(4), 188–205.

Andon, M. B. and Anderson, J. W. (2008) State of the art reviews: the oatmeal-cholesterol connection: 10 years later. *American Journal of Lifestyle Medicine*, **2**(51), 51–57.

Asp, N. G. and Bryngelsson, S. (2008) Health claims in Europe: New legislation and PASS-CLAIM for substantiation. *Journal of Nutrition*, **138**(6), 1210S–1215S.

Australian Government (2003) Food for Health – Dietary Guidelines for Australians [online]; available: http://www.nhmrc.gov.au/guidelines/publications/n29-n30-n31-n32-n33-n34 [last accessed 9 May 2013].

Bae, I. Y., Kim, S. M., Lee, S. and Lee, H. G. (2010) Effect of enzymatic hydrolysis on cholesterol-lowering activity of oat β-glucan. *New Biotechnology*, **27**(1), 85–88.

Beer, M. U., Wood, P. J., Weisz, J. and Fillion, N. (1997) Effect of cooking and storage on the amount and molecular weight of $(1{\rightarrow}3)(1{\rightarrow}4)$-β-D-glucan extracted from oat products by an in vitro digestion system. *Cereal Chemistry*, **74**(6), 705–709.

Behall, K. M. and Hallfrisch, J. G. (2006) Effects of barley consumption on CVD risk factors. *Cereal Foods World*, **51**(1), 12–15.

Brummer, Y., Duss, R., Wolever, T. M. S. and Tosh, S. M. (2012) Glycemic response to extruded oat bran cereals processed to vary in molecular weight. *Cereal Chemistry*, **89**(5), 255–261.

Cheickna, D. and Hui, Z. (2012) Oat beta-glucan: its role in health promotion and prevention of diseases. *Comprehensive Reviews in Food Science and Food Safety*, **11**(4), 355–365.

Codex Alimentarius (2011) Guidelines for Use of Nutrition and Health Claims, [online]; available: http://www.codexalimentarius.org/standards/list-of-standards/en/?no_cache=1 [last accessed 9 May 2013].

Deutsche Gesellschaft fur Ernahrung e.V. (2005) 10 guidelines of the German Nutrition Society (DGE) for a wholesome diet [online]; available: http://www.dge.de/modules.php?name=Content&pa=showpage&pid=16 [last accessed 15 May 2013].

EFSA (European Food Safety Authority) (2010) Scientific Opinion on the substantiation of health claims related to whole grain (ID 831, 832, 833, 1126, 1268, 1269, 1270, 1271, 1431) pursuant to Article 13(1) of Regulation (EC) No 1924/20061 [online]; available: http://www.efsa.europa.eu/en/efsajournal/pub/1766.htm [last accessed 9 May 2013].

EFSA (European Food Safety Authority) (2011a) Scientific and technical guidance for the preparation and presentation of an application for authorisation of a health claim (revision 1) [online]; available: http://www.efsa.europa.eu/en/efsajournal/pub/2170.htm [last accessed 9 May 2013].

EFSA (2011b) Scientific Opinion on the re-evaluation of lutein preparations other than lutein with high concentrations of total saponified carotenoids at levels of at least 80%. *EFSA Journal*, **9**(5).

EFSA (2011c) Scientific Opinion on the substantiation of a health claim related to barley beta-glucans and lowering of blood cholesterol and reduced risk of (coronary) heart disease pursuant to Article 14 of Regulation (EC) No 1924/2006. *EFSA Journal*, **9**(12).

EFSA (2011d) Scientific Opinion on the substantiation of health claims related to arabinoxylan produced from wheat endosperm and reduction of post-prandial glycaemic responses (ID 830) pursuant to Article 13(1) of Regulation (EC) No 1924/2006. *EFSA Journal*, **9**(6).

EFSA (2011e) Scientific Opinion on the substantiation of health claims related to beta-glucans from oats and barley and maintenance of normal blood LDL-cholesterol concentrations (ID 1236, 1299), increase in satiety leading to a reduction in energy intake (ID 851, 852), reduction of post-prandial glycaemic responses (ID 821, 824), and "digestive function" (ID 850) pursuant to Article 13(1) of Regulation (EC) No 1924/2006. *EFSA Journal*, **9**(6).

EFSA (2011f) Scientific Opinion on the substantiation of health claims related to betaine and contribution to normal homocysteine metabolism (ID 4325) pursuant to Article 13(1) of Regulation (EC) No 1924/2006. *EFSA Journal*, **9**(4).

EFSA (2011g) Scientific Opinion on the substantiation of health claims related to lutein and protection of DNA, proteins and lipids from oxidative damage (ID 3427), protection of the skin from UV-induced (including photo-oxidative) damage (ID 1605, 1779) and maintenance of normal vision (ID 1779, 2080) pursuant to Article 13(1) of Regulation (EC) No 1924/2006, *EFSA Journal*, **9**(4).

EFSA (2011h) Scientific Opinion on the substantiation of health claims related to resistant starch and reduction of post-prandial glycaemic responses (ID 681), "digestive health benefits" (ID 682) and "favours a normal colon metabolism" (ID 783) pursuant to Article 13(1) of Regulation (EC) No 1924/2006. *EFSA Journal*, **9**(4).

EFSA (2011i) Scientific Opinion on the substantiation of health claims related to rye fibre and changes in bowel function (ID 825), reduction of post-prandial glycaemic responses (ID 826) and maintenance of normal blood LDL-cholesterol concentrations (ID 827) pursuant to Article 13(1) of Regulation (EC) No 1924/2006. *EFSA Journal*, **9**(6).

Electronic Code of Federal Regulations (2012a) Health claims: fiber-containing grain products, fruits, and vegetables and cancer [online]; available: http://www.ecfr.gov/cgi-bin/text-idx?c=ecfr;sid=502078d8634923edc695b394a357d189;rgn=div8;view=text;node=21%3A2.0.1.1.2.5.1.7;idno=21;cc=ecfr [last accessed 9 May 2013].

Electronic Code of Federal Regulations (2012b) Health claims: fruits, vegetables, and grain products that contain fiber, particularly soluble fiber, and risk of coronary heart disease., [online]; available: http://www.ecfr.gov/cgi-bin/text-idx?c=ecfr;sid=502078d8634923edc695b394a357d189;rgn=div8;view=text;node=21%3A2.0.1.1.2.5.1.8;idno=21;cc=ecfr [last accessed 9 May 2013].

Electronic Code of Federal Regulations (2012c) Health Claims: Soluble fiber from certain foods and risk of coronary heart disease (CHD)., [online]; available: http://ecfr.gpoaccess.gov/cgi/t/text/text-idx?c=ecfr;sid=502078d8634923edc695b394a357d189;rgn=div8;view=text;node=21%3A2.0.1.1.2.5.1.12;idno=21;cc=ecfr [last accessed 9 May 2013].

European Commission (2011) Commission Regulation (EU) No 1160/2011 of 14 November 2011 on the authorisation and refusal of authorisation of certain health claims made on foods and referring to the reduction of disease risk Text with EEA relevance [online]; available: http://eur-lex.europa.eu/LexUriServ/LexUriServ.do?uri=CELEX:32011R1160:EN:NOT [last accessed 9 May 2013].

European Commission (2012) Commission Regulation (EU) No 432/2012 of 16 May 2012 establishing a list of permitted health claims made on foods, other than those referring to the reduction of disease risk and to children's development and health Text with EEA relevance, [online]; available: http://eur-lex.europa.eu/LexUriServ/LexUriServ.do?uri=CELEX:32012R0432:EN:NOT [last accessed 9 May 2013].

Fardet, A. (2010) New hypotheses for the health-protective mechanisms of whole-grain cereals: what is beyond fibre? *Nutrition Research Reviews*, **23**(01), 65–134.

Fitzsimmons, R. (2012) Oh, what those oats can do. Quaker Oats, the Food and Drug Administration, and the market value of scientific evidence 1984 to 2010. *Comprehensive Reviews in Food Science and Food Safety*, **11**(1), 56–99.

Food Standards Australia New Zealand (2013) Wholegrains and coronary heart disease – FSANZ consideration of a commissioned review [online]; available: http://www.foodstandards.gov.au/consumerinformation/nutritionhealthandrelatedclaims/reviewsforhighlevelc3090.cfm [last accessed 9 May 2013].

Food Standards Australia New Zealand (2007) Preliminary Final Assessment Report – Proposal P293, Nutrition, Health and Related Claims – Attachment 5: Technical Report: Diet-disease relationships [online]; available: http://www.foodstandards.gov.au/foodstandards/proposals/proposalp293nutritionhealthandrelatedclaims/p293preliminaryfinal3502.cfm [last accessed 9 May 2013].

Food Standards Australia New Zealand (2013) Standard 1.2.7 – Nutrition, Health and Related Claims [online]; available: http://www.comlaw.gov.au/Series/F2013L00054.

Gordon, D. T. (2003) Strengths and limitations of the U.S. whole-grain foods health claim. *Cereal Foods World*, **48**(4), 210–214.

Hawkes, C. (2004) *Nutrition Labels and Health Claims: the global regulatory environment*. World Health Organization, Geneva, Switzerland.

Health Canada (2007) Eating Well with Canada's Food Guide [online]; available: http://www.hc-sc.gc.ca/fn-an/food-guide-aliment/index-eng.php [last accessed 9 May 2013].

Health Canada (2009) Guidance Document for Preparing a Submission for Food Health Claims [online]; available: http://www.hc-sc.gc.ca/fn-an/legislation/guide-ld/health-claims_guidance-orientation_allegations-sante-eng.php [last accessed 9 May 2013].

Health Canada (2010) Oat Products and Blood Cholesterol Lowering-Summary of Assessment of a Health Claim about Oat Products and Blood Cholesterol Lowering [online]; available: http://www.hc-sc.gc.ca/fn-an/label-etiquet/claims-reclam/assess-evalu/oat-avoine-eng.php [last accessed 9 May 2013].

Health Canada (2012a) *Summary of Health Canada's Assessment of a Health Claim about Barley Products and Blood Cholesterol Lowering*. Health Canada, Ottawa, ON.

Health Canada (2012b) *Summary of Health Canada's Assessment of a Health Claim about Whole Grains and Coronary Heart Disease*. Health Canada, Ottawa, ON.

Howarth, N. C., Saltzman, E. and Roberts, S. B. (2001) Dietary fiber and weight regulation. *Nutrition Reviews*, **59**(5), 129–139.

Ikegami, S., Tomita, M., Honda, S., *et al.* (1996) Effect of boiled barley-rice-feeding in hypercholesterolemic and normolipemic subjects. *Plant Foods for Human Nutrition*, **49**(4), 317–328.

International Association for Cereal Science and Technology (ed.) (1998) *Method 166: Determination of beta-glucan in barley, oat and rye*. International Association for Cereal Science and Technology, Vienna, Austria.

Ippolito, P. M. and Mathios, A. D. (1991) Health claims in food marketing: Evidence on knowledge and behavior in the cereal market. *Journal of Public Policy and Marketing*, **10**(1), 15–32.

Izydorczyk, M. S., Storsley, J., Labossiere, D., *et al.* (2000) Variation in total and soluble β-glucan content in hulless barley: Effects of thermal, physical, and enzymic treatments. *Journal of Agricultural and Food Chemistry*, **48**(4), 982–989.

Keenan, J. M., Pins, J. J., Frazel, C., *et al.* (2002) Oat ingestion reduces systolic and diastolic blood pressure in patients with mild or borderline hypertension: a pilot trial. *The Journal of family practice*, **51**(4), 369.

Keogh, G. F., Cooper, G. J. S., Mulvey, T. B., *et al.* (2003) Randomized controlled crossover study of the effect of a highly β-glucan-enriched barley on cardiovascular disease risk factors in mildly hypercholesterolemic men. *American Journal of Clinical Nutrition*, **78**(4), 711–718.

Kim, H. J. and White, P. J. (2010) *In vitro* bile-acid binding and fermentation of high, medium, and low molecular weight β-glucan. *Journal of Agricultural and Food Chemistry*, **58**(1), 628–634.

Kim, S.-Y., Nayga, R.M. Jr., and Capps, O. Jr. (2001) Food label use, self-selectivity and diet quality. *The Journal of Consumer Affairs*, **35**(2), 346–363.

Kim, H., Behall, K. M., Vinyard, B. and Conway, J. M. (2006) Short-term satiety and glycemic response after consumption of whole grains with various amounts of β-glucan. *Cereal Foods World*, **51**(1), 29–33.

Kivelä, R., Gates, F. and Sontag-Strohm, T. (2009a) Degradation of cereal beta-glucan by ascorbic acid induced oxygen radicals. *Journal of Cereal Science*, **49**(1), 1–3.

Kivelä, R., Nyström, L., Salovaara, H. and Sontag-Strohm, T. (2009b) Role of oxidative cleavage and acid hydrolysis of oat beta-glucan in modelled beverage conditions. *Journal of Cereal Science*, **50**(2), 190–197.

Kivelä, R., Sontag-Strohm, T., Loponen, J., *et al.* (2011) Oxidative and radical mediated cleavage of β-glucan in thermal treatments. *Carbohydrate Polymers*, **85**(3), 645–652.

Kivelä, R., Henniges, U., Sontag-Strohm, T. and Potthast, A. (2012) Oxidation of oat β-glucan in aqueous solutions during processing. *Carbohydrate Polymers*, **87**(1), 589–597.

Malaysia MoH (2010) Guide to Nutrition Labelling and Claims [online]; available: http://fsq.moh.gov.my/v3/images/filepicker_users/5ec35272cb-78/Perundangan/Garispanduan/Pelabelan/GuideNutritionLabel.pdf [last accessed 9 May 2013].

Marchonie, M. (2009) Fat in fiber's clothing? Nutrient-spiked foods top shopping list. *USA Today*, 20 August 2009.

Newman, R. K. and Newman, C. W., ed. (2008) *Barley for Food and Health: Science, Technology, and Products*. John Wiley and Sons, Inc., Hoboken, NJ.

Paul, G. L., Ink, S. L. and Geiger, C. J. (1999) The quaker oats health claim: A case study. *Journal of Nutraceuticals, Functional and Medical Foods*, **1**(4), 5–32.

Pereira, M. A., O'Reilly, E., Augustsson, K. and Brown, M. M. (2004) Dietary fiber and risk of coronary heart disease. *Evidence-Based Eye Care*, **5**(4), 226–227.

Pick, M. E., Hawrysh, Z. J., Gee, M. I., *et al.* (1996) Oat bran concentrate bread products improve long-term control of diabetes: a pilot study. *Journal of the American Dietetic Association*, **96**(12), 1254–1261.

Pins, J. J., Geleva, D., Keenan, J. M., *et al.* (2002) Do whole-grain oat cereals reduce the need for antihypertensive medications and improve blood pressure control? *The Journal of Family Practice*, **51**(4), 353–359.

Regand, A., Tosh, S. M., Wolever, T. M. S. and Wood, P. J. (2009) Physicochemical properties of Beta-glucan in differently processed oat foods influence glycemic response. *Journal of Agricultural and Food Chemistry*, **57**(19), 8831–8838.

Regand, A., Chowdhury, Z., Tosh, S. M., *et al.* (2011) The molecular weight, solubility and viscosity of oat beta-glucan affect human glycemic response by modifying starch digestibility. *Food Chemistry*, **129**(2), 297–304.

Statistics Canada (2009) Mortality, Summary List of Causes [online]; available: http://www5.statcan.gc.ca/access_acces/alternative_alternatif.action?l=engandloc=http://www.statcan.gc.ca/pub/84f0209x/84f0209×2009000-eng.pdfandt=Mortality,%20Summary%20List%20of%20Causes [last accessed 9 May 2013].

Swain, J. F., Rouse, I. L., Curley, C. B. and Sacks, F. M. (1990) Comparison of the effects of oat bran and low-fiber wheat on serum lipoprotein levels and blood pressure. *The New England Journal of Medicine*, **322**(3), 147–152.

Tapola, N., and Sarkkinen, E. (eds) (2009) *Oat Beta-Glucan*. CRC Press, Boca Raton, FL.

Tosh, S. M., Brummer, Y., Wolever, T. M. S. and Wood, P. J. (2008) Glycemic response to oat bran muffins treated to vary molecular weight of β-glucan. *Cereal Chemistry*, **85**(2), 211–217.

Tosh, S. M., Brummer, Y., Miller, S. S., *et al.* (2010) Processing affects the physicochemical properties of β-glucan in oat bran cereal. *Journal of Agricultural and Food Chemistry*, **58**(13), 7723–7730.

UK NHS (2011) The Eatwell Plate [online]; available: http://www.nhs.uk/livewell/goodfood/pages/eatwell-plate.aspx [last accessed 9 May 2013].

USDA (US Department of Agriculture) (2011) My Plate [online]; available: http://www.choosemyplate.gov/food-groups/ [last accessed 9 May 2013].

US Department of Health and Human Services/Food and Drug Administration and Nutrition/Center for Food Safety and Applied Nutrition (2009) Guidance for Industry: Evidence-Based Review System for the Scientific Evaluation of Health Claims – Final [online]; available: http://www.fda.gov/Food/GuidanceRegulation/GuidanceDocumentsRegulatoryInformation/LabelingNutrition/ucm073332.htm [last accessed 15 May 2013].

US FDA (Food and Drug Administration) (1999) Health Claim Notification for Whole Grain Foods [online]; available: http://www.fda.gov/Food/IngredientsPackagingLabeling/LabelingNutrition/ucm073639.htm [last accessed 15 May 2013].

US FDA (Food and Drug Administration) (2003) Health Claim Notification for Whole Grain Foods with Moderate Fat Content [online]; available: http://www.fda.gov/Food/LabelingNutrition/LabelClaims/FDAModernizationActFDAMAClaims/ucm073634.htm [last accessed 15 May 2013].

Webster, F. H. and Wood, P. J. (eds) (2011) *Oats: Chemistry and Technology*, 2nd edn. AACC International Press, St, Paul, MN.

WHO (2012) Cardiovascular Diseases [online]; available: http://www.who.int/mediacentre/factsheets/fs317/en/index.html [last accessed 9 May 2013].

Zank, G. M. and Kemp, E. (2012) Examining consumers' perceptions of the health benefits of products with fiber claims. *Journal of Consumer Affairs*, **46**(2), 333–344.

17

Oh, What Those Oats Can Do: Quaker Oats, the US Food and Drug Administration, and the Market Value of Scientific Evidence 1984–2010

Robert Fitzsimmons

Harvard College, Cambridge, MA, USA

> *We made many a "bran new" theory of life over a thin dish of gruel, which combined the advantages of conviviality with the clear-headedness which philosophy requires.*
>
> Henry David Thoreau, *Walden* (1854)

17.1   Introduction

> *Oats supply what brains and bodies require.*
>
> Quaker Oats advertisement (1880)

In 2005, readers of magazines like *Good Housekeeping* and *Cooking Light* were witness to the opening of a bold new frontier. Advertisements offered the public a luscious, chewy oatmeal cookie, warm from the microwave. Oozing with chocolate chips, the cookie beckoned those with a sweet tooth everywhere to a healthy breakfast. No longer would lovers of a morning pastry struggle with guilt, Quaker Oats proclaimed, "Your childhood dreams have come true; you can have

Oats Nutrition and Technology, First Edition. Edited by YiFang Chu.
© 2014 John Wiley & Sons, Ltd. Published 2014 by John Wiley & Sons, Ltd.

Now he has another reason to smile!

The FDA confirms the first food specific health claim

Soluble fiber from oatmeal, as part of a low saturated fat, low cholesterol diet, may reduce the risk of heart disease.

Quaker Oatmeal. Oh, what those Oats can do.™
http://www.quakeroatmeal.com

Figure 17.1 Quaker celebrates its historic FDA approval with a full-page notice. (Quaker advertisement: *New York Times*, 1997.) Reproduced by permissions of The Quaker Oats Company.

a chocolate chip cookie for breakfast. Indulge responsibly with Quaker's Oatmeal Chocolate Chip Breakfast Cookies. Made with whole-grain Quaker Oats and sprinkled with chocolate chips, Quaker Breakfast Cookies are a good source of iron, calcium, and fiber. Your mouth will think it's a chocolate chip cookie, but your body will know better" (Quaker Oats, 2012).

Whether or not one believes these advertising claims (see Figures 17.1 and 17.2), why is nutrition being used to entice buyers of *chocolate chip cookies*?

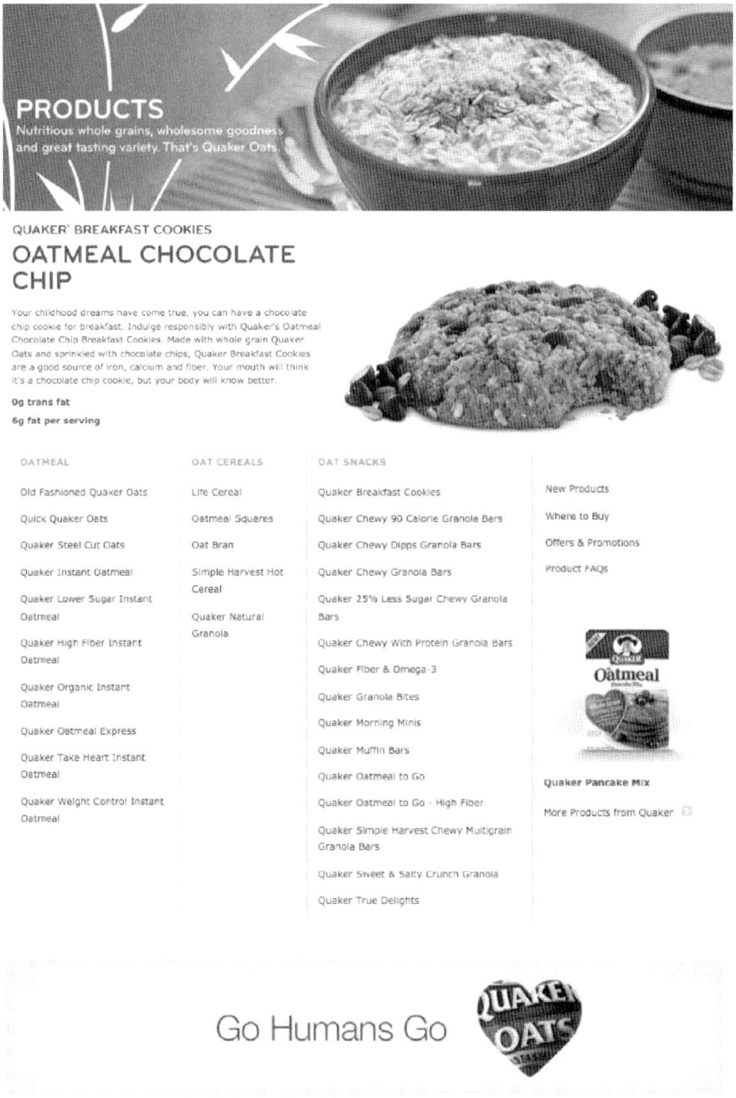

Figure 17.2 Setting forth Quaker brand identity, the full range of value-added options, and a new health-focused advertising campaign in a single web page (Quaker website, 2010). Reproduced by permissions of The Quaker Oats Company.

As a quantitative, reductionist approach to food, nutrition allows scientists to discuss food in terms of discrete, experimentally verifiable components that can be linked to health. At the same time, contemporary marketers have found scientific expertise to be an especially convincing promotional tool. However, to consider only the breakfast cookie is to miss a more radical shift: the underlying association between oats and heart health that makes Quaker's advertisement copy credible to consumers. This chapter attempts to chronicle the company's groundbreaking 20-year translation of nutrition science into successful consumer marketing.

The story of oatmeal, the first government-certified "health food," illustrates the ways that modern society approaches the interactions of commerce. The claim showed a new way to value a food product: it utilized scientific evidence to convince regulators and scientific language to sell to consumers. Quaker's claim added market value to sell more oatmeal and oat products. However, it also became important in the overall corporate strategy, positioning the manufacturer as socially responsible and responsive to consumer needs, generally as "wholesome" as its products. The claim's impact on the actions and attitudes of industries, policymakers, and even academics illustrates the power of two paramount American cultural values, scientific expertise and capitalist commercialism, combined.

Many American consumers are amazed to find that food products have only borne mandatory nutrition labels since 1994. They are even more surprised to learn that what the US Food and Drug Administration (FDA) terms "health claims" did not formally exist until 1997. Health claims are intended to be concise summaries of the nutritional research surrounding a food's effect on health, to "characteriz[e] the relationship between a food nutrient and the risk of a disease or health-related condition" (Wansink and Cheney, 2005). Manufacturers voluntarily print claims on their labels after completing a thorough scientific review by the FDA. What makes health claims so valuable—and controversial—is that the information goes beyond advertising nutrients and attempts to assign a prophylactic role to the product, such as protection against osteoporosis or heart disease. The greater specificity achieved with health claims is a clear market advantage for businesses, but critics insist that it overstates the certainty of scientific knowledge and misleads consumers.

Today, health claims are ubiquitous. For example, in 1998, more than 25% of the 11 000-plus new products were marketed based on their nutritional attributes (Nestle, 2003). Sales have continued to increase; global revenues for nutrition-marketed food grew by an average of 15.8% per year between 2002 and 2007. This far outpaced overall growth of food sales of 2.9% per year, indicating an important growth area for the slow-growing, highly competitive food industry (*The Economist*, 2009).

Beyond the current frequency of claims today, another reason that consumers may be accustomed to health information on their food is the longstanding belief linking food to health in popular discourses. This broad cultural appreciation allowed advertisers throughout the twentieth century to make oblique references to "wholesome" or "hearty" products without violating the law.

Food health claims were prohibited through the Pure Food and Drug Act of 1906.[1] This first federal law, like those that followed, resulted from a very old understanding from both consumer advocates and industrialists that regulation was necessary to gain consumer trust and ensure a fair marketplace. The law aimed to prevent consumers from being misled by scientific (or

---

[1] This law had been the result of a legacy of poor quality and adulteration in consumer products, in particular worthless and often dangerous "patent medicines," and given urgency with the publication of Upton Sinclair's meat-industry expose *The Jungle* earlier that year.

pseudoscientific) information; changes occurred in how scientists understood the properties of food and the degree to which that knowledge was transmitted to the public.

Ancient though the dietetic tradition may be, however, the modern history of diet and health knowledge is strikingly new. Mainstream consensus on the components of a healthy diet—expressed in measurable quantities of specific nutrients—was neither attained nor disseminated to the public until the late 1970s. Today, most nutritionists agree that Americans consume too much. Overconsumption is the cardinal dietary sin; saturated fat, sodium, refined carbohydrates, and other harmful nutrients are often particularly demonized. As both critics and company spokespeople will readily admit, the key to continued growth in the marketplace is increased sales: eating *more*, not less. Thus, the direction to eat more of a healthy food—containing fiber, monounsaturated fat, or any of a plethora of antioxidants, for example—has been embraced.

Historian Rima Apple notes that, "commercial firms were and are well aware of the power of scientific rhetoric in American culture," though longstanding regulation made it difficult to use health claims in selling food (Apple, 1996). These conventions broke down in the early 1980s due to savvy collaboration between a cereal company (Kellogg) and a government agency (the National Cancer Institute, or NCI). Kellogg's All-Bran advertised a claim in line with epidemiological evidence of the time, but in blatant violation of FDA rules. A pro-industry White House resolved the controversy by lifting the ban on health claim marketing. Apple believes that "the cultural authority of science sells," and marketers were quick to utilize scientific language to promote products from cookies to margarine once restrictions were lifted (Apple, 1996). Yet, this powerful promotional tool held potential for abuse through misrepresentation of scientific evidence or active ingredient content. Consumers expressed frustration, or even distrust, toward both industry and government regulators (Levenstein, 1993).

To show how health claims function, this paper attempts to show the historical progression of claims throughout the last 25 years. Health claims emerged as commercially viable—and legal—in the late 1980s with cereal advertisements but the deregulated environment led to a crisis in 1988–1990, as unscrupulous advertising touted the anticholesterol benefits of oat bran. The bran craze eventually crashed but its effects persist in the Nutrition Labeling Education Act (NLEA) of 1990, which severely restricted the scope of health labeling but institutionalized certification of those claims that the FDA deemed most valid. This certification process proved exceptionally valuable for those food marketers who could muster the scientific evidence to win approval. Quaker Oats, the first company to gain FDA certification, serves as a case study for how health claims can be leveraged for market advantage.

This approval was momentous for Quaker's public image and the company's overall strategy, establishing the brand as a leader in nutrition-focused foods. The claim's effects on shoppers' nutrition knowledge, and consequently in improving the national diet, are more difficult to discern. However, in examining the arguments and assumptions present in debates over nutrition and the proper role of commerce in public health, observers can begin to understand the beliefs that may underlie their own modern attitudes. We may come to understand how

American society values scientific expertise, and how businesses thus derive commercial value from that faith.

Merely reacting to science could be chaotic; positive, prominent studies could build brands and pad profits, or destroy them if reported negatively in the media. On the other hand, the food industry saw that science that was properly interpreted and leveraged could be a promotional tool of unprecedented market power. Ernest Dichter, a designer of product packages, notes, "If one wants to act rationally, one must, at all costs, find a reason which makes the irrational seem rational. Some of what appears on the package – especially the words – are there to reassure consumers that their impulsive choice was also a sensible one. Thus, nutritional information on the package won't be emotionally neutral. Anything that's put on a package—even a bunch of scientific names and numbers—can trigger feelings as well as thoughts" (Hine, 1997).

The debates that surrounded food regulation in the late twentieth century and that came to a head in the campaign Quaker spearheaded for oats and heart health essentially came down to unease over the possibility that scientific testimony's ability to sway consumer opinion might constitute added market value.

Newly added value did not necessarily entail a totally novel marketing approach. Nutrition marketing has occurred practically since the advent of national food advertising. Here again, Quaker had been the pioneer. The company's very first newspaper advertisement in 1879—also the national cereal industry's first—was a health pitch. "One pound of Quaker Oats builds as much muscle and bone as three pounds of beef," it read. "Oats [...] supply what brains and bodies require" (Sivulka, 1997). Despite the more stringent regulations enacted after 1906, the company continued to allude to the heartiness and good nutrition of its product in marketing. These advertising approaches were key to Quaker's growth but too vague to be influential in modern advertising.

Two mutually reinforcing trends made a shift toward specificity possible in the 1990s. Firstly, scientifically based health claims in the late 1990s were not only allowed but explicitly certified by the federal government. Manufacturers may have strained to meet the FDA's stringent standards of proof but the credibility that an approved health claim gives was highly useful. This phenomenon reinforced the second trend, an expanding public awareness and concern for nutrition. Consumers' preoccupation with dietary health was made manifest in everything from newsweeklies to popular cartoons. The confluence of these two factors created what journalist Michael Pollan has referred to as the "age of Nutritionism" (Pollan, 2008).

If this is, indeed, the age of Nutritionism, it is worth exploring the values and desires that have made nutrition knowledge so important today. Surety about health and nutrition has become important to consumers, mainly due to rapid acceleration in the dissemination of (often contradictory) scientific evidence. In order to satisfy this felt need, the government stepped in to *certify* the science conducted on constituent parts of foods and their health effects. This reflected a belief by all parties concerned—consumers, companies, and regulators—that a review of science would produce an objective judgment on the interaction of food and health. Apple noted that, "The rhetorical power of science in our culture is incredibly potent, so potent that many people [have] wrapped themselves in the flag of science. [...] Partisans of these controversies argue with contemporary science, with contradictory and contested science" (Apple, 1996).

The science may be controversial but the results are not. Those products that are certified because of this process receive "added value" in the eyes of consumers and, accordingly, manufacturers. What is quickly apparent, however, is that the objectivity of such science is as much a rhetorical tool (albeit an effective one) as it is a fact. Decisions on health claims' validity are driven by expectations of American consumers, regulators, and companies regarding food, health, and how information linking two should be communicated. The importance of consumers' choice to heed health information, in particular, has been overlooked in most prior histories of food regulation.

This chapter attempts to show that, although the attitudes and laws surrounding health information about food have existed in flux for the last 100 years, health claims did not become used as effective, legal marketing until the 1980s. Until this time, consumer values, regulatory priorities and paradigms, and corporate structure and strategy had not coalesced to produce an optimally receptive environment. It will become apparent that the attitudes and institutional arrangements surrounding scientific evidence are crucially important in nutrition science. The incredibly complex nature of biochemistry and physiology means that no conclusion can be intuitive, certain, or final. These seemingly exogenous factors therefore make scientific knowledge actionable in the here and now. In recognizing the often-ignored role of *all* participants in the debate in shaping consensus and applications of knowledge, it is hoped that this chapter will add a more realistic account of how corporate and government entities are shaped by public demand, as well as how these institutions' actions shape public understandings of science.

The French gourmand Antoine Anthelme Brillat-Savarin once proposed, "dis-moi ce que tu manges, je te dirai ce que tu es;" – tell me what you eat, and I will tell you what you are (Brillat-Savarin, 2000). How much more revealing, in our modern "age of nutritionism," to understand *why* we choose to eat what we do? If food is, as it has been called, our most tangible link to the material world, the emergence of nutrition as a method of understanding our own bodies' interactions with aliment, science has made unprecedented progress toward the organization of our lives. Additionally, if the use of this knowledge (whether correct or incorrect) can effectively sell products, nutrition labeling constitutes a use of science at the crossroads of the modern human condition – nonliving food turning into living being, individuals forming market demographics, inert information taking on monetary value. In a small, commonplace food label, science figures powerfully in everyday lives. Modern consumers purchase and consume science for breakfast.

17.2   Wild oats: The oat bran craze 1988–1990

What God has joined together, no man should put asunder. Put back the bran!
Sylvester Graham, *Lectures on the Science of Human Life* (1854)

In the late 1980s, millions of Americans joined the oat bran craze; it was, by all accounts, the most widespread health fad the nation had ever known. "Talk about buzzwords. It spawned hundreds of new products, tens of millions of dollars in sales, and created a national shortage of a grain once used to feed horses. All a company had to do was sprinkle a few flakes on its bread, doughnuts, potato chips, or even beer, slap the magic words 'oat bran' on the label, and watch the cash roll in" (Liebman, 1990).

Astoundingly, a prominent scientific study and a best-selling diet book were credited with nothing short of a revolution in American eating. Whereas US citizens bought less than 4 lbs of oats each in 1987, the average would be 7 lbs in 1989 (Moser, 1989). While the craze profoundly impacted national eating habits, the ephemeral nature that helped it grip the country so universally was also its downfall. Americans found it easy to abandon the extra roughage when a Harvard School of Public Health study questioned bran's health effects in January 1990.

To understand the oat bran craze in its historical context, one must understand that no mere diet book or journal article caused millions of Americans to double their consumption of oat bran. As historian Harvey Levenstein notes, "to be accepted, new ideas about food must also fit in with people's social and economic aspirations" (Levenstein, 1988). Throughout the twentieth century, society increasingly valued scientifically produced nutrition knowledge over more traditional dietary guidance. With hunger now relatively rare, they worried about chronic diseases of overabundance like atherosclerosis and cancer. At the same time, adoption of broader societal value placed on affluence, self-sufficiency, and personal choice led many Reagan-era Americans to believe they could make prophylactic decisions for their health through diet and fitness.

In many ways, oatmeal is the traditional food made modern. The oat was particularly amenable to analysis under emerging scientific and medical paradigms. Its virtues are well represented in reductionist, statistical, biomedical research: a specific, unique fiber measurably improved an easily administered blood test. Oats were not controversial. Their nutrient values were esteemed in dietary paradigms that nutrition science codified: low-fat, all-natural, plant-based, high in protein and fiber. Oats were nutritionists' "health food" long before regulators certified it so.

17.3 Brantastic voyage: Oats through dietetic history

Johnson: Oats [are] a grain, which in England is generally given to horses, but in Scotland appears to support the people.
Elibank: Yes, and where else will you see such horses and such men?

Boswell, *Life of Johnson* (1791)

The history of oats, and that of the Quaker corporation, is entwined with historical transitions in nutrition knowledge. Oats have traditionally been considered to have many of the health-promoting qualities that were attributed to it by

Figure 17.3   Quaker's message adapted for weight-conscious "Negative Nutrition" (Quaker advertisement: LIFE, 1967). Reproduced by permissions of The Quaker Oats Company.

scientists in the late 1980s. The attribution of health properties to food certainly predates the oat bran craze, and even modern medicine (see Figure 17.3). "It is probably not possible," notes historian J. Worth Estes, "to differentiate foods from drugs with mutually exclusive definitions. The ambiguity, troublesome as it may be in some contexts, has deep historical roots" (Estes, 2000). An understanding of the current controversies benefits from a long-standing perspective on beliefs about health properties of food.

For centuries, oatmeal's value was established in traditional dietetic practice (and the even-more-resilient common sense that some term "the wisdom of mom") (Pollan, 2009). For nearly two millennia, the guiding nutritional wisdom of the learned Western world was embodied in dietetics, a system of knowledge originating in the writings of classical physicians Hippocrates (460–370 BC) and Galen of Pergamon (AD 129–217). This tremendously persistent ancient paradigm managed "humours," bodily fluids which corresponded to the elements comprising the Aristotelian universe and which therefore had to be kept in balance for health.[2] Foods were assigned elemental affiliations based on sensory qualities and played a key part in this balance.

Such valuations even withstood the Scientific Revolution of the sixteenth and seventeenth centuries, when other Aristotelian paradigms were challenged and supplanted by experimental theories. The nineteenth century, however, marked the end of the dietetic tradition in the professional discourse. Noted German biochemist Justus von Liebig (1803–1873) announced that he had broken aliment down to its ultimate constituents: protein, carbohydrate, and fat. These same compounds made up the human body: We are literally what we eat. "The production of the constituents of blood cannot appear more surprising," he believed, "than the occurrence of the fat of beef and mutton in cocoa beans, of human fat in olive oil, of the principal ingredient of butter in palm oil, and of horse fat and train oil in certain oily seeds" (Liebig, 1843). Food no longer needed to be thought of as a mystery, Liebig argued. It was observable, reducible, understandable with science.[3]

Wilbur Atwater (1844–1907), an American, improved upon Liebig's findings by creating a "calorimeter" in his laboratory at the Wesleyan University. This machine was able to quantify the energy produced by the combustion of food (a rough allegory for human digestion). Popular magazines carried Atwater's articles extolling the virtues of a rational, scientific diet for health and economy. Because "the distinction between cheap food and expensive food disintegrated under chemical analysis," oats were particularly prized in Atwater's system. A sound nutritional source that provided quality protein and carbohydrates for energy, oats sold at prices far lower than refined wheat flour (Harper and LeBeau, 2003).

Atwater here embodied a paradigm shift that historian Harvey Levenstein has referred to as the "New Nutrition" (Levenstein, 1988). Atwater believed he had utilized science to identify those foods most beneficial to growth and health, based on adequate calories, fat, carbohydrate, and protein. Atwater and his associates sought to introduce scientific efficiency to the rhythms of work and family life. "These chemists," Levenstein reports, "recommended that people select their food foods on the basis of their chemical composition, rather than taste, appearance, or other considerations. In other words, they were telling people to

---

[2] For a more detailed discussion of ancient dietetics, see Innocenzo Mazzini, "Diet and Medicine in the Ancient World" in *Food: A Culinary History from Antiquity to the Present* (eds Jean Louis Flandrin *et al.*) Columbia UP, New York, 1999, pp. 141–152.

[3] Michael Pollan has isolated Liebig's discoveries as the birth of the new field of "nutritionism," a pathological discipline that he believes has abstracted human relationships with food to an unhealthy degree (Pollan, 2008).

eat 'what was good for them' rather than 'what they liked'" (Levenstein, 1988). New Nutritionists aimed to supersede tradition with experimental fact.

If oatmeal was a favored food in the Atwater system, its stock rose higher with the discovery of various "vitamines" in the 1920s and 1930s, isolated organic compounds essential to life. Vitamin-based "Newer Nutrition" (to again use Levenstein's terminology) reached a high-modernist apex in the Second World War and its aftermath. Due to strict rationing of foodstuffs needed at the front, an emphasis was placed on maximizing nutrition with limited resources.[4] By far the most important dietary legacy of the war, however, was the federal government's first attempt at prescribing a national diet. The Department of Agriculture's first attempt at dietary guidelines was released in 1943 at the behest of President Franklin Roosevelt. The "Basic Seven" (later "Basic Four"), in place until 1979, stressed whole grains, and offered oatmeal for breakfast as a perfect way to start the day (USDA, 1943).

The abundance of the postwar period greatly impacted Americans' views on food; the great increase in availability also led to an increase in consumption of rich foods like red meat, with the American meal of steak and potatoes newly available to a range of consumers. Yet a growing medical consensus was forming that blamed dietary fat for the increase in American heart disease cases. The nation closely followed the results in the media throughout the 1960s and 1970s. "The Fat of the Land," as one article was titled, had gone from a celebratory idiom to an object of deep anxiety (*Time*, 1961).

The growing consensus in the medical community eventually spurred Senator George McGovern (D-SD), chair of the Select Committee on Nutrition and Human Needs, to call for a re-evaluation of the government food guidelines when it released *Dietary Goals for the United States*, which advocated that Americans reduce their fat and cholesterol intakes and increase consumption of complex carbohydrates and fiber (US Senate, 1977). Foods low in fat were now favored as the new "health" foods. Oats and other starches, as the basis of responsible low-fat meals, received the biggest boost of all.[5]

The McGovern report contributed to an attitude that Harvey Levenstein has termed the "Negative Nutrition," in contrast to the more positive "Newer Nutrition" dominant in the first half of the twentieth century (Levenstein, 1994).[6] Negative Nutrition is focused on the role of overnutrition and certain "bad" foods in the etiology of chronic disease, particularly of cardiac conditions.[7] In the words of Ronald Reagan's 1984 presidential campaign, the 1980s were "Morning in

---

[4] Margery Vaughn, the government's senior rationing nutritionist, advocated in the *American Journal of Nursing* that Americans start the day with a full cup of oatmeal, as the richest source of "vegetable protein" available at breakfast, a responsible way to do one's "war job" by efficiently using food for energy (Vaughn, 1943).

[5] Indeed, they even had a presidential seal of approval: Eisenhower was a well-known oatmeal devotee, eating it daily following a heart attack in 1955 (Taubes, 2007).

[6] Levenstein credits fellow food historian Warren Belasco with the original coining of the term.

[7] McGovern's report shocked the USDA into action, and since 1979 the agency has been federally mandated to produce new dietary guidelines every 5 years.

America"—an era with worries and values distinct from earlier decades. Economic growth empowered personal choice, and a failing Soviet Union allowed Americans to focus their anxieties domestically, even internally, for the first time in years. Levenstein finds it "difficult to think of a society in which [diet obsession] was more pronounced than the United States of the 1980s" (Levenstein, 1994). America's weakened health now possessed widespread attention—and profit potential.

Dietetics began as qualitative judgments, ascribing relationships between certain foods and states of disease. Throughout three millennia, these relationships were preserved in professional medical discourse as well as popular tradition. However, with the ascendancy of experimental science in the nineteenth century, professional quantitative judgment took priority. Foods would be reduced to their constituent parts for study, with nutrition's role in the etiology of disease left to demonstrable causation (as in vitamin deficiencies). The growth of nutritional epidemiology and intense public interest around the discipline led to a re-introduction of qualitative information, linking food to a variety of illnesses humans suffered. No matter the system, valuations of oats as a healthful, hearty food remained remarkably stable, though expressed in new terms.

17.3.1   The oatmeal epic: A brief history of Quaker Oats

Oats have arrived! [...] A race of capable millers and inventors appear! [...] There are vast resources! Startling New Methods! It is the dawn of Big Business in the world's strongest nation! And the Quaker Mill at Ravenna, by reason of a priceless birthright, has a star role in The Oatmeal Epic.
Richard Ellsworth Day, *Breakfast Table Autocrat* (1946)

The Quaker Oats Company that encountered the oat bran boom of the late 1980s had roots stretching back to the mid-nineteenth century. As the first major national cereal corporation, Quaker pioneered its business in the United States. Quaker was "modern from its birth, free to pioneer and innovate," in the words of Arthur Marquette's 1967 corporate history. The company "changed the breakfast habits of the nation, revolutionized food marketing, created national brand food advertising, and, in passing, demonstrated to industry that diversification of [...] product lines is the life of trade" (Marquette, 1967).[8] In the process of making breakfast cereals familiar and appealing to Americans, Quaker was at the vanguard of mass production, branding, and advertising.

As might be expected from any century-old organization, Quaker's history is full of notable events, from the world's first cereal-box prizes[9] to the creation of the iconic Cap'n Crunch.[10] However, Quaker's methods of creating a powerful

---

[8] The grandiloquence of an earlier style of history likely strikes most readers as humorous today.

[9] In 1901, Quaker commissioned a silver-plated, engraved cereal spoon from the Oneida flatware company. Consumers could send away their box-top coupon to receive the spoon in the mail (Edmonds, 1958).

[10] Cap'n Crunch is the product of one of the most bizarre collaborations in business history. In 1963, the cereal itself was specially developed to stay crunchy in milk by venerable consulting firm Arthur

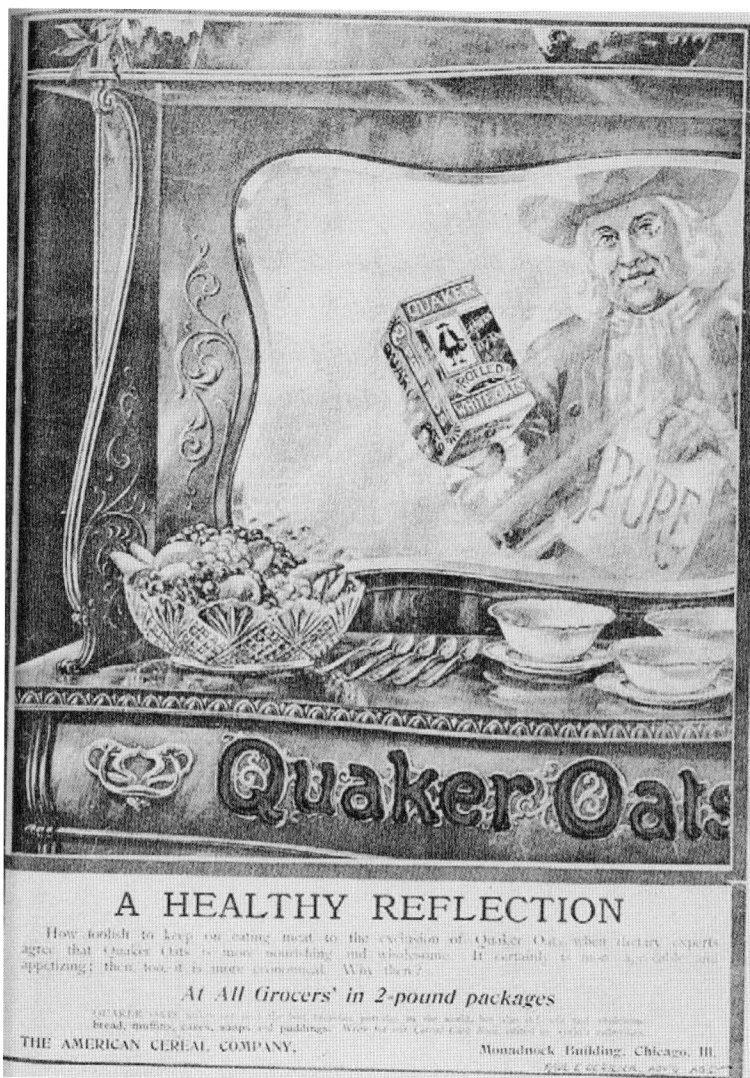

Figure 17.4   The *Saturday Evening Post's* first full-page, two-color advertisement featured a Quaker health appeal (Quaker advertisement: *Saturday Evening Post*, 1880). Reproduced by permissions of The Quaker Oats Company.

brand are of most import to this discussion of selling oat science in the late twentieth century. For more than 125 years, the company has both responded to and shaped consumer mores, blending the modern-scientific and the rustic-traditional identities of oats in order to sell more canisters of its namesake product (Figure 17.4). These tactics are here traced throughout Quaker's corporate history to

---

D. Little, while Jay Ward, animator of the classic Rocky and Bullwinkle series, created the character (Bruce and Crawford, 1995).

show the persistent success this approach has brought, and its different manifestations in separate social and scientific circumstances.

The group that eventually formed Quaker Oats began as a number of independent mills throughout the Midwest, owned and operated by millers of German and Scottish origin. The oldest mill was that of Ferdinand Schumacher (1822–1908), a German immigrant whose *Jumbo* mill in Akron, Ohio, was the nation's largest since its founding in 1850 (Bruce and Crawford, 1995). However, in 1881 Schumacher's rivals formed a trust to combat his market dominance. Harry Parsons Crowell (1855–1943), an ambitious 26-year-old who in 1881 had purchased a bankrupt mill in Ravenna, Ohio, led the group. Crowell became the outright leader when Schumacher, humbled by a devastating mill fire in 1886, joined the Quaker group.[11]

Whereas Schumacher had succeeded by selling large volumes of commodity oats, Crowell believed in the future food would be sold with intelligent marketing. Expanding the market for Quaker's product beyond traditional oatmeal-eating Scots, Irish, and Germans would require an unprecedented appeal directly to the consumer, making the case for oats as a better way to eat breakfast. Crowell recognized that the group's oats needed to be separated from the commodity crops that grocers scooped out of bulk barrels. Rock-bottom margins and brutal price competition meant that grain producers lived and died with the next crop. Crowell believed that a product with a stable identity of quality and purity could escape this cycle of "commodity hell."[12]

The "Quaker" name, chosen to convey the brand's "frugality, thrift, neatness, orderliness and integrity," was Crowell's chief asset in the merging of the various mills (Marquette, 1967).[13] Certainly, the mark was innovative; when it was registered in 1877, no other national company had marketed a breakfast cereal. Juliann Sivulka remembers Quaker as "marketing's first success with brand-name, packaged goods" of any kind (Sivulka, 1997).[14] She credits the company's eminence to two factors that turned oats from agricultural commodity to powerful brand. Firstly, Crowell focused on turning oats from a bulk good into a packaged one, removing worries about contamination and creating "a more desirable product by packaging them in a cardboard box printed with the picture of the Quaker man and a recipe" (Sivulka, 1997).

Secondly, and most crucial, was the way this novel packaged food product was promoted. Crowell left no avenue unexplored in his quest to sell more oats. "The pioneering cereal carried its message of health through a variety of forms: newspapers, magazines, streetcar signs, billboards, booklets, samples, cooking demonstrations, store displays, premiums inside the carton, calendars, cookbooks, and

---

[11] Schumacher would continue to be an active member of the group, however, engaging in multiple power struggles with Crowell over the years (Day, 1946).

[12] "The abyss called commodity hell" is the industrial fear of lacking product differentiation. It is a relatively recent, though illustrative term for a longstanding problem, likely coined by General Electric CEO Jeffrey Immelt (Hof, 2004).

[13] The previous owners had little relationship to the actual Quaker religious group, but rather chose the name while "searching the encyclopedia for a virtuous identity that would instill buyer confidence" (Hof, 2004, p. 31).

[14] Although Quaker moved to the familiar canister in 1915, one sees the same basic design today.

picture cards. Within a few years, the Quaker Oats trademark character became familiar nationally" (Sivulka, 1997).

Quaker thus became emblematic of the rise of branded products: Manufacturers were discovering for the first time to "add value." This concept, common parlance in the food industry, refers to the provision of advantages like better taste, convenient preparation, attractive packaging and branding, and (most recently) better nutrition through processing of raw food ingredients, in order to produce a product consumers appreciate and for which they will pay a premium. Crowell's attempts to add value through good branding led to Quaker being perhaps "the most promoted product ever" at the turn of the twentieth century (Hine, 1997).

Many historians of marketing argue, "Henry Crowell was one of the strongest forces in the creation of modern advertising" (Marquette, 1967). His long career of high profile, innovative campaigns (from the 1880s to the 1940s) had broad influence on both industry and eating habits. Harvey Levenstein points to "the emergence of large corporate entities profiting from mass markets for their mass produced foods" in the "mounting of large campaigns designed to change food habits through persuasion" (Levenstein, 1988). Quaker may have been hawking oatmeal but it changed American's attitudes toward food. Eating became an important issue of consumer *choice*.

These precepts served Quaker well throughout the twentieth century. The company sought to compete in the increasingly competitive cereal market by emphasizing innovation and research in both marketing and nutrition science. Its product advertisements never failed to stress oats' health image, and a willingness to spend for maximum impact meant that these health messages reached audiences with impact. The nation's first color advertisement, for example, hawked Quaker on the back cover of the *Saturday Evening Post* in 1899: "A Healthy Reflection: How foolish to keep on eating meat to the exclusion of Quaker Oats when dietary experts agree that Quaker Oats is more nourishing and wholesome. It certainly is more agreeable and appetizing; then, too, it is more economical. Why then?" (Marquette, 1967).

The marketing underlined actual research. In 1923, Quaker acquired the very first license for a scientific process to artificially enrich food, negotiating $60 000 in annual royalties with biochemist Harry Steenbock and the University of Wisconsin in order to advertise oats' vitamin D content (Apple, 1989). By 1946 research showed that oats "had just about everything – Vitamin B1, [...] available iron, phosphorus, and protein for body building! Also, Riboflavin and Calcium! If the Quaker scientists continue to find new blandishments, who can foretell what will be reported by 1950?" (Day, 1946).

Many believe that with such a strong precedent to follow, competitors like Kellogg and General Mills consciously modeled their marketing on Quaker's innovations (Bruce and Crawford, 1995).[15] These two competitors pushed Quaker to third place by the mid-twentieth century by more quickly and intelligently diversifying product lines and businesses. While Quaker foundered with investments

---

[15] With more than $12 million in sales by 1899, Quaker was America's best-selling breakfast during the early twentieth century. Indeed, it was so popular that Oliver Wendell Holmes called the company "the autocrat of the breakfast table."

in dog food, industrial chemicals, and Fisher-Price toys, its competitors domi-
nated the breakfast market with ready-to-eat cereals and Pop-Tarts.[16] By the late
1980s, the Quaker Oats brand was more than 100 years old, overextended in poor
product lines and unprofitable side businesses. Worse, it seemed to have run out
of interesting things to say. "It seemed that Quaker could not attract the attention
of American consumers," said members of the company's oatmeal group; it was
obvious the company as a whole was searching for identity (Nordhielm, 2006).
The bran craze, however, once again made consumers oat-conscious.

17.3.2   The muffin and the mania:[17] Bran breaks through

Ay, sir, they be ready: the oats have eaten the horses.
William Shakespeare, The Taming of the Shrew (1594)

Historically, the FDA had outlawed language on product labels linking foods to
specific health conditions since 1906, when the Pure Food and Drug Act prohibited "false or misleading" claims under penalty of seizure and prosecution
(US Congress, 1906).[18] As the FDA has historically been a small, underfunded
agency, George Kurian believes "the basis of the law rested on the regulation
of product labeling rather than pre-market approval" (Kurian, 1998). Monitor-
ing claims after they reached the market was not ideal but enabled the FDA to
correct blatant abuses.

The FDA had been particularly focused on misleading food claims. Under the
Food, Drug, and Cosmetic (FD&C) Act of 1938, a food that claimed effectiveness
against diseases was considered a drug under FDA rules, and would have to file
a new drug application; they were also prohibited from reaching the market until
the FDA gave its approval. Hilts explains the bureaucratic hurdles entailed, "the
product [...] would have to go through the same painstaking process as any drug
would" (Hilts, 2003). The time, expense, and high burden of proof involved acted
as an effective deterrent.

In 1984, however, one of Quaker's competitors challenged the status quo. Kellogg's All Bran, citing recent epidemiological research, advertised the purported
ability of dietary fiber to fight cancer.[19] All-Bran reaped enormous benefits: an

---

[16] Quaker made attempts to catch-up with products like Cap'n Crunch and Chewy granola bars
but remained a minority player in these markets. Some say, of course, that such diversification was a
deviation from the health of whole grains, adulterating them with sugar and other ingredients. (see
Bruce and Crawford 1995 for an entertaining look at earlier twentieth century competition).

[17] (Burros, 1988).

[18] This broad definition was clarified to distinguish food from drug in the 1938 Food, Drug, and
Cosmetic Act (FDandC), which remains in effect today in a modified form.

[19] Although dietary fiber is not considered a controversial topic in the popular discourse, it is worth
noting that the research that supported a relationship between fiber and cancer has been repeatedly
challenged in the medical literature, as many have criticized the observational studies as showing, in
fact, the benefits of a diet low in refined carbohydrates and high in fruits and vegetables (both are
correlated with high fiber intake), not specifically of a fibrous diet. This underscores a continuing
ambiguity about the actual biological mechanisms of any fiber, including oats' β-glucan (Taubes,
2007).

"astonishing" 47% increase in market share within the first six months of the campaign showed unequivocally that "health claims sold products" (Nestle, 2003).
Surprisingly, the FDA took no action on Kellogg's health claim.[20]

The claim's success perhaps owes more to savvy navigation of government
bureaucratic structure, rather than scientific rigor. Kellogg developed the claim
with the National Cancer Institute, the FDA's sister agency within the Department of Health and Human Services (HHS). This intradepartmental end-around
left the FDA powerless. Although Dr. Sanford Miller, the FDA's head of Food
Safety and Applied Nutrition, considered the product claim "incorrect because
there's no evidence that this kind of fiber can help," the agency was unable to
publicly contradict an NCI statement.

In the eyes of those higher up in the Reagan administration, "the FDA's job
[was] to encourage this trend [...] by cooperating with industry to make healthful
products available," as FDA head Dr. Frank Young explained (Food Chemical
News, 1985). The agency announced a "cautious green light [...] to use health
claims in promotions" until new rules could be drafted (Colford, 1985). Sud-
denly, minor changes to a product label could add a powerful new marketing
incentive. Within five years of the All-Bran campaign, an estimated that 40%
of all new food products advertised their health appeal (Hilts, 2003). "What
has All-Bran wrought?" wondered *Consumer Reports*, even as the venerable
magazine revamped its coverage to compare cereals' fiber content (*Consumer
Reports*, 1986).

Claim labels allowed average consumers to consider food in a quantitative,
reductionist framework. Dietary information was disseminated along a clear specialist hierarchy. Peer-reviewed scientific media served to convert the popular
media, who then translated expert scientific testimony into more easily digestible
information. Along the way, valuable quantitative context detailing everything
from statistical power to experimental design was completely excised.[21] Only
then, it was believed, could the lay public understand these recommendations,
and spend their money accordingly.

Whether health claims sold consumers on sound nutritional practice was far
less clear. The FDA found in 1987 (and again in 1997) that health claims' chief
benefit was rather dubious: a quick shopping heuristic that shoppers could use
to decide upon a product without consulting the back of the label for quantitative nutrition information (Levy and Stokes, 1987, Levy *et al.*, 1997). Economists
at the Federal Trade Commission (FTC) presented a more positive interpretation focused on the measurable effects of the advertisements: consumption
changes, which were positively influenced in favor of high-fiber options (Ippolito

---

[20] For a (politically-slanted) report about the incident, see United States Cong., House, Committee
on Government Operations, *FDA's Continuing Failure to Prevent Deceptive Health Claims for Food*,
101st Cong. 2nd Sess. (14 November, 1990).

[21] Interestingly, in retaining some of the numerical trappings of nutrition science—grams of fiber,
cholesterol readings in milligrams/deciliter—these simplified accounts actually increased the mystique
and power of the science they described. "Science has methods to offer to achieve longevity, and you
can use them," journalist Robert Kowalski wrote in 1987. "You *can* cut the risk of heart disease and
other life-shortening diseases. That's a promise" (Kowalski, 1987).

and Mathios, 1989).[22] Both had no choice but to admit that the conflicting data caused ambivalence and confusion.

The rapid increase in health claims as a result of the All-Bran fiasco created a precedent for oat bran in 1988. A correspondent for *Fortune* magazine remembered the more proximate causes for the mania:

> What started the craze was a California medical writer, Robert E. Kowalski, whose book *The 8-Week Cholesterol Cure* was published in 1987. The cure's magic bullet: oat bran. The book climbed onto the best seller list and sits there still. Then, the following spring, the *Journal of the American Medical Association* published a report on the virtues of reducing cholesterol through diet, including oat bran. Right away, grocers across the land heard the nonstop patter of feet in the cereal aisle (Moser, 1989).

Fortune credits the influence of peer-certified scientific testimony, combined with the accessibility of a popular source, in swaying public opinion.

Oat bran was launched into the public consciousness with Kowalski's *8-Week Cholesterol Cure*, a diet plan that promised to help the reader control not only their weight but also their heart disease risk. Kowalski claimed to have lowered his cholesterol 115 points in two months by combining a low-fat diet with regular exercise and riboflavin, a B vitamin. Yet the centerpiece element of Kowalski's program was oat bran, part of the cereal in "that familiar cylindrical container with the smiling Quaker face that today seems so wholesome" (Kowalski, 1987). The skeptical reader might wonder how this worked. Kowalski explained, "Just what can be expected by making oat bran a part of the daily diet[?] Oat bran significantly lowers both total cholesterol and LDL cholesterol while not at all lowering the protective HDL levels. As a side effect, oat bran helps maintain a normal glucose level in the blood of diabetic patients. [However,] no one knows for certain how oat bran works" (Kowalski, 1987).

In addition, the fiber in oat bran would keep dieters from being hungry and provide them with essential vitamins and minerals. The familiar Quaker breakfast staple was imbued with so many mysterious health effects that if oat bran was not the panacea consumers took it to be, the confusion was understandable.

Kowalski, though not a medical doctor or researcher, invoked two forms of expertise that have proved enormously effective with the lay public, despite not satisfying scientific standards. Firstly, the author was a journalist with graduate-level training in physiology who had spent years covering scientific discoveries for national newspapers (Kowalski, 1987). In addition, Kowalski had worked in the pharmaceutical, medical, and food industries (in effect, every industry that would be interested in his health phenomenon). Kowalski's professional and academic experience combined technical knowledge and the savvy to bring bran to the masses.

---

[22] Whereas the FDA was charged with regulation of food constituents and labels, advertising for food products fell under the jurisdiction of the FTC. The FTC is focused, like the FDA, on consumer protection, but also has the encouragement of competitive business practices in its mission. Thus, the FDA and FTC not only collected conflicting data but also fell into their habitual regulatory philosophies when asked to interpret these complexities.

Secondly, however, and likely just as important as any professional affiliation, was the fact that Kowalski was a two-time heart attack survivor. Repeatedly, he communicated to readers that he understood their concerns. Like any good parent, Kowalski knew he "had so much to live for! I adored my very young children [...] Never before had I wanted so badly to live" (Kowalski, 1987). The combination of an "everyman" story with Kowalski's claims to understand scientific truth made for exceptionally powerful rhetoric. The *8-Week Cholesterol Cure* was a number-one *New York Times* bestseller, and remained on the list for an unprecedented 115 weeks (McDowell, 1989). More than two million copies were sold; 8% of the United States population believed Robert Kowalski enough to buy his book. The fad was underway.

Kowalski's personal experiences and the testimonies his doctors gave sold a great many diet books and the existing scientific literature Kowalski could cite was voluminous. Nestle reports that "11 studies [...] by 1988 had demonstrated reductions in blood cholesterol ranging from 3% to 26% as a result of eating from 1 to 5 ounces of oatmeal or oat bran daily" (Nestle, 2003). While these studies provided a wealth of background reading, oat bran became a phenomenon with the well-timed publication of a review, entitled "Cutting into Cholesterol: Cost-Effective Alternatives for Treating Hypercholesterolemia," which appeared several months after Kowalski's book entered the bestseller lists (Kinosian and Eisenberg, 1988). The authors, Dr. Bruce Kinosian and John Eisenberg, had medical credentials and business experience. Relying on previous studies' empirical findings, the study compared the effectiveness of oat bran to that of two prescription medicines, cholestyramine resin and colestipol.[23] While all three interventions worked well, oat bran was *far* cheaper. The authors estimated that a significant reduction in cholesterol (85 mg/dL) would cost about $20 000 per person with oat bran; prescription drugs would cost between seven- and ten-times more. While individual consumers could certainly appreciate these savings, Kinosian and Eisenberg believed that the societal cost reduction—in health and more traditional monetary assets—would be enormous. Equally large were the study's effects on bran consumption: "Even though there was no new evidence in the article about the usefulness of oat bran in lowering cholesterol, by the time the information had made its way into the press, oat bran had become the magic bullet consumers were looking for" (Burros, 1989).

Although bran was far cheaper than prescription drugs, this certainly did not mean that it was an unprofitable business. Products with oat health claims were selling like (fibrous, whole-grain) hotcakes. Journalist Penny Ward Moser noted that the craze had the feeling of a riotous collective binge, "When I went to the store myself, I was astounded: There they all were, yuppies in Izod shirts or Fila joggers, reading cereal boxes as if they were bodice rippers. The only oatmeal left was apple-cinnamon in a green box (my husband had thought it was lime because he didn"t have his glasses). Like any good consumer in a frenzy, I bought it" (Moser, 1989).

---

[23] The first statin drugs had only recently been approved for sale (1987) and thus were not included in the analysis. However, the authors estimated that oat bran would be equally effective as the new drugs while saving around $11 000.

Moser believed oats were "traditionally good as animal feed or a breakfast that tasted like wallpaper paste," but as "a way to battle killer cholesterol" they had rapidly regained their place in the American diet and Americans' conscious (Moser, 1989).

The numbers supported Moser's experience. A 900% increase in oat demand in 1989 drew both giants like Quaker and General Mills but also upstarts like Health Valley, which quickly became one of the nation's largest health-food corporations as a result (Moser, 1989; Nestle, 2003). In the fiscal year 1988–1989, sales of oats and bran grew 35.6% to $2.26 billion, representing an astounding 31.4% of total cereal sales and outpacing growth in the industry as a whole by more than 20 percentage points (Erickson and Dagnoli, 1989a). "If you look at the market for hot cereal, which is primarily an oat-based product, dollar sales were up 25% last year after years of stagnating," noted one industry expert in 1989. "People have really started to eat oat bran like horses. If this keeps up, there won't be enough left for the horses and we might have to start feeding them doughnuts and cake" (Kleinfeld, 1989).

Quaker Oats is the brand name nearly synonymous with oats in America, and has been since it was the first registered trademark for a cereal in 1877. It might be wondered, however, why it is an exemplar of changing American notions of diet and health in the late twentieth century. Even its primacy in the oat bran craze requires explanation: Certainly, one company does not a craze make. What made the craze so astounding was its gold-rush feeling: it seemed every food company in existence (and many that hadn't existed before) was churning out a bran-filled offering. The craze mattered most, however, to Quaker, the company that owed its name and its continued market competitiveness to *Avena sativa*.

Whereas Quaker's oat bran was primarily found in its traditional offering, hot cereal, its rush to bring new products to the market was somewhat behind other food companies. These rivals sprinkled bran on potato chips, beer, brownies, and other counterintuitive products. Many consumer groups were incensed at the liberal use of bran: "when doctors talk about lowering your cholesterol, I don't believe they're talking about doughnuts with a smidgen of oat bran in it," protested Bonnie Liebman, director of nutrition for Center for Science in the Public Interest (CSPI) (Kleinfeld, 1989). It hardly seemed to matter, however. Consumers were buying oat-bran products and it seemed everyone was clamoring to answer the demand. In 1988, more than two hundred brand-new oat bran products jostled for shelf space (Nestle, 2003).

The fact that it cost up to $30 million to introduce a food product in 1989 (and that 90% of those new products failed to become permanent offerings) gives an indication of the enormous profit potential of bran (Shapiro, 1990). "Food marketers—acting without government guidance or regulation—are facing a once-in-a-lifetime opportunity" for profit (Erickson and Dagnoli, 1989b). Enormous corporate accounts would be thrown at those advertisers savvy enough to capitalize on health. Quaker spent $357 million on advertising in 1990, and General Mills $471 million (*Advertising Age*, 1990). Quaker's McKinney believed that advertisers and industry needed to catch up to the other players in health, "This isn't that difficult to solve. We've advanced a lot in dietary options and the medical profession's perception of the relationship between diet and health. Now we need to resolve the relatively simple matter of communicating those ideas" (Erickson and Dagnoli, 1989b).

When these advertisements were memorable and effective (like Quaker's long-running "Do the Right Thing" campaign with mustachioed pitchman Wilford Brimley), profits beckoned.

The sheer size of the shift in consumer demand managed to catch major oat producers unprepared. The chaos was such that Americans actually purchased all of the oat bran Quaker produced for domestic consumption. "We're out of capacity, people are screaming, everyone starts chasing this thing because there's this roaring demand going," remembered Polly Kawalek, the vice president in charge of Quaker's hot cereal division (*Chicago-Booth Magazine*, 2000). Quaker brought in bran from overseas and ran its plants around the clock (Moser, 1989).

The craze was noticeably impacting the priorities of all of agriculture. Mary Shelman, director of the Harvard Business School agribusiness program, worked in the rice industry in the late 1980s and recalls a priority effort made by rice producers to nutritionally analyze their bran for health benefits (Shelman, personal communication). The potential to be "the next oat bran" was alluring—if consumers were willing to pay top dollar for an agricultural byproduct normally fed to cattle, why would producers protest? Certainly, humans paid a bit more for their bran than the livestock had.

The sheer number of companies rushing to fill the oat bran demand created a wealth of options for consumers. However, this was not always advantageous. Without the rigid consumer protection that had been previously provided by the FDA, consumers were left to guess which products fairly represented their nutritional content. Even if one accepted the evidence underlying that oat bran craze, how could it be known that any given purchase was a quality source of the coveted ingredient?[24] Confusion over the science was only exacerbated by insufficient FDA regulation.

Consumers' frustrations were magnified for Quaker, which considered itself unfairly discriminated against by the FDA's lack of involvement. The company had always prided itself on its status as a nutrition leader, extensively financing proprietary and academic research. However, the oat bran mania challenged this model. It allowed unscrupulous corporations to benefit from Quaker's own work, even as they damaged the industry's hard-won reputation. McKinney vented in the *New York Times*: Quaker Oats had scientists on its staff, carried out serious research, and adhered to rigorous scientific standards. The company made use of its own work and that of science in general to make claims for its products. But most companies, [McKinney] said, had little interest in science; "They just sprinkle a little oat bran on their product and ride this train without any backing for their claims."

---

[24] Although nutritional labels were not mandated for all foods in the late 1980s, the oat bran craze demonstrated the inadequacy of a purely nutrient-based label. Oat bran sold so well because it claimed to be better than other forms of fiber in heart disease prevention; a label that merely labeled all fiber in the same way would not answer consumers' concerns. Bran doughnuts were probably not the best source, of course. One might need to eat more than 100 bran cookies, which commonly contained 0.5 g of bran each, to get the AMA's recommended daily dose.

Clearly, these abuses could not stand in the future; the FDA needed to be empowered for the protection of America's citizens and its businesses.

17.3.3   Suing wild oats: A bran backlash

The two made one crop of wild oats, for which he was heartily sorry, and he could not see that those oats are of a darker stock which are rooted in another's dishonour.

E.M. Forster, *Howards End* (1910)

Figure 17.5   Ambivalence about bran mirrored other anxieties of the late 1980s. Copyright © Mark Allen Stamaty. Reproduced with permission from the artist.

The bran fad's demise was due in large part to the publication of yet another scientific study in a prestigious journal, calling bran's cholesterol-lowering effects into question (Figure 17.5) (Swain *et al.*, 1990).[25] The press was abuzz: "Has the oat bran bubble burst?" asked the CSPI (Liebman, 1990). The answer soon became clear; by January 22, the craze appeared to be "on the wane" as consumers lost faith in product claims (Liesse and Dagnoli, 1990).

Oat bran's headlines did not necessarily represent a more easily comprehensible scientific consensus. Indeed, "[1989] has been a record year for sitting in front of a television screen, a bag of oat-bran potato chips in one hand, a bottle of oat-bran beer in the other, watching late breaking reports on menacing foods" (Dullea, 1989). No conclusion was safe; even though it doomed oat bran

---

[25] Twenty subjects were fed 90 g of either oat bran or refined white flour daily for six weeks. In the end, the authors found no significant cholesterol-lowering benefits for the fiber-fed subjects.

donuts, the Harvard study design was harshly criticized within the medical community (Roubenoff and Roubenoff, 1990).[26] "The authors have managed to confuse the American public further with a poorly designed and underpowered trial that draws erroneous conclusions," wrote one dissenting reader (Burris, 1990). Conclusive data were hard to find.

Businesses were equally vexed; many had only seen the craze continuing. Given that Americans had consumed seven pounds at the craze's height and only four the previous year, analysts were far too bold when they "expect[ed] that number to go to ten by 1991" (Moser, 1989). Quaker lost more than $20 million on oats alone in 1990, as most of its new bran products had hit the market in December 1989. With two years of work and planning wasted, "we were disgraced. Everyone else left," remembered Quaker veteran Polly Kawalek (*Chicago-Booth Magazine*, 2000).

Many critics have targeted the rise and precipitous fall of oat bran as an exemplar of the harmful effects of too-simplistic scientific media coverage. Epidemiologists argue that the media errs in assuming that any conclusions were, indeed, conclusive. When "each little jigsaw piece is picked up by the media and made into a message," as one doctor complained in 2006, the public gets "the impression that nutritional advice changes every day of the week" (Trivedi, 2006). The difficulty and uncertainty of thorough, large-scale nutrition studies mean that almost any finding is only a partial answer in an overall nutritional understanding.

Bonnie Liebman, nutrition director at CSPI, maintained that most people were not adequately interested or qualified to interpret original scientific information. "You can't expect Joe Six-Pack to read six reports on diet and cholesterol," she believed, expressing exasperation with news reports that "typically portray food scientists as befuddled figures in white coats who are constantly scratching their heads and reversing themselves" (Dullea, 1989). If the media could not report science accurately the average consumer lacked both the interest and the education to check facts, the situation seemed bleak (Liebman, 1990).[27]

It seemed that academic nutrition science was not attuned to the public's expectations. Jonathan Pizer, an ordinary resident of Chicago's north side, fretted: "[Now] they tell us oat bran really doesn't reduce cholesterol. [...] What are we supposed to do?"[28] A nationwide Gallup poll conducted in 1989 confirmed that Mr. Pizer's frustrations were by no means isolated. An overwhelming majority (82%) of consumers wished to see more products advertised as having health benefits and 47% had purchased such foods in the previous thirty days. At the same time, however, nearly half stated that they did not believe that such products would reduce cholesterol (Erickson and Dagnoli, 1989b). Thus, even though these consumers may have listed Quaker's oatmeal among the healthiest foods

---

[26] The study was conducted on a small population of dieticians, who were already engaged in a heart-healthy lifestyle.

[27] Liebman insisted that expert opinion had to be the only standard, but that such expertise could not be held to timelines: "[I]f you're beginning to feel like a volleyball, bouncing back and forth with each new headline about cholesterol, don't despair. Researchers sometimes take years to get to the final answer. Once you accept that, it's easier to take the next bounce of the ball." This sort of relativism, obviously, was hard for many consumers to swallow (Liebman, 1990).

[28] Dowd, "No Perrier?" "It's Reagan's fault," Mr. Pizer added. "I don't know how, but it is."

they purchased (alongside apples and yogurt), they remained skeptical of any health message. Public trust was severely damaged.

While the federal government continued to debate the best regulatory response, state-level governments invoked the constitutional right to regulate commerce within their jurisdictions. The National Association of Attorneys General coordinated an all-out assault of litigation on offending cereal producers. Kellogg faced lawsuits in Iowa and New York for claiming that its "Rice Krispies" were an excellent B-vitamin source. General Mills' "Benefit" cereal, which used psyllium fiber (the base of Metamucil) to claim anticholesterol powers, was challenged and pulled from the market in 1989 (Nestle, 2003).[29]

By far the most bitter, protracted legal battle was Texas attorney general Jim "Mad Dog" Mattox's suit against Quaker. Mattox sought a temporary restraining order against the company, claiming "[the] oat bran craze in this country was primarily started by Quaker in order to sell its products," and that "consumers have been duped" (*Business Week*, 1989). Quaker strenuously objected, and countersued. "With the states jumping in, [food labeling] has become like playing baseball with 50 referees: It's hard to play competitively and on a level playing field," said one of Quaker's lawyers. "In fact, it's hard to play at all" (Erickson and Dagnoli, 1989b). Although federal regulations would be a "mixed blessing," Quaker argued that a unified national policy would be a significant improvement over uneven state-by-state standards (Erickson and Dagnoli, 1989b).

Mattox v. Quaker Oats was particularly poignant because those who did *not* believe the studies showing bran's promise also invoked science. Studies showing oat bran's effectiveness would be rebutted with contradicting research, critiques of experimental design and size, or allegations that corporate sponsorship had adulterated findings. Historian Rima Apple has observed a similar phenomenon in the decades-long debate over vitamin supplementation. "Lack of evidence did not undermine faith in a scientific solution to the vitamin controversy. Both supporters and opponents considered science to be the arbiter of the question. [...] did not provide clear, unambiguous results. Consequently both sides used the very lack of scientific consensus to defend their own stands and to accuse their opponents of scientific ignorance. Consumers themselves used the uncertainty of orthodox science to defend their right to choose" (Apple, 1996).

Oat bran, like so many scientific skirmishes, became a battle over beliefs, with empirical evidence used as a rhetorical weapon.

Whereas vitamins have been more or less a niche product for the 90 years since their discovery, the oat bran fad marked the first time in national memory that a large percentage of consumers simultaneously adopted a dietary change.[30]

---

[29] The suit was originally filed in Texas, but when the federal government took up the case, "Big G" abandoned the claim and 50 tons of Benefit, which aided the gastrointestinal regularity of livestock in South Dakota (Bruce and Crawford, 1995).

[30] Perhaps the clearest manifestation of the oat-bran-as-archetype phenomenon in the subsequent 20 years has been the low-carbohydrate diet craze of the early 2000s. The "low-carb" concept was nothing new. Indeed, the first modern weight-loss diet, William Banting's *Letter on Corpulence, Addressed to the Public* (1863) had been based on avoiding bread sugary puddings. Dr. Robert Atkins had made bestseller lists intermittently since the 1970s with his antiestablishment "Diet Revolution."

Although bran sales eventually stopped crackling, calls for tougher standards did not. Bonnie Liebman of CSPI grew bolder: "Maybe [lawsuits will] teach companies like Quaker Oats not to plaster exaggerated claims about heart disease on their labels before the science is more certain" (Liebman, 1990).

The *New York Times* took a more positive viewpoint as *Mattox v. Quaker Oats* came to an end in February 1990 (in anticipation of federal labeling law changes later that year). Editorials called on the FDA to rewrite its regulations for the good of both customers *and* producers when the Congressional bill was finalized. "The delay [to reinstate rules] is doubly unfortunate. Advertising has great power to mislead [… but] can also convey valuable nutrition information, which consumers increasingly seek. […] The confusion over food claims hurts manufacturers, too. Investments promoting foods with oat bran may have suffered now that new studies dispute oat bran's value in reducing cholesterol. It's a proper role of government to determine where consensus exists and then let food makers inform the public" (*New York Times*, 1990).

The chaos that the craze became had ironically left united the public and industry awakened to new possibilities and frustrated at the abuses of a totally open market.

17.3.4   Conclusion

The oat bran craze speaks powerfully to the public's views on health, to food producers' calculations of profitable promotions, and to government's struggles to regulate this intersection of science and business. Michael Pollan, perhaps the most prominent critic of the phenomenon, regards oat bran as an archetype. "The Year of Eating Oat Bran—also known as 1988—served as a kind of coming-out party for the food scientists, who succeeded in getting the material into nearly every processed food sold in America. Oat bran's moment on the dietary stage didn't last long, but the pattern had been established and every few years since then a new oat bran has taken its turn under the marketing lights" (Pollan, 2008).

Oat bran was not the first indication that consumers cared about health; it was however, the first case of a sudden change in food habits that captured national attention and powerfully affected corporate bottom lines. With oat bran "science was removed from the laboratory and placed squarely in the center of American culture and commerce." The possibilities of large-scale dietary change were readily apparent to industry.

Bruce and Crawford (1995) believe that "during the 1980s, the Reagan administration's laissez-faire economic and social policies allowed corporate America to gorge itself at a breakfast of the vanities." In reality, negative consequences

---

In the 2000s, however, low-carb went mainstream. It should not surprise that the craze closely resembled oat bran. It had its "Kowalski" (bestselling popular-science author Gary Taubes), contradicting studies in *JAMA* and the *New England Journal of Medicine* and high-profile media attention in national sources like the *New York Times* and *Time*. It was also highly profitable: the craze was estimated to sell $10 billion in products in 2003, with Atkins Nutritionals earning $200 million that year, before being acquired by a Goldman Sachs-led private equity group. The eventual crash, unsurprisingly, was similarly disastrous for those corporations on the bandwagon.

were apparent to all, though preferred responses varied. Total deregulation had been a failure, and new laws were needed. "Congress is considering legislation to hurry the FDA along. The oat bran imbroglio [...] attest[s] to the liveliness of the experiment the agency started with All-Bran—and the urgency of completing it" (*New York Times*, 1990). Reinvigorated legislation would not be easy to create or to enforce. Oat bran, however, convinced the government that it was the best way to protect both consumers and business.

17.4   Gruel intentions: The NLEA and Quaker's health claim 1990–1997

Spare your breath to cool your porridge.
                    Miguel de Cervantes, *Don Quixote de la Mancha* (1605)

The nation could not choke down bran beer forever. By early 1990, with new scientific studies calling bran's effect on cholesterol into question, the craze came crashing down like a poorly constructed grocery store display. Many consumers felt confused by scientific authority, abandoned by regulators, and cheated by food manufacturers. Corporations, too, felt that their rights to fair competition were ignored by totally unregulated advertising and labeling. The bran controversy demonstrated a need for clear, enforceable law to substantiate health claims.

"Food is becoming the pharmacy of the 1990s. People don't take vitamins, they eat different foods" (Kleinfeld, 1989). The public decided which products were successful, and those with health messages—valid or not—were selling. The power of marketing, if properly channeled and controlled, offered great potential to educate the public. Unregulated claims, by contrast, hurt as often as they helped. The FDA's inability to control the bran craze led directly to an act of Congress, the NLEA of 1990.

One of the principal supporters for stronger legislation was Quaker Oats. Accordingly, Quaker was the first to successfully navigate the health-claim process created by the NLEA and thus gained the market-viable certification of FDA approval, a unique and significant advantage for oatmeal. Quaker's experience transformed the way corporations viewed the possibilities of nutrition information; rather than a neutral or negative experience, working with the FDA could provide unprecedented opportunities.

17.4.1   Regulating regularity: The NLEA of 1990

To the consuming fire we consign thee – BRAN. Ten thousand dieticians have wept over thee in vain.

                    USDA Dietician Margaret Sawyer (1927)

The passage of the NLEA is viewed as a triumph for the FDA but can more accurately be termed a carefully considered compromise. The complexity of food industry regulation was readily apparent in the passionate debates nationally and in Congress over the bill. Marion Nestle believes that "separate lines of regulatory proposals for food labeling [and] health claims on food labels [...] became thoroughly entangled with each other" (Nestle, 2003). The ever-present difficulty of philosophically distinguishing food from drug reached new levels in lawmaking.

The NLEA became law on 8 November 1990, ending a six-year regulatory moratorium. Foods that made health claims would once again need to be pre-registered with the FDA. However, they would not be held to the high standard of proof that had once applied under the FD&C Act of 1938 (US Congress, 1990). The NLEA, therefore, constituted an effort to strike a balance between these all-or-nothing approaches, to allow viable claims.

The FDA had suspended its regulation of claims with All-Bran in 1984 as "an experiment" but critics were highly ambivalent about the agency's success. "Food health claims have since proliferated, but the agency's experiment remains unfinished," the editors stated, because "the FDA has delayed writing urgently needed rules to clarify what claims may be made and what degree of scientific consensus should support them" (*New York Times*, 1990). Consumer groups like the CSPI agreed, insisting "good, detailed labeling could reduce disease" (Hilts, 2003). Well-defined rules would be responsible health policy.

Many within the food industry also supported the proposed legislation, albeit for different reasons. Research-based firms sought to differentiate their products from those with unreasonable claims, to protect consumers and company profits. Quaker in particular was strident in its call for reforms: the company found itself disadvantaged against less-scrupulous companies that had no compunction in making questionable claims. Luther McKinney called cereal "a bad business" and said Quaker would "welcome some responsible regulation" (Hilts, 1989). Corporations and trade groups filed petitions with the FDA to request a formal inquiry into the matter (Nestle, 2003). With representatives of both sides of the market aligning for change, it was clear that the government would have to re-evaluate its policies.

Regulators were fed up as well. Dr. Louis Sullivan, director of the US Department of Health and Human Services, called for "sweeping changes;" he believed, "The grocery store has become a Tower of Babel, and consumers need to be linguists, scientists, and mind readers to understand the many labels they see. Vital information is missing, and frankly some misleading health claims are being made" (Hilts, 1990a). New FDA commissioner David Kessler was blunt in a speech at a food industry convention: his agency "*must* stand for, it *must* embody, strong and judicious enforcement [...] Let me remind all of you neither to underestimate the rigor of this agency nor the strength of its resolve" (Hilts, 2003).[31]

Resolve aside, however, the FDA struggled to reassert its authority, to harsh criticism from elected officials. Representative Ted Weiss (D-NY) called the moratorium "an unhealthy idea" and worried that "the public can be harmed and, at the very least, can be wasting their money on things that say, 'Eat this product and prevent X disease'" (*New York Times*, 1988). Finally, spurred to

---

[31] Emphasis original.

action by oat bran chaos, Congressional subcommittees stepped in to debate the components of a new label law in October of 1989.[32] Henry Waxman (D-CA), a noted consumer advocate, sponsored the bill that would eventually become law. The influential liberal did not mince words: "In today's market, the industry has an incentive to make exaggerated and inaccurate health claims on foods. I hope the FDA will finally begin prohibiting health claims unless the agency first finds they have a sound scientific basis" (Nestle, 2003). The pressure on the FDA was intense.

Drafting actionable everyday regulations from Congressional guidelines was not nearly as simple as it appeared. In particular, many in the media believed the FDA's plans ran afoul of the Bush administration's deregulation sympathies. The media was deeply skeptical of the influence of appointed politicians, who they believed were biased by lobbyists and contributions, over the supposedly more neutral career bureaucrats and specialists at the FDA (*New York Times*, 1992).[33]

Accusations of political interference were particularly aimed at the White House. Its Office of Management and Budget, "to which every agency would have to apply if it wanted to implement regulation," was excoriated as a pro-business conservative cabal (Hilts, 2003).[34] Facing this criticism, in late 1992, Bush approved a modified version of the FDA's plan, with less stringent definitions for front label statements like "low fat." The Bush administration's about-face was something of a reversal from the Reagan administration's pro-industry stance. However, many in the food industry realized that "litigation and uncertainty for food companies and consumers" would be the only result of further delay, before rules had to be ratified. Lacking a solid reason to oppose the current proposals (and with the agency's chief, David Kessler, threatening to resign if the FDA was slighted), the White House relented (Hilts, 2003). Philip Hilts, a *New York Times* correspondent covering FDA at the time, later wrote that the NLEA compromise produced the "most important food regulation ever" (Hilts, 2003).[35]

FDA's final regulations focused on packages' front *and* back panels. A standardized quantitative back panel label listed the nutrient content of the food inside and attempted to fit the food into the daily diet with percentages of the recommended daily allowance. The label was an updated, universal version of the information already required on foods that made health claims since 1984

---

[32] A less-exhaustive bill, HR3028, actually reached the House Committee on Energy and Commerce in August 1989 but was incorporated into the later, more comprehensive bill.

[33] The *Times* flatly declared, "Agriculture Secretary Edward Madigan [a Bush political appointee] put the interests of the meat industry ahead of the health of Americans."

[34] A House subcommittee found that OMB interfered to such a degree in the proposal process that the FDA's regulatory powers had been "neuter[ed]" (Hilts, 1990c).

[35] Although the NLEA was passed in 1990, it was acknowledged that the FDA would require time to develop its policies. The original legislation gave the agency until November 8, 1992 to establish its final revised rules [HR 3562: 2(a)(4)(B)(i).] This deadline was given urgency by the stipulation that if the deadline was not met, the rules as laid out in the law would go into effect without change (though in practice this was not enforced). After internal FDA politics and the President's decision were finalized, regulations were published on January 6, 1993 (Nestle, 2003).

(about 30% of products). Billions of labels would have to be changed, at an estimated cost of $2 billion to producers (Bruce and Crawford, 1995).

The front of the label offered significant profit potential. Regulators at the FDA braced themselves in anticipation (ultimately correct) that "this provision would ultimately produce the most serious controversy" (Hutt, 1995). Health claims offered the credence of the FDA's scientific imprimatur, combined with a specific linkage to a disease. Health claims occupied the middle ground between scientific information and advertisement copy, so the FDA intended to enforce stringent standards and expert panel review, so that only solid consensus would suffice.[36] The FDA "had twice proposed regulations to govern these matters prior to enactment of the [NLEA]," former FDA chief counsel Peter Hutt believes. The new regulation would be especially divisive because "it reflected some of the same criteria that had been proposed by FDA and strongly criticized by the food industry (Hutt, 1995).

Congressional law and FDA proposals only represented one perspective; they would be significantly shaped by compromises with industry and consumers. The FDA was inundated with more than 47 000 comments from corporations, scientists, and the public filed in the months leading up to its final rulings on 6 January 1993 (US DHSS/FDA, 1997). Food labels were a matter of intense interest among both consumers and producers. Most seemed to agree that the prior regime had been highly dysfunctional. Consumer groups and public health officials, for their part, generally were pleased with the increased scope of FDA power. The food industry was more circumspect, reserving judgment to see how friendly enforcement would be.

By 1993, the FDA hoped it again had a system in place that would allow fair product evaluation. The agency had to cede its ambitions of totally restricting health claims in return for effective, though circumscribed, regulatory power. Food marketers were granted the possibility of official certification and credibility with consumers. In exchange, they were held to greater accountability and saddled with the significant cost and, thus, inconvenience of label changes on all products. Consumers, for their part, received the boon of standardized and inspected information to protect against the gravest mistruths; how effectively these labels would be produced and implemented was unclear.

Consumer theory, a branch of microeconomics, attempts to relate individual choices to the iron rules of supply and demand. Economists working with the FDA and FTC examined the NLEA's potential impact using the "economics of information" developed by George Stigler, a University of Chicago economist (Stigler, 1961). Consumers prefer goods with better information, so in the mindset of the food industry claims would "add value" because many would pay more for a certified-nutritious product. When combined with Columbia economist Kelvin Lancaster's "product characteristics theory," marketers and regulators

---

[36] The standard of "Significant Scientific Agreement" (SSA) was identified as ". standard of "the (1) totality of (2) published evidence, from (3) well-designed studies, (4) conducted in a manner consistent with standard methods, about which (5) significant scientific agreement must exist (6) among qualified experts" (Nestle, 2003).

"started looking at products such as foods [...] not just as consumption commodities, but as a bundle of attributes" that consumers searched (Drichoutis *et al.*, 2006). Knowledge was (market) power.

Foods' appearance and taste were "search" (readily apparent when making a choice) and "experience" (evaluated after purchase and use) attributes, respectively. Nutritional characteristics, while vitally important to many consumers, were categorized as "credence attributes," those values that could not be easily observed or experienced but had to be taken on faith. The obvious problem lay in finding trustworthy information, and thus the FDA-approved label provided the solution (Caswell and Mojduszka, 1996). Labels took specialized and inaccessible knowledge, quantified it, and related it to an individual's health needs. Credence goods would become search goods, and public health would benefit.

Labels exerted pressure upon both the food industry and the consumer to make better choices. The grocery shopper who considered health due to the legislation and consciously chose nutritious options also encouraged the manufacturer to produce healthier goods in more variety and volume. Economists Caswell and Mojduszka forecasted that "in the long run producers are likely to respond by developing or purchasing new ingredients to replace less healthful nutrients without significantly changing the taste of products" (Zarkin and Anderson, 1992).

Food companies and industry trade groups, however, were ambivalent about the regulations. A lack of public trust hurt all companies, and specific corporations whose competitive advantages were compromised (like Quaker) were among the NLEA's strongest supporters. However, many found themselves unable to endorse stricter regulation as a matter of short-term cost as well as principle. The costs for industry were predicted to be enormous – on the order of $2400–10 800 per product (Drichoutis *et al.*, 2006). Thomas Hine remembers 1993 as a "managerial nightmare" but a "busy and prosperous year" for package designers (Hine, 1997).[37] The profound changes had the potential to remake the industry; apprehension stemmed from uncertainty over who would best grasp the new opportunity.

Quaker, too, was initially unsure of how it should use the new regulations to its best advantage. "It's a mixed blessing," said Luther McKinney, then Quaker's vice president of law and corporate affairs. "We are obviously pleased they are finally getting this out," McKinney noted, but he remained concerned about overly stringent standards of proof for claims (Hilts, 1990b). In another interview, McKinney noted, "This is a good thing for consumers and a good thing in terms of resolving who's in charge of food labeling that the company would be dealing with a single federal government rather than the more pugnacious local and state regulators" (Liesse and Colford, 1990). By the mid-1990s, however, with cereal sales mired in a post-oat bran slump, innovative new strategies were in order (Berry, 1990). The organization sought to leverage any competitive advantage available and a potential health claim was an attractive opportunity.

Without a doubt, the NLEA was perceived as a major victory for the FDA because its passage marked the successful completion of years' work for the

---

[37] In fact, the costs were minimized due to the foreknowledge of claims approaching, giving manufacturers a chance to work the changes into their natural package design "life cycle."

agency.[38] It became a "popular measure that earned the agency credit for working to make reforms that industry could live with" (Hilts, 2003). Hilts believes the FDA accomplished its goals because its final proposal was "a balance of rules that protected consumers, ensured that ample information was available to them, and encouraged businesses to compete on the basis of health and nutrition" (Hilts, 2003). For the food industry, the NLEA meant increased accountability and the expense of new labeling, but these negative effects were easily counterbalanced by new marketing opportunities and increased credibility. Consumers were given the empowering information to choose a responsible diet, while maintaining the ability to buy—and eat—as they pleased. In many ways, the NLEA was not a perfect compromise but it *was* actionable. The law brought order to a method of claims-making that became a uniquely powerful appeal to the modern consumer.

17.4.2 Totality of the evidence: Scientific specificity in health claims

Bran does not irritate, it titillates!
> Dr. J.H. Kellogg, *The Itinerary of A Breakfast* (1920)

The NLEA presented challenges for food manufacturers, and also exciting new opportunities. Both hinged on the new level of specificity mandated by the legislation. As changes in the regulatory environment encouraged food companies to seek approval for health messages, debates over what was being specified in any prospective claim—indeed, over what *could* be known or specified—intensified. Health claims manifested a concept of products' uniqueness in promoting health, mediated through peer review studies and other expert testimony but certified with a stamp of government approval. The broad public credibility that a claim garnered was invaluable in a skeptical market. There could be no question: companies sought FDA certification because it now had market value.

The type of health claims possible under the NLEA leveraged specific information in an attempt to satisfy a number of different constituents. Expert bodies like the FDA appreciated the rhetorical weight that the scientific language of specificity provided. The public would also benefit, consumer advocates argued, from certified claims served to educate about good dietary practices. Most crucial for food companies was the ability to create maximum market value from a claim that benefited their product over their competitors.

---

[38] The NLEA did not end controversy, however. Rather, a reinvigorated enforcement of the law would foment legal battles and legislative maneuvers. The 1994 Dietary Supplement Health and Safety Act (DSHEA) created the less-regulated "supplement" category, which enabled manufacturers to make unapproved claims with a disclaimer The 1997 FDA Modernization Act (FDAMA) shortened the FDA's deliberation period to only 120 days and required the agency to approve health claims based on consensus statements published by the US government or the National Academy of Sciences. Finally, the 1999 Supreme Court ruling in *Pearson v. Shalala* established that the "significant scientific agreement" requirement stifled free speech by requiring more agreement than is typical except for the most conservative claims. The cumulative effect of these decisions prevented the most of blatant untruths but weakened FDA regulatory capacity.

The eventual 875-page release explicitly spelled out regulatory policy under the new NLEA labeling regime, from standardized serving sizes to label formatting (US DHSS/FDA, 1993). In anticipation of health claim petitions, the FDA did its best to be ready, outlining its preferences for paperwork formatting and evidence submission (US DHSS/FDA, 1993). Most important for the future of Quaker oatmeal were 375 pages detailing official policies on "use of health claims that characterize the relationship of a substance to a disease or health related condition on the labels and in labeling of foods" (US DHSS/FDA, 1993). These included "pre-approved" exemplar claims that the FDA had ratified at the suggestion of the original Congressional legislation (Nestle, 2003). These claims, which included such food–disease relationships as "fruit and vegetable consumption and a decreased cancer risk," were generally unobjectionable. These claims were meant partially to codify conventional scientific wisdom and establish a template for the phrasing of a summary of current scientific evidence in later, petitioned label claims.

Interestingly, in its initial rulings, the FDA determined that Congress' suggestion of a soluble fiber-heart disease relationship did not meet its standards of proof. "The evidence is not sufficient to attribute the reduction in risk to soluble fiber or to a specific type or characteristic of soluble fiber" (US DHSS/FDA, 1993). For decades, however, health commentators had considered the bran (and thus, fiber) of grains to be healthful and decried the nutritional worth of modern refined grains.[39] Indeed, the medical establishment's faith in the worth of fiber in cancer prevention legitimized All-Bran's illegal claim in 1984. There was little disagreement that fiber was beneficial, such a diverse category needed to be better defined.

Much of the problem, it seemed, was a lack of specificity. The FDA repudiated overgeneralization: "wheat bran, oat bran, and rice bran (all heterogeneous mixtures of fibers) are not similar in composition. It is also very difficult to analyze dietary fiber chemically, and thus it is hard to correlate the role of specific fiber components to health effects" (US DHSS/FDA, 1993). The agency ended its literature review by leaving the future open: "If, however, additional information becomes available [...] then the FDA encourages manufacturers to petition for a health claim for their particular product" (US DHSS/FDA, 1993). Clearly, to attain greater claim specificity, a petitioner would need to assemble more specific and impressive data.

Soon after the FDA revealed its claims-making methodology in 1993, it became clear that Quaker had been hard at work marshaling support for oatmeal and its β-glucan. The company had sponsored a meta-analysis, a statistical paper combining the results of several previous studies to arrive at a more powerful estimation of a variable's effect, in 1992. This controversial analysis examined 20 studies and its researchers concluded that bran had a "modest" effect on blood cholesterol (about 5 "points," measured in mg/dL) (Rispin *et al.*, 1992).[40]

---

[39] It was these whole grains that were to pave the "Road to Wellville" promoted by John Harvey Kellogg and C.W. Post, the early kings of health claims at the turn of the twentieth century.

[40] While 5% may not appear to be a significant reduction, the reduction in risk of heart attack can be as much as 10%, depending upon one's prior cholesterol score.

Quaker trumpeted the results as a validation of its namesake product's health value. Skeptics, predictably, cried foul, suspecting Quaker's sponsorship had purchased the results. Such studies, however, are not generally considered a breach of medical ethics; corporate sponsorship is common and Quaker's support was clearly disclosed in the report. Vitamin historian Rima Apple notes that commercial companies and nutrition scientists have had a mutual understanding for decades. "While advertisements were quick to support their products with the latest scientific research, the needs of commerce also helped shape that very research. Scientific researchers, unsurprisingly, were not oblivious to the concerns of manufacturers" (Apple, 1996).

The NLEA finally went into effect on May 8, 1994. Quaker's petition was submitted to the FDA less than one year later, on 22 March 1995, citing support from 37 studies from 1980 to 1995 (US DHSS/FDA, 1996). The petition effectively collated the pre-existing evidence "suggesting that oat fiber reduces blood cholesterol levels by 5–10% [that] has appeared since the early 1960s" (Nestle, 2003). It was hoped that this targeted and highly bounded claim petition would provide exactly the specificity that the FDA had called for earlier when it rejected the "whole grain" claim in 1993. Here was a specific food, and even a specific type of fiber, β-glucan, that appeared to be oatmeal's "active ingredient."[41]

One might read gratitude into the FDA's statement: "The petitioner stated that while current research may not demonstrate that β-glucan is the only component of oats that affects blood lipids, it does suggest that it is an excellent marker for cholesterol reduction potential" (US DHSS/FDA, 1996). Instead of further whole-food headaches, the scientific evidence could be focused around a reductionist view of a single ingredient. Further, Quaker was able to provide a dosage recommendation to show a measurable effect; it speculated that 3 g of β-glucan daily would reduce cholesterol 5% in most subjects (US DHSS/FDA, 1996). Here, ironically, was a food being discussed in terms usually reserved for drugs, so as *not* be held to the stringent standards of a drug.

The claims-making process revealed that not all epidemiological evidence was equal in the eyes of the FDA. The agency thoroughly critiqued the studies that Quaker submitted, evaluating each study according to how well it fulfilled agency goals for accuracy and scope in data collection, statistical analysis, and proper methodology (US DHSS/FDA, 1996). The agency eventually rejected 20 of the 37 initial studies submitted. Some studies simply did not show a conclusive anticholesterol benefit. Others suffered from poor design and could not be generalized (US DHSS/FDA, 1996).[42] Despite these weaknesses, the FDA's reviewers determined that a significant scientific case remained when less-than-satisfactory studies were removed.

---

[41] The specific mechanism underlying β-glucan's anticholesterol properties remains uncertain. However, the most popular theory holds that the fiber absorbs water in the intestines (hence "soluble" fiber), creating a slush that removes some of the cholesterol in food, as well as digestive bile made from cholesterol, as it moves through the digestive tract. More bile must be created, but the body must draw cholesterol from the blood, as β-glucan has also trapped cholesterol from food.

[42] Some used only subjects with high cholesterol and no control group or did not record fiber grams, providing no dosage relationship.

The final FDA ruling appeared on 23 January 1997, after nearly 2 years of deliberation. "Based on its review of evidence [...], the agency has concluded that the type of soluble fiber found in whole oats, i.e., beta (β)-glucan soluble fiber, is primarily responsible for the association between consumption of whole oats [...] and an observed lowering of blood cholesterol levels. [...] Therefore, the FDA has decided to make the subject of the health claim "soluble fiber from whole oats" and has concluded that claims on foods relating the consumption of soluble fiber from whole oats to reduced risk of heart disease are justified" (US DHSS/FDA, 1997).[43]

The ruling had limitations, however. The claim would apply only if a single serving of the food product contained at least 0.75 g β-glucan, no more than 3 g fat, and no more than 1 g saturated fat per serving.[44] Quaker originally requested an unqualified claim, without any caveats like "as part of a diet low in saturated fat and cholesterol," citing nine studies that showed a blood-cholesterol benefit without any change in diet. Removing a "lifestyle change" injunction would have made the marketing pitch yet stronger.

However, the FDA was not inclined to return to the oat bran sensationalism of the late 1980s. Dr. David Kessler, the FDA commissioner from 1990–1997, believed that era's labels had been "so opaque that only consumers with the hermeneutic abilities of a Talmudic scholar can peel back the encoded layers of meaning" (Nestle, 2003). The agency insisted upon a qualification based on a low-fat, low-cholesterol diet appearing in the label language, as it represented public health policy consensus of the time. "Absolute claims about diseases affected by diet are generally not possible," the ruling maintained, "because such diseases are almost always multifactorial" (US DHSS/FDA, 1996). The FDA wished to curtail interpretive marketing hyperbole; it viewed these health claims primarily as an educational tool and source of responsible dietary information.

17.4.3   Why is this man smiling? Oats approved for first FDA claim

He receives comfort like cold porridge.

William Shakespeare, *The Tempest* (1610)

The significance of the health claim approval was certainly not lost on the media. Nutrition and health were already popular topics in most general-interest publications, but the Quaker story combined health with a significant legal landmark

---

[43] β-glucan is a relatively rare type of fiber; oats represent the dominant source in the average American's diet. However, barley is another grain rich in the fiber, and in 2006 barley manufacturers succeeded in extending the claim to their products. The willingness of such a small trade group to engage in the lengthy and expensive FDA process shows a certified claim's appeal.

[44] These limitations severely hampered Quaker's ability to promote processed products like breakfast bars and cookies using the claim; the strategy thus far has thus been to establish less-processed oats' healthfulness and to count on translated associations and more vague invocations of health (without the FDA claim) in advertising such products.

and multibillion dollar business opportunities. The decision was touted as "cause for elation [...] for executives at Quaker" (Stout, 1997).

The possibilities of approved claims opened a multitude of options for product promotion. "An approval would not only breathe life back into the flat [...] cereal industry but would "revitalize ad spending and new products" across the category (Haran, 1996). Marketers were particularly excited because the FDA claim would not only make food-specific claims acceptable but would also allow them to use plain, simple language (provided it was not untrue). An industry consultant believed that "The only manufacturers who won't be increasing their ad spending as a result of this claim will be those asleep at the switch. All the smart marketers" would be "trying to take advantage of a long overdue approval by the FDA" (Haran, 1996). As exciting as the claim was for cereal manufacturers, the repercussions for the entire food industry made the approval the subject of massive anticipation.

Both before and after the claim approval, the FDA found itself under heavy criticism. Regulators spent all of 1996 processing comments that flowed in response to the agency's proposal. The claim proposal elicited "approximately 1450 letters, each containing one or more comments, from consumers, professional organizations, government agencies, industry, trade associations, and health care professionals" (US DHSS/FDA, 1997). Many comments took issue with terminology: Quaker's proposal had used the term "oatmeal" to refer to all whole oat products, which risked consumer confusion. Other commenters also perceived a Quaker bias for its own signature product: studies were submitted claiming similar benefits for other grains as sources of β-glucan; other writers advocated for other soluble fibers and denied β-glucan's uniqueness (US DHSS/FDA, 1997).

The more serious reservations, however, had to do with the possibilities for consumers to be misled even if Quaker made a legitimate case for its claim (Figure 17.6). Some opposed the FDA's ruling out of a worry that a health claim would "mislead consumers in that it creates the impression that consumption of certain foods (oat bran and oatmeal) alone will protect against CHD [coronary heart disease]" (US DHSS/FDA, 1997). This "magic bullet" reservation did not dissipate after the claim was approved. Representatives of the consumer health group CSPI came out strongly against the health claim. The center's director of nutrition said of the FDA position that oats supplemented a cholesterol-lowering diet and regular exercise: "It's like saying that oats and aspirin will get rid of your headache." CSPI maintained that the FDA's approval "had less to do with nutrition than the oat industry's lobbying campaign to recover from lagging sales after the oat fad of a decade ago" (Stout, 1997).

Certainly, no one argued that the claim was without market value. An FDA-certified health food would appeal to consumers increasingly consumed by what Michael Pollan has termed the American Paradox: "a notably unhealthy people obsessed by the idea of eating healthily" (Pollan, 2004). Many of these nutrition-conscious Americans were also growing highly skeptical of food-industry messages, making an independently certified health claim important for credibility. More Americans were basing their buying behavior on health every year. A 1998 study found that "almost 80% of Americans believed nutrition to be

Why is this man smiling?

FDA Proposes
First Food-Specific Health Claim:

Diets high in oatmeal and low in saturated fat and
cholesterol may reduce the risk of heart disease.

Figure 17.6 Full-page advertisement published for more than a year before the final FDA decision (Quaker advertisement: *New York Times*, 1996). Reproduced by permissions of The Quaker Oats Company.

'very important' in their purchase decisions and 70% of American consumers read nutrition labels on food products" (Nordhielm, 2006). Legitimated health claims were becoming a consumer mandate.

In the aftermath of Quaker's successful petition, its executives aggressively sought to leverage oatmeal's new certification. "It was obvious that Quaker should capitalize on the FDA's recent approval [...]. A huge segment of American society was already affected" by heart health worries (Nordhielm, 2006). The market potential was massive. While the FDA's earlier decision to proclaim plant-based foods protective was a scientifically conservative, prudent decision, it was far from viable advertisement copy. Quaker's claim added specificity to make oatmeal seem *uniquely* healthy. Obviously, the ability to affect sales was much greater with an implicit recommendation to "eat more oats" rather than a general exhortation towards more fruit, vegetables, and whole grains.

Here, Quaker stayed true to an old advertising maxim: that the most successful products need a "unique selling proposition" or USP. The USP was a theory of prominent mid-century advertising executive Rosser Reeves (1910–1984), which stated a product needed a singular reason for why it needed to be bought or was better than its competitors (Sivulka, 1997). Reeves was highly critical of so-called brand image advertising, warning that a brand was a result of quality and could be complementary to, but did not replace, superior products.

Quaker's experience, in many ways, mirrored Reeves' theory. For more than 100 years, it had one of the longest-lived and strongest brands in consumer products. "The Quaker trademark was registered in 1877," Michman and Mazze (1998) report, "and Quaker was the first firm in the breakfast cereal industry to have high identification with consumers." To further grow the market for a mature product, however, new approaches would be needed to return to product differentiation in and to increase oatmeal sales. "Admakers relied less on the old ways of advertising to reach this savvy audience [of health-conscious baby boomers], which questioned the sincerity of advertisers' pitches" (Michman and Mazze, 1998). Government certification of Quaker's claim created an authentic point of differentiation would make oatmeal attractive to the cynical but highly lucrative health consumer of the 1990s.

Skeptics wondered why Quaker expended the time and resources it did for a health claim, whereas others would simply take advantage. It was "a long, arduous process to get the FDA approval [...] And, if approval is given, all other oatmeal marketers could make the same claim" (Pollack, 1996). While this was technically true, the strength of Quaker's brand meant that the company was uniquely positioned to take advantage of an oat health claim. With 60% share of the hot cereal market in 1993, any expansion of the segment would be predominantly in Quaker's favor (Galbraith, 1993). If a product helped all of oatmeal, it would help Quaker exponentially more.

If shareholders outside of Quaker's management had not already grasped the significance of the FDA claim, the company's 1997 annual report was quick to make it clear: the front cover announced, "We've turned the corner" (Quaker Oats, 1989–2000). As a key to that transition, "Quaker generated excitement in 1997 with the approval of the first-ever food-specific health claim by the Food and Drug Administration. We made sure that the word got out [... and] consumers listened, as evidenced by another four consecutive quarters of volume growth" (Quaker Oats, 1989–2000) Polly Kawalek, the head of Quaker's oatmeal division, was completely convinced. She "firmly believed that publicizing these results [...] could form the foundation for an effective marketing campaign to encourage regular consumption of oatmeal by adults" (Nordhielm, 2006). With a solid appeal to nutrition-focused adults (particularly mothers), the entire market size would increase, as these consumers typically bought for families.

In addition to driving growth, the claim would help to reinforce the company's overall identity as "a leading producer and marketer of wholesome foods and beverages" (Quaker Oats, 1989–2000). Quaker sought to cement this brand image in consumer minds through extensive spending on announcements for the general public as well as the medical community. The familiar Quaker man beamed up from the pages of the *New York Times* as the copy asked, "Why is this man smiling?" Quaker ran full-color advertisement on 14 January 1996, the day of the FDA proposal's release in the *Federal Register* and an entire year before the claim would be officially sanctioned (Quaker Oats, 1996). Quaker clearly had a sense of history as it boasted of the "first food-specific health claim."

One year later, the 22 January 1997 advertisement in the *New York Times* again featured the familiar Quaker, looking rather pleased with himself, "Now he has another reason to smile! The FDA approves the first food-specific health claim: Soluble fiber from oatmeal, as part of a low saturated fat, low cholesterol diet, may reduce the risk of heart disease. Quaker Oatmeal. Oh, what those oats can do" (Quaker Oats, 1997).

Quaker had been the first to capitalize on major transition in regulatory priorities for the FDA. As the agency's outlook changed toward a positive promotional agenda, so too did the attitude of industry.[45] No longer was the relationship with the government merely one of mitigating harm to profit potential. The savviest marketers would shift from developing the most ambitious claims that the FDA would *not* ban, to working *with* the agency to make unique, certified product statements and reach consumers. Approval could be a major boost to credibility and a market advantage. "Does oat fiber lower blood cholesterol levels? Of course it does—conditionally," Nestle qualifies. It is "the potential marketing benefits—not the science—that explain companies' persistent attempts to obtain FDA authorization for health claims" (Nestle, 2003).

17.4.4   Conclusion

When Quaker's *New York Times* advertisement ran in January 1997, the Quaker had any number of reasons to smile. FDA regulation, lapsed since the 1980s, had been reinstituted, but in a form that reflected American consumers' desires for healthy, honestly advertised food. The NLEA changed the regulatory environment, making claims more difficult but also imbuing them with greater significance. To be scientifically defensible, food had to be broken into constituents, tested, and quantified. A natural product had to fit into the reductionist nutritional paradigm. The shift provided Quaker with a competitive advantage, as it had one of the few products with a long history of peer-review studies showing a health effect. Oats were certified with the specificity to make the product seem *uniquely* healthy, a new and unique claim concerning a major health risk.

In response, the food industry developed an appreciation for the market value of FDA approval. A shift in perspective occurred, from working around

---

[45] In 1996, the FDA approved a claim for sugar alcohols on behalf of the National Chewing Gum Manufacturers' association, certifying that these sugar replacements prevented cavities when replacing sugared gum. The insistence on priority from Quaker, however, is indicative of its valuation of the claim.

regulation to make effective marketing pitches, to working with or even *through* the FDA to differentiate a product. Certification meant market share, millions of dollars, even a revitalized corporate identity. Quaker's experience, if it proved successful, would be at the vanguard of a new way to sell products. The implications of using such certified scientific knowledge as a promotional tool were profound. The food industry certainly saw the commercial possibilities, but on a larger level; the FDA's labeling decisions meant that science now entered every consumer's life on an unprecedented level. As consumers faced rows of labels hawking health claims, many could hardly resist thinking of their food in biomedical, disease-oriented terms. Science had been inextricably linked with a basic, universal experience – the purchase and consumption of food.

The labeling controversy combined several of the most contentious debates in professional discourse: scientific, political, and high-stakes financial conflict. Throughout these controversies, however, no party lost faith in the explanatory (or rhetorical) power of science. Each used its own interpretation of scientific studies, or invoked the scope of scientific surety to show why its position was valid. Sound, scientific nutrition was the universal good, the precept that was so difficult to achieve consensus around despite common rhetoric. When a notion is so uncontroversial yet at the center of so much debate, one wonders how many different motives are represented in the concept. The ability to reach compromise and achieve the interests of government, industry, and the consumer speak to that multiplicity.

17.5 Cash crop: Leveraging scientific evidence 1997–2010

Yours for Health, Wealth, and a Good Breakfast…
It maketh a merry heart.

Quaker advertisement (1896)

The 1997 FDA decision to approve a health claim for β-glucan fiber's relationship to heart disease represented an unprecedented opportunity for Quaker and a watershed moment for the food industry. As with any historic opportunity, however, company executives found themselves under pressure to take advantage of the new market opportunities. Quaker integrated nutrition into traditional consumer appeals like taste, convenience, and its strong brand image. Science constituted new "added value" for its products.

When vice president Polly Kawalek assumed executive control of Quaker's oatmeal division in 1996, she envisioned a new direction for Quaker products that brought its health appeal to as many consumers as possible. Kawalek's direct orders from the executive office were "to cater to oatmeal consumers at every stage of life and to introduce products that were in keeping with Quaker's overall brand profile—products that consumers felt comfortable relying on for good nutrition and pleasing taste" (Nordhielm, 2006). The FDA health claim was an exceptionally powerful tool to shape consumer preferences in this way.

With increasing competition both within the hot cereal industry and from other breakfast options, the company needed new appeal. Through its health claim publicity, Quaker sought to justify higher prices for what was essentially commodity grain (Nordhielm, 2006).[46] Quaker's products, although not expensive in absolute terms, cost on average 56% more than similar private-label brands.[47] The company's efforts were, therefore, focused on creating unique, premium options. This new identity would most effectively leverage the Quaker brand and separate it from the commodity competition management feared.

The Quaker group's actions from 1997 to 2000 would not prevent a takeover by the massive multinational PepsiCo in 2001. However, they would create an archetypal approach for integrating America's increased nutrition consciousness into effective marketing and product design. Kawalek obsessed over finding a "breakthrough message that provides a reason for consumers to eat oatmeal," and found it by combining health with other consumer needs like convenience, taste, brand appeal, and value. Nutrition reinvented Quaker oatmeal in the twenty-first century, reimagining a simple grain to reflect how modern consumers understood their food. The health claim cemented Quaker's brand power, which was "characterized by the distinctive nature of its brand personality, by the appeal and relevance of its image, by the consistency of its communication, by the integrity of its identity, and by the fact that it has stood the test of time [and its ability to] convey subtly different messages to different consumer groups within the same market" (Keough, 1994).

Even before Quaker effectively leveraged its claim, other manufacturers were hard at work developing cases for health messages of their own. In the 4 years before PepsiCo acquired Quaker, major manufacturers had 10 claims approved by FDA review, eight of which attempted to leverage a heart-health appeal like oats (Nestle, 2003). If a new era of opportunity dawned for the food industry in January 1997, Quaker oatmeal was served for breakfast.

17.5.1   Gains from grains: Health claims in Quaker's business plan

I had as like you would tell me a mess of porridge.
                    William Shakespeare, *The Merry Wives of Windsor* (1602)

In order to understand Quaker Oats' urgency in applying for and receiving the first food-specific health claim, it is helpful to consider where the company stood in the early 1990s. When its claim petition was submitted in 1994, Quaker Oats faced an uncertain future in a highly competitive marketplace. However, the rapid changes following the institution of the NLEA in 1993 offered a novel

---

[46] Consumers paid, on average, $3.48 per box of Quaker oats and $2.23 per box of generic oats.

[47] Oatmeal has one of the highest value-added rates of an agricultural product, second only to corn syrup, with only 7% of the value of oats sold in store being "farm value" (the price paid to the farmer), As much as 73% of consumers' price goes to pay for production, distribution, and promotion (Nestle, 2003).

opportunity to shape the new regulatory and retail environment. Quaker's oat-meal division was severely challenged: it was fighting a loss of pricing power due to new competition, the industry's highest promotional costs, and seri-ous questions about its brand's relevance to modern consumers. With a claim, the company hoped to differentiate its products, reach out more effectively to a segmented market, and revitalize its brand image through a breakthrough health claim.

"The last decade of the twentieth century is as critical as it is unique," industry forecasters claimed at the time. "It is critical because of the extreme pervasive-ness of environmental and technological change affecting the structure of the food industry [… and] unique inasmuch as sophisticated technological changes along with social changes have created many more opportunities for firms to develop differential advantages" (Michman and Mazze, 1998).

Quaker was particularly well suited to develop just such an advantage. As the dominant maker of oatmeal, a product with solid scientific support, it could offer consumers an FDA-certified health food. Just as the FDA and FTC inac-tion on Kellogg's All-Bran had broken open a Pandora's (cereal) box in the late 1980s, a successful Quaker campaign would be the trendsetter for a new era of government-mediated health claims.

Quaker Oats was in rapid transition during the FDA's deliberations. When it acquired the Gatorade line from the small Stokely-vanKamp group in 1983, Quaker unwittingly became a new company. Gatorade became phenomenally successful and took control of the sports drink category, which was growing expo-nentially due to the same interest in functional food that had fueled the oat bran craze. Beverages became the fastest-growing part of Quaker's line, making up 42% of sales by 2000 (Quaker Oats, 1989–2000). By 1998, CEO Bob Morrison felt confident enough to label Quaker "primarily a beverage concern," with Gatorade the "gem in our North American portfolio" (Balu, 1998).[48] Quaker's most rapid growth was outside its historical core business.

Although hot cereal products were not the segment in Quaker's portfolio with cachet in the 1990s, many analysts like those at Bernstein Group, a leading mar-ket research firm, still considered its historical business key to future success. Although Gatorade's meteoric rise had captivated management and media atten-tion, "We believe that the performance of the much more mundane (and, on a relative basis, profitable) hot-cereal business will ultimately determine whether Quaker consistently achieves its targeted goal of 7% real earnings growth. While the growth prospects for Gatorade are clearly superior to those of Quaker's oatmeal business, the margins in hot cereal (20%+) remain among the best in the industry" (Galbraith, 1993). Hot cereal would be challenging but oatmeal remained the heart and soul of Quaker Oats' image, as well as its profitability.

The company's market position varied widely across categories. In addition to its namesake, Quaker owned market-leading powerhouse brands in Gatorade, Rice Cakes, Chewy granola bars, and Rice-a-Roni (Nordhielm, 2006). In the

---

[48] This was hyperbole: beverages accounted for 38% of sales in that year, and dry food remained the dollar sales-leader throughout the company's independent history (Quaker Oats, 1998).

overall breakfast market, the group was a distant third behind behemoths Kellogg and General Mills (Michman and Mazze, 1998). Yet Quaker had always been "king of the hot cereal market," and held a commanding 67% by 1993 (Galbraith, 1993; Michman and Mazze, 1998). The margins for oatmeal were superb. In 1992, Quaker enjoyed 20% profit margins for its hot breakfast offerings (Galbraith, 1993). The company jealously defended its position from other name brands, flooding the market with promotions and price cuts to drive out any incursions by other brand names (Nordhielm, 2006).

Despite its dominance, Quaker had reason to worry about the hot cereal segment. Firstly, hot cereal was not a growth area. Following the peak of the oat bran craze, hot cereal sales declined or held even every year thereafter (Quaker Oats annual report, 1993). Analysts believed "while Quaker is the market leader in this segment, the size of the market will probably diminish over time" if conditions did not change (Michman and Mazze, 1998). Indeed, the fact that Quaker had actually increased its market share in the early nineties had little effect on analysts' projections, so dismal was the category outlook.

The category as a whole failed to connect with a number of consumers in the basic "value added" framework of food-processing logic. Following Marion Nestle, one might define the industry's traditional "marketing imperatives" as taste, cost, and convenience (Nestle, 2003).[49] In the view of Quaker executives, hot cereals were struggling because other options were equally cost effective and often as tasty but far less time consuming (Nordhielm, 2006). To make matters worse, hot cereals were increasingly feeling price pressure from a new source. Private label ("store brand") products formed a rapidly growing category in the early 1990s, as consumers "continue[d] to discover that private label products provide[d] high quality at significantly lower price points than their branded counterparts" (Galbraith, 1993). By 1993, private label offerings held 10% market share and were rapidly gaining ground (Galbraith, 1993).

Quaker's frustrations with private label competition stemmed from the fact that the branded product was sold at a markup more than one-half over generic (one of the highest in consumer products), although there were very few ways that their product was differentiated from cheaper alternatives (Galbraith, 1993). The general feeling in the industry was that "hot cereal makers are, for the most part, [. . .] selling a commodity product for value-added prices" (Galbraith, 1993). Without some new appeal, analysts believed Quaker's brands would likely experience a "gradual decline in cereal margins as the company is forced to lower the price points of its less unique brands [. . .] in order to compete with less expensive private label alternatives" (Galbraith, 1993).

The emergence of competitive, quality private label products also challenged the Quaker ethos of extensive promotional spending to propel their products to market success. In 1993, Quaker spent more than $983 million to promote its

---

[49] Levenstein believes these marketing imperatives have stayed relatively constant throughout time; he believes cereal manufacturers at the turn of the twentieth century focused on convenience, purity, and brand appeal (though he omits taste). Health was an appeal at this time, as well; however, Levenstein believes these vague, unregulated claims were merely "slogans that implied everything but promised nothing" (Levenstein, 1988).

products (Quaker Oats annual report, 1994). These were the highest promotional costs of any producer in the food industry, relative to food sales.[50] Instead of building a brand with "pull-side" consumer appeals to make consumers desire the product, Quaker spent 65% of its promotional budget on so-called "push tactics" or "trade greasing." This meant paying retailers to put the brand's products on sale and increase in-store visibility to "push" sales off store shelves (Twitchell, 1996).

Yet the overwhelming focus on price promotions was not proving terribly effective. These promotions encouraged precisely the wrong sort of buyer behavior; "customers responded to price deals by stocking up during the promotion with enough oatmeal to last until the next promotion" (Nordhielm, 2006). By 1993, promotion-cost growth had outpaced actual sales growth by 6% over the prior decade (Galbraith, 1993). More money thrown at cost promotions produced diminishing returns, in addition to its expense.

In light of these challenges, many at Quaker thought a paradigm shift was in order.[51] Polly Kawalek advocated a return to Quaker's roots and the institution of a more aggressive "pull-side" strategy to stimulate consumer demand through intelligent consumer marketing. This smarter allocation of promotional energies would work synergistically to drive demand for Quaker products, unrelated to their price. Kawalek noted that "price was of some concern to consumers, but less so than product benefits and also less so than intangible benefits derived from the brand name itself" (Nordhielm, 2006). So long as the brand's benefits were effectively communicated to the appropriate audiences, management was confident the Quaker name would continue to be powerful.

Effective advertising and marketing would be critical to creating and reinforcing perceptions of Quaker quality through nutrition. Although the health claim as eventually approved would apply to all qualifying oat products, regardless of brand, Quaker stood to gain most from the resultant product differentiation. Analysts conceded that "the hot-cereal category has evolved into essentially a two-player game [Quaker and generic];" Quaker applied for the health claim with full awareness of the fact that it was near synonymous with "oats" in the mind of the American consumer (Galbraith, 1993).[52]

If all went to plan, the oatmeal health claim would constitute nothing less than a re-imagination of oatmeal's place in the diet of the average consumer. Michman and Mazze (1998) believed "brand revitalization [...] can be approached by market expansion" and this became a corporate quest as Quaker began to

---

[50] Fully 20% of revenues went directly back to promotions, more than double the percentage outlay of Kellogg, the cereal market leader (Galbraith, 1993).

[51] A number of options for improving hot cereal profitability were pursued by Quaker, in addition to the health claim petition. Bernstein Research's 1993 report on Quaker catalogues these attempts, which included more technology-driven inventory management, new value-priced ready-to-eat (RTE) cereal lines in bags. New product lines attempted before the health claim largely foundered, however, suggesting a synergy between the claim and successful product targeting.

[52] Quaker correctly anticipated that it would be better able to take advantage of health appeal than generic products, which competed only through low price; some (such as Wal-Mart's "Great Value" brand) still have not added the claim to their labels more than a decade later.

30–Day
Dietary Intervention Program
to Help Lower Cholesterol

The Smart Heart Challenge

Real people. Real results.
One hundred residents of Lafayette, Colorado took the Smart Heart Challenge in February, and 98 lowered their total cholesterol.

- The program focuses on one simple change: eating a good-sized bowl of oatmeal daily for 30 days, as part of a low saturated fat, low cholesterol diet.

- The Challenge serves as a catalyst for people to make further dietary and lifestyle changes. In fact, 95 percent of the Lafayette participants said the Challenge inspired them to continue on a course toward a heart-healthy lifestyle.

- People prove to themselves after 30 days that dietary changes can make contributions toward the fight against heart disease.

Scientific basis: More than 37 independent clinical studies have documented the cholesterol-lowering effect of soluble fiber (beta-glucan) from oats, the only major grain shown to lower cholesterol. Some studies have attributed a 4–8% decrease in total cholesterol to the addition of oatmeal to a healthy diet. Other studies included oatmeal in a specific low-fat diet which resulted in a 10–12% decrease in total cholesterol.

To order a free Smart Heart Challenge guide and patient education materials, call toll-free (888) 328-6287

Challenge.

© 1998 The Quaker Oats Company

How does oatmeal lower cholesterol?

Oat soluble fiber mixes with cholesterol-based bile acids in the intestines and prevents them from being absorbed. The oat-fiber then carries them out of the body. In response, the liver pulls cholesterol out of the bloodstream to replace these bile acids. Cholesterol levels then drop.

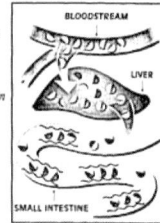

CHOLESTEROL   CHOLESTEROL-BASED BILE ACIDS   OAT SOLUBLE FIBER

Low Density Lipoprotein Cholesterol (LDL-C) Cereal Study

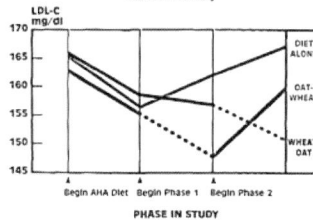

Begin AHA Diet   Begin Phase 1   Begin Phase 2

PHASE IN STUDY

Dotted lines indicate periods in which the subjects added 3.5g beta-glucan per day to their diets. This produced significant reductions in both total and LDL cholesterol when compared to wheat cereal or a fat-modified (AHA Step 1) diet alone.

Source: Keenan, Joseph, M.D., University of Minnesota Medical School, Department of Family Practice and Community Health, 1991.

Figure 17.7 This advertisement contained no straightforward "pitch" but was styled like a research report and used scientific terms to detail the Lafayette, CO, study (Quaker advertisement: *Journal of the American Dietetic Association*, 1999). Reproduced by permissions of The Quaker Oats Company.

leverage its health claim in 1998. They clarified that "[brand revitalization] can mean finding new uses for the brand or new users of the brand, or getting users to buy more of the brand." Health claims could make the crucial difference in all three senses, adding a new "use" for oatmeal in preventing heart disease, which would interest more users and motivate them to eat it more often, even daily for the cholesterol benefit (Figure 17.7). An answer to the company's largest marketing conundrum—how to shape consumer behavior—might be handed down in a single regulatory decision.

When the FDA finalized that decision in January 1997, Quaker's situation had become yet more complex. In 1994, the company attempted to diversify its beverage business with the $1.7 billion purchase of the Snapple iced tea brand. "Even now," Harvard Business School professor John Deighton claims, "mere mention of Quaker Oats' acquisition of Snapple causes veteran dealmakers to shudder" (Dreighton, 2002). The brand was at an absolute peak when it was purchased and quickly lost momentum as Quaker bungled marketing and distribution. Quaker divested the brand in 1997 for a paltry $300 million but the losses were significant beyond cash; the company had also folded several divisions to buy Snapple (Nordhielm, 2006). The blow to stockholder confidence was so severe that CEO William Smithburg, who had been in power since 1979, lost his job (Waters, 1997).

When new CEO Bob Morrison took over in 1997, he immediately sought to revitalize the group, eliminating senior management and reorganizing the company structure to highlight the prime "value-driving businesses." Gatorade and oatmeal were to lead the way. More important than internal structure was the public image Morrison cultivated, however: "Quaker positioned itself as the manufacturer that offered consumers food and beverage products with 'healthy nutritional profiles'" (Nordhielm, 2006). Quaker was the pre-eminent maker of oats but needed to make oats matter to consumers. Armed with an FDA claim, Quaker sought to differentiate oats from competing products and categories, more intelligently segment markets, and effectively leverage one of the nation's most trusted brands.

17.5.2   Accounting for taste: Tailoring health products for market segments

Truly, a peck of provender: I could munch your good dry oats.
    William Shakespeare, *A Midsummer Night's Dream* (1595)

Because of the FDA's approval, oats were effectively positioned as a food with unique health benefits. The scientific implications of such official approval did not concern business people, however. In a way, the FDA's approval of a health claim exclusive to oat fiber created a "brand" around oats. Consumers historically had a variety of feelings about the oat but government certification created near-universal acceptance as a uniquely healthy natural food. Let academics bicker over proof; the business people at Quaker were occupied with promoting a new-found brand property. "It is possible," wrote leading food marketing researchers, "to influence consumers to see this utilitarian benefit as the most salient related to the food" (Wansink and Cheney, 2005). Quaker believed it possessed an appealing product and could overcome major shortcomings (like relatively bland taste and longer preparation time) by reframing oats in this nutrition paradigm.

The standard practice with a successful brand, of course, was to press it for further market advantage through "brand extension." The creation of new offerings with the desirable characteristics of the original product, but with adjustments made for differing consumer tastes and lifestyles, expands the brand's appeal.[53] Brand extensions can make a company more forward-looking as well; they help by "accessing new sources of revenue, by revitalizing a brand in the eyes of the consumer, or by helping a business to respond to a significant change in the market" (Miller and Muir, 2004).

Extension of the new oat identity would mean reconstructing oats and their "goodness" as ingredients, which could add appeal to various Quaker products and the corporation as a whole. Minimally processed oats had been the context of research and federal approval but Quaker recognized that its consumers would be interested in—and willing to pay a premium for—convenient, tasty oat products like new flavors, instant cups, and breakfast bars (Figure 17.8). Oats as a health-promoting *identity*, not only as a single-component natural food, would be key to effectively leveraging the claim to maximum benefit. Thus, even products which did not meet the FDA's minimum criteria for the heart health claim would have their nutritional appeal boosted, merely through being part of the Quaker brand, which would be linked with oats and their health benefit.

Food processors believe the elements which consumers consider when purchasing food are price, taste, and convenience (Nestle, 2003). Marketing theorists Michman and Mazze (1998) stress that a brand's innate appeal constitutes an important factor as well, citing multiple studies to support this observation. Over the past 50 years, however, as consumers have enjoyed increased discretionary income, prices of relatively inexpensive staples (like oats) have become less important (Nordhielm, 2006). In 2003, marketing publication *Brand Week* noted that "decades ago, when times were simpler consumers were interested mostly in the product's taste and value. Packaging was a secondary consideration, other than throwing in special offers to tempt kids. But these days, with meal occasions boiled down to their bare essentials, packaging and delivery have emerged as key weapons in the cereal marketer's arsenal" (Reyes, 2003).

Many consumers now defined "value" as much by the box (health claims and convenient preparation) as what was *in* the box (taste and price). Competitive advantage in the cereal category could be attained through the health, convenience, and brand characteristics a cereal possessed (Nordhielm, 2006).

As new boxes of oatmeal were stocked on store shelves in early 1997, it was hard to miss the fact that Quaker deemed health an added value. A new label loudly proclaimed oats' certified health value. "Soluble fiber from oat bran," the label states (in large type), "as part of a low saturated fat, low cholesterol diet" (in smaller type), "may reduce the risk of heart disease" (in a red heart symbol). Predictably, oatmeal's role is highlighted while the relative importance of overall diet is de-emphasized. This is hardly surprising; the claim is only part public service announcement but all advertisement.

Far more revealing of the deep impact of an FDA-mediated claim is the way in which Quaker has marketed *all* of its oat products, including those that do

---

[53] Brand extension is a routine strategy for other reasons as well. It diversifies to mitigate risk that one poorly-received product will impact the bottom line and pushes competitors off supermarket shelf space (Michman and Mazze, 1998),

QUAKER OATS

The Great American Breakfast

Voted So by Millions of Women, by Culinary Experts and Dietetic Authorities, on These Important counts

Deliciousness—Steaming, flavory and wonderful, no other hot breakfast compares. Rich, plump oats, milled under the watchful scrutiny of Quaker experts. All that rare "Quaker" flavor is embodied—a flavor to be found in no other kind of oats.

Rich in Nutriment—A breakfast that "stands by" you through the morning. Contains more protein than any other cereal. Rich in essential carbohydrates. And when served with milk, combines the necessary vitamines.

Quick Quaker cooks in 3 to 5 minutes—That's faster than plain toast. No cooking bother, no kitchen mess. A rich, hot breakfast in a jiffy.

Why Quaker Oats "stands by" you through the morning

DO YOU feel hungry, tired, hours before meals? Don't jump to the conclusion of poor health. Much of the time you'll find it is largely brought on by an ill-balanced diet.

To feel right you must have well-balanced, complete food. At most meals you can get it. That is, at luncheon and dinner. But the great dietetic mistake is usually made at breakfast—a hurried meal, often badly chosen.

That is why Quaker Oats is so widely urged today. The oat is the best balanced of all cereals grown.

Contains 16% protein, food's greatest tissue builder; 58% carbohydrates, the great energy element; is well supplied with minerals and vitamines. Supplies, too, the roughage essential to a healthful diet that makes laxatives seldom needed!

Few foods have its remarkable balance. That is why it stands by you through the morning.

Why go on with less nourishing breakfasts?

THE QUAKER OATS COMPANY

★

Hot oats and milk is the dietetic urge of the day

The Quaker on a label means the world's standard in cereal products . . . a symbol of the finest grains that grow, of the finest milling known.

November 1926 Good Housekeeping

Figure 17.8 Expert and lay testimony of oats' taste, health, and convenience (Quaker advertisement: *Good Housekeeping*, 1926). Reproduced by permissions of The Quaker Oats Company.

not meet the claim criteria. From granola bars to breakfast cookies, the company promoted "whole grain Quaker oats parents insist on" as an assurance of quality and responsible eating. "Your mouth may think it's a cookie," Quaker promises, "but your body will know better". The more generic "goodness of oats" may not be tied to any scientific or federal authority but the unstated associations that those authorities have certified with the claim certainly matter to consumers.

17.5.2.1 Brand Quaker's image was based upon a commitment to a good-faith effort for its customers. The company had been a pioneering national advertiser since Henry Parsons Crowell's national campaigns in the 1880s made Quaker a household name. As early as the 1920s, company chairman John Stuart had valued Quaker's "brands, trademarks, and goodwill" more highly than its substantial "bricks and mortar" equity. These mores formed the "core of [Quaker's] management philosophy" (Marquette, 1967). Brand associations were becoming yet more important in the 1990s; in 1992, marketing theorist Jean Noel Kapferer asserted, "In [the] future, the brand will be the most important asset of the firm" (Kapferer, 1992). Image was everything for Quaker, fighting hard for space on consumers' breakfast tables even more than on the store shelf. Because oatmeal was historically at the center of Quaker's corporate identity, marketers believed that a "failure to maintain and support these important [brand] assets can lead to failure of the company itself," bolstering oatmeal's image through nutrition was critical to the entire corporation (Stobart, 1994).

"For over 130 years," Quaker advertisements read in 2010, "we've been inspired by the power and wholesome goodness of the amazing oat" (Quaker Oats, 2010b). The new FDA claim was important to this image, as it gave oats cachet in the present and projected toward a future of good health through tasty, nutritious products certified by science. Quaker sought to take ownership of nutrition as its "brand authority," the area of expertise which consumers believe a company is a leader. This entailed a circumspect, long-term strategy from the oatmeal division. Camillo Pagano, a Nestle executive in the early 1990s, noted, "The authoritative brand must reflect the culture of the company that makes it, the culture of the product category it belongs to, and the culture of the countries in which it is sold and should be managed from three perspectives: what it has stood for in the past, what is needed from it today, and what is expected from it tomorrow" (Pagano, 1994).

Quaker's emphasis on the "goodness of oats" and its unique and longstanding relationship with this nourishing natural product informed its entire product portfolio, and its corporate image.

Quaker had been one of America's most trusted brands since it became one of the country's *first* brands in the late nineteenth century. Certainly, the Quaker enjoyed an enviably positive reputation. "Brand equity was bulletproof nutritionally and evoked wonderful memories," noted Quaker oatmeal division chief Polly Kawalek (Nordhielm, 2006). In poll after poll, Quaker was identified as one of the companies consumers trusted most. "Clearly, there is a leverageable opportunity," said Prudential Securities analyst Jeff Kanter. "Very few brands mean wholesome and better-for-you snacks more than Quaker" (Thompson, 2006).

However, as corporate executives reviewed the company's performance in the mid-1990s to revise long-term strategy, the mark's familiarity had actually become something of an impediment. "Consumers thought they already knew all about it," Kawalek observed (Nordhielm, 2006). Quaker's research showed the market effects of that boredom—consumers bought oatmeal out of habit or family tradition, giving the brand one of the food industry's highest household penetration percentages; 80% of homes had Quaker oatmeal in the pantry (Nordhielm, 2006). Nearly every home had the familiar round Quaker Oats canister. Yet when it came time to eat breakfast, the smiling Quaker that had appealed in the supermarket lost out to other options.

The reasons varied; for many consumers, more convenient breakfasts that could be made quickly eased a hectic morning. Others were attracted by more indulgent breakfasts laden with more fat, sugar, and salt than Quaker's relatively bland options. Quaker needed to give consumers a fresh reason to think of its oats. If consumers remembered oatmeal for heart health and ate it daily, consumption would accelerate and demand would increase without needing to buy more advertisements or lower prices. Brand appeal would move its premium oatmeal into shopping carts, and the new health appeal would make Quaker appealing out of the supermarket, as well. With oatmeal moving off store shelves *and* cupboard shelves at a faster pace, Quaker would be making money in sales and reducing its promotional costs in one stroke.

17.5.2.2 Convenience In the decades after World War II, Americans, especially women, increasingly worked outside the home. With 59.5% of women (and 77% between the ages of 25 and 54 years) in the work force by 1997, time to cook meals was shrinking in proportion to the increase in spending power two-income households enjoyed (Hayghe, 1997). This meant that working families would seek – and pay for – food that fit with their more hectic lifestyles. Products' ease and speed of preparation thus became increasingly important concerns.[54]

"Never underestimate the appetite of American consumers for convenience;" Polly Kawalek made this a mantra (Nordhielm, 2006). Yet no matter how "quick" or even "instant" Quaker oatmeal had become over the past century, heating still required time and an appliance.[55] Cold cereal was traditionally the convenient option but foods that could be eaten as breakfast substitutes while commuting, such as shakes, bagels and muffins, and granola bars were taking market share from cereals, hot and cold alike. Nordhielm believes that, "Quaker regarded these [convenience] barriers to consumption to be of greater danger to [its] volume goals than any direct market-share threat from competitors" (Nordhielm, 2006).

The company's oatmeal had always struggled with the way Americans in the late twentieth century eat breakfast but recent failures in the mid-1990s magnified the problem, as attempts at making oats a convenience food actually backfired and hurt conventional oatmeal sales. For example, Quaker's "Quick and Hearty" microwaveable oatmeal was "virtually indistinguishable" from the familiar old-fashioned Quaker tube. The old-fashioned oats' promotional budget was entirely shifted to Quick and Hearty, and both products suffered. Quick and Hearty was discontinued in 1996 (soon after its introduction) but promotional funding did not return to the classic oatmeal, and thus sales did not either (Nordhielm, 2006). Without a major strategy change, oatmeal's re-entry into the modern breakfast would be challenging.

---

[54] Whether the free time gained by processed food outweighed the disadvantages of less-nutritious food is a matter of contentious debate (Pollan, 2009).

[55] The company had actually introduced one of the first "convenience" foods with its "Quick Oats" in 1921, which were specially steamed and flattened, reducing cooking time from 20 to 5 minutes. Further milling reduced cooking time to the familiar 1-minute "instant" version introduced in 1966 (Marquette, 1967).

The paradigm—and the pitch—shifted in the wake of the 1997 claim approval. The government-approved promotional boost was ecstatically received at head-quarters. Although "oatmeal never achieved parity with substitute products in terms of perceived convenience," nutrition offered a chance to redefine precisely what product was being sold, to Quaker's advantage. Oatmeal was thus *more* than a breakfast food; it was a daily commitment to good health. When one was making a prudent, healthy choice, an extra minute seemed more justifiable.

The oatmeal division's efforts after 1996–1997 attempted to circumvent the negative associations with preparing oatmeal. "Express" cups, instant oatmeal sold in a microwaveable cup, were well received as a time and effort-saving inno-vation, easily carried into the car or on the subway for the morning commute. Express cups satisfied both time-pressed consumers and Quaker executives. The cups were an exceptional "value added" product—retailing for about $1.75 in 2009, they sold for four times the equivalent amount of flavored instant packets (about 42¢), or six times the same amount of plain Quaker oats (less than 30¢). Yet, margins had never been at issue—Quaker needed to increase its volume. It seemed that the key to increasing Americans' consumption for the long term lay in options that broke out of the traditional heat-and-eat model entirely.

The rapidly increasing market share of the "breakfast bar" category was pred-icated on its portability and out-of-the-package readiness. In 2008, Quaker intro-duced its most convenient "oatmeal" yet.[56] "Oatmeal to Go" bars removed all preparation entirely—these microwaveable bars, which resemble oat cakes with a drizzle of frosting, might be thought of as an oat-laden version of that child-hood breakfast favorite, Kellogg's Pop-Tart. Pre-baked and bound together with a liberal helping of fat and sugar, these soft and chewy products boast "the same health benefits as a bowl of Quaker Instant Oatmeal in a convenient breakfast bar" (Quaker Oats, 2010a).

Importantly, this product specifically avoids using FDA claim language, as Oat-meal to Go bars fall far outside the agency's fat and sugar guidelines for the oat claim. The competitive advantage of Quaker's convenience offerings was "all the goodness of oats." Within this paradigm, the product was providing better nutri-tion quickly, instead of just satiating hunger. Quaker sought to end oatmeal's dissonance with modern lifestyles by creating a *new* lifestyle product—health on the go.

17.5.2.3 Taste

As thoroughly as Quaker, like any other major food manu-facturer, considered every aspect of its products, an appealing taste was uni-versally recognized as a prerequisite to gaining and retaining customers. A food product required significant thought and expenditure in research and development, advertising, branding, packaging, distribution, and countless other

---

[56] Although Quaker's "Chewy" granola bars were an iconic brand that could be eaten for breakfast, granola had strayed from its 1960s counterculture associations, and by the 1990s many consumers considered it a candy-like product. Quaker executives specifically stated that they did not want oat fiber to turn into "another granola" trend (Belasco, 1989).

considerations. What made these expenses justifiable was an appetizing product consumers reached for repeatedly. Taste built brands.

If the 1980s marked the advent of nutrition-conscious consumers, the 1990s launched a somewhat contradictory trend. "Though they knew that their food should be healthy, Americans were increasingly guided by taste," notes Nordhielm (2006). Many health-conscious consumers also expressed a preference for rich, indulgent food and new tastes. Indeed, industry studies in 1995 and 1998 asked consumers to list their top criteria for choosing a product and showed the two trends existing in parallel. Consumers replied that taste and nutrition were their top purchase motivations, with 79–81% listing health and 80–86% choosing hedonically (Nordhielm, 2006). Quaker executives saw the opportunity inherent in the numbers. A product that consumers saw as healthy but which also had a satisfying taste would appeal to a massive constituency.

The increased demand for oatmeal products allowed Quaker to reassess its products' taste in the overall brand strategy. With new customers buying oatmeal, Quaker attempted to understand what each customer wanted from their morning bowl of oats. The 1980s saw the broad adoption of market segmentation, a technique whereby marketers attempted to identify distinct "segments," or demographics to target. These lifestyles could be made important by age, sex, income, marital status, and any number of other characteristics. "The ultimate aim of this new wave of marketing is to reach different groups with specific messages about how certain products tie into their lifestyles," advertising theorists claimed (Turow, 1997).

Manufacturers hoped to distill the needs of a narrow section of American eaters, and then to address them with perfectly on-pitch advertising and product design. Michman and Mazze (1998) viewed market segmentation as highly important for cereal manufacturers, and specifically noted the effectiveness of "strategies that target consumer segments that are nutrition oriented, desire a high-fiber cereal, are diet oriented, desire a natural-ingredients cereal, or are concerned about cholesterol."[57,58] Effectively segmenting the market meant that Quaker could sell value-added products with precisely the "values added" that each consumer desired. These concepts were not new in 1997, yet "adult segments demonstrated unique consumption patterns that had not been recognized by Quaker in the past," reported Rick Gomez, director of marketing for the oatmeal division. Whereas the company traditionally had "focused primarily on two consumer groups: kids and adults, with little differentiation of the unique needs of various cohorts within these populations," the company looked hard at its marketing assumptions as it sought to most effectively to take advantage of oats' nutrition appeal (Nordhielm, 2006).

---

[57] The authors asserted the unique value of market segmentation for the health-conscious market, but noted that price pressure would be important in years to come. (However, an insistence on price was something of a fringe position among cereal-marketing experts at this time.)

[58] Quaker had actually created an ambitious market-segmentation campaign in 1990; Quaker Direct attempted to target 18 million consumers with personalized coupons. When the expensive campaign failed quickly, Quaker had shied away from complex marketing techniques but was attempting segmentation again seven years later (Turow, 1997).

While marketers found that shoppers shared universal concerns around convenience and similar associations with the Quaker brand, taste was another matter. The 1990s produced several widespread taste trends. Exoticism, rich and indulgent foods, and sweet flavors were shaping American breakfasts (Stanton and George, 1997). However, separate market segments manifested even these broad values in distinct ways. Quaker's attempts to understand the various shoppers who picked up its canister—and to learn something about those who didn't—represented an attempt to align its corporate priorities with those of the public. After identifying and targeting specific consumer groups, Quaker got to work designing and marketing products that took advantage of its claims and their associated health appeal.

Key to Quaker's strategy was the targeting of young consumers. The key was "to market to two audiences simultaneously—moms and kids—with two different messages," Quaker's focus groups determined. "Communicate the health and nutrition benefits to moms and convince kids that oatmeal tastes good and is fun to eat" (Nordhielm, 2006). Moms were informed of oats' health benefits and lower sugar content. Kids, for their part, were introduced to formulations like "Sea Adventures" (which turned the oatmeal blue when heated) and "Dinosaur Eggs" (with sugar eggs that melted when microwaved to reveal candy dinosaurs).[59] Quaker had aligned mothers, who it regarded as "nutritional gatekeepers," with their children's tastes.

Quaker's market research confirmed the suspicion that children were not the only consumers with a serious sweet tooth and hedonist tendencies. As they hurried to longer hours at higher-paying, more stressful jobs, consumers increasingly reached for highly sweetened breakfast snacks. These products already accounted for 31% of adults' breakfasts in 1995 and grew another 4% by 1998 (Nordhielm, 2006). It was clear that traditional oatmeal needed updated products to compete for these new tastes. With the added advantage of its heart health claim, Quaker figured it could appeal to both customers' rational and indulgent natures. The result was a "Deluxe Collection," developed in 1997. Beginning with Baked Apple and French Vanilla, instant oatmeal expanded to include no fewer than fourteen flavors by 2010 (Quaker Oats, 2010b).

Quaker marketed to distinct consumer groups' taste buds but also attempted to tailor health appeals to specific groups, presented in tasty products. In perhaps the ultimate manifestation of nutrition as a market value, in 1999 Quaker released "Nutrition for Women" instant flavored oatmeal, enriched with iron, calcium, folic acid, and soy protein (Nordhielm, 2006).[60] These nutrients were selected based on specific health conditions that concerned women, such as bone health. Natural foods proponents may have shuddered at the notion of enriching

---

[59] Sea Adventures tested well in focus groups but was quickly pulled from the market following the discovery the blue dye in the oatmeal permanently stained children's clothes. (Somewhere, Michael Pollan was apoplectic.) Dinosaur Eggs, on the other hand, continue to be phenomenally successful.

[60] Quaker here was shrewdly taking advantage of the cutting edge of health claims; soy protein was approved for an FDA health claim in 1999, and its positive effects on postmenopausal estrogen levels were becoming widely known.

a pure food with artificially derived nutrients based on inconclusive evidence but the large female customer base received the product enthusiastically. Encouraged, Quaker added other popular targeted nutrition products, such as "Take Heart" (enriched with potassium to lower blood pressure), "High Fiber" (with inulin, a chicory-root extract, to provide extra roughage), and "Weight Control" (low-calorie, artificially sweetened oatmeal with extra fiber and protein for dieters). With consumer segments demanding personalized products, Quaker took health marketing to an entirely new level by tailoring its marketing explicitly to higher-frequency oatmeal eaters like women and dieters.

"It was obvious" to marketing professionals at the time that "Quaker should capitalize on the FDA's recent approval." The claim would cement Quaker products as an essential part of a healthy modern diet, as "a huge segment of American society was already affected by or had fears of being affected by high cholesterol" (Nordhielm, 2006). No matter what niche Quaker's products attempted to fill, Kawalek's team ensured that nutrition was a centerpiece of the advertising message. After the initial boost in publicity from the claim being approved in 1997, the company sought to keep oats' benefits for heart health firmly at the center of its product and corporate identity.

17.5.3   Sweet deal: Quaker merges with PepsiCo

If this business were to be split up, I would be glad to take the brands, trademarks and goodwill, and you could have all the bricks and mortar, and I would fare better than you.

Former Quaker chairman John Stuart (1920s)

To understand Quaker Oats and its approach to nutrition today, it is necessary to see the role Quaker plays for PepsiCo, the multibillion-dollar conglomerate that Quaker merged with in 2001. After operating independently for nearly 125 years, the company that had innovated national distribution and branding in America found itself "a relatively small player in food" at the end of the twentieth century (Leonhart, 1998). Though the company remained profitable, its total revenues of $5.04 billion in 2000 (up 7% from the year before) lagged far behind the leaders of an "industry of giants" (Waters, 1997; Quaker Oats, 1989–2000). Larger companies like PepsiCo recorded $23 billion in sales that year from a high-profit, low-nutrition combination of soft drinks and Frito-Lay salty snacks (PepsiCo Inc., 2000–2008).

Quaker possessed several characteristics that kept its value high. First was the fact that in the late 1990s Quaker seemed highly attuned to the modern consumer. Gatorade was a high-growth phenomenon and the company's ability to promote simple oats in the same function-focused manner made the food industry take notice. As a result, Quaker was doubly attractive for its profitability. Every one of Quaker's business units was growing and making money at the turn of the millennium.

Quaker's brand had developed a reputation for health promotions as well as its own fiscal health, putting it squarely at the center of larger businesses' acquisitive consciousness. *Forbes* summarized why other brands coveted Quaker's products and reputation:

> Quaker had a built-in advantage in selling its granola bars: It was hard to imagine that the guy on the oatmeal box could sell you anything that wasn't going to do your body good. Pepsi and [other successful brands] have the opposite effect on most people: Those brands make one think of summer barbeques and junk food (Herper and Schiffman, 2001).

Financial stability made Quaker a safe acquisition, but "brands, trademarks, and good will" continued to be most its valuable assets. Buying Quaker meant that the purchaser had the chance to adopt its hard-won nutritional legitimacy.

After several years of speculation, a bidding war began between PepsiCo, Coca-Cola, and France's Groupe Danone in 2000. PepsiCo finally made the winning bid on its second attempt, after a low offer was rebuffed. The company would pay $13.4 billion in stock and assume $760 million in Quaker's debt (Sorkin and Winter, 2000). Gatorade, which held overwhelming market share in the sports drink category and was worth about $2 billion in revenues annually at the time, was particularly seen as a key to long term expansion into health-focused food for these purveyors of salt and sugar. "Gatorade is expected to give Pepsi a nearly 80% market share of the sports drink space," *Forbes* reported. "And that's a deal sweeter than sporty sugar water" (Herper and Schiffman, 2001). However, had PepsiCo overpaid for the Gatorade brand, given that it was now in possession of a significant food operation outside its core soda-and-salt competencies?

Most analysts did not expect Quaker's oat business to form part of the PepsiCo group permanently; most believed the company would divest the food businesses after a mandatory waiting period imposed by the Securities Exchange Commission. If PepsiCo did decide to retain Quaker's food brands, *Forbes* went so far as to ask whether the soda giant had "wasted" its $14 billion investment by taking on more than it needed in the view of many analysts, Quaker's food business was a "a slow-growth liability that caused Coca-Cola to spit Quaker out like so much unsweetened oatmeal." If the company was, indeed, holding cereal businesses it did not understand, *Forbes* humorously suggested that "PepsiCo Chief Executive Steve Reinemund must feel a bit like Dustin Hoffman's character at the end of the 1967 film, *The Graduate*, looking at his newly stolen bride and wondering: 'Okay, now what?'" (Herper and Schiffman, 2001).

Other analysts, including Bank of America's Bryan Spillane, believed it would not benefit PepsiCo to quickly divest Quaker's food lines. They disagreed with the characterization of PepsiCo as a beverage-maker first. Just as Quaker had evolved into a health brand with the oat health claim and Gatorade, they argued, Pepsi had diversified; the health appeal of Quaker's oat businesses would complement the indulgence of Frito-Lay snacks to give Pepsi a more complete portfolio. "Given current trends in beverages and the aging expansion at Frito-Lay North America, this acquisition represents [...] a preemptive action to fortify and potentially accelerate growth rates over the long term," Spillane wrote in a research note (Herper and Schiffman, 2001). The debate continued after Quaker formally joined PepsiCo in August 2001.

In the end, PepsiCo elected to retain not only Gatorade or those snacks which complemented Frito-Lay's competencies but the entirety of Quaker. The company thus became one of the four operating divisions in the PepsiCo corporation. Though by far the smallest group in terms of revenues (only about 4% in 2009), Quaker Foods North America became the cutting edge of PepsiCo's realigned corporate image, focused on nutrition marketing as an effective public relations tool but also a driver of major profits company-wide through an improved Pepsi image (PepsiCo Inc., 2009).[61] Quaker immediately performed well, posting a 21% improvement in profitability within its first year with PepsiCo. But "it was at the operating profit level that the synergies with Quaker were most clearly seen" for Pepsi as a whole, which rose 11% for the year to $5.3 billion" (Nutraingredients USA, 2003).

It was a confluence of all of Quaker's "marketing imperatives"—taste, convenience, brand, and nutrition—that made it such an attractive acquisition for PepsiCo. These imperatives had propelled Gatorade's success, which in turn motivated the takeover. However, it was their role in the creation of a health-food icon, Quaker Oats, which made operating the entire company a sound decision for PepsiCo. Today, PepsiCo is America's largest food company.[62] In 2007, it was recognized as "the world's biggest functional food company" poised "at the cutting edge of developments in health" (Piribo Group, 2007). Defined by soda and potato chips only six years before, PepsiCo has become the leader of an explosively growing market. Analysts believe "its strategy has seen it put health and wellness at the centre of everything it has done since the turn of the century" (Piribo Group, 2007). It is no coincidence that this commitment to health-conscious products and marketing began in 2000. PepsiCo's health kick started with the acquisition of Quaker's famous brands, trademarks, and good will.

17.5.4   Standard and porridge: Assessing the value of health claims

Poor fellow never joyed since the price of oats rose.
                           William Shakespeare, *Henry IV Part I* (1597)

The profound implications of an FDA-approved health claim for Quaker's brand have been discussed at length in this chapter. However, the effects of such a change on Quaker's real-world bottom line cannot be ignored. As early as 1993, industry analysts were noting that, "hot cereals may hold the key to Quaker's earnings outlook" for the future (Galbraith, 1993). Quaker's ability to maintain or expand consumer interest a highly lucrative product category—boasting

---

[61] Interestingly, in 2005 the Pepsi company reorganized its structure to include Quaker as the leader of its health brands group, known as Quaker Tropicana Gatorade (QTG), with the belief that the importance of these brands should be protected with management and marketing synergies.

[62] Internationally, Pepsi stands behind only Nestlé (Switzerland) and Unilever (UK); its health focus appears to have spurred both to commit publicly to marketing nutrition more extensively.

almost unheard-of 30% profit margins—would be a clear indicator of its continued competitiveness across the board (Baldwin and Soudakov, 2008).

The FDA claim was approved at a time when Quaker was engaged in an internal debate over the company's future identity. Gatorade signaled that nutritional products would be key to Quaker's long-term plan. "Was oatmeal a product that could fulfill the changing needs of Quaker consumers," management wondered, "or was it a relic of the past?" (Nordhielm, 2006). The FDA claim offered an answer to these concerns: Oatmeal could be a thoroughly modern breakfast, shaped by a faith in nutrition science.

However, Quaker's results were initially inconclusive. The 1997 year "generated excitement" and featured strong growth in hot cereal (Quaker Oats, 1989–2000). However, the claim's possibilities had not been used fully; many of Quaker's new value-added oat products had yet not been released. The following year disappointed, however, in part because conditions during the fiscal year 1998 were far from ideal. Quaker's sales declined sales growth declined 7% as the warmest winter in recent memory cooled down consumers' appetites for hot cereal (Quaker Oats, 1989–2000). Even as Quaker announced that, "we are focused on driving innovation and growth" to leverage their claim, Kawalek was in CEO Robert Morrison's office lobbying for patience to do just that (Nordhielm, 2006). The company's oldest product was still its most profitable and Quaker's annual report was quick to note, "the brand continues to be fresh and relevant to consumers" (Quaker Oats, 1989–2000). Kawalek was granted another year.

The oatmeal team worked full time to revamp the product line. Sales in 1999, with established marketing, new products, and better weather conditions, were record-breaking. Oatmeal revenues totaled $485 million, a 12.5% improvement. These numbers would have encouraged any investor, but Quaker insiders were ecstatic: the average sales growth over the five previous years had only been 1.6% (Baldwin and Soudakov, 2008). Quaker had satisfied its hopes for the claim.

17.5.5 Conclusion

The market for Quaker's oats continued to grow after the PepsiCo acquisition. At present, hot cereal is the fastest-growing segment of the cereal industry, with Quaker predictably claiming the largest share of the spoils. Mintel, a leading market research firm, reported that "growth in hot cereals has outpaced growth in the larger cold cereal segment [...] as consumers 'rediscovered' hot cereal for its positive health benefits," and credits the change to "the health properties associated with oatmeal along with the convenience factor of instant varieties" (Mintel Group, 2010). Quaker offerings are prominent within PepsiCo's top ten bestselling brands worldwide and helped Pepsi generate more than $1.9 billion in North American sales in 2008 (PepsiCo, 2000–2008). Solid growth and extremely positive consumer perceptions indicate that Quaker rightly anticipated a claim's appeal to the public, and thus its value to the brand.

These results indicate several important developments in "health" food that continued into the twenty-first century. Faced with declining growth, Quaker recognized the trend of nutrition consciousness that was now bypassing cost. However, management cleverly enhanced their products' appeal by promoting synergies between health and other consumer priorities like taste, convenience, and brand appeal. Even outspoken critics like Marion Nestle, who decry functional foods as "more about marketing than health," recognize the effectiveness of Quaker's approach to give "[a food] already classified at the base of the [USDA's Food Pyramid] a bit more of an edge" (Nestle, 2003).

In keeping with the principles of market segmentation, the latest nutrition science was tailored to consumer lifestyles. Claims in this new age differentiated products not only through *authenticity*, but also through *specificity* to particular lifestyles; better living through biochemistry. By intelligently marketing products with scientifically worded claims, Quaker reflected consumer values for its own benefit. As much as producers like Quaker are maligned, and consumers painted as victims, they are still working within a framework of consumer choice; those products that reflected lifestyles as well as health values sold.

In these new appeals, Quaker's health appeal became its brand. Indeed, in 2009, the group formally realigned its public identity to highlight its trademark grain: "Nutritious whole grains, wholesome goodness, and great tasting variety. That's Quaker Oats" (Quaker Oats, 2010b). This shift "helps us elevate and communicate the power of this surprising super grain—the oat—to meet the needs of the growing number of health conscious consumers," said Mark Schiller, the current president of Quaker Oats. "We feel that Quaker has a great opportunity as a market leader in health and wellness to leverage our product portfolio under one powerful platform" (Quaker Oats, 2010c).

17.6 Conclusions

Glegg paused from his porridge and looked up, not with any new amazement, but simply with that quiet, habitual wonder with which we regard constant mysteries.

George Eliot, *The Mill on the Floss* (1860)

Shortly after its claim was approved in 1997, Quaker Oats prepared an unprecedented advertising campaign (see Figure 17.9). That year, 100 residents of the tiny town of Lafayette, Colorado, were recruited to take the "Quaker Smart Heart Challenge," eating a daily bowl of oatmeal. Ninety-eight lowered their cholesterol. In 1998, television and print advertisements told the stories, "In one TV spot, a police officer named Dan is pictured in front of a sign showing a speed limit of 30 mph, detailing how his cholesterol dropped by 29 points. Others show residents against backdrops of numbers—bingo signs, price markers—that equal the number of points their cholesterol fell" (Pollack, 1998).

Before long, these ordinary people achieved an immortality usually reserved for star athletes and cartoon characters: they graced the front panels of Quaker boxes. In this new era of FDA health claims, science was an officially certified promotional tool. Selling products in this new way would impact not only

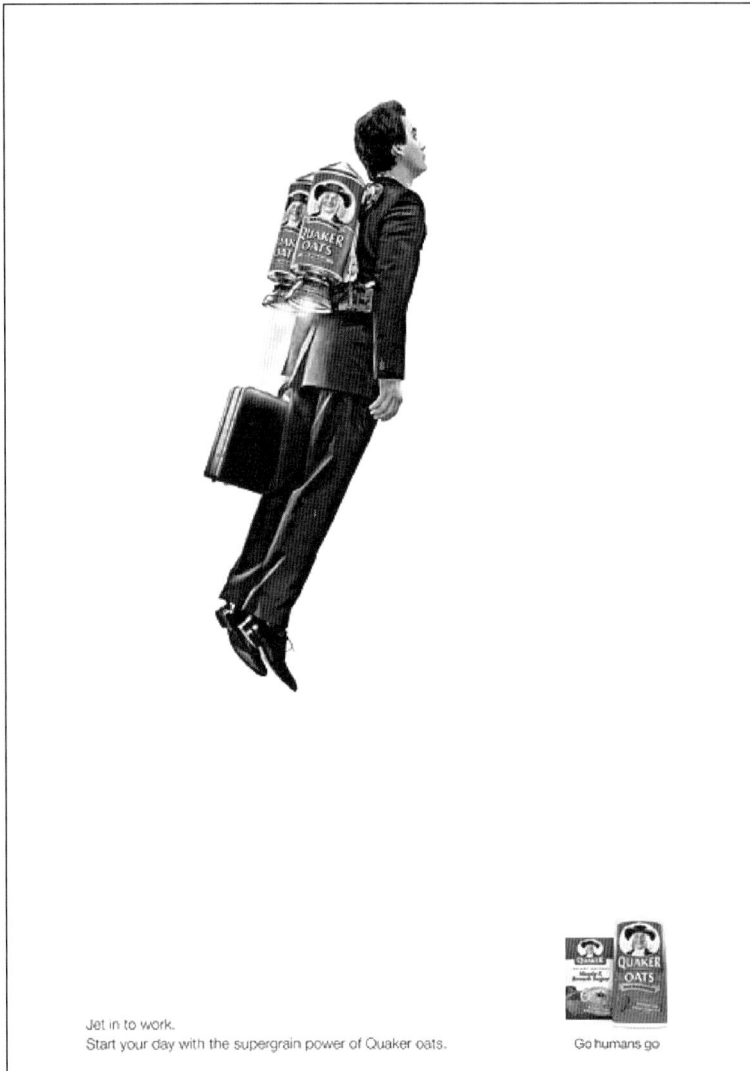

Jet in to work.
Start your day with the supergrain power of Quaker oats.   Go humans go

Figure 17.9   Quaker's "Go Humans Go" Campaign promoted the oat as "a super grain that powers your day, [. . .] knowing you did something good" (*Quaker advertisement*: BrandWeek, 2009). Reproduced by permissions of The Quaker Oats Company.

corporate bottom lines but also everyday American lives.[63] To properly leverage its claim, Quaker needed more than scientific evidence, regulatory approval, or

---

[63] Lafayette has hosted an annual Oatmeal Festival sponsored by Quaker every year since 1997, featuring the world's largest oatmeal breakfast (200 gallons of oatmeal and 140 toppings), a health fair with free cholesterol screenings, a "Quicker Quaker" 5k run and the "Smart Heart Start:" Free oatmeal and cholesterol screenings for 30 days in exchange for a pledge to eat Quaker every morning to test Quaker's 5-point cholesterol reduction promise.

even marketing dollars. The engagement and credence of consumers gave claims their value.

As the health claim trend accelerated, many observers sought to identify the power structure inherent in this information exchange. Scientists identified (supposedly) health-promoting compounds, while marketers translated science into persuasive claims. Regulators integrated experimental and commercial perspectives, evaluating scientists' studies as testimony in evaluating corporate claims. The public, while often portrayed as a passive recipient of knowledge and advertising, played a powerful role. As anthropologist Mary Douglas cautions, "theories of consumption which assume a puppet consumer, prey to the advertiser's wiles [...] or lemming consumers rushing to disaster, are frivolous, even dangerous" (Douglas, 2006). A growing awareness of nutrition motivated corporations to push claims and the FDA to evaluate (and allow) them. No party was powerless.

Information on product labels reflected and reinforced a change in the products inside, and thus a change in the nation's diet. The health publicity that Quaker received because of the 1997 claim opened opportunities for value-added products aimed at the specific health needs of key demographics. The cutting edge of commercial food became so-called "nutraceuticals" or "functional foods," value-added products marketed from a health standpoint. With these foods, historian J. Worth Estes notes, the industry "seems to sidestep even the vague definition to which the FDA is tethered" rendering it "not possible to differentiate foods from drugs" (Estes, 2000). Quaker's familiar canister of plain oats, a product as close to nature as one might find on grocery store shelves, was only the first iteration of these new possibilities.

As Quaker expanded its own line with nutrient-enhanced oats and convenience foods, other corporations sought their own claims. Critics like Michael Pollan remember the FDA's decision as the triumph of nutrition marketing in America, "Nutritionism had become the official ideology of the Food and Drug Administration; for all practical purposes the government had redefined foods as nothing more than the sum of their recognized nutrients. Adulteration had been repositioned as food science [... and the change would allow] hundreds of "traditional foods that everyone knows" to begin their long retreat from the supermarket shelves and for our eating to become more "scientific." [...] Hyphens sprouted like dandelions in the supermarket aisles: *low-fat, no-cholesterol, high-fiber*" (Pollan, 2008).[64]

New foods, developed with claims in mind, included margarines enriched with plant sterols believed to eliminate cholesterol, soy-infused snack bars claiming breast cancer risk reduction, and other novel components. Sanctioned "health food" began with a simple grain but the future lay in more sophisticated, and less familiar, formulations based on natural health-promoting components.[65]

---

[64] This chapter has attempted to show that Pollan's rhetoric is too simple regarding the FDA's regulatory attitude. Yet his frustration stems from an objective phenomenon, proliferation of claims.

[65] Even FDA-approved claims affirming the value of β-glucan have become more commercialized in the past decade. Joining Quaker's claim are officially sanctioned trademark products like Betafiber® and Oatrim®, extracts which may be used to gain an oat label claim.

From a purely promotional point of view, health claims have been highly effective. As economists Drichoutis and colleagues reported, "Health claims in the front of the package have been found to create favourable judgements about a product. For example, when a product features a health or nutrient content claim, consumers tend to view the product as healthier and are then more likely to purchase it, independent of their information search behavior." (Drichoutis *et al.*, 2006).

However, consumer advocates like the CSPI charged that the FDA's lenience in permitting industry-petitioned health claims on front labels undercut its own quantitative back labels.

Manufacturers countered that front label health information was tantamount to public service. Grams and percentages were precise but they were the language of specialists. Everyday shoppers, on the other hand, were making split-second decisions at the store; front labels that connected nutrients to health conditions would resonate without becoming confusing. Companies argued claims were "a good solution to an important issue and demonstrate what we've been saying all along: that the interests of consumer and the food industry are not at odds" (*Advertising Age*, 1990).

However, the simplicity of front label health claims which attempted to tie nutrients to health conditions reintroduced subjectivity—a disquieting notion for scientists but far more comfortable for creative marketers. Nestle recognizes this tension, noting, "Like any other kind of science, nutrition science is more a matter of probabilities than of absolutes and is, therefore, subject to interpretation. Interpretation, in turn, depends on point of view. Government agencies invoke science for regulatory decisions. [...] Advocates invoke science to question the safety of products perceived as undesirable. In contrast, scientists and food producers, who might benefit from promoting research results, nutritional benefits, or safety, tend to view other-than-scientific points of view as inherently irrational" (Nestle, 2003).

Whatever one's position, the use of scientific evidence and wording has proved rhetorically essential. However, a recent phenomenon in the popular discourse has been the emergence of eloquent, bestselling critics who question the undue influence and liberal scientific interpretations of the food industry. These skeptics manifest a range of responses to scientific uncertainty, however.

Michael Pollan, a University of California-Berkeley professor, is among the best-known food industry critics. He decries "nutritionism," the reductionist approach to food that he believes the food industry has exploited to profit from processed products marketed with health appeals. "The implicit message" of nutrient labels and health claims, Pollan claims, "was that foods, by comparison, were coarse, old-fashioned, and decidedly unscientific things – who could say *what* was in them really?" Claims, on the other hand, "gleamed with the promise of scientific certainty" (Pollan, 2008). Pollan warns against the "hopelessly corrupt [...] "pseudoscientific beauracratese" gracing product labels today, which are "official FDA euphemism[s] for 'all but meaningless'" (Pollan, 2008). He advises that consumers avoid all foods bearing claims on their labels, as "for the most part it is the products of food science [that bear claims] founded in incomplete and often erroneous science" (Pollan, 2008). In denying the ability of corporate interests to utilize science, Pollan denies wholesale the power of experimental knowledge about food.

Marion Nestle, a professor of public health at New York University, brings a sociological perspective and explores the "politics" of the laws underlying nutrition claims. For depth of research, there is no equal in the secondary literature. Like Pollan, Nestle decries "misappropriated" science used to sell processed food, but she supports the USDA Food Pyramid and does not reject science entirely (Nestle, 2003). She does, however, adamantly distrust the ability of the food industry seek both health and profit. She believes that "unless we are willing to pay more for food, relinquish out-of-season produce, and rarely buy anything that comes in a package or is advertised on television, we support the current food system every time we eat a meal" (Nestle, 2003). This broad anticonsumerism seems misplaced in a discussion of high-level public health policy and impractical for many Americans.

As perhaps the most respected nutritionist in the world today, Professor Walter Willett of the Harvard School of Public Health wields considerable influence with both government and industry. Willett absolutely respects the role of the food industry in creating a healthier diet but maintains that stringent scientific standards are necessary. "The fact that [food companies] are linking health and what people eat should help create awareness," he observes, "if the claims are right" (Willett, personal communication). To this end, Willett recently took his message directly to food industry executives, telling them "an unregulated market is doing for human health what it has done for the US economy" (Kasdon, 2009). Personally, however, he prefers that academics work on the side of regulation and policy, and refuses to work with manufacturers on a paid basis, so as not to compromise his scientific objectivity.

To varying degrees, then, America's foremost thinkers on diet and health each reject the opportunity to work more closely with industry. Yet it is undeniable that the food industry's influence on public perceptions of diet is powerful; Nestle estimates that food marketing budgets for single products are often 10- to 50-times larger than USDA spending to promote the Food Pyramid (Nestle, 2003). Levenstein goes so far as to credit the combination of the two approaches with the making of modern diet, "The right social and intellectual climate was not, in itself, enough. In order for new nutritional ideas to be adopted by mainstream Americans and for national food habits to change, it was necessary for the changes to be actively fostered by the two most powerful institutions in society, namely, government and giant food corporation. The former, working through formal and informal information systems, and the latter, with their influence and advertising in the mass media, have been the only forces with the necessary resources to spread the message on a mass basis" (Levenstein, 1988).

There is no doubt that such spending can not only profoundly influence the public's priorities in food choice but also feelings about what food *is* and can *do* for their lives.

However it is also clear, as Willett notes, that food producers are not entirely forcing nutrition on the public but rather "responding to an interest" on the part of consumers, and marketing can have a positive effect by ensuring that better nutrition remains in the public eye (Willett, personal communication). Each author recognizes, to a greater or lesser degree, that the food industry is responding to consumer interest in diet and health, even as it contributes to escalating anxieties and confusion. If this is true, perhaps a more appropriate response is recognition that the food industry can be a positive force while making a profit, should nutritionists apply their pressure and expertise in collaboration as well as antagonism.

A 2008 study conducted by USDA, however, perhaps revealed the most important trend that Quaker's claim had wrought. The authors found that consumers in 2005–2006 used food labels less frequently than they had in 1995–1996. Consultation of all nutritional indicators, from calories to fat to sodium and ingredient lists, declined between 3–11% (Todd and Variyam, 2008). From a purely objective standpoint, the FDA's labeling efforts were wasted on a large percentage of the public. The results showed that, when shoppers did check back labels, they did not consult the fat, cholesterol, or sodium information which had been important a decade before.

Indeed, the *only* nutrient consumers checked with greater frequency in 2005–2006 was dietary fiber, by about two percentage points. The USDA report recognized the unique appeal of marketers and the popular media in raising public awareness: they believed, "The increase in the use of fiber information (and, to a lesser extent, sugar) among label users [...] is a notable exception. The role of adequate fiber in promoting good health has received much attention in the press recently. [...] Perhaps similar exposure to information about other nutrients in food will lead to increased label use" (Todd and Variyam, 2008).

The coordinated efforts of marketing and manufacturing professionals for Quaker's health-claim push combined public health and profit, with explicit regulatory sanction. This unprecedented campaign may have impacted consumer priorities to an extent that government and academic promotional efforts could not accomplish alone. Marketing and media use of scientific evidence helped "sell" the public on a particular perception of a healthy diet. More importantly, shoppers attempted to conform to this effectively marketed health message, as opposed to other nutrition messages that were met with acknowledgement, but indifference.

Though it has been difficult to quantitatively *prove* the effectiveness of claims, certainly Quaker is renowned for its nutritional savvy. Experts on health marketing cite Quaker's experience as the quintessential FDA claim success story. Cornell professor Brian Wansink posits that "a higher level of understanding [about food and health] occurs" when health claims are present. "Because the consumer knows that a certain type of food contains nutritional properties and that these properties produce certain health benefits, the consumer is able to conclude that this food will lead to certain health benefits" (Wansink, 2004). Awareness of the connections between diet and health motivate consumers to eat better.

Wansink thus believes the oat claim's success is proof that good marketing and good nutrition can exist in tandem. Five factors made it particularly effective. Quaker targeted a specific group (children with Dinosaur Eggs, seniors with Wilford Brimley), sought media attention (advertisements celebrated the claim in national newspapers and medical journals), co-promoted the claim as a corporate partner to government agencies (supporting the USDA's focus on whole grains), focused on quantifiable or observable results (reducing cholesterol score points), and focused on a disease many consumers have a personal relationship with (heart disease, America's most common cause of death). He posits that that one "might call [efforts like Quaker's campaign] education, public service, or simply good parenting," not mere marketing alone (Wansink, 2004). In the future, Wansink concludes, those who want to sell products for profit but also for the public good would do well to emulate Quaker's approach to oats.

The fact that Quaker's claim seems to have been influential, however, by no means should imply that it is going unchallenged. In 2007, the CSPI scored a victory when it threatened to sue the company for "exaggerated" claims that portrayed oats as a "unique" health food that "actively finds" and removes cholesterol capable of lowering cholesterol more than any other nutrition choice. "Oatmeal is a healthy food, but that's no excuse to give people the impression that it will miraculously remove cholesterol from your arteries or to otherwise exaggerate its benefits," stated Steve Gardner, litigation director for CSPI (Center for Science in the Public Interest, 2007). Quaker settled out of court, agreeing to redact the claim in favor of more circumspect language.

The company came under fire again in 2009, when the FDA questioned PepsiCo's participation in an industry-wide "Smart Choices" campaign. The program seeks to further make essential front *and* back label scientific information into a green check mark, to "help shoppers easily identify smarter food and beverage choices." Critics, including Willett and Nestle, contend its lenient criteria allow "horrible choices" like sugary granola bars to be advertised as health food (Neumann, 2009). They see "Smart Choices" and similar efforts as attempts by the industry to utilize the trust the public now has in FDA-certified front labels, without true federal review. Both USDA and FDA have indicated the industry's actions will be reviewed in the drafting of new, likely more stringent label standards in 2010 (*The Economist*, 2009). Clearly, label claims will remain controversial.

Perhaps the only certain lesson from health-claim labels is the tremendous market value of scientific knowledge, both in selling products and, in a larger sense, in building a competitive modern brand. Quaker Oats was by no means America's largest food company throughout its history, but rather established itself through innovation, quality, and consumer loyalty—"brands, trademarks, and goodwill"—which its founders had trusted. The FDA claim Quaker acquired and leveraged into valuable brand property attracted those large multinational corporations that had always struggled to build brands fitting the public's new standards of nutrition.

The Quaker brand invoked nostalgia when it submitted its label claim petition in 1994; soon, it became clear that Quaker was a sign of things to come. By 2009, analysts tracking the $78 billion industry recognized that "there is a sound commercial logic behind [a] shift towards health and nutrition," noting the frantic efforts of executives at Unilever, Kraft, and Nestle, PepsiCo's main worldwide competitors, to make themselves into nutritional powerhouses. Experts projected that the market for nutraceuticals would continue to outpace traditional food, earning $128 billion by 2013 (*The Economist*, 2009). Quaker asked consumers, "Why is this man smiling?" while it awaited its claim in 1996. The answer was clear by 2009; Peter Hutt, the FDA's former chief counsel, knew that Quaker smirked because "they'd just made a gazillion dollars" (P Hutt, personal communication).

To understand the role of science as a key to developing trust in the modern marketplace, this chapter has attempted to define precisely what a health claim "meant" to consumers, regulators, and corporations throughout the last 25 years. A key component to any history of science is that what we regard as "fact" is never neutral, nor is it stable. The scientific evidence brought to bear in the modern history of health claims has been a powerful rhetorical and promotional tool.

β-glucan was transformed from cattle feed to "superfood" in the late 1980s when scientifically demonstrated health properties lent it quantitative credibility and a connection to disease prevention. Yet without authoritative certification, scientific dissent deflated the craze. In the context of Quaker's 1994 appeal, a multitude of studies and quantitative data proved sufficient to convince an expert panel at the FDA, which approved the unprecedented claim. This certification expanded possibilities for Quaker's profitable product portfolio. Today, Quaker's health appeal makes it a key part of PepsiCo, and the model of long-term strategy for corporations attempting to provide for their customers' needs and wants in the twenty-first century.

Historian Harvey Levenstein has speculated as to the modern rules that govern eating for mainstream Americans and believes them to state, "That taste is not a true guide to what should be eaten; that the important components of food cannot be seen or tasted, but are discernable only in scientific laboratories; and that experimental science has produced rules of nutrition which will prevent illness and encourage longevity" (Levenstein, 1988).

The scientific approach embodied in Levenstein's food rules, when applied to the humble oat, elevated it to a status as *the* iconic "health food." However, without the control and certification of regulators, health appeal proved ephemeral. Quaker would require legal intervention in the NLEA of 1990, scientific approval in the FDA's 1997 decision, and business savvy thereafter to translate health appeal into a more effective way to do business. In the end, Quaker's claim paved the way not only to profit for a particular product or company but to a new appreciation for the persuasive power of scientific evidence in modern society.

Oh, what those oats can do.

References

Advertising Age (1990) The 100 leading national advertisers. *Advertising Age* 26 September.

Apple, R. (1989) Patenting university research: Harry Steenbock and the Wisconsin Alumni Research Foundation. *Isis* **80**, 375–394.

Apple, R. (1996) *Vitamania: Vitamins in American culture.* Rutgers UP, New Brunswick, NJ.

Baldwin, C.Y. and Soudakov, L. (2008) *PepsiCo's bid for Quaker Oats.* Harvard Business Publishing, Boston, MA.

Balu, R. (1998) Business brief: New CEO views Quaker Oats as primarily a beverage concern. *Wall Street Journal* 24 January, 1.

Belasco, W. (1989) *Appetite for Change: How the Counterculture Took on the Food Industry*. Pantheon, New York.

Berry, K. (1990) All about breakfast cereal; the snap has turned to slog. *New York Times* 18 November.

Brillat-Savarin, J.A. (2000) *The physiology of taste, or, meditations on transcendental gastronomy* (trans M.F.K. Fisher). Counterpoint, Berkeley, CA.

Bruce, S. and Crawford, B. (1995) *Cerealizing America: The unsweetened story of American breakfast cereal*. Faber and Faber, New York.

Burris, J. (1990) Letter to the editor: Oat bran and serum cholesterol. *New England Journal of Medicine*, **320**, 1748.

Burros, M. (1989) De Gustibus: 'Oat bran' may be on the label, but how much is in the box? *New York Times* 22 March, C4.

Business Week (1989) Snap, crackle, stop: States crack down on misleading food claims. *Business Week* 25 September.

Caswell, J.A., and Mojduszka, E.M. (1996) Using informational labeling to influence the market for quality in food products. *American Journal of Agricultural Economics* **78**, 1248–1253.

Center for Science in the Public Interest. (2007) *Quaker agrees to tone down exaggerated health claims on oatmeal* [Online]. Available: http://www.cspinet.org/new/200704171.html (last accessed 12 May 2013).

Chicago-Booth Magazine (2000) Profile: Polly Kawalek. *Chicago-Booth Magazine* Spring.

Colford, S.W. (1985) Food marketers let health claims simmer. *Advertising Age* **19**, 12.

Consumer Reports (1986) What has All-Bran wrought? *Consumer Reports* October, 638–639.

Day, R.E. (1946) *Breakfast table autocrat: The life story of Henry Parsons Crowell*. Moody Press, Chicago, IL.

Douglas, M. (2006) *The world of goods: Towards an anthropology of consumption*. Routledge, London.

Dreighton, J. (2002) How Snapple got its juice back. *Harvard Business Review* **1**.

Drichoutis, A.C. *et al.* (2006) Consumers' use of nutritional labels: A review of research studies and issues. *Academy of Marketing Science Review* **9**.

Dullea, G. (1989) What to eat? Confusion is the main course. *New York Times* 3 December, 172.

Economist, The (2009) The unrepentant chocolatier. *The Economist* 29 October.

Edmonds, W.D. (1958) *Oneida: The First Hundred Years 1848–1948*. Oneida Ltd, Oneida, NY.

Erickson, J.L. and Dagnoli, J. (1989a) Healthy food pace quickens, leaving regulators behind. *Advertising Age* 25 September, 3.

Erickson, J.L. and Dagnoli, J. (1989b) Wary consumers want more health ad info. *Advertising Age* **12**, 12.

Estes, J.W. (2000) Food as medicine. In: *The Cambridge World History of Food* (eds K.F. Kiple and K.C. Ornelas), pp. 1534–1553. Cambridge UP, Cambridge, UK.

Food Chemical News (1985) Continued FDA-FTC split on food claims suggested at NFPA meeting. *Food Chemical News* **6**, 28–29.

Galbraith, S. (1993) *Bernstein research: Quaker Oats Company*. Bernstein Global Wealth Management, New York.

Haran, L. (1996) Oat brands set to reheat cereal sales. *Advertising Age* **29**, 12.

Harper, C.L. and LeBeau, B. (2003) *Food, society, and environment*. Prentice Hall, Upper Saddle River, NJ.

Hayghe, H.V. (1997) Developments in women's labor force participation. *Monthly Labor Review* **120**, 41–46.

Herper, M., and Schiffman, B. (2001) Pepsi bought Quaker. Now what? *Forbes* [Online]. Available: http://www.forbes.com/2001/08/02/0802topnews.html {last accessed 12 May 2013).

Hilts, P.J. (1989) FDA is preparing new food rules to curb label claims. *New York Times* 30 October, A1.

Hilts, P.J. (1990a) In reversal, White House backs curbs on health claims for food. *New York Times* 9 February, A1.

Hilts, P.J. (1990b) U.S. plans to make sweeping changes in labels on food. *New York Times* 8 March, A1.

Hilts, P.J. (1990c) Panel says White House delayed effort to curb food health claims, *New York Times* 15 November, B14.

Hilts, P.J. (2003) *Protecting America's health: The FDA, business, and one hundred years of regulation*. Knopf, New York.

Hine, T. (1997) *The total package: The secret history and hidden meanings of boxes, bottles, cans, and other persuasive containers*. Back Bay, New York.

Hof, R.D. (2004) Building an idea factory. *Business Week* 11 October.

Hutt, P. B. (1995) A brief history of FDA regulation relating to the nutrient content of good. In: *Nutrition labeling handbook* (ed. R. Shapiro), pp. 1–28, Marcel Dekker, New York.

Ippolito, P.M. and Mathios, A.D. (1989) *Health claims in advertising and labeling: A study of the cereal market*. Federal Trade Commission, Washington, DC.

Kapferer, J.N. (1992) *Strategic brand management*. Free Press, New York.

Kasdon, L. (2009) Goodbye trans fats; Now it's salt's turn: Top nutritionist looks to change how America eats. *Boston Globe* [Online]. Available: http://www.boston.com/lifestyle/food/articles/2009/03/04/goodbye_trans_fats_now_its_salts_turn/ (last accessed 12 May 2013).

Keough, D. (1994) The importance of brand power. In: *Brand power* (ed. Paul Stobart). New York UP, New York.

Kinosian, B. and Eisenberg, J. (1988) Cutting into cholesterol: Cost-effective alternatives for treating hypercholesterolemia. *Journal of the American Medical Association* **259**, 2249–2254.

Kleinfield, N.R. (1989) Catching the anti-cholesterol fever. *New York Times* 16 April, 31.

Kowalski, R. (1987) *The 8-week cholesterol cure: How to lower your blood cholesterol up to 40 percent without drugs or deprivation*. Harper and Row, New York.

Kurian, G. (1998) *A historical guide to the U.S. Government*. Oxford UP, New York.

Leonhart, D. (1998) Stirring things up at Quaker Oats. *Business Week* 30 March, 42.

Levenstein, H. (1988) *Revolution at the table: The transformation of the American diet*. California UP, Berkeley, CA.

Levenstein, H. (1993) *Paradox of Plenty: A Social History of Eating in Modern America*. Oxford UP, New York.

Levenstein, H. (1994) *Paradox of plenty: A social history of eating in modern America*. Oxford UP, New York.

Levy, A.S. and Stokes, R.C. (1987) Effects of a health promotion advertising campaign on sales of ready-to-eat cereals. *Public Health Reports* **102**, 398–403.

Levy, A.S. *et al.* (1997) *Consumer tmpacts of health claims: An experimental study*. Food and Drug Administration, Washington, DC.

Liebig, Justus (1843) *Familiar Letters on Chemistry: In Its Relations to Physiology, Dietetics, Agriculture, Commerce, and Political Economy*. Taylor, Walton and Maberly, London.

Liebman, B. (1990) Has the oat bran bubble burst? *Nutrition Action Newsletter* 1 January.

Liesse, J. and Dagnoli, J. (1990) America's oat bran craze on the wane; Controversial research fuels product fallout. *Advertising Age* 22 January, 1.

Liesse, J. and Colford, S. (1990) Marketers like new label law. *Advertising Age* 29 October, 67.

Marquette, A.F. (1967) *Brands, Trademarks and Good Will: The Story of the Quaker Oats Company*. McGraw-Hill, New York.

McDowell, E. (1989) Top selling books of 1988: spy novel and physics. *New York Times* 2 February, C24.

Michman, R.D. and Mazze, E.M. (1998) *The food industry wars: Marketing triumphs and blunders*. Quorum, Westport, CT.

Miller, J., and Muir, D. (2004) *The Business of Brands*. John Wiley & Sons Ltd, Chichester, UK.

Mintel Group (2010) Breakfast cereal (US) September 2009: Hot cereal [Online]. Available http://oxygen.mintel.com/display/418754/ (last accessed 15 May 2013).

Moser, P. (1989) How I made $812 in the oat bran craze. *Fortune* 9 October [Online]. Available: http://money.cnn.com/magazines/fortune/fortune_archive/1989/10/09/72556/index.htm (last accessed 12 May 2013).

Nestle, M. (2003) *Food politics: How the Food Industry Influences Nutrition and Health*. California UP, Berkeley, CA.

Neumann, W. (2009) For your health, Froot Loops. *New York Times* 4 September, B1.

New York Times (1988) F.D.A. faulted for shift on food label rules. *New York Times* 13 April, C4.

New York Times (1990) Foodbusters. *New York Times* 2 February, A30.

New York Times (1992) Honest food labels? Fat chance. *New York Times* 18 November, A26.

Nordhielm, C.L. (2006) *Quaker Oats's oatmeal division*. Northwestern University Kellogg School of Management, Evanston, IL.

Nutraingredients USA (2003) Quaker merger boosts PepsiCo profits. *Nutraingredients USA* [Online], Available: http://www.nutraingredients-usa.com/Consumer-Trends/Quaker-merger-boosts-PepsiCo-profits (last accessed 15 May 2013).

Pagano, C. (1994) The management of global brands. In: *Brand power* (ed. P. Stobart), pp. 53–64. New York UP, New York.

PepsiCo, Inc. (2000–2008) *Annual report*. PepsiCo Inc., Purchase, NY.

PepsiCo, Inc. (2009). Form 10-K: Fiscal year 2009. In: *Edgar search. Mergent online*. Harvard Business School Baker Library, Boston, MA [Online]. Available: http://www.sec.gov/Archives/edgar/data/77476/000119312509033126/d10k.htm (last accessed 12 May 2013).

Piribo Group. (2007) *PepsiCo: The World's Biggest Functional-Food Company. 10 Case Studies in its Strategies in Sports Drinks, Fruit Drink, Snack Innovation and Cereals*. New Nutrition Business, London.

Pollack, J. (1996) Nutraceuticals take healthy food to a new level. *Advertising Age* 2 December, 54.

Pollack, J. (1998) Ordinary people star in new fall Quaker effort: Ads document cholesterol drop among residents of Colo. town. *Advertising Age* 17 August, 42.

Pollan, M. (2004) Our national eating disorder. *New York Times Magazine* 17 October, MM64.

Pollan, M. (2008) *In Defense of Food: An Eater's Manifesto*. Penguin, New York.

Pollan, M. (2009) Rules to eat by. *New York Times Magazine* 6 October, MM64.

Quaker Oats. (1996) Why is this man smiling? Advertisement. *New York Times* 14 January, A40.

Quaker Oats. (1997). Now he has another reason to smile! Advertisement. *New York Times* 23 January, A15.

Quaker Oats (1998) *Annual Report 1997*. Quaker Oats Company, Chicago.

Quaker Oats (1989–2000) *Annual Reports*. Quaker Oats Company, Chicago, IL.

Quaker Oats (2010a) Quaker oatmeal to go [Online]. Available: http://www.quakeroats.com/products/oat-snacks/oatmeal-to-go-bars/brown-sugar-and-cinnamon.aspx   (last accessed 12 May 2013).

Quaker Oats (2010b) Quaker products [Online.] Available: http://www.quakeroats.com/products.aspx (last accessed 12 May 2013).

Quaker Oats (2010c) Quaker Oats announces re-positioning of business to focus on power of the whole grain oat [Online]. Available: http://www.pepsico.com/PressRelease/Quaker-Oats-Announces-Re-Positioning-of-Business-t.html (last accessed 12 May 2013).

Quaker Oats (2012) Quaker Oats oatmeal chocolate chip [Online]. Available: http://www.quakeroats.com/products/cookies/breakfast-cookies/oatmeal-chocolate-chip.aspx (last accessed 12 May 2013).

Reyes, S. (2003) What Will Become of the Box? *Brand Week* 27 January.

Rispin, C.M. *et al.* (1992) Oat products and lipid lowering: A meta-analysis. *Journal of the American Medical Association* **267**, 3317–3327.

Roubenoff, R.A., and Roubenoff, R. (1990) Letter to the editor, Oat bran and serum cholesterol. *New England Journal of Medicine* **320**, 1746–1747.

Shapiro, E. (1990) New products clog food stores. *New York Times* 29 May, D1.

Sivulka, J. (1997) *Soap, sex, and cigarettes: A cultural history of American advertising.* Wadsworth, Belmont, CA.

Sorkin, A.R. and Winter, G. (2000) PepsiCo said to acquire Quaker Oats for $13.4 billion in stock. *New York Times* 4 December, A23

Stanton, J.L. and George, R.J. (1997) *21 Trends in Food Marketing for the 21st Century.* Raphel, Atlantic City, NJ.

Stobart, P. (1994) Preface. In: *Brand power.* (ed. P. Stobart). New York UP, New York.

Stigler, G. (1961) The economics of information. *The Journal of Political Economy* **69**, 213–225.

Stout, D. (1997) FDA moves back oatmeal and a test for drug use. *New York Times* 22 January, A10.

Swain, J.F. *et al.* (1990) Comparison of the effects of oat bran and low-fiber wheat on serum lipoprotein levels and blood pressure. *New England Journal of Medicine* **322**, 147–152.

Taubes, G. (1997) *Good Calories, Bad Calories: Fats, Carbs, and the Controversial Science of Diet and Health.* Knopf, New York.

Time (1961) The fat of the land. *Time* 13 January, 48–52.

Todd, J.E. and Variyam, J.N. (2008) The decline in consumer use of food nutrition labels, 1995–2006. Economic Research Report No. 63, Economic Research Service, United States Department of Agriculture, Washington, DC.

Thompson, S. (2006) New Quaker snacks aim to capitalize on wholesome image. *Advertising Age* 7 October, 6.

Turow, J. (1997) *Breaking Up America: Advertisers and the New Media World.* Chicago UP, Chicago.

Trivedi, B. (2006) The good, the fad, and the unhealthy. *New Scientist* **191**, 42–49.

Twitchell, J.B. (1996) *Adcult USA.* Columbia UP, New York.

Wansink, B. (2004) *Marketing Nutrition: Soy, Functional Foods, Biotechnology, and Obesity.* Illinois UP, Champaign, IL.

Wansink, B. and Cheney, M.M. (2005) Leveraging FDA health claims. *Journal of Consumer Affairs* **39**, 386–397.

Waters, J. (1997) What new CEO must do to craft Quaker's revival. *Crain's Chicago Business* 27 October.

US Congress (1906) *Pure Food and Drug Act. United States Statutes at Large.* 59th Cong., 1st Sess., Chp. 3915, Sec. 1.

US Congress (1990) *To Amend the Federal Food, Drug, and Cosmetic Act to Prescribe Nutrition Labeling for Foods, and for Other Purposes*. 101st Cong. 2nd Sess. US Government Printing Office, Washington, DC.

USDA (1943) *National Wartime Nutrition Guide* (NFC-4), Washington, DC.

US DHSS/FDA (United States Dept. of Health and Human Services/Food and Drug Administration) (1993) Food labeling; General requirements for health claims for food. *Federal Register* **58**, 2487.

US DHSS/FDA (United States Dept. of Health and Human Services/Food and Drug Administration) (1996) Food labeling: Health claims; Soluble fiber from whole oats and risk of coronary heart disease; Proposed rule. *Federal Register* **61**, 308.

US DHSS/FDA (United States Dept. of Health and Human Services/Food and Drug Administration) (1997) Food labeling: Health claims; Soluble fiber from whole oats and risk of coronary heart disease; Final rule. *Federal Register* **62**, 3584.

US Senate (1977) *Dietary Goals for the United States*. 95th Cong. 1st Sess. US Government Printing Office, Washington, DC.

Vaughn, M. (1943) Good nutrition under rationing. *The American Journal of Nursing* **43**(11), 1002–1006.

Zarkin, G.A. and Anderson, D.W. (1992) Consumer and producer responses to nutrition label changes. *American Journal of Agricultural Economics* **74**, 1202–1207.

Part VI
Future Recommendations

18

Overview: Current and Future Perspectives on Oats and Health

Penny Kris-Etherton

Department of Nutritional Sciences, The Pennsylvania State University, University Park, PA, USA

18.1 Chapter summaries

The nutritional composition of oats has been characterized and there is a good understanding of the β-glucans, starch, and types and amounts of many antioxidants. β-glucans account for the majority of oat soluble dietary fiber (10% of the dry weight of oat bran) and have unique functional properties (e.g., solubility, molecular weight, and gelation properties) compared with β-glucans from other cereal grains. The β-glucan content of oats varies by cultivar and is affected by the environment, as well as isolation, purification, and detection methods. Oat starch is a slowly digestible, resistant starch. Approximately 40–65% of the weight of groats (hulled grain of oats) is starch, which is located primarily in the endosperm. Many phytochemicals have been identified in oats. The main lignan in oats is syringaresinol, which is a precursor of mammalian lignans. Recent interest has also centered on avenanthramides, which are soluble phenolic metabolites that are unique to oats. Avenanthramides are a group of alkaloids that have high antioxidant activity. Their content can vary by more than twofold in oats [from 300–600 parts per million (0.03–0.06%)] and is affected by cultivar and processing conditions. Although present in small quantities, avenanthramides serve as the primary source of the antioxidant and anti-inflammatory properties of oats. Oats also contain various vitamins, minerals, and small amounts of saponins, flavonoids, and prostaglandin inhibitors.

Heat processing generally decreases the amount of avenanthramides, even though some are more heat stable than others. Other important lignans in oats are pinoresinol (194 μg/100 g) and lariciresinol (183 μg/100 g). Total lignan

---

content in oats is higher than in wheat, barley, and millet but lower than in flaxseed, rye, and buckwheat. The total lignan content of five different spring oat cultivars has been shown to vary appreciably (i.e., 820–2550 µg/100 g). Phytosterols in oats include β-sitosterol, sitostanol, campesterol, campestanol, and others. Phenolics, carotenoids, and vitamin E also are found in oats. Ferulic acid is the predominant phenolic acid in oat flour, accounting for about 76% of the total phenolic acids in oat flour. Oat flakes (whole grain) have lower levels of phenolic acids (472 mg/kg) than oat bran (651 mg/kg). Some of the flavonoids identified in oats include apigenin, luteolin, tricin, kaempferol, and quercetin. It is evident that oats contain many nutrients and bioactive compounds that vary in a manner that reflects differences among cultivars, growing conditions, and processing conditions. An interdisciplinary approach among various science disciplines is (and will be) important to optimize nutrient and bioactive compound profiles of oats, as well as to identify and develop optimal processes for isolating compounds in oats that are of both nutritional and commercial interest.

Epidemiological and clinical studies have shown that oats have many benefits on major chronic diseases, including cardiovascular disease, diabetes, and obesity. In addition, they have favorable effects on immune function, gut health, and skin health. It has been suggested that components of oats may decrease the risk of certain cancers based on epidemiologic associations. For example, dietary fiber has been shown to be inversely related to breast, prostate, and colorectal cancers in different population studies. The weight of evidence linking oats with risk of these various chronic diseases differs, with the most robust for cardiovascular disease.

Many studies have evaluated the underlying mechanisms by which oats and their different bioactive compounds confer health benefits. As is always the case, this has advanced our understanding of the health effects of oats and oat components but has also raised numerous questions about the specific physiological effects of these compounds, what molecular species within a bioactive family mediate the effects on risk factors for chronic diseases, and how all of these effects might be modulated by oat milling and processing.

In the Nurses' Health Study (Liu *et al.*, 1999), consumption of two to four servings of cooked oatmeal per week decreased risk of coronary heart disease by approximately 30%. The effects of whole grain oats on cardiovascular disease risk factors, including elevated total cholesterol (TC), low-density lipoprotein-cholesterol (LDL-C), and blood pressure (BP), have been extensively studied. Meta-analyses have shown that daily consumption of soluble fiber from oats ($\geq$ 3 g) lowers TC and LDL-C levels 1.3–1.8% per gram β-glucan, with greater reductions reported for hypercholesterolemic individuals. In addition, high-molecular weight β-glucans have been shown to reduce LDL-C levels to a greater extent than low molecular weight β-glucan. Plant sterols, another bioactive compound in oats, lower TC and LDL-C levels by inhibiting cholesterol absorption.

Several prospective cohort studies have reported an inverse relationship between blood pressure and dietary intake of whole grains and/or fiber. In the Physicians' Health Study with 13 368 male participants (Kochar *et al.*, 2012), whole grain cereal intake of more than seven servings per week lowered the relative risk of hypertension by 19% and 20% [when body mass index (BMI) was less than 25 kg/m^2 or greater than 25 kg/m^2, respectively]. Two meta-analyses of 24 or 25 randomized, controlled trials reported that a mean supplemental fiber intake of 7.2–18.9 g/day reduced systolic BP (SBP, 1.1–1.2 mm Hg) and diastolic BP (DBP, 1.3–1.7 mm Hg). Soluble fiber produced a greater effect on SBP (–1.3 mm Hg) and DBP (–0.8 mm Hg) than insoluble fiber (–0.2 and –0.6 mm Hg, respectively). The controlled clinical studies with oats do not consistently show a BP-lowering effect, although there is some evidence that they do have a hypotensive effect, especially in individuals with high BP. Likewise, the clinical studies that evaluated the effects of oat bran and β-glucan on BP have reported mixed results.

Prospective epidemiological studies suggest that habitual consumption of whole grains is inversely associated with BMI and body weight gain. There is evidence that consumption of fermentable dietary fiber increases appetite-suppressing gastrointestinal (GI) hormones, which may mediate a reduced energy intake that would account for the lower BMI and attenuated weight gain over time. However, clinical studies designed to evaluate the effects of oats or oat fiber (principally β-glucan) have reported mixed results for appetite, food intake, and body weight loss.

Whole grain consumption, including whole grain oats, is associated with reduced risk of type 2 diabetes. In the Nurses' Health Study (Liu et al., 2000), consumption of cooked oatmeal five to six times per week reduced the risk of type 2 diabetes by 39%. Research has shown a beneficial effect of a low glycemic index diet that emphasizes large flake oatmeal, oat bran, beans, nuts, bulgur, flax, and other low glycemic index foods on glycated hemoglobin A1c (HbA1c), a measure of long-term plasma glucose levels. HbA1c decreased by 0.50% in the low glycemic index group (Thomas and Elliot, 2009). In an analysis of 15 studies, increased dietary fiber intake lowered HbA1c (by 0.26%) and fasting blood glucose levels (by 0.85 mM) in participants with type 2 diabetes mellitus (Post et al., 2012). However, the clinical studies specifically with oat products demonstrated mixed effects on long-term blood glucose control in individuals with diabetes.

Contrary to longer-term clinical studies, acute studies have consistently demonstrated a beneficial effect of oat bran on postprandial glucose levels (Sadiq Butt et al., 2008). This review noted that oat bran lowers postprandial glucose and insulin levels and the effect has been attributed to β-glucan, which influences glycemic and insulinemic responses. Oat β-glucan develops high viscosity and slows both starch digestion and the rate of gastric emptying. Although these effects are realized in the upper GI tract, there are effects in the colon that are related to oat β-glucan being a fermentable fiber that is converted to acetate, propionate, and butyrate.

Oats contribute dietary fiber, resistant starch, and oligosaccharides to the diet, which affect gut health. The fiber in oats may help reduce the rate of gastric emptying and promote satiety. The viscosity of oats appears to have important physiological properties, including the regulation of satiety-related gut hormones. Oat fiber remains essentially intact during transit through the small intestine. Both total and soluble fiber in oats increase stool weight. Undigested and unabsorbed carbohydrates are fermented in the large intestine, where they may function as prebiotics. Nondigestible carbohydrates in oats (dietary fiber and resistant starch) may positively affect the gut immune system by altering the number and composition of intestinal microflora and affecting gut-associated lymphoid tissue. While there is little information about the immune effects of prebiotics, it is clear that they affect gut microbial composition. These changes in gut microbes may improve resistance to pathogens by increasing the numbers of bifidobacteria and lactobacilli, which decrease gut pH and adversely affect pathogen populations. This area of science is emerging, and much remains to be learned about the specific effects of oats and oat bran on gut health.

The nutritional composition of oats is more "complete" than other whole grains, such as corn, wheat, and rice. On a 100 g-serving basis, oats are a good source of many nutrients: thiamin, folate, iron, magnesium, copper, zinc, and potassium, and have a low sodium/potassium ratio. The amino acid profile of oats is higher in leucine, lysine, and phenylalanine than other whole grains (Chapter 4).

Oats are used to treat a variety of dermatologic conditions. Avenanthramides, metabolites that are unique to oats, have anti-inflammatory, antipruritic, antioxidant, and antifungal properties, which confer benefits for atopic dermatitis, contact dermatitis, pruritic dermatoses, sunburn, drug eruptions, and other skin conditions. Colloidal oatmeal is used to protect and relieve minor skin irritations and itching due to eczema; it also has many beneficial dermatologic properties and functions as a cleanser, moisturizer, and skin protectant. The flavonoids in oats protect against ultraviolet A radiation and tocopherol protects against both inflammation and photodamage. There is some histopathological evidence of improved wound healing after topical treatment with oat extracts.

Although there is not a chapter on oats and cancer, there is some epidemiological evidence discussed about oat constituents and their anticancer properties. For example, oat compounds that may be protective against cancers of the GI tract include enterolactone and lignans. The studies conducted to date provide only suggestive information about a protective association between oat bioactive compounds and risk of certain GI cancers. Thus, further research is needed to gain a better understanding about whether oats are protective against GI cancers.

The last part of the book addresses nutritional and food health claims. Health claims are important because they educate consumers about informed food choices that can improve their health. In addition, they provide regulatory guidance for the food industry in appropriate use of claims on product labels. Approved health claims for oat products around the world (i.e., Canada, the United States, the European Union, and other countries) are based on extensive research necessary to substantiate oat health claims according to Codex guidelines. The claims vary among countries and the evidence required to substantiate a health claim also varies. One chapter presents a detailed summary of the steps involved for the health claim that was issued for oats in the United States, including the legal aspects and economic benefits. This chapter provides a road map for seeking additional health claims that could be related to other diseases/conditions beyond cardiovascular disease.

18.2   Relevance to the nutrition and dietetic communities and the medical profession

Given the many health benefits ascribed to oats in this book, it is evident that this information must be communicated to consumers by the nutrition and dietetic communities and other health professionals. The standards of practice in nutrition, dietetics, and medicine are guided by an extensive evidence base, which is used to formulate dietary recommendations that are made by professional organizations, such as the Academy of Nutrition and Dietetics (through their Evidence Analysis Library, in part), the American Heart Association (AHA), and the American Diabetes Association, among others. In addition, dietary recommendations are issued by federal agencies. Relative to oats, dietary recommendations by all of these organizations have been made to consume fiber-rich whole grains (one-half or more of total grain consumption should be from whole grains) for health promotion (Consistent with those made by USDA/DHHS, 2010) and for prevention of major chronic diseases (Bantle *et al.*, 2008; Lloyd-Jones *et al.*, 2010).

Members of the healthcare professions are well aware of the major chronic diseases that affect world populations. As of 2010, ischemic heart disease (number 1 ranking) and stroke (number 3 ranking) were two of the top 12 world health problems that could be favorably affected by oat consumption (Cohen, 2012; Lim *et al.*, 2012). Important risk factors recently highlighted by the Global Burden of Disease Study that could be affected by oats include high BP, high BMI, and high fasting blood glucose levels (Cohen, 2012; Lim *et al.*, 2012), as well as an elevated LDL-C level, as noted by the AHA (Roger *et al.*, 2012).

The impetus for the many dietary guidelines that have been issued is based on recognition of their importance in promoting health and well-being, and for reducing risk of chronic diseases. A recent meta-analysis of 23 studies with 11 085 randomized patients reported that lifestyle modification programs were associated with reduced all-cause mortality, cardiac mortality, cardiac readmissions, and nonfatal reinfarctions (Janssen *et al.*, 2012). Implementation of lifestyle modification programs, including a healthy diet and program of physical activity, beneficially affected many risk factors associated with heart disease. The extensive evidence base showing benefits of a healthy diet that includes whole grains and oats for the world's top health problems must be coupled to programs that are effective in modifying consumer health behaviors to reduce the burden of disease. A large integrated effort is needed that spans the spectrum from producing healthy food to identifying effective strategies that encourage consumers to adopt healthy diet and lifestyle practices. This will require major inputs from the agricultural and food processing industries as well as from the various health professional communities. For this to be effective on a global scale, it must be coordinated by federal, state, and local agencies, as well as various health organizations.

18.3 Future needs and recommendations

A common theme in the book is the "need for more research and innovations" to produce the healthiest oats and oat products possible for optimal health outcomes and to increase the wider global usage and consumption of whole grain oat and oat products. To address the needs identified, future considerations and actions should be prioritized.

(i) Stop the decline in oat production and reverse the trend of oats as an orphan crop: Programs must be put in place to encourage growers to increase oat production. This may require novel and creative long range, public–private strategic collaborations to increase production and yields in North America that are sustainable and profitable.

(ii) Explore new tools to identify oat varietals of specific interest to increase production: Key tools that are being used now (and will be in the future) for developing the best oat varieties are using "-omics" technologies, including metabolomics and genome sequencing (Rasmussen *et al.*, 2012). These technologies are invaluable tools for large-scale geno- and phenotyping of breeding populations for developing new oat cultivars that are optimized for both oat production and human health. This approach may allow farmers to plant the varietals of interest, providing them with an economic incentive to increase output.

(iii) Identify and characterize oat bioactive compounds for health effects: It is clear that different constituents of oats affect many major diseases. Questions remain about what specific bioactive compounds (including isoforms) in oats account for the observed health effects. Recent interest has also focused on avenanthramides, a group of soluble phenolic metabolites that are unique to oats. These must be identified so that future varieties of oats can be "engineered" to have the desired traits.

(iv) Understand the impact of processing on oat nutritional and functional benefits: How processing affects oat chemical composition and bioactive functionality needs further study, especially the impact on health outcomes.

(v) Explore new tools for mapping oat modes of action and dietary customization: Nutrigenomics is a discipline that offers the promise of many new exciting breakthroughs in understanding the mechanisms by which diet, dietary patterns, nutrients, and bioactive compounds affect nutrition-related diseases. Nutrigenomics provides in-depth information about how nutrients interact with genes and how genes influence metabolism (Zeisel, 2011). As we gain a better understanding of single nucleotide polymorphisms that account for the variability in nutrient/bioactive metabolism, we may be better positioned to modulate these by diet. With respect to oats, it may be possible to modify their nutrient/bioactive profiles that have the greatest benefit on a population basis to reduce risk of chronic diseases. In addition, it may be feasible to custom "design" the composition of oats to be of benefit to specific diseases/conditions. Thus, the use of nutrigenomics may guide the development of "enhanced foods" that are of greatest benefit to specific population groups and individuals.

(vi)   Explore further research on oats and specific health areas:

GI cancer: Further research is needed to gain a better understanding about whether oats are protective against human GI cancers in both the upper and lower gut.

Children and women: Additional research is needed to understand the benefits of incorporating/increasing whole oat intake in the diets of children and women over a long period.

Food intake: A better understanding of oats and macronutrient interactions (e.g., proteins, fats) on satiation and satiety may be important to assess potential longer-duration benefits to caloric and weight control in different age and gender cohorts, including young children, teens, and the elderly

Skin health: Further investigation and research to better understand the benefits of oat and oat constituents on skin health, for example, repair skin dryness and hydration.

Blood pressure: Additional studies are needed to better define the role of whole oats, β-glucan, and other oat constituents on BP. Although several prospective cohort studies have reported an inverse association between whole grain and fiber intake and BP, intervention trials with whole oats or concentrated β-glucan (3.0–7.7 g/day β-glucan) did not demonstrate consistent reductions in BP except in individuals with hypertension and/or increased adiposity.

(vii)   Increase oat consumption: Although whole oats and oat products have beneficial health effects, worldwide consumption of oats remains low compared to other grains. In 2011/2012, oats accounted for only 1% of global whole grains consumed, surpassed by corn, wheat, rice, and barley. Oat production at 22.6 million metric tons is markedly lower than the 696.4 million metric tons of wheat (USDA, 2013). Many factors contribute to the decrease in oat production, including lack of profitability for farmers to plant oats in North America versus other whole grains. Another reason may be the low usage of oats in staple products such as breads and crackers. To increase oat production, some of the new tools identified previously in item (iii) could be applied. To increase consumer daily intake of whole grain oats, good tasting, familiar, and portable whole grain oat products with high sensory appeal and affordability should be made readily available and accessible. Priority should be placed on education that is targeted to children and teens, using repetitive, simple messages on the nutrition and health benefits of whole grain oat products. Innovative education programs linked to food labels should be explored further. Paul and colleagues documented the effectiveness of food labels on consumer food choices (Paul *et al.*, 1999). New health claims related to whole oats and/or oat bioactive compounds and chronic disease risk factor reduction beyond cholesterol must be explored.

(viii) Focus on smart and sustained communication and education: The nutritional and health benefits of oats must be conveyed at multiple levels over sustained periods. These efforts should include scientists, healthcare professionals, regulators, and the public. Early education of consumers will provide a lifetime understanding for the use of oats in their diets.

(ix) Collaborate on clinical trials: A major challenge going forward is to increase support for the research community that meets societal needs. It is evident that oats have many health benefits; however, additional research is needed to further the goals of developing the best oats and oat products possible, issuing specific dietary recommendations for oat consumption, and identifying strategies that increase consumption as a means to benefit health. Many longer trials are necessary to understand the long-term benefits of oat products on chronic disease outcomes or relief of symptoms. Since these trials are expensive, creative and strategic public–private partnerships and collaborations should be considered.

Oats are a grain with a unique chemistry and nutritional profile that may play a critical role in reducing risk of many chronic diseases and favorably affecting skin, gut, and immune health. It is the first food to be approved by the US Food and Drug Administration in 1997 for a health claim to reduce cholesterol when consumed as part of a healthy diet. Today, a health statement for oats and cholesterol reduction has also been approved by the European Food Safety Commission and Health Canada. Despite this fact, consumption of oats still trails behind other whole grains, such as wheat, corn, and soybeans. The decline in oat production in North American has made oats an orphan crop, becoming the single most important threat to increasing oat intake and usage as a staple in the diet. Solutions to meeting the stated challenges will require creative longer-term planning and strategic multistakeholder collaboration. It will also require long-term funding to encourage oat and health research.

References

Bantle, J.P. *et al.* (2008) Nutrition recommendations and interventions for diabetes: A position statement of the American Diabetes Association. *Diabetes Care* **31**, S61–S78.

Cohen, J. (2012) A controversial close-up of humanity's health. *Science* **338**, 1414–1416.

Janssen, V. *et al.* (2012) Lifestyle modification programmes for patients with coronary heart disease: A systematic review and meta-analysis of randomized controlled trials. *European Journal of Preventive Cardiol*ogy 28 September [Epub ahead of print]. doi: 10.1177/2047487312462824.

Kochar, J. *et al.* (2012) Breakfast cereals and risk of hypertension in the Physicians' Health Study I. *Clinical Nutrition* **31**, 89–92.

Lim, S. *et al.* (2012) A comparative risk assessment of burden of disease and injury attributable to 67 risk factors and risk factor clusters in 21 regions, 1990–2010: A systematic analysis for the Global Burden of Disease Study 2010. *Lancet* **330**, 2224–2260.

Liu, S. *et al.* (1999) Whole-grain consumption and risk of coronary heart disease: results from the Nurses' Health Study. *American Journal of Clinical Nutrition* **70**, 412–419.

Liu, S. *et al.* (2000) A prospective study of whole-grain intake and risk of type 2 diabetes mellitus in US women. *American Journal of Public Health* **90**, 1409–1415.

Lloyd-Jones, D.M. *et al.* (2010) Defining and setting national goals for cardiovascular health promotion and disease reduction: the American Heart Association's strategic Impact Goal through 2020 and beyond. *Circulation* **121**, 586–613.

Paul, G.L. *et al.* (1999) The Quaker Oats health claim: A case study. *Journal of Nutraceuticals, Functional and Medical Foods* **1**, 5–32.

Post, R.E. *et al.* (2012) Dietary fiber for the treatment of type 2 diabetes mellitus: a meta-analysis. *Journal of the American Board of Family Medicine* **25**, 16–23.

Rasmussen, S. *et al.* (2012) Metabolomics of forage plants: a review. *Annals of Botany* **110**, 1281–1290.

Roger, V.L. *et al.* (2012) Executive summary: Heart disease and stroke statistics – 2012 update: A report from the American Heart Association. *Circulation* **125**, 188–197.

Sadiq Butt, M. *et al.* (2008) Oat: unique among the cereals. *European Journal of Nutrition* **47**, 68–79.

Thomas, D. and Elliott, E. J. (2009) Low glycaemic index, or low glycaemic load, diets for diabetes mellitus. *Cochrane Database of Systematic Reviews* 1 (Art. No. CD006296). doi: 10.1002/14651858.CD006296.pub2.

USDA (US Department of Agriculture) (2013) Grain: World Markets and Trade. Foreign Agricultural Service Circular Series FG 05-13. http://www.fas.usda.gov/psdonline/circulars/grain.pdf (last accessed-16 May 2013)

USDA (US Department of Agriculture)/DHSS (Department of Health and Human Services) (2010) *Dietary Guidelines for Americans*, 7th edn. US Government Printing Office, Washington DC.

Zeisel, S.H. (2011) Nutritional genomics: Defining the dietary requirement and effects of choline. *Journal of Nutrition* **141**, 531–534.

Index

Agri-Food Canada (AAFC or AAC), 11
  Cereal Research Center (CRC), 11
appetite, 271–4
  effects of oats on, 271–4
  fiber and, 271–4
  β-glucan and, 271–4
  physiological mechanisms, 274–6
  stimulating hormones, 272–4
  suppressing hormones, 272–5
  visual analog scale, 271
avenanthramides, 5, 172, 180–81, 195–6,
        255–6, 312–3, 315–6, 316–7, 317–8,
        321, 429, 432, 434
  analysis methods, 201
    liquid chromatography, 201
    mass spectrometry, 201
  biosynthesis, 197, 201–6
    CoA: hydroxyanthranilate
      N-hydroxycinnamoyl transferase
      (HHT), 201, 204, 209–11, 211–4,
      214–6
    environmental effects on, 207–9
    in false malting, 219–21
    hydroxycinnamoyl/benzoyl-
      CoA:anthranilate
      N-hydroxycinnamoyl/
      benzoyltransferase (HCBT), 211–14
    localization, 216–8
    metabolites, 201–2, 203
    pathways, 203
  chemistry, 197–201, 211–4
    cloning, 211–4
    Crown Rust resistance, 205–206,
      206–9

  metabolism, 201–2
  processing of, 429–30
  solubility, 200–201
  stability, 197–9
  victorin sensitivity, 206–7
composition of, 180, 195–6, 312–3
epidemiological data, 255, 256, 260
function, 219, 256–61, 313–8, 429, 432
  antifungal, 256, 317–8
  anti-inflammatory, 258–61, 313–6,
    314
  antioxidation, 199–200, 257–8, 314,
    317
  antiproliferative, 258–61
  antipruritic, 314, 316–7
health effects, 4–5, 180–81, 256–8,
    258–61, 313–8, 321–22
nomenclature, 196–7, 256–8
phytoalexin, 206–7, 211–4, 222
plant defense activators, 218–9
  Actigard™, 218
  benzo (1,2,3)-thiadiazole-7-
    carbothioic-S-methyl ester (BTH),
    218–9
properties, 199–201
  antioxidant, 199–200
  solubility, 200–201
Avena sativa, 172, 311, 313, 315, 321, 324,
    326, 376

blood pressure, 4–5, 186, 239, 240, 430
  dietary fiber and, 240–46, 430
  dietary patterns and, 240, 430
  effects on, 4–5, 186, 240–46, 246–51, 430

Oats Nutrition and Technology, First Edition. Edited by YiFang Chu.
© 2014 John Wiley & Sons, Ltd. Published 2014 by John Wiley & Sons, Ltd.

blood pressure (*Continued*)
Health Professionals Follow-up Study,
240
mechanisms of, 246
Physicians' Health Study, 430
see health benefits, blood pressure, 4–5,
186, 240–46, 246–51
Women's Health Study, 240
body mass index, 4–5, 265, 431
effects on, 4–5, 268–9, 271, 431
breeding, 9–11, 19–22, 24–9, 34–36, 36–41,
42, 44–5, 59–60
challenges, 10, 15, 19, 23–24, 28–30
goals, 10, 30, 31
independent culling, 25–6
index selection, 27–8
multiple traits, 19–25
pairwise association, 19–22
three-way association, 23–5
programs, 38–42, 44
single traits, 11–9
β-glucan concentration, 17
grain yield, 12–5
groat percentage, 16
kernel weight, 16
oil concentration, 17–8
protein concentration, 18–9
test weight, 15–6
strategies, 25–8

Canada's Food Guide, 35
Canadian Grain Commission Inspection
Division Grade Standards, 34
carbohydrate metabolism, 281
clinical studies of, 284–9
dose response, 289–91
oat bran, 286–8
oat-derived β-glucan preparations,
289
whole oats, 284–6
digestion of, 282–4
epidemiology of, 281–2
glucose lowering, effects on, 282–4
long term, 291–2
cardiovascular disease, 3, 171–2, 175–6,
178, 179, 181, 184, 185–6, 189, 229,
232, 255–6, 335, 340, 341, 343, 345,
346, 347, 362, 391
cholesterol lowering in, 230–31, 231–3,
374, 375, 388–90, 391, 413, 430, 436

effects on, 3, 171–2, 175–6, 178, 179,
181, 184, 185–6, 189, 229, 335, 340,
341, 343, 345, 346, 347, 362, 391,
430
lipoproteins, 229, 230–31, 231–3, 255–6,
258
see health benefits, epidemiological
evidence, cardiovascular disease,
171–2, 175–6, 178, 179, 181, 184,
185–6, 189, 255–6, 362, 391, 430
triglycerides, 230, 233
and whole grains, 255–6
carotenoids, 172, 181–2
composition of, 181
health effects, 181–2
cereal grains, 73, 77, 303, 429
barley, 78, 171, 172
corn, 73, 76, 78, 171, 172, 255
health effects of, 305, 306
nutrition of, 78–80, 80–81, 81–3, 83–91
oats, 74, 76, 78, 171, 172, 255, 257, 260,
303
rice, 73, 76, 78, 171, 172, 255
rye, 78, 171, 172, 179, 303, 305
wheat, 73, 76, 78, 172, 255, 303, 305
cholesterol, 4–5, 41, 43, 46, 56, 144, 144–7,
147–52229, 230–31, 231–3, 374, 375,
388–90, 391, 413, 430, 436
effects on, 4–5, 41, 43, 46, 56, 144, 144–7,
147–52, 229, 230–31, 231–3, 374,
375, 388–90, 391, 413, 430,
436
see also health benefits,
cholesterol-lowering, 4–5, 144,
144–7, 147–52, 229, 230–31, 231–3,
374, 375, 388–90, 391, 413, 430, 436
colloidal oatmeal, 5, 311–2, 432
avenanthramides in, 312–3
effects of, 5, 432
skin health, 318
skin pH, 318
properties, 313, 313–8, 432
antifungal, 314, 317–8
anti-inflammatory, 313–6, 314
antioxidant, 314, 317
antipruritic, 314, 316–7
hydration, 318
structure, 312–3
composition, 312–3
processing, 312

dehulling equipment, 38–9, 39–40, 59
  Codema LLC laboratory dehuller
    LH5095, 39
  Streckel & Schrader KG Laboratory
    hulling machine BT459, 39
diabetes, 152, 171–2, 178, 183, 187, 255,
    256, 281–2, 286, 291, 292, 335–336,
    430–31
  effects on, 152, 183, 187, 255, 256, 282,
    291, 292, 335–6, 430–31
  see Health benefits, diabetes, 152, 172,
    178, 187, 255, 256, 282, 291, 292,
    335–6, 430–41, 433
Dietary Goals for the United States, 367
digestive health, 299–301
  bowel habits, 299–300, 302
  diarrhea, 299–301
  effects on, 299–301
  fiber and, 299–301, 301–2
  irritable bowel syndrome (IBS), 300–301
  large bowel, 302
  laxation, 299, 302
  mechanisms of, 303–6
  see Health benefits, epidemiological
    evidence, digestive health, 188
  whole grains and, 299–301, 302, 305–6

Eastern Cereal and Oilseed Research
    Center (ECORC) of Agriculture,
    11–12
European Food Safety Authority, 281, 286

FDA, 3–4, 41, 56, 311–2, 319, 358, 360, 372,
    373, 382–7, 387–90, 391, 391–4,
    395–6, 401, 401–2, 411–2, 413–5
fiber, 4–5, 10, 41–2, 43–4, 49, 54, 55, 73, 75,
    75–8, 147, 230–31, 299–301, 301–2
  fatty acids and, 301–2
  fermentation, 301–2, 302, 303
  health effects, 4–5, 10, 35, 37, 41, 43–4,
    52, 56, 230–31, 364, 367, 372, 373,
    377, 380, 387–90, 394, 395, 402, 415,
    418
  hypertension and, 240–46
  insoluble, 43–4, 55, 75, 76, 78
    arabinoxylan, 43, 44
    cellulose, 43, 45, 74, 75, 78
    noncellulosic polysaccharides, 43
  measurement methods, 35, 42,
    43–4

soluble, 35, 41, 43–4, 52, 55, 75, 78
  arabinogalactan, 43
  arabinoxylan, 43, 44
  cholesterol-lowering effects, 230–31
  effects on cholesterol production,
    230–31, 388, 389, 394, 402
  β-glucan, 76, 78, 123–4, 131, 389,
    391
total dietary, 41–2, 43–4
flavonoids, 5, 172, 174–6, 429–430, 432
  composition of, 174–5
  health effects, 5, 175–6
Food, Drug, and Cosmetic (FD&C) Act of
    1938, 372, 383
food pyramid, 413, 417

GGE biplot analysis, 11–4, 15, 16, 17, 18,
    19, 20, 21, 26, 27
  genotype (G), 11, 13
  genotype-by-environment (GE), 11, 13
  G/[G+GE] ratio, 11, 14, 15, 16, 17, 18
  β-glucan, 3–5, 10, 17–8, 22–4, 27, 30–31,
    35–6, 39, 41, 42–3, 43–44, 47, 50–52,
    54, 55–6, 60, 123–4, 131, 172, 176–8,
    256, 429–31, 435
  composition of, 123–4, 176–7
  extraction, 129–31, 131–2, 133–5
    purification, 133–5
    quantification, 132–3
  fermentation, 301–2, 302, 303
  health benefits, 41, 42–3, 56, 60, 143, 146,
    147, 155, 159, 256, 265–6, 281, 282,
    283–4, 284, 291–2, 336, 339, 340–42,
    343
  body weight, 266–71
  carbohydrate metabolism, 281, 282–4,
    284, 291–2
  cholesterol-lowering, 144, 144–7,
    147–52, 229, 231–33, 256, 336, 339,
    340–42, 343, 348, 350, 388–390
  clinical trial data, 144, 147, 150–52,
    157–8
  digestive health, 188, 299–301
  energy intake, 144, 147, 156
  glucose-lowering, 144, 145, 147, 152–6,
    256, 282, 282–4, 284, 291–2, 336,
    339, 340–42
  hypertension, 186, 246–51
  immunological effects, 156–7
  insulin attenuation, 147, 149, 152–6

GGE biplot analysis (*Continued*)
  matrix effects on, 144, 147, 149,
    149–52
  physicochemical properties, effects
    on, 155
  satiety, 144, 156, 256
  location, in oats, 126, 131–2
  nutritional properties, 144
    colon, 144, 146
    fermentation, 144, 146
    gastrointestinal, 144–7
    small intestine, 145–6
    stomach, 145
  physicochemical properties, 4, 39, 42–3,
    123, 143, 144, 146, 147, 147–8,
    149–52, 299
    cholesterol-lowering, 145–6
    components of, 124, 129
    measurement methods, 42, 43
    molecular structure, 124–31
    molecular weight, 124, 127, 128, 129,
      132, 134
    solution of, 124, 129, 132, 133, 135–33
    viscosity, 144, 145, 145–6, 148, 149–51,
      151–2, 299, 306
    see Rapid visco analyzer, 42, 43, 112
    see also Fiber, 4–5, 10, 41–42, 43–4, 49,
      54, 55, 429–33
  solution properties, 135
    conformation, 135–9
    rheology, 139–43
  sources of, 123, 127, 128, 156, 157–8
glucose levels, 4–5, 43, 144, 145, 147, 152–6
  effects on, 4–5, 43, 144, 145, 147, 152–6
  see also Health benefits,
    glucose-lowering, 4–5, 43, 144, 145,
      147, 152–6
grain, 9–10, 12–5, 20, 21, 22, 23, 24–7, 28,
    30–31, 34, 35, 36, 39–42, 46, 54,
    58–9, 60, 73, 74, 171, 172, 173
  components of, 37–9, 74, 172
    groat, 36, 39, 74, 177, 180
    hull, 37–9, 74
  quality characteristics, 10–11, 34, 35, 36
    groat breakage, 36, 39
    kernel size, shape, 34, 36, 39–40
    nutrition, 35, 36, 41–2
  weight, 34, 36, 40–41
  structure, 74
  yield, 10–11, 12, 14, 15, 22, 23, 24, 28, 29,
    30–31

groat, 10–11, 16, 22, 23, 24, 28, 29, 30–31,
    36, 37–9, 177, 180
  breakage, 36, 39
  components, 36, 171, 180
    bran, 36, 171
    germ, 36, 171
    starchy endosperm, 36, 171
  β-glucan in, 16, 22, 23, 24, 28, 29

health benefits, 4–5, 144, 144–7, 147–52,
    152–6, 229, 239–40, 255–6, 258–61,
    265–66, 282, 291, 292, 299–301,
    303–6, 313–7, 317–8, 335–6, 362, 374,
    375, 388–90, 391, 413–4, 430–33, 433
  blood pressure, 4–5, 186, 240–46,
    246–51
  body weight, 266–71, 431
  cholesterol-lowering, 4–5, 144, 144–7,
    147–52, 176, 177–8, 185, 186, 188,
    229, 230–31, 231–3, 374, 375,
    388–90, 391, 413, 430, 436
  diabetes, 152, 172, 178, 187, 255, 256,
    282, 291, 292, 335–6, 430–31, 433
  digestive health, 299–301
  epidemiological evidence, 171, 179,
    185–9, 255, 256, 260, 362, 391,
    430–31
    cancer, 255, 260–61
    cardiovascular disease, 171–2, 175–6,
      178, 179, 181, 183, 185–6, 189,
      255–6, 362, 391, 430
    diabetes, 187, 255, 256, 430–31
    digestive health, 188
    hypertension, 186
    obesity, 187–8, 430–31
  glucose-lowering, 4–5, 43, 144, 145, 147,
    152–6, 177, 187, 256, 430–31
  skin health, 318–23
Health Professionals Follow-up Study,
    240

lignans, 172, 178–9, 429–30, 432
  composition of, 178
  health effects, 179

market, 33–7, 359, 362, 370–72, 372–3, 376,
    384, 387, 393–4, 395, 397–9, 401–3,
    412
  feed, 34, 37
  milling, 34, 37, 41
  value, 360, 362, 387, 391, 394, 408, 419

milling, 34, 37
  bran, 55–8
    health benefits, 56
    properties, 55–8
    quality, 58
    yield, 56
  flakes, 47–9
    processing treatment, 47–9
    properties, 47–9
    quality, 47–8
  flour, 49–52
    pasting, 50–52
    processing treatment, 51
    properties, 50–52
    quality, 50–52
    use in bread, 49, 53–55
    use in extruded products, 49, 55
    use in pasta, 49–50, 52–3
  β-glucan, 42, 50–51
  groats, steel-cut, 49
    properties, 49
    quality, 49
  mycotoxins, 58–9
    Fusarium, 58–9
      deoxynivalenol, 58
      zearalenone, 58
    Penicillium, 58–9
      ochratoxin A, 59
    tolerance levels
MyPlate, 35

National Cholesterol Education Program,
      267–9, 270
Nationwide Oat Test, 11, 12, 19, 29
nutritionism, 362–3, 366, 415–6
Nutrition Labeling Education Act, 361,
      382, 382–387, 387–90, 394, 396,
      420

oats
  agronomic traits, 10, 42, 46–7
    fertilizer, 46–7
    management techniques, 46–7
  breeding, single vs. multiple traits,
      11–25
  clinical applications, 318–23
    atopic dermatitis, 318–20
    contact dermatitis, 320–21
    drug eruptions, 323
    sunburn, 322
    wound healing, 321–2

  components of, 4–5, 75–91, 429–32
    dietary fiber, 41–2, 43–44, 75–8, 177,
        429–32
    fat, 46, 80–81
    β-glucan, 4–510, 17–8, 22–4, 27, 30–31,
        176–8, 255–6, 429–31
    minerals, 83–92, 172, 429
    phytochemicals, 172–85, 429–30
    protein, 45–46, 78–80
    starch, 44–5, 429, 431–32
    vitamins, 81–3, 172, 182–4, 256, 258,
        429–30
    volatiles, 57–8
  consumption, 4–5
  as a crop, 9–10, 11, 28–9
  cultivar, 9–11, 12, 14–5, 19, 23, 25, 26,
      27–8, 29, 30, 31, 35–6, 37–8, 40,
      42–3, 60, 429–30, 434
  effects on
    appetite, 271–4
    blood pressure, 240–46, 246–51
    body weight, 266–71
    carbohydrate metabolism, 281
    cholesterol, 229, 230–31, 231–3
    diabetes, 152, 187, 255, 256, 282, 291,
        292, 335–6, 430–31
    digestive health, 299–301
  β-glucan, 35–6, 39, 41, 42–3, 43–4, 47,
      50–52, 54, 55–6, 60, 123–4, 131
  health claims, 4–5, 41, 42–3, 56, 60,
      336–8, 338, 339–46, 346–8, 348–9,
      349–51, 382, 387–90, 391–94,
      396–401, 411–2, 432–33
    dietary recommendations and, 339–46
    labeling, 339–46, 382, 382–387, 387–90
    nutritional information and, 348–9,
        360, 361–2, 363, 373, 377, 381, 382,
        391, 432
    research on, 338, 339–46, 349–51, 360,
        371, 372, 377, 388, 416
    substantiation, 228, 349–51, 382
    value assessment, 360, 363, 364, 365,
        371, 381, 385, 395, 402, 407, 411–2,
        415
  history of, 363–72
    dietetics
      oat bran craze, 363–4, 364–8
      Quaker Oats and, 368–72
  milling, 34, 37
  nutrients, 4–5, 41–2, 75–91
  production, 5, 9, 29, 73–4

oats (*Continued*)
  protein, 78–80
    amino acid composition, 80
  quality, 33–4, 36
    appearance, 36–7
    aroma, 56–8
    assessment, 35–6
    color, 36–7
    flavor, see Oats, quality, taste, 34, 47, 56
    groat content, 37–9
    hullability, 38, 59
    milling yield, 37
    nutrition, 41–2
    physical, 36, 39–40
    processing, effects on, 34–5, 37, 43, 48–50, 51, 57
    taste, see Oats, quality, flavor, 34, 47, 56
  shelf life, 34, 57, 58
  side effects, 323–6
    allergy, 325–6
    contact dermatitis, 323–5
    percutaneous sensitization, 325–6
    urticaria, 323–5
  source of, 172
    dietary fiber, 41–2, 43–4
    fat, 46
    protein, 45–6
    starch, 44–5
  starch, 95–6, 102
    characteristics, see Starch, characteristics of, 97–101, 107, 113, 114, 115–6
  value chain, 33, 36, 59, 362, 387, 391, 394, 408, 419
    breeders, 33, 34, 36, 59
    exporters, 33
    food companies, 33, 34
    grain companies, 33, 34, 59
    growers, 33, 59
    processors, 33, 34, 36, 44, 59
    quality, 33–4, 36
obesity, 265–6
  effects on, 266–71
  fatty acids and, 275–6
  fiber, effects on, 266–71
  Nurses' Health Study, 265–6
  physiological mechanisms, 274–6
  satiety, 271–4
  weight management, 266–71, 276

phenolics, 172–6
  composition of, 172–73, 173–4, 174–5
  health effects, 175–6
Physicians' Health Study, 430–31
phytoalexins, 206–7, 212–3, 222
phytochemicals, 172–85, 256
phytosterols, 184–85
  composition of, 184
  health effects, 185
prebiotics, 303–6
  carbohydrates, 304
  effects of, 303–6
  fermentation, 303–4, 305–6
  gut and, 303–5
  immunity and, 304–5
  microflora, effects on, 306
  whole grains, in, 305–6
Pure Food and Drug Act of 1906, 360, 372

Quaker Oats, 368–72, 382, 396–401
  advertising, 357, 358, 359, 365, 369, 392, 394, 400, 403
  brand, 359, 361, 370, 372, 376, 393–94, 395, 397, 397–401, 401–5
  business plan, 396–401
    competition, 361, 371–3, 375, 376, 379–380, 387, 388, 396, 397, 399, 407
    PepsiCo merger, 409–11
  history, 368–72
  legal battles, 378–81
  marketing plan, 362, 411
    convenience, 405–6
    health claims, 382, 387–90, 391–94, 396–401, 411–2
    taste, 406–9
    value, 411–2
  media coverage, 370–71, 373, 374–5, 377, 378, 379–381, 381
  products, 371, 372, 379, 395, 398, 402
    bars, 372, 390, 397, 401–3, 405–6, 410, 419
    cereal, 368, 370, 374–5, 376–377, 380, 405, 412
    cookies, 358, 390, 403
    nutrition enhancements, 386, 401, 402, 405, 406, 408, 410, 415
    oatmeal, 370, 375, 379, 388, 390–94, 395, 397–401, 402, 405, 412, 412
Quaker Smart Heart Challenge, 413
Quaker Uniform Oat Nursery, 21

rapid visco analyzer, 42, 43, 112

skin health, 318
  colloidal oatmeal, 311–2
starch, 95–6, 102
  amylopectin, 96
  amylose, 96, 105–6
  characteristics of, 97–101, 107, 113, 114,
    115–6
    amylose:amylopectin ratio, 97–8
    molecular structure, 98–101
    molecular weight, 98–101
  components of, 104–7
    ashes, 105
    extraction, 106–7
    isolation, 106–7
    lipids, 105–6
    phosphorus, 105
    proteins, 105–6

gelatinization, 107–9, 111, 112
industrial uses of, 115–6
methods of analysis, 97–101
nutrition, 98, 115–6
pasting, 112–5
retrodegradation, 112–3, 115
supramolecular structure, 102–3
  crystallinity, 103–4
  granules, 102–3

USDA, 175, 367, 382, 413, 417–8,
  419
US dietary guidelines, 73, 171

vitamin E, 182–4
  composition of, 182–3
  health effects, 183–4

Women's Health Study, 240